Stephen J.

THE UNIFIED CYCLE THEORY

How Cycles Dominate the Structure
of the Universe and Influence Life on Earth

Outskirts Press, Inc.
Denver, Colorado

The opinions expressed in this manuscript are solely the opinions of the author and do not represent the opinions or thoughts of the publisher. The author has represented and warranted full ownership and/or legal right to publish all the materials in this book.

The Unified Cycle Theory
How Cycles Dominate the Structure of the Universe and Influence Life on Earth
All Rights Reserved.
Copyright © 2009 Stephen J. Puetz
V2.0

Cover Photo © 2009 JupiterImages Corporation. All rights reserved - used with permission.

This book may not be reproduced, transmitted, or stored in whole or in part by any means, including graphic, electronic, or mechanical without the express written consent of the publisher except in the case of brief quotations embodied in critical articles and reviews.

Outskirts Press, Inc.
http://www.outskirtspress.com

ISBN: 978-1-4327-1216-7

Library of Congress Control Number: 2008943454

Outskirts Press and the "OP" logo are trademarks belonging to Outskirts Press, Inc.

PRINTED IN THE UNITED STATES OF AMERICA

Table of Contents

Chapter 1 — 1
Introduction to the Unified Cycle Theory

Chapter 2 — 7
Definitions and Reference Points

Chapter 3 — 17
Markets Oscillate in Cycles

Chapter 4 — 25
Cycles & Sub-Cycles

Chapter 5 — 33
The Forces Behind Economic Cycles

Chapter 6 — 37
How Physical Cycles Affect Humans

Chapter 7 — 45
Classification of Cycles

Chapter 8 — 61
The Turning Point Distribution Principle

Chapter 9 — 79
Financial Leverage & Cycles

Chapter 10 — 89
Electromagnetic Cycles

Chapter 11 — 131
Gravitational & Motion Cycles

Chapter 12 — 141
EUWS Cycles

Chapter 13 — 155
Giga-Year EUWS Cycles

Chapter 14	165
2.46 Gyr, 821 Myr, and 274 Myr EUWS Cycles	
Chapter 15	185
91.3 Million Year EUWS Cycle	
Chapter 16	199
30.4 Million Year EUWS Cycle	
Chapter 17	211
10.1 Million Year EUWS Cycle	
Chapter 18	219
3.38 Million Year EUWS Cycle	
Chapter 19	225
1.13 Million Year EUWS Cycle	
Chapter 20	231
376 Thousand Year EUWS Cycle	
Chapter 21	237
125 Thousand Year EUWS Cycle	
Chapter 22	247
41,728 Year EUWS Cycle	
Chapter 23	261
13,909 Year EUWS Cycle	
Chapter 24	267
4,636 Year EUWS Cycle	
Chapter 25	271
The 1545-Year EUWS Cycle	
Chapter 26	283
The 515-Year EUWS Cycle	
Chapter 27	315
The 172-Year EUWS Cycle	
Chapter 28	339
The 57-Year EUWS Cycle	
Chapter 29	359
The 19-Year EUWS Cycle	
Chapter 30	391
The 6.36-Year EUWS Cycle	
Chapter 31	403
The 2.12-Year EUWS Cycle	
Chapter 32	415
The 258.11-Day EUWS Cycle	
Chapter 33	419
The 86.04-Day EUWS Cycle	

Chapter 34 425
The 28.68-Day EUWS Cycle
Chapter 35 433
Summary, Analysis, and Conclusions
Chapter 36 449
Summary of the Unified Cycle Theory
Chapter 37 455
Predictions Based on the EUWS Cycles
Appendix 459
References 475

Chapter 1
Introduction to the Unified Cycle Theory

This book presents a new theory, the *Unified Cycle Theory*, which describes the source and timing of naturally occurring oscillations that dominate both the fabric of life on Earth and many functions within our universe. Throughout the last century, the scientific community progressed tremendously in realizing that our environment fluctuates more often than not. Unfortunately, theories advanced to explain cyclic behavior generally failed to consistently match the historical record left by nature. The *Unified Cycle Theory* fills that void.

Documented cycles range widely in diversity. They include fluctuations in the stock market, commodity prices, economic activity, wars, civilizations, global climate, ice-ages, geological formations, and abundance of life on Earth. This book examines all of these cycles, plus more.

In addition, the *Unified Cycle Theory* makes predictions about our universe. If you're interested in history, science, geology, physics, astronomy, climatology, biology, mathematics, psychology, sociology, philosophy, economics, or investment theory, then read on. The theory includes portions of all of these academics. Furthermore, this book covers cycles ranging as short as 27 days to ones spanning billions of years. The evidence comes from a variety of scientific, academic, and business sources – which this book methodically details.

The Unified Cycle Theory

In spite of the great advances in identifying cycles, this increased knowledge yielded little in the way of providing a consensus about why they occur. This book attempts to correct that shortcoming. These chapters step back, dispose all preconceptions, reexamine the wealth of information available, and use deductive processes to formulate a new theory.

The first few chapters start with my initial interest in cycles and the methods used to determine the forces behind oscillations. Chapter 2 contains a reference of technical terms and acronyms used in later chapters. While this book takes a scientific approach to analyzing cyclic behavior, it is written as prose. As much as possible, it avoids technical jargon often used in scientific abstracts.

Middle chapters switch to identifying specific cycles and describing their characteristics. Except for many cycles with frequencies above one million years, various individuals and research teams previously discovered almost all cycles reviewed in this book. These chapters simply re-examine and re-interpret the original works. The middle portion becomes especially critical to the theory. Full understanding of the attributes of individual cycles paves the way for conclusions reached during the final chapters.

The book concludes with exercises in logic. Attributes of cycles, along with numerous correlations, permit dabbling in the always dangerous chore of determining cause-and-effect. In reviewing many previously discovered cycles, incorrect cause-and-effect assumptions led to erroneous conclusions. Re-examination yields some quite surprising results. An analysis of the entire cyclic spectrum shows that a large number of frequencies, previously thought to be independent, convincingly link themselves together as a cohesive group. This demonstrated unity immediately eliminates the cause-and-effect propositions outlined in all older theories, and it provides the basis for the *Unified Cycle Theory*.

Now you have an overview about what this book entails. If you're fascinated with markets, social behavior, and the universe, then you will be anxious to get started. But at this point, it's necessary to step back in time and review the thought process that lead to the theory. This basis later evolved into something quite unexpected.

For most of my adult life, I've been involved in market analysis. The equity markets captured most of my attention; however, considerable time was also devoted to analyzing derivatives, credit markets, commodities, housing, and other markets. During my first few years of research, the realization hit that most investors failed to anticipate market tops and bottoms. Almost always, analysts and investors became wildly bullish at tops, and excessively bearish at bottoms.

Because of the inability of the vast majority of financial experts to correctly predict the future state of markets, alternative methods of analyzing markets became attractive. The concept of contrary opinion provided a good starting point. But after extensive investigation, serious shortcomings emerged. Even though analysts' and investors' opinions reached lopsided extremes at major turning points, sentiment gradually shifted as markets moved. Sometimes, lopsided sentiment persisted a while before encountering a market reversal.

The lifetime of both bull markets and bear markets can range from as short as a few months to as long as several decades. As a market rises, sentiment becomes more bullish. However, as a move progresses, it's next to impossible to tell exactly when sentiment becomes too lopsided. Some bull markets peak with 70% bulls, others at 80%, still others at 90%, on so on. And the time-frame to get that last 5% or 10% bullish can take many months (or even years). So while contrary opinion can be a useful indicator, it has definite flaws as a reliable timing device.

Eventually, another alternative caught my attention – cyclic analysis. Similar to contrary opinion, using cycles to predict markets has many limitations. Proponents claim hundreds of financial cycles exist. As my awareness of cycles increased, it became hard to tell which cycles were valid, and which were not. Of the many alleged cycles, validation methods were often either questionable or completely missing. Nonetheless, quite compelling evidence surfaced every once in a while. One of the early contributors to the field, Edward Dewey, coauthored his hallmark book, *Cycles: The Science of Prediction,* presenting the case for cycles. [Dewey and Dakin, 1947]

Dewey's analysis always contained thorough, thought-provoking methods. Most importantly, he explained cyclic concepts in easy, understandable ways. He continued his research through the *Foundation for the Study of Cycles* where he wrote many excellent articles on the subject.[Foundation for the Study of Cycles, 2008] The ideas Dewey introduced stuck with me. His concepts always held a place in my mind – as I struggled to understand why markets moved in a constant state of flux. Yet, in spite of the great works performed by the early pioneers, many problems remained.

The list of issues included....... What is the source of cyclic behavior in the markets? Whatever the source, how are humans influenced? Which cycles are valid? Which ones aren't?

With these nagging questions in mind, I went on to gather hordes of information about cycles. Using this information, the search began for a plausible explanation of cyclic behavior in the markets. Over a period of years, as the database grew, an explanation finally emerged that even surprised me. Some questions remain unanswered. Regardless, the research progressed to the point where construction of the *Unified Cycle Theory* became possible.

As the name implies, this theory organizes, unifies, explains, and clarifies the many cycles already discovered by a variety of scientists, researchers, and analysts. In addition, the theory predicts a set of characteristics for one set of cycles with a mysterious source. And it also predicts important turning points in mankind's social and economic future. A brief, general outline follows.

> 1) Markets (in particular) and human activities (in general) aren't efficient. Rather, they oscillate in cycles. They over-shoot reasonable norms at tops, and then sink below fair

values at bottoms.

2) As the *Elliott Wave Theory* suggests, cycles develop in waves that consist of ever smaller sub-waves. And conversely, every cycle serves as a sub-component of a larger wave. Hence, large waves exist that span millions, billions, and trillions of years. At the opposite end of the spectrum, extremely small cycles exist that repeat with a frequency of a few seconds, or less.

3) Many of the cycles already discovered are valid. However, insufficient data, combined with inadequate theories about their operation, caused slight miscalculations of their frequencies. The *Unified Cycle Theory* provides a method for estimating the length of some of these long-term cycles down to the day.

4) The force behind all of the major cycles predominantly originates from the interaction of the Earth with various physical forces from both inside and outside our universe. These external forces create a series of electromagnetic-type fluctuations. Some of these oscillations can be quite insignificant, while others can be literally Earth-shaking.

5) The constantly changing force fields surrounding Earth affects the way some people feel and act. Hence, these changes affect the collective mood of the world's population. These forces affect humans by causing our electro-chemical nervous systems to operate differently as the forces fluctuate. When enough people begin to feel differently, they alter their behavior. These altered actions then reflect themselves in social interactions, including oscillating markets and fluctuating economic activity. The *Unified Cycle Theory* **does not suggest** that these external forces cause people to think differently – only that it causes them to **feel differently**. These emotional oscillations result in manias, depressions, herd instinct actions, and other social behaviors. But man's ability to use rational thought remains unaffected during these emotional transitions. Markets fluctuate around a "fair value" norm related to rational thought. Yet, markets seldom sit at "fair values" for extended periods.

6) Three types of cycles affect activity on Earth:
   - Geomagnetic cycles that result from magnetic field interactions among the Sun, Moon, and Earth. The most influential geomagnetic cycles repeat in periods of less than 15 years.
   - Gravitational cycles that result from the interaction of the Sun, Moon, Earth, and planets. These cycles can span thousands of years.
   - A set of more dominant, but mysterious, cycles that probably originate from outside our universe.

7) These dominant, mysterious cycles exhibit harmonics of three. That is, take the period of any of these cycles, multiply it by three, and the next larger cycle results. Conversely, take any cycle belonging to this sequence, divide its frequency by three,

and the next smaller cycle appears. From this point forward, the book refers to this set of cycles as the Extra-Universal Wave Series (EUWS). The reason for this naming convention will become apparent as the book progresses.

8) The EUWS cycles exhibit a Turning Point Distribution (TPD) Principle. This principle describes how the EUWS cycles reverse direction at tops and bottoms. Variations from theoretical turning points distribute themselves in a non-random manner. A normal distribution indicates random behavior around a mean. If a EUWS cycle misses its theoretical turning point, alternatives include theoretical turning points of its sub-cycles. This tendency produces non-random, stair-step distributions.

9) When a EUWS cycle deviates from its expected turning point, it may continue to deviate in a similar manner for several more cycles. When this happens, the cycle appears to shift. But eventually, EUWS cycles always return back to their expected theoretical oscillations. This book includes extensive examples of cyclic-shifts.

10) In financial markets, decision-making becomes more emotional and pronounced when speculators buy with borrowed money. During periods of mass speculation, cyclic patterns become more visible than normal. After a major top, falling markets often force speculators out of leveraged holdings. Once the deleveraging process removes speculators from the marketplace, cyclic visibility returns to a more obscure state. This tendency explains why cycles appear stronger near market tops, and why they either disappear or becomes less visible after large deflations.

The remainder of this book provides evidence supporting the above statements. It offers explanations, and it predicts future events for both mankind and the universe. The web-site for *Cycles Research Institute* accurately describes the nature of cycles, and thus the challenges involved in determining their source:

> "Cycles may be defined as any phenomena repeating after fairly regular time intervals. Generally cycles are found in anything to which numerical measurements may be assigned at intervals in time. Edward Dewey of the *Foundation for the Study of Cycles* said that every field studied had been found to have cycles.... The cause of cycles is generally that some restorative force exists whenever any system is displaced from its equilibrium. This applies equally to the motion of a pendulum, the orbiting of planets and the state of the economy, because it doesn't matter whether the system and the restoring force is physical or not.... It may be confidently said that although cycles are found in everything, they are *never perfectly regular*, but the period of the cycles also varies in cycles. In the case of the planets, they have elliptical orbits causing small variations in motion, but the disturbance of other bodies also causes other variations over varied time scales.... Because everything in the universe affects everything else, all vibrations, oscillations and disturbances reverberate about forever, continuously interacting. The wave nature of the universe is

attested to by all modern scientific theories and the *study of cycles* is a wonderful way to study how influences move from one place to another, one time to another and one field of study to another. This can be done because cycles have many specific periods present that can be recognized, like fingerprints, when they appear somewhere else." [Cycles Research Institute, 2008]

Chapter 2
Definitions and Reference Points

This book presents material in a way to accommodate both the general public and the scientific community. To achieve that goal, this chapter contains terms, abbreviations, acronyms, labels, phrases, and geological periods commonly used by research specialists. While reading, if you come across a term you're unfamiliar with, refer back to this section. If you're already an expert in these areas, you may wish to skim over this chapter.

The final portion of this chapter outlines the charting conventions used in this book. Some of these conventions are non-standard, so everyone should review that portion of the chapter.

1) Acronyms, abbreviations, and terms.

- **AP**. After Present – Refers to a timescale in the distant future. These future points can be thousands, millions, billions, or trillions of years from now.

- **BAO**. Baryon Acoustic Oscillations – Sound waves originating from the creation of baryonic matter in the early stages of the universe, shortly after the Big Bang. BAO provides astrophysicists a standard ruler for measuring the age of various universal events.

- **BP**. Before Present – Refers to a timescale of events far in the past. These past events occurred thousands, millions, billions, or trillions of years ago. In this book, the year 2000 represents the present.

- **CRF**. Cosmic Ray Flux – Fluctuations in the rate of flow of cosmic rays within the universe or a galaxy.

- **ENSO**. El Nino Southern Oscillation – Temperature fluctuations in surface waters in the Eastern Pacific.

- **EUWS**. Extra-Universal Wave Series – A theoretical group of waves with no definitive beginning or end. The length of each child-wave can be calculated by dividing the length of its parent wave by 3. The length of each parent-wave can be calculated by multiplying the length of its child wave by 3.

- **GISP**. Greenland Ice Sheet Project – A joint research effort by Denmark, Switzerland, and the United States in evaluating past Greenland climate conditions by studying its ice core.

- **GISP2**. Greenland Ice Sheet Project, Phase 2 – A 1993 follow up study to GISP. The ice core drilled for GISP2 was deeper and better than the original GISP core from 1971.

- **GRIP**. Greenland Ice Core Project – A European research project that evaluated changes in Greenland's climate over the past 100,000 years by studying the Oxygen-18 content of different layers of its ice core.

- **Ga**. Giga-Years Ago – A billion years ago.

- **Geomagnetic AA Index**. A measure of Earth's magnetic field, calculated daily for the past 150 years.

- **GPI**. Geomagnetic Paleointensity – A measure of the magnitude of the Earth's magnetic field. The paleointensity scale (GPI) measures the field over millions of years by analyzing the magnetic properties of ancient rocks. This differs from Geomagnetic AA which measures daily changes.

- **Gyr**. Giga-Year Timescale – This represents billion year intervals.

- **IAE**. Ice-Age Epoch – An extended period of cold climate, resulting in large increases in polar ice caps and extensive formation of glaciers.

- **Ka**. Kilo-Years Ago – A thousand years ago.

- **Kyr**. Kilo-Year Timescale – This represents thousand year intervals.

Definitions and Reference Points

- **LACC**. Low Altitude Cloud Cover – Clouds that shield the Earth from solar radiation, thus helping cool global temperatures if they form in large numbers.

- **LMC**. Large Magellanic Cloud Galaxy – A galaxy close to the Milky Way.

- **MW**. Milky Way Galaxy – The galaxy where our solar system resides.

- **Ma**. Mega-Years Ago – A million years ago.

- **Myr**. Mega-Year Timescale – This represents million year intervals.

- **NASA**. National Aeronautics and Space Administration – The U.S. Government agency responsible for launching rockets and satellites.

- **NOAA**. National Oceanic and Atmospheric Administration – A U.S. government organization, operating within the Department of Commerce, assigned various environmental responsibilities.

- **Paleo**. Involving ancient conditions. Paleo is often used as a prefix to other word. For example, paleoclimate refers to ancient climate conditions.

- **SFR**. Star Formation Rate – Stars form at an uneven rate. SFR quantifies the rate of their creation.

- **SN**. Supernova – A large star with a mass at least 10 times greater than the Sun's.

- **Ta**. Tera-Years Ago – A trillion years ago.

- **Tyr**. Tera-Year Timescale – This represents trillion year intervals.

- **TPD**. Turning Point Distribution – Actual turning points of the Extra-Universal Wave Series cycles aren't evenly distributed. Instead, reversal points concentrate around theoretical turning point of its largest sub-cycles.

- **Unconformity**. A geological gap leaving no trace of Earth's history during the time spanned by the gap.

- **VADM**. Virtual Axial Dipole Moments. VADM measures paleointensity. Paleointensity refers to the intensity of the Earth's magnetic field measured over time.

- **WMAP**. Wilkinson Microwave Anisotropy Probe – A satellite mission launched in June 2001 with the purpose of measuring the temperature of the radiant heat left over from the Big Bang. The project is a joint venture between NASA's Goddard Space

Flight Center and Princeton University. It's scheduled to be completed during September 2009.

2) Geological Timescale.

The geological timescale changes as better information becomes available. These changes continue to be made year after year. Geological intervals consist of eons, eras, periods, and epochs – listed in the following section.[Gradstein *et al.*, 2005] & [National Science Foundation, 2008]

Supereon	**Years Covered**
Pre-Cambrian	4.6 to Ga to 0.542 Ga

Eons	**Years Covered**
Hadean	4.6 Ga to 3.8 Ga
Archean	3.8 Ga to 2.5 Ga
Proterozoic	2.5 Ga to 0.542 Ga
Phanerozoic	0.542 Ga to present

Eras	**Years Covered**
Eoarchean	3800 Ma to 3600 Ma
Paleoarchean	3600 Ma to 3200 Ma
Mesoarchean	3200 Ma to 2800 Ma
Neoarchean	2800 Ma to 2500 Ma
Peleoproterozic	2500 Ma to 1600 Ma
Mesoproterozoic	1600 Ma to 1000 Ma
Neoproterozoic	1000 Ma to 542 Ma
Paleozoic	542 Ma to 251 Ma
Mesozoic	251 Ma to 65.5 Ma
Cenozoic	65.5 Ma to present

Periods	**Years Covered**
Siderian	2500 Ma to 2300 Ma
Rhyacian	2300 Ma to 2050 Ma
Orosirian	2050 Ma to 1800 Ma
Stathrian	1800 Ma to 1600 Ma
Calymmian	1600 Ma to 1400 Ma
Ectasian	1400 Ma to 1200 Ma
Stenian	1200 Ma to 1000 Ma
Tonian	1000 Ma to 850 Ma
Cryogenian	850 Ma to 630 Ma
Ediacarian	630 Ma to 542 Ma
Cambrian	542 Ma to 488 Ma

Definitions and Reference Points

Ordovician	488 Ma to 444 Ma
Silurian	444 Ma to 416 Ma
Devonian	416 Ma to 359 Ma
Carboniferous	359 Ma to 299 Ma
Permian	299 Ma to 251 Ma
Triassic	251 Ma to 200 Ma
Jurassic	200 Ma to 145 Ma
Cretaceous	145 Ma to 65.5 Ma
Paleogene	65.5 Ma to 23.0 Ma
Neogene	23.0 Ma to present

Epochs (Recent)	**Years Covered**
Paleocene	65.5 Ma to 55.8 Ma
Eocene	55.8 Ma to 33.9 Ma
Oligocene	33.9 Ma to 23.0 Ma
Miocene	23.0 Ma to 5.33 Ma
Pliocene	5.33 Ma to 1.81 Ma
Pleistocene	1.81 Ma to 0.115 Ma
Holocene	0.115 Ma to present

3) Blytt-Sernander Classification of Northern European Climate Periods.

The Blytt-Sernander classification refers to climate conditions in Northern Europe during the last 14,000 years. The classification system was based on a study of Danish peat bogs.[Ipedia.net, 2008]

Geological Epoch	**Blytt-Sernander**	**Years Spanned**
Late Pleistocene		
	Older Dryas	14.0 Ka to 13.6 Ka
	Allerod	13.6 Ka to 12.9 Ka
	Younger Dryas	12.9 Ka to 11.5 Ka
Holocene		
	Boreal	11.5 Ka to 8.9 Ka
	Atlantic	8.9 Ka to 5.7 Ka
	Subboreal	5.7 Ka to 2.6 Ka
	Subatlantic	2.6 Ka to present

4) Sunspot Periods.

Over thousands of years, various solar periods have been associated with unusually high or low sunspot numbers. These noteworthy periods hold special names, listed below.[NASA, Earth Observatory, 2001], [U.S. Geological Survey, 2000], and [Usoskin *et al.*, 2007]

Solar Period Name	Years Spanned	Associated Climatic Event
No name, minimum	3370-3300 BC	Cool
No name, maximum	3160-3130 BC	Warm
No name, minimum	2890-2830 BC	Cool
No name, maximum	2530-2510 BC	Warm
No name, maximum	2250-2230 BC	Warm
No name, maximum	2090-2050 BC	Warm
No name, maximum	1800-1780 BC	Warm
No name, minimum	1410-1370 BC	Cool
No name, minimum	810-720 BC	Cool
No name, maximum	465-425 BC	Warm
No name, minimum	390-330 BC	Cool
No name, minimum	650-720	Cool
Oort Minimum	1010-1050	Cool
Medieval Maximum	1100-1200	Medieval warm period
Wolf Minimum	1280-1340	Cool
Spoerer Minimum	1420-1530	Cool
Maunder Minimum	1645-1715	Little Ice-Age
Dalton Minimum	1795-1825	Cool
Modern Maximum	1920-present	Global warming

5) Common Methods for Dating Paleo Events and Paleo Materials.

Scientists have developed hundreds of different dating methods for measuring the age of various stages of Earth's progression. This book often refers to these methods. To ensure you're familiar with the subject, the following section briefly explains some of the more common dating methods.

- **Organic Dating – Carbon-14.** All living organisms contain a small percentage of an unstable isotope of carbon, Carbon-14. With a half-life of 5730 years, the amount of Carbon-14 decay tells how long ago an organism lived. However, Carbon-14 dating has a significant limitation. Because of its relatively short half-life, it only proves effective in estimating the death of creatures that lived within the past 60,000 years.

- **Inorganic Dating – Beryllium-10.** As cosmic rays strike the atmosphere, they create radioactive Beryllium-10. This isotope holds a half-life of 1.51 million years. Researchers use Beryllium-10 ratios for dating ice cores and rocks, as well as other inorganic materials.

- **Inorganic Dating – Potassium-40.** With an extremely long half-life of 1.3 billion years, this potassium isotope proves effective for determining the age of rocks more than one billion years old. By studying the ratio of Potassium-40 to Argon-40, the age of old rocks can be determined fairly accurately.

Definitions and Reference Points

- **Ice-Ages and Air Temperature – Oxygen-18**. Researchers analyze the ratio of two oxygen isotopes to estimate temperatures in the past. These isotopes are Oxygen-16 and Oxygen-18. Over 99% of the Earth's oxygen consists of Oxygen-16, while the Oxygen-18 portion only amounts to about 0.2%. Structurally, Oxygen-18 contains 10 neutrons and 8 protons, making it slightly heavier than Oxygen-16 with 8 neutrons and 8 protons. Because of their relative lightness, ocean waters containing Oxygen-16 molecules evaporate into the atmosphere more readily than water with Oxygen-18. During ice-ages, the process of evaporation at the equator and rainfall in the Polar Regions combine to transfer unusually heavy concentrations of Oxygen-16 into the polar ice shelves. As global temperatures rise, the ice shelves melt, and the frozen Oxygen-16 molecules transfer back into the oceans. In paleoclimatology, researchers take samples from ice-cores and marine sediments. By analyzing these samples to determine the ratio of Oxygen-18 to Oxygen-16, scientists calculate ancient temperatures reasonably well. In this book, charts labeled with the ratio of Oxygen-18 to Oxygen-16 appear as d Oxygen-18 (‰) – delta Oxygen-18 in parts per thousand. This ratio correlates to global climate conditions inversely. That is, as delta-Oxygen-18 falls, temperatures increase; and the opposite holds true as the ratio rises. Because of this inverse relationship, these ratios appear on the y-axis of charts with smaller numbers at the top of the y-axis, and larger numbers at the bottom of the y-axis.

6) Age of Important Events in our Universe and Solar System.

- Age of the Universe   13.73 ±0.12 Ga [Hinshaw *et al.*, 2008]
- Age of the Milky Way   13.60 ±0.80 Ga [Del Peloso *et al.*, 2005]
- Age of the Sun   4.57 ±0.11 Ga [Bonanno *et al.*, 2002]
- Age of the Earth   4.54 ±0.04 Ga [Dalrymple, 1991]

7) The 4 Basic Physical Forces.

The four basic physical forces determine how all known structures in our universe interact with each other, by either attraction or repulsion. These forces act as the primary candidates for regulating cycles. These physical forces are described below, in order of weakest to strongest.[Georgia Perimeter College, 2008]

- **Gravitational Force** – This is the weakest of the four forces. It is an attractive force that affects all types of particles. However, as compensation for its weakness, it operates over an infinite range of space. Gravity dominates the motions of large-scale objects such as planets, stars, and galaxies.
- **Electromagnetic Force** – Electromagnetic fields are much stronger than gravity, but not as intense as either of the nuclear forces. This force can be either attractive or repulsive, depending on the type of electrical particles it interacts with. Similar to gravity, electromagnetic fields operate over an infinite range. Electromagnetic forces affect objects that range in size from atoms to all living organisms.

- **Weak Nuclear Force** – The next-to-strongest influence is the weak nuclear force. It is not important in the structure of the universe; however, it affects all types of particles and the properties of those particles from a short range. The weak nuclear force is neither attractive nor repulsive.
- **Strong Nuclear Force** – The strongest force is an attractive one, but it only operates from a very short range. It is the dominant force for small particles, up to the size of the nucleus of an atom. The strong nuclear force affects quarks, but not electrons or neutrinos.

8) The Electromagnetic Spectrum.

Electromagnetic fields surround all bodies in the Universe. In addition, they affect all living organisms in some manner. That fact alone elevates the importance of considering electromagnetic fields in cyclic studies. NASA's Goddard Space Flight Center describes the electromagnetic spectrum.

> "We may think that radio waves are completely different physical objects or events than gamma-rays. They are produced in very different ways, and we detect them in different ways. But are they really different things? The answer is 'no'. Radio waves, visible light, X-rays, and all the other parts of the electromagnetic spectrum are fundamentally the same thing. They are all electromagnetic radiation.... Electromagnetic radiation can be described in terms of a stream of photons, which are massless particles each traveling in a wave-like pattern and moving at the speed of light. Each photon contains a certain amount (or bundle) of energy, and all electromagnetic radiation consists of these photons. The only difference between the various types of electromagnetic radiation is the amount of energy found in the photons. Radio waves have photons with low energies, microwaves have a little more energy than radio waves, infrared has still more, then visible, ultraviolet, X-rays, and ... the most energetic of all ... gamma-rays."[Newman, NASA, 2008]

The electromagnetic spectrum follows.
  Radio waves
  Microwaves
  Infrared rays
  Visible light
  Ultraviolet light
  X-rays
  Gamma rays

9) Timescales Used in the Charts.

In this book, the charts contain non-conventional methods. To avoid confusion, the following rules apply.

Definitions and Reference Points

a) For timescales covering events within the last 10,000 years, Gregorian calendar dates are used.
- Negative numbers on the x-axis represent years BC (before Christ), or BCE (before Common Era).
- Positive numbers on the x-axis represent years AD (anno Domini), or CE (Common Era).
- Dates appear with a format of year-month-day.

b) For timescales covering events spanning more than 10,000 years, the year 2000 represents the present.
- Negative numbers represent the number of years BP (before present).
- Positive numbers represent the future, or years AP (after present).

c) Regardless of the timescale used, all times appear on charts on the x-axis.
- Time always advances from left to right.
- Events on the left side of the charts represent times in the past.
- Events on the right side of the charts represent either recent times, or times in the future.
- Tick marks on the x-axis generally represent "theoretical peaks" of EUWS cycles.

Chapter 3
Markets Oscillate in Cycles

Markets oscillate in cycles. They have fluctuated for at least 2,000 years, and they will certainly do so in the future. That's the history of markets and the history of the universe. Just about everything moves in cycles. Coming chapters heavily document this cyclic activity. And they show how these oscillations consistently link themselves to naturally occurring physical cycles.

Yet, in spite of this long cyclic history, consumers complain when retail items rise in price, and investors cry for government action when security prices fall. In general, humans prefer price stability. This preference for stability shows not only in views of how markets operate, but in views of nature and philosophy. Looking at the history of human change, it reveals deep-rooted belief systems that inhibit cyclic interpretations. Especially for older people, change comes with greater difficultly than it does with youth. Old ideas, even incorrect ones, can survive for centuries. And new ideas, no matter how strong the supporting evidence, aren't easily accepted.

As an example of shifting concepts, over the centuries, the idea that a stable Earth resided at the center of the universe gradually (and grudgingly) gave way to the realization that Earth accounts for an insignificant portion of our universe. And within our universe, everything oscillates in cycles. These oscillations sometimes result in great collisions, which in turn, temporarily cause further instability.

The Unified Cycle Theory

While the concept of continual oscillation has gained acceptance by astrophysicists in their view of our universe, the same cannot be said of economics. Especially in the USA during the past fifty years, economists generally agree that government planning can control the business cycle. Politicians promise spending projects to keep the economy expanding. Our Federal Reserve central bank lowers interest rates at the first sign of recession. And big businesses leverage their balance sheets during times of economic weakness to avoid terminating long-term employees. Optimists cite these actions by our political, financial, and business leaders as assurances that the United States will prosper indefinitely.

Yet, in the long-run, history shows that central planning never works well in the marketplace. The obvious short-term benefits of government interventions aimed at halting business cycle recessions seriously mask the long-term damage these actions create. Central planning clearly failed in the former Soviet Union. And economic central planning will certainly fail in the United States as well. Government interventions create unnatural distortions in the short-term, with devastating consequences in the long-term.

Carrying this logic further, for every action, there is an equal and opposite reaction. All of the positive short-term economic fixes utilized by our political leaders create offsetting negative long-term liabilities. But alas, the long-term always arrives, and usually at a time least expected. In the long-run, cyclic forces always overwhelm government attempts to promote unnatural, sustained, economic expansion. At that critical point, after it's too late to undo the damage, citizens finally realize the folly of fruitless interventions. Misguided interventions then lead to lack of trust in government. With trust in government destroyed, free-market pricing finally returns. Later, this book presents evidence that the USA reached a major inflection point during January 2007. For the next few decades, cyclic forces will likely trump all attempts to manage our economy. From a cyclic viewpoint, the world entered an extended bear market during 2007.

The short-term spending sprees so kindly provided by legislators during times of economic weakness create revenue deficits and long-term debts that need to be paid back in one form or another (either by willing debt repayment, or erosion of purchasing power through inflation, or by default). All three of these debt readjustment mechanisms are associated with economic recessions and depressions. Artificially low interest rates "fixed" by our Federal Reserve central planners discouraged vitally need savings while encouraging debt accumulation in our country. Now, leveraged balance sheets threaten corporate liquidity. Because of this leverage, even mild economic downturns put corporate survival at risk.

For the bearish camp, these long-term concerns became increasingly evident with each passing year. Debts, deficits, leverage, and instability progressively worsened every year. The leverage extremes encountered prior to the Great Depression were easily exceeded during the 1970s and 1980s. Even with these excesses in place, the stock market crash of October 1987 failed to topple leveraged speculators. After a few years of hesitation, the speculative fever reemerged stronger than ever during the early 1990s.

The '90s gave birth to the technology bubble, and stock markets soared worldwide. Debt levels expanded even further. Like the '87 crash, the bursting of the technology bubble in 2000 did little to slow speculative fever. With its monetary spigot wide open, easy credit continually flowed from the Federal Reserve. Between 2002 and 2006, the giant government-sponsored mortgage lenders (along with private mortgage bankers) assisted by making loans available to anyone and everyone – regardless of their credit history and financial condition. As a result, financial bubbles formed in housing, structured finance, derivatives, Wall Street alchemy, commercial real estate, hedge funds, commodities, and foreign equities.

All of this booming activity led the majority to believe that the United States had indeed found the recipe for permanent prosperity. This boom served as convincing evidence for the majority. For most economists and investors, the only relevant history amounted to economic statistics following the Great Depression. Economists often viewed pre-Depression history as ancient, unimportant records. After the Depression, economic recessions surfaced occasionally, but our country avoided a repeat of the 1930s' crisis.

In fact, since 1982, recessions became shallower and less frequent. The majority noticed and followed these trends. With only two mild recessions within a 25-year period, by early 2007, the vast majority became quite convinced that government leaders had learned how to properly manage the economy. And the majority firmly believed that our financial leaders could easily manage any type of downturn, without allowing it to turn into something more serious. Most people viewed the possibility of a deflationary depression as extremely remote. But for me, the enormous credit-expansion of the past 25 years created angst instead of comfort. Excesses expanded alarmingly fast. The list of concerns grew long.

- How far can our financial system go beyond the extremes encountered during 1929? By the early 1990s, in order to lessen the strain on large commercial banks, the Federal Reserve effectively eliminated reserve requirements, thus allowing almost unlimited lending. From 1985 to 2007, what made borrowers so willing to leverage themselves to an extreme more than two times greater than the worst part of the Great Depression of the 1930s? See **Chart 3A**.

- How far can mortgage debt expand relative to the market-value of the homes securing them? By late 2007, the collective mortgage debt in the US exceeded 50% of the market value for all homes. In contrast, this same ratio ranged from only 5% to 15% during the 1920s, 1930s, and 1940s, and from only 20% to 25% during the 1960s, 1970s, and 1980s. From 1985 to 2007, what made lenders willing to radically change their loan standards by virtually eliminating down-payments and lending up to 100% of a home's value? See **Chart 3B**.

- Why did so many real estate appraisers willingly overestimate assessed home values?

- Why were rating agencies (S&P, Moody's, Fitch, etc.) so eager to stamp AAA ratings on bundled packages of sub-prime loans?

- Why did hoards of home-buyers fib about their incomes in order to qualify for home purchases? And why were lending institutions readily accepting mortgage applications without verifying applicants' incomes?

- Why did the major commercial and investment banks hide so many assets off of their balance sheets?

- Why were government regulators so lax in enforcing prudent lending guidelines for Fannie Mae, Freddie Mac, and other mortgage lenders?

- Why were major investment banks so eager to give up standard collateral protections while loaning massive amounts of money to private equity firms using "covenant lite" agreements?

- Why were so many hedge funds willing to leverage their investments many times over with large amounts of illiquid investments? (Especially in light of the LTCM collapse in 1998.)

The point has been made. As our national system of restraints, oversight, and financial ethics broke down, the economy expanded further, and the bubbles grew ever larger – and completely out of control. With this underlying financial deterioration spreading like cancer, confidence had clearly mushroomed to the point of tremendous over-confidence. Reckless risk-taking had replaced conservative assessments of risk-reward ratios. The extent of the bullishness and over-confidence can be seen from the **Chart 3C**.

As measured by *Investors' Intelligence*, over a period of about 4 years, financial advisors became more bullish toward the stock market during early-2007 than at any other time in the fifty year history of this index.[Investors Intelligence, 2008] The other times sentiment reached comparable extremes were in 1966, 1977, and 1987. Nasty bear markets followed each of these sentiment peaks.

The questions kept coming to my mind. How much longer can this financial insanity prevail? How far beyond the 1929 extremes can our financial system expand before it reaches the breaking point? How much more bullish can investors become with disastrous long-term liabilities looming over markets?

And that's where the *Unified Cycle Theory* and cyclic analysis comes into play. In my earlier years, I had made the mistake of assuming that the 1929 extremes would never again be exceeded. I had assumed that it was next to impossible for bankers and lenders to extend credit more freely than they did prior to the Great Depression. I had assumed that they would learn from history and not repeat the mistakes of their grandparents. But I was wrong – dead wrong. Mankind doesn't seem to learn much from the financial mistakes of earlier generations. As a group, we keep making the same mistakes over and over. Legislative actions in recent years serve as an example. Congress and businesses found ways to circumvent the regulatory safeguards put in place following the depression of the 1930s.

As confidence expands, people tend to throw away safeguards and protections. And that's what the financial system in the United States did in a big way from 1971 through 2007. First, in 1971, Nixon closed the gold window to foreign central banks, thus eliminating the federal government's need to control credit growth. Next, rules were relaxed for buying homes, thus allowing the quasi-government mortgage lenders Fannie Mae and Freddie Mac to grow out of control and dominate the home-loan marketplace. The Federal Reserve constantly lowered reserve requirements throughout this period. Requirements for holding government securities were effectively eliminated. And during 2007 and 2008, the Federal Reserve agreed to hold non-government securities as collateral for troubled banks in need of cash.

All of these actions, combined with the expanding confidence, allowed the 1929 excesses to be exceeded by a wide margin. As awareness of these extremes became glaringly apparent, my search intensified for a cyclical explanation. Was the financial insanity that took place from 1991 to 2007 a one-shot deal? Or, did history provide a record of equivalent expansions?

Fortunately, with the help of excellent research already conducted by a variety of individuals and organizations, the answers appeared readily. As the research proved fruitful, the *Unified Cycle Theory* began to grow into a theory explaining the cyclical tendencies for both mankind and our universe. Chapter 4 provides a more detailed explanation about the logic behind the theory.

Chart 3A

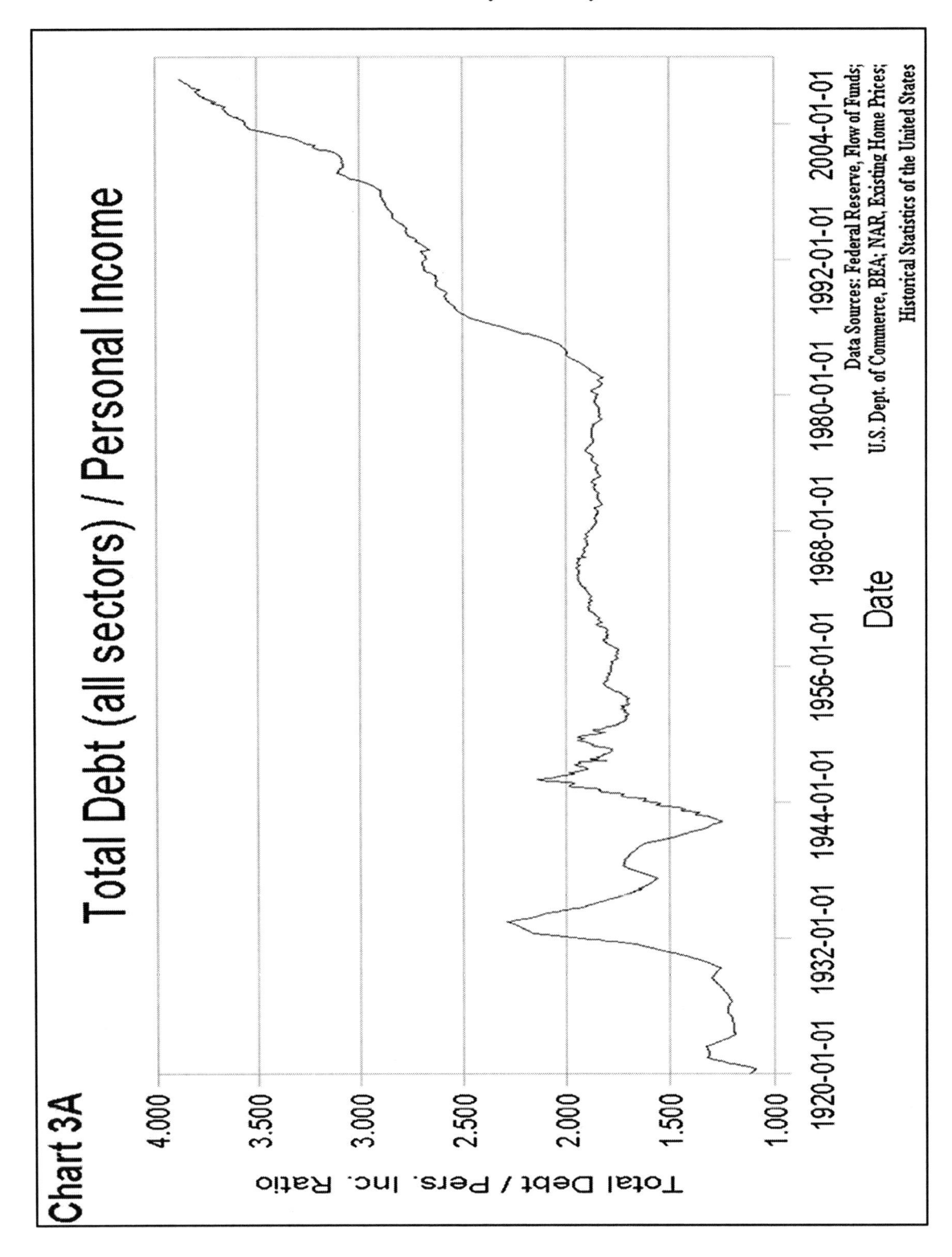

Total Debt (all sectors) / Personal Income

Data Sources: Federal Reserve, Flow of Funds; U.S. Dept. of Commerce, BEA; NAR, Existing Home Prices; Historical Statistics of the United States

Markets Oscillate in Cycles

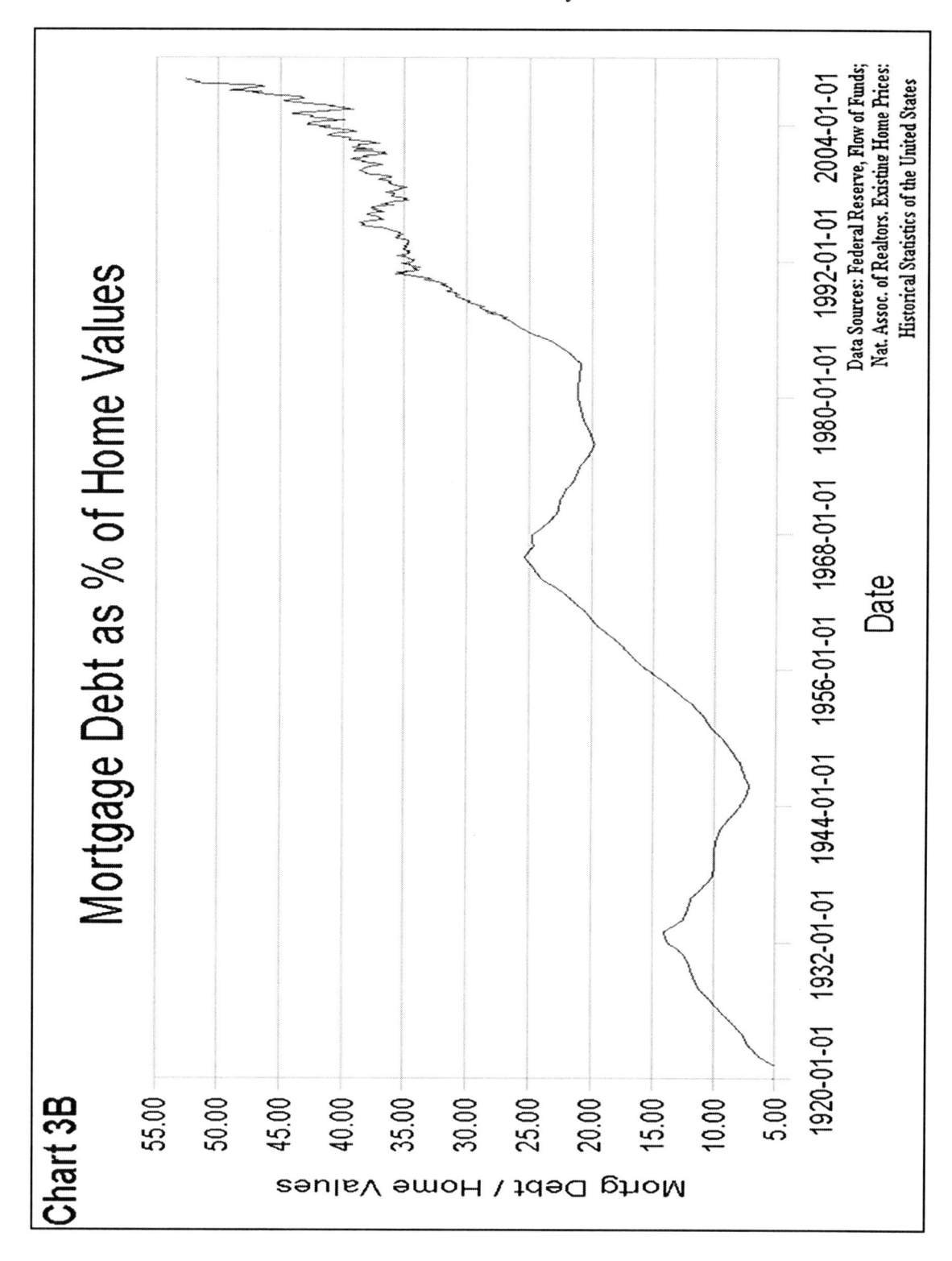

Chart 3B

Mortgage Debt as % of Home Values

Data Sources: Federal Reserve, Flow of Funds;
Nat. Assoc. of Realtors. Existing Home Prices:
Historical Statistics of the United States

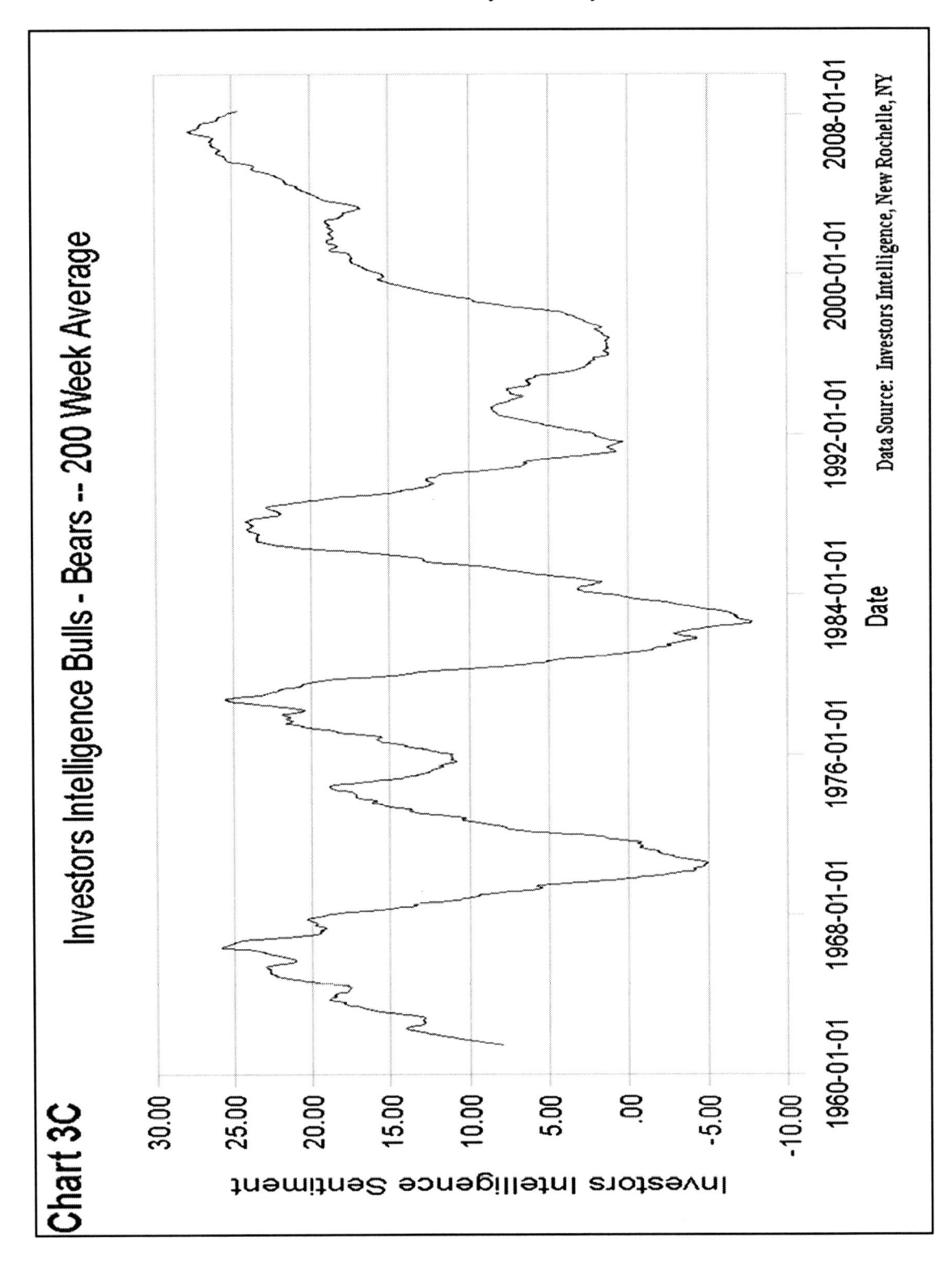

The Unified Cycle Theory

Chart 3C — Investors Intelligence Bulls - Bears -- 200 Week Average

Data Source: Investors Intelligence, New Rochelle, NY

Chapter 4
Cycles & Sub-Cycles

The idea that economies fluctuate in waves has been around for a while. However, a clear concept of how cycles influence consumers has befuddled researchers. At the same time, Federal Reserve officials and Treasury economists persuaded citizens that government policy can control any potential cycles. As a result, fiscal stimulus packages abound. However, the idea that business cycles can be controlled only becomes believable from the relatively short-term perspective of the last sixty years in the United States.

By examining financial history over many centuries, equivalent periods of extended economic prosperity actually take place on occasion. And like night follows day, each of these multi-decade booms ends with a prolonged bust. From this perspective, the global prosperity of the past few decades should be viewed as an anomaly that will end with great economic disaster.

The basis for this research started with my discovery of a 2.1-year cycle that dominated stock market activity over the past thirty years. I wondered if this cycle could be connected to various cycles already discovered by others. The core of the *Unified Cycle Theory* arose from three ideas. First, the concept that all waves appear as components of larger waves came from the *Elliott Wave Theory*. Second, Edward Dewey's observation that cycles often appear in harmonics of three was employed. Third, the theory used cycles already identified by the *Foundation for the Study of Cycles*. In combination, these ideas produced the content for **Table 4.1**.

Table 4.1 – Previously Discovered Social and Economic Cycles.

Cycle Frequency (years)	Cycle's Influence
2.1 years	Stock Prices
6.4 years	Stock Prices
19.0 years	Stock Prices
50 to 60 years	Kondratieff Wave
57 years	International battles
170 years	Civil Wars
510 years	Civil Wars

Since the mid-1970s, Kondratieff Wave behavior captivated my imagination. The cycle's name comes from its discoverer, Russian economist Nikolai Kondratieff (1892-1938). He worked at the Agricultural Academy and Business Research Institute in Moscow. However, Kondratieff's ideas didn't sit well with Soviet Premier Joseph Stalin. Stalin jailed him in 1932, and later executed him before a firing squad in 1938. With his book *On the Notion of Economic Statics, Dynamics, and Fluctuations,* Kondratieff reported the discovery of a long-wave in commodity prices that repeated every 50 to 60 years.[Kondratieff, 1924]

Residing a step below the Kondratieff Wave, Dewey wrote extensively about a 19-year stock market cycle. Actually, various researchers reported this frequency anywhere from 17.7 to 20 years, with 19 residing close to the midpoint. This cycle heavily influenced business and real estate trends. In addition, the *Foundation for the Study of Cycles* reported cycles with periods of 6.4, 170, and 510 years.

If an unknown force linked all of these cycles by a harmonic of three, then small errors existed in their reported frequencies. By making minor adjustments of less than 1% to each cycle, a perfect harmonic sequence emerged that produced remarkable fits to existing datasets. **Table 4.2** shows the resulting sequence of cycles. This set of six cycles provided the basis for the *Unified Cycle Theory*. And it set the stage for the discovery of 19 other cycles in the Extra-Universal Wave Series.

As mentioned in Chapter 1, alleged business cycles often lack validation – with undesirable methods employed in estimating the frequencies. For example, many annual cycles have good reasons for their precise, one-year frequency. However, no logical reason exits to indicate that frequencies for multi-year cycles should equal integers. Yet, publications usually report multi-year cycles as integers. This either indicates a great deal of uncertainty, or a little bit of laziness, or both.

Table 4.2 – Adjustments to Conform with the EUWS Cycles.

EUWS Calculated Period	Adj. to Original Cycle	Cycle's Influence
2.12 years	1.0 % higher	Stock prices
6.36 years	0.6% lower	Stock prices
19.08 years	0.4% higher	Stock prices & real estate
57.24 years	0.4% higher	Wars & commodity prices
171.72 years	1.0% higher	Civil wars
515.16 years	1.0% higher	Civil wars

Equally important, human activity measurements can never be precise. The measurements themselves contain systemic errors. The error associated with the cycle's length should always be reported with the error.

Cyclic turning points also hold great value. Yet, researchers often avoid assigning dates to reversal points. For a practitioner, knowing that a 10.4-year cycle exists becomes meaningless without also knowing its theoretical peaks and troughs.

As far as the turning points for the EUWS cycles, except for the 515.16-year rhythm, all of the cycles peaked on January 24, 2007. The expected peak of the 515.16-year cycle comes on October 10, 2178. The remainder of this book reinforces these turning points. Scores of case studies show how past oscillations match with these ideal inflection points. For your reference, the **Appendix** provides a list of all theoretical turning points for all EUWS cycles.

Concerning their error, the EUWS frequencies contain accuracy within 2%. This estimate was arrived at in an unconventional manner – by trial and error. First, the estimate could not be calculated for each cycle individually. The EUWS cycles act as a group. If one frequency is adjusted, they all must be. That's because they link themselves by a harmonic of three. Hence, the error had to be estimated collectively. To make a long story short, the match between reversal points in historical data and theoretical peaks began to break down significantly by adjusting the EUWS cycles more than 2%. In fact, a slight breakdown occurred at 1%, but not enough to confidently claim a 1% error threshold. Nonetheless, more than likely, the EUWS frequencies published in this book contain accuracy within 1%.

Now, back to the adjustments made to get these cycles to match. Every cycle in the EUWS series relates to its neighboring cycle precisely by a factor of three. Some may argue that adjusting these frequencies in perfect alignment creates a charade. But that argument lacks merit. None of these cycles were precise to begin with. Estimates made from limited sets of data produced the original frequencies. Furthermore, all of these cycles came with no credible theory explaining their fluctuations. Their original discoverers simply stated their statistical significance, without providing a credible cause.

As time passed, more data became available. Additional data allowed new estimates with more precision. But remember, in spite of their inaccuracies, the original estimates proved to be correct within 1%. That accuracy shows the original discoverers performed fairly good work. This same process of making adjustments continues throughout the book. New data, along with a new theory, allows a large number of previously estimated frequencies to be fine-tuned to attain greater precision.

As a final rebuttal, if this core group of six cycles didn't behave as a group, then their predictions should have failed miserably. Bad theories produce dismal predictions. As a testimonial to its validity, the *Unified Cycle Theory* predicted an indefinite number of cycles with 19 of them already verified. Starting with Chapter 12, this book reviews each of the 19 predicted (and discovered) EUWS cycles, as well as the core group of six.

As time passes, more data will become available. This additional data will make future estimates even more reliable. Hence, some of the information in this book will soon become obsolete. Someone will come along to better estimate all EUWS frequencies. The cycle lengths will be slightly modified, and the dates of the turning points will surely be modified as well. However, don't expect major changes.

The next potential point of contention comes from the fact that these cycles represent different activities. Thus, it may be argued that they cannot originate from the same cyclical source. At first glance, fluctuations in stock prices, commodity prices, real estate, international wars, and civil wars represent completely different human activities. So how can they be linked? The answer lies with the timing of their turns. If these cycles acted on their own, their peaks and valleys should appear randomly. But they don't. Instead, the turning point of every third EUWS sub-cycle precisely matches with its parent cycle's inflection point. This continual, persistent behavior indicates these cycles share a common link. As the investigation moves into the longer-term EUWS cycles, the range of events broadens to include fluctuations in global climate, oscillations in evolution, and changes in the geological time scale. But for now, attention remains on markets.

Historically, stock and commodity quotes only became available on a daily basis during the last 125 years. Prior to about 1880, statisticians only recorded stock prices on a limited basis. Sporadic quotes from Europe exist back to the year 1700, after business corporations first formed. Commodity prices records actually go back thousands of years. However, the reliability and frequency of the data becomes somewhat suspect with age.

In spite of these obstacles, abundant knowledge about ancient civilizations becomes increasingly available each year. New archaeological finds continually add to the data collection. In addition, the scientific community has constantly improved dating methods. Today, we know the age of ancient events with accuracy unimaginable just a few decades ago. With precise indicators of ancient business activity still largely unavailable, a combination of data about commodity prices, architecture, art, food, trade, monetary systems, and wars provide significant clues about the state of ancient cultures and civilizations.

When it comes to cycles, longevity provides the ultimate validation tool. That's why ancient records become so critical. For example, the 19.08-year cycle leaves its mark through 15 cycles since 1720. However, this cycle becomes even more significant because of its appearance in ancient records. This fact minimizes modern monetary systems as the primary cause of the 19.08-year cycle. Furthermore, a cycle that repeats for 15 periods attains significant reliability. One that spans hundreds of periods becomes virtually undeniable.

Establishing the link between the EUWS cycles became critical to the *Unified Cycle Theory*. Hence, digging into historical archives became a requirement. The details come in later chapters. But for now, some examples will better illustrate how one part of the *Unified Cycle Theory* connects various human activities.

The 6.36-year EUWS cycle corresponds to an economic recession cycle. It becomes useful in predicting stock market reversals, as well as economic downturns and expansions. For example, the last three theoretical peaks for the 6.36-year cycle coincided very closely with actual stock market tops. The theoretical peak on May 7, 1994 came about the same time as the stock market downturn that year, which arrived just ahead of Orange County's default and the Mexican debt crisis. The next theoretical peak came on September 12, 2000, coinciding with the NASDAQ bubble top. Then, on January 24, 2007, the latest theoretical peak of the 6.36-year cycle arrived as the bubble burst for sub-prime lending.

Exactly three times greater than the 6.36-year cycle, the 19.08-year cycle comes next in the sequence. This cycle correlates closely with financial panics and severe bear markets. From this point forward, this oscillation will be referred to as the 19-year cycle. The 19-year cycle exhibits characteristics similar to its 6.36-year brethren. However, the bear markets, the financial losses, and the economic downturns reach a notch higher in severity. The last three peaks of the financial panic cycle came on November 24, 1968 – just days before an extensive 6-year bear market began; December 23, 1987 – within two months of the infamous October 1987 stock market crash; and January 24, 2007 – which coincided almost exactly with the bursting of the sub-prime lending bubble.

Next in the list of ever-intensifying cycles comes the 57.24-year economic depression cycle. From this point forward, this oscillation will be referred to as the 57-year cycle. The last three cyclic peaks arrived on July 31, 1892 – which coincided with the start of the 1892-1896 depression; October 26, 1949 – this represented a significant deviation, with the Great Depression of the 1930s starting almost twenty years earlier (an explanation of this deviation comes later); and on January 24, 2007. This 57-year cycle suggests that a new financial crisis began with the sub-prime crash. Eventually, it should turn into something exceeding the Great Depression.

The 171.72-year cycle comes next in the sequence. From this point forward, this oscillation will be referred to as the 172-year cycle. This cycle triggers economic depressions so severe that the resulting weakness subjects nations to both external attack and massive civil discontent. The previous peak of the 172-year cycle came on May 5, 1835. This top coincided almost exactly with the start of the longest bear market in the history of the United States. After the bubble in railroad stocks burst in 1835, stock prices declined seven straight years until finally hitting bottom in 1842. During the down-phase of this cycle from 1835 to 1921, the United States endured its only civil war from 1861 to 1865; fell into economic depression twice – 1835-42 and 1892-96; and encountered repeated financial panics about every 19 years. On a global basis, the outbreak of World War I came near the bottom of the cycle, covering the years 1914 through 1918. It ended with 40 million casualties. January 24, 2007 marked the latest peak of the 172-year war cycle. Hence, the down-phase is just beginning. Nonetheless, the intensity of the ongoing financial crisis indicates this crash will turn into a depression more severe than the one during the 1930s. As the downturn intensifies, another civil war among competing factions within the United States will become a distinct possibility.

This example ends by ratcheting up to 515.16-years – which represents the next step in cyclic frequency and strength. From this point forward, this oscillation will be referred to as the 515-year cycle. Peaks in this cycle heavily coincide with golden eras of history's great civilizations. Once a civilization peaks, it normally takes 100 to 300 years before the subsequent collapse completes its course. Over the past 5,000 years, 515-year theoretical peaks consistently coincided with the start of civilization collapses for the ancient Egyptians, Greeks, Romans, and other noteworthy civilizations. Details come in Chapter 26.

It's important to continue the discussion about the link between cycles in the stock market, economic depressions, wars, and the ebb-and-flow of civilizations. Ancient civilizations reveal their history through their art, architecture, agriculture, technology, geography, and war. But perhaps the most important item of all is often overlooked – a civilization's monetary system. A nation's monetary system reflects its strength. Monetary systems seldom remain stable. Instead, they fluctuate with their economy. During times of growth, the content of coinage remains stable and honest. However, once a civilization passes its prime, monetary authorities debase their coinage (or print paper notes in lieu of coins) to sustain spending programs. In modern times, central bank credit creation equates to the coinage debasement practiced by ancient, declining civilizations. In both cases, monetary systems disintegrate in ways that provided short-term benefits at the cost of long-term stability.

In its early stages, a growing civilization conquers new lands. Territory expands. Spoils from the victory of war enrich the national treasury. However, once territorial expansion halts, an empire reaches a plateau in prosperity – often called the golden era of the civilization. But lavish living standards, tax shortfalls, revolts, and wars all contribute to a national loss of savings. At that point, rather than cutting back on expenditures (corresponding to the loss in income) political leaders of great civilizations often continue to spend excessively. Several monetary symptoms reveal the weakness that results from over-spending.

These symptoms include debasement of coins (practiced by the Roman emperors during their declining years), printing paper money as currency (practiced by the Mongolians as Kublai Khan's rule weakened), or the issuing of credit as a replacement for money (used by modern central banking authorities). Excessive spending can last for decades. However, relying on accumulated wealth (savings) rather than current income to sustain living standards eventually leads to national impoverishment. As a great civilization approaches the poverty stage, financing a large, effective army becomes next to impossible. And without a sufficient army, a dominant military power soon becomes vulnerable. A complete civilization collapse usually follows shortly after the poverty stage.

Regardless of the type of monetary system a nation employs, it's hard to disguise debasement, inflation, over-issuance of credit, lack of savings, and over-consumption. When these symptoms appear, it means needed spending cuts failed to materialize. Civilization collapse usually follows rather quickly.

Never underestimate the condition of a nation's monetary system when examining its history, strength, and longevity. Stock market swings, credit market conditions, government spending, wars, and the costs associated with maintaining civil-order all interact closely as the fabric of civilization. They all move together as part of the same social behavioral cycle. A critical aspect of these social cycles relates to the magnitude of their associated EUWS periods. The general rule is: the greater the EUWS cycle, the larger the associated boom-bust movement.

Using this concept as it relates to the current global economic situation, it brings us back to the reasons why the financial insanity from 1982-to-2007 went so far beyond the excesses encountered in 1929. The answer lies with the fact that government leaders, investors, and consumers all followed their herd instincts. They jumped in together to enjoy the intoxicating effects of the final blow-off stage of a naturally occurring 172-year cycle in economic activity. The down-phase for this cycle, which just began in January 2007, will likely cause more financial pain and social conflict than the Great Depression of the 1930s. Until it reaches its trough in 2092, the impoverishing effects from the 172-year cycle will intensify with time. The likelihood of another world war or the use of nuclear bombs in battle conditions will increase as social conditions worsen. Only time will tell, but the immediate prospects appear dim.

Chapter 5
The Forces Behind Economic Cycles

The first four chapters served as an introduction to the smaller components of the EUWS. But what causes these cycles to materialize? No theory holds much worth if it cannot explain the sequence of events that it attempts to predict. To answer that question, the remaining chapters identify numerous correlations between these cycles and a variety of ancient events.

However, a few words of caution are necessary. It's somewhat easy to correlate one event with another. But it's much more difficult to determine cause and effect. Two events could be correlated by mere coincidence. Here's an example. For about thirty years, the NFL conference of the Super Bowl winner correlated highly with the performance of the stock market the following year. If the National Conference won, stocks went up. If the American Conference won, stocks declined. Yet, I doubt if very few people seriously believed that the winner of a football game affected the stock market's performance the following year.

No matter how incredible a correlation appears, it pays to be careful about jumping to conclusions about cause-and-effect. Especially early in the book, simply note the correlations. Along with the evidence, this book quotes various experts. These experts often provide opinions about cause-and-effect. Pay attention to their reasoning, but also be aware that their opinions might be erroneous.

If factor A correlates with factor B, then four possible explanations exist:

The Unified Cycle Theory

1. A causes B
2. B causes A
3. A third, unknown factor causes both A and B
4. Their correlation happens by mere chance

In this book, the third possibility (that a third factor causes the correlation) may come into play quite often. Chapter 35 reviews the entire set of evidence one final time – concluding with cause-and-effect assessments.

Now, the search begins for candidates causing these cycles. As I became increasingly convinced that financial cycles resulted from naturally occurring oscillations, identifying the source became critical. Without an answer, the whole theory seemed suspect.

The effort actually became somewhat tricky. As the analysis deepened, its diversity spread. Many components emerged, with many different sciences involved. The range of academic fields included economics, finance, biology, history, astronomy, physics, geology, climatology, psychology, sociology, and different subsets of these fields.

But at its core, astronomy and physics played the key role. From one perspective, these two sciences seem nearly identical. Astronomy deals with large objects outside of our planet Earth arena, while physics concentrates on very small objects (from the human point of view) such as the composition of matter and the forces that interact with this matter. However, out of necessity, astronomers require proficiency in both sciences – hence the term astrophysicist.

Addressing how order is maintained in our known universe, Stephen Hawking wrote the following in his book, *A Brief History of Time*:

> **"A well-known scientist ... once gave a public lecture on astronomy. He described how the earth orbits around the sun and how the sun, in turn, orbits around the center of a vast collection of stars called our galaxy. At the end of the lecture, a little old lady at the back of the room got up and said: 'What you have told us is rubbish. The world is really a flat plate supported on the back of a giant tortoise.' The scientist gave a superior smile before replying, 'What is the tortoise standing on?' 'You're very clever, young man, very clever,' said the old lady. 'But it's turtles all the way down!'"** [Hawking, 1988]

This comical reference leads us to the paradox of the composition of matter and the size of the known universe. While being more comfortable and understandable, the concept of finite existence opens the <u>turtles all the way down</u> paradox on both ends of the physical scale.

From the perspective of large things, if our universe contains finite mass, then what lies beyond the end of our universe? Does a large vacuum extend beyond the end? Or does some type of wall prevent matter from going beyond the edge of our universe? If so, what's the composition of this wall? And equally perplexing, from the viewpoint of small things, if quarks represent the smallest component of matter, then what are quarks made of?

From a logical standpoint, the concept of infinity in both directions seems more realistic than the concept of finiteness. The known order of physical composition, from smallest mass to largest mass follows: Quarks, atoms, moons, planets, stars, solar systems, galaxies, super-galaxies, and our known universe. As matter lumps together in ever larger masses, different bodies form. In essence, the size of a collection of mass determines its properties.

Now, back to the question: what are quarks made of? Do they consist of vibrating strings, as described by the relatively new String Theory? If so, what are strings made of? This questioning could go on endlessly to reveal ever smaller entities.

At the other end of the spectrum, why does the known universe have to be the end of mass-collections? Could it be possible that other universes cling together in groups to form universe clusters? And this type of possibility could go on indefinitely to produce ever-larger collections of mass, all residing within a much larger multiverse.

Based on a large number of observations, this concept of never-ending cycles (and infinity in both directions) became the core of the *Unified Cycle Theory*. It's important to interject that physicists generally reject the concept of infinity on both ends of the spectrum. However, the question still attracts considerable attention, as this discussion from astrophysicist Stephen Hawking indicates:

> "I believed strongly that the universe came into being, at a finite size, and I felt that implied that the universe now, was still of finite size, or closed. However, after Neil [Turok] gave a seminar on open inflation in Cambridge, we got talking. We realized it was possible for the universe to come into existence, at a finite size, but nevertheless, be either a finite, or an infinitely large universe now."
> [Hawking, 1998]

Basically, the concepts of both <u>infinity</u> and <u>nothing</u> amount to mathematical representations that create a conundrum when applied to the real world. For the time being, the infinity debate will be set aside – mostly because all solutions seem inadequate.

However, the topic can't be completely avoided. That's because the set of cycles outlined in **Table 4.2** bring the issue to the forefront. The six cycles join as a series, with a harmonic of three. But why stop at six? I suspected the EUWS could extend the harmonic in both directions. At a minimum, it seemed worth a try! And that summarizes the thought process involved in formulating the theory, and estimating the EUWS frequencies.

Next, to arrive at a list of causative candidates, basic physics provided help. Section 6 of Chapter 2 summarizes the four fundamental forces in the known universe. The short list consists of the gravitational force, the electromagnetic force, the strong nuclear force, and the weak nuclear force. The search for the source of the EUWS cycles involved identifying which of these four forces caused (either directly or indirectly) the array of fluctuations.

In reviewing these forces, the electromagnetic force looks the most promising, although none of the other forces can be completely ruled out. Electromagnetic fields affect all living organisms. For example, the movement of bacteria in water and the navigational skills of pigeons are both linked to the structure of the Earth's magnetic field. Hence, electromagnetism surfaces as a good starting point for analyzing how cycles affect human behavior. Surprisingly, however, gravity also plays an indirect role in some cycles that affect mankind. This association will be explained later.

Next, the weak nuclear force may also affect human behavior. This force only acts strongly from a very short range – at the sub-atomic level. Nonetheless, in action, it's a stronger force than either gravity or electromagnetism. It affects all quarks, and it's the only force to have a significant impact on neutrinos. A later chapter presents evidence correlating neutrino cycles to two EUWS cycles.

Finally, a link between the strong nuclear force and oscillations in human activity cannot be found. However, the strong nuclear force acts as a primary agent in the reactions that allow the Sun to radiate light and energy to Earth – thus allowing mankind to survive. Hence, the strong nuclear force cannot be totally excluded either.

Chapter 7 classifies known cycles by their associated physical forces. This classification helps in the analysis that follows later in the book.

To conclude, the forces behind the cycles that affect humans, as well as a variety of other events, are well known. They originate from the four basic physical forces. The extent to which they affect mankind remains to be determined. However, before proceeding with the classification system, focus shifts to the manner in which these forces influence human behavior.

Chapter 6
How Physical Cycles Affect Humans

As my market-related interests expanded, claims about celestial events correlating to market movements came to my attention. Examples included markets that followed sunspot activity, phases of the moon, and eclipse cycles. At first, I met these claims with skepticism. After all, how could sunspots cause the stock market to fluctuate? The idea seemed just as radical as the notion that the Super Bowl winner influences stock market trends.

In spite of the skepticism, curiosity got the best of me. After a few years of studying these relationships, it became apparent that these events correlated too closely to be the product of mere chance. Endless nights of contemplation finally resulted in a plausible explanation.

Gravity became the first candidate for investigation. Because of their close proximity, Moon's gravitational field significantly interacts with Earth's. In combination with gravity from the Sun, ocean tides provide the clearest example. However, lunar gravity exerts a negligible influence on individuals. This fact makes it unlikely that gravitational cycles influence market oscillations.

Furthermore, the Moon wasn't alone in correlating with market fluctuations. Market moves often correlated with sunspot activity and seasonal cycles as well. Unlike the tides, these cycles have no direct connection to gravity. Whatever causes these behavioral cycles, a common cause held more credibility than a separate cause for each cycle. Gravity failed to fit the requirement

The Unified Cycle Theory

of a common cause.

The search for a force associated with both the Sun and Moon led to the investigation of the Earth's magnetic field. That decision proved successful. While fluctuations in geomagnetism cannot fully explain market cycles, they provided an important first step.

As an example of a geomagnetic cycle affecting market behavior, the theory illustrates how sunspot activity influences market behavior. This happens through an extensive chain of events. The hypothesized sequence follows: The sunspot cycle reaches a maximum. Sunspots act as a barometer of dramatic changes taking place on the Sun's surface. Sunspots appear when massive solar eruptions take place. These eruptions intensify the solar wind's electrical charge. The solar wind continually spews from the Sun. After traveling through space for about 3 or 4 days, the wind hits the Earth's atmosphere. These solar wind strikes cause our atmosphere to become highly ionized. The Aurora Borealis (or Northern Lights) appear. The Aurora Borealis serves as another symptom. The solar wind bombardment causes the Earth's geomagnetic field to change. As the final step in this process, these geomagnetic fluctuations affect the electro-chemical nervous systems in all humans. These changes in nervous system operation affect how people feel. This causes the collective human mood to fluctuate. In response, a large number of individuals change their investment strategies. And finally, economic and market conditions change accordingly. As amazing as this sequence seems, existing academic studies support each step of the process. Details follow in a few paragraphs.

In addition to the electromagnetic cycles, other physical cycles correlate with and affect humans in similar sequences. However, few in the academic community have investigated these areas. In an attempt to explain social cycles, some competing theories propose that economic and behavioral cycles sustain lives of their own. Those theories suggest that human activity cycles play out as an inherently natural part of human social existence. Others claim that economic cycles result from loss of memory – making mankind destined to continually repeat the mistakes of our ancestors. The *Unified Cycle Theory* completely rejects those ideas.

Occurring naturally in our universe, identifiable physical cycles cause all major human activity cycles. The remaining portions of this book present the correlations, research abstracts, and explanations that help isolate the various physical forces. In addition to affecting humans, these cycles play a dominant role in life for all species, global climate, and even the operation of our universe.

Now, it's necessary to slow down and contemplate this idea. In its entirety, the *Unified Cycle Theory* amounts to an entirely new concept; however, broken down into its separate components, most of the key concepts were previously theorized by others. But to enhance credibility of the *Unified Cycle Theory*, this book does more than just theorize. Instead, it presents chapters of evidence supporting each critical part of the theory.

Getting back to the solar influence on markets, it's necessary to break down the theorized string of events to see if it really does make sense. The *Sun and Sunspots* web page for NOAA's

National Weather Service explains in greater detail how sunspots affect Earth:

> "Coronal Mass Ejections and solar flares are extremely large explosions on the photosphere. In just a few minutes, the flares heat to several million degrees Fahrenheit and release as much energy as a billion megatons of TNT. They occur near sunspots, usually at the dividing line between areas of oppositely directed magnetic fields. Hot matter called *plasma* interacts with the magnetic field sending a burst of plasma up and away from the Sun in the form of a flare. Solar flares emit x-rays and magnetic fields which bombard the Earth as *geomagnetic storms*. If sunspots are active, more solar flares will result creating an increase in geomagnetic storm activity for the Earth. Therefore during sunspot maximums, the Earth will see an increase in the Northern and Southern Lights and a disruption in power grids and radio transmissions." [NOAA, National Weather Service, 2008]

And the *Solar Physics* web page for the National Aeronautics and Space Administration (NASA) further explains how the eruptions associated with sunspots transfer energy to Earth:

> "The solar wind is not uniform. Although it is always directed away from the Sun, it changes speed and carries with it magnetic clouds, interacting regions where high speed wind catches up with slow speed wind, and composition variations. The solar wind speed is high (800 km/s) over coronal holes and low (300 km/s) over streamers. These high and low speed streams interact with each other and alternately pass by the Earth as the Sun rotates. These wind speed variations buffet the Earth's magnetic field and can produce storms in the Earth's magnetosphere." [NASA, Solar Physics, 2008]

The only possible point of contention with the idea that sunspot cycles act as an indicator of stock market trends lies with the final part of the sequence – the proposition that human beings respond to fluctuations in the Earth's magnetic field. But even here, researchers have linked geomagnetic fluctuations with human behavior.

In a British Journal of Psychiatry article entitled *Geomagnetic Storms: Association with Incidence of Depression as Measured by Hospital Admission,* R.W. Kay wrote:

> "The hypothesis that geomagnetic storms may partly account for the seasonal variation in the incidence of depression, by acting as a precipitant of depressive illness in susceptible individuals, is supported by a statistically significant 36.2% increase in male hospital admissions with a diagnosis of depressed phase, manic-depressive illness in the second week following such storms compared with geomagnetically quiet control periods.... Effects on cell membrane permeability, calcium channel activity, and retinal magneto-receptors are suggested as possible underlying biochemical mechanisms."[Kay, 1994]

Some may argue that this study only relates to mental patients. However, many seasoned investors will counter that with the observation that insanity is a prerequisite for investing. Humor aside, Kay makes three important observations that mirror my research. First, geomagnetic activity correlates with human activity. Second, this correlation operates on a delayed basis. And third, a seasonal cycle in geomagnetic activity appears. Chapters 7 and 10 contain extensive details about sunspots, geomagnetic cycles, and their delayed impact on mankind.

In an article from the University of Minnesota's Institute for Rock Magnetism, entitled *Out, Damned Spot!* Senior Scientist and Facility Manager, Mike Jackson, describes other examples of links between geomagnetism and human activity, including, astonishingly, birth rates!

> "There is a surprisingly extensive literature on correlations between solar-driven geomagnetic activity and a wide variety of biomedical phenomena. At least it came as a surprise to me, though in retrospect it probably shouldn't have. After all, since Galvani's frog experiments we've known of the importance of tiny electrical impulses in the nervous system, and it is plausible, at least in a vague way, to imagine subtle effects related to magnetic storms. The plasma streams from coronal mass ejections (CMEs) are made up of fast-moving, highly-ionized particles, some fraction of which are able to penetrate through the Earth's magnetic field.... Numerous related studies follow a similar empirical approach, identifying (or challenging previously-claimed) statistical correlations between geomagnetic indices and the occurrence of various medical conditions. Some of these seem generally plausible (e.g., migraines, epileptic seizures, and cardiac failures), and others are more unexpected. For example Randall & Moos published a 1993 article in the *International Journal of Biometeorology* ...entitled 'The 11-year cycle in human births.' They compiled national and regional birth data for the period 1930-1984 for the U.S., England and Wales, New Zealand, Japan, Switzerland, New South Wales Australia, and Baden-Württemberg Germany. They then compared these statistics with the AA geomagnetic activity index and related parameters including the Wolf sunspot index and temperature. Astonishingly, the strongest bivariate correlation was between births and sunspots (Spearman rank correlation coefficient 0.86, p<0.01)."[Jackson, 2003]

As far as the actual mechanism causing this correlation, Jackson ends the article leaving us guessing:

> "Three conceivable mechanisms for DC magnetic field bio-effects in humans – including magneto-reception – were discussed (and dismissed) by Sastre *et al* [2002]. First, and most obvious, is the torque exerted on the permanent magnetic moment of single-domain magnetites, which may act as tiny switches for opening/closing trans-membrane ion channels [e.g., Kirschvink *et al.*, 2001] or produce other changes in neural cell activity. Sastre *et al* [2002] conclude that their EEG measurements during controlled field reversals should show evidence

of such activity if it occurs, but do not. Similarly they argue against induction related to motion in a static field, and field control of chemical reaction rates that involve electron spin states, based on lack of observed EEG response. Yet they acknowledge several caveats and remaining puzzles, including one of their own experimental results, which did show significant (p<0.01) changes resulting from a sudden steepening of the ambient field inclination to a near vertical orientation. The question remains open." [Jackson, 2003]

Of course, the suggestion that geomagnetic cycles influence economic activity opposes the mainstream view. In the United States, most people believe that government leaders, central bankers, commercial bankers, and business leaders determine the state of economic affairs. Especially in the political arena, citizens cast their votes for candidates they believe will improve economic conditions by "fixing" any and all problems.

In the capital markets, the prevailing view goes something like this ... as economic and monetary conditions change, investor sentiment changes correspondingly. However, according to the *Unified Cycle Theory*, just the opposite happens. In his book *Socionomics: The Science of History and Social Prediction*, Robert Prechter, Jr. describes the relationship between market action and investor sentiment this way:

"Mood impels action. An increasingly positive social mood causes people to buy stocks and expand businesses... [while] an increasingly negative social mood causes people to sell stocks and contract businesses... Actions do not affect mood - it's the other way around." [Prechter, 2003]

Prechter's observation matches the thesis outlined by the *Unified Cycle Theory*. Further details about magnetic fields help in understanding how they affect mood swings and the decision-making process. Electrochemical processes serve as building blocks for the human nervous system. And electromagnetic fields affect all living organisms. Scientists have known this fact for decades.

A World Health Organization report entitled *Environmental Health Criteria for Magnetic Fields* describes various creatures with observable effects from geomagnetism:

"Some organisms possess sensitivity to static magnetic fields with low intensities comparable to that of the geomagnetic field (about 50 µT). Phenomena for which there is substantial experimental evidence of sensitivity to the earth's field include: (a) direction finding by elasmobranch fish (shark, skate, and ray); (b) orientation and swimming direction of magnetotactic bacteria; (c) kinetic movements of mollusks; (d) migratory patterns of birds; and (e) waggle dance of bees." [World Health Organization, 1987]

However, the fact that the Earth's magnetic field helps certain organisms navigate doesn't provide evidence of an unconscious effect on human moods. But the report does state the intensity needed for magnetic fields to affect human health:

> **"It can be concluded that available knowledge indicates the absence of any adverse effects on human health due to exposure to static magnetic fields up to 2T. It is not possible to make any definitive statements about safety or hazard associated with exposure to fields above 2T. From theoretical considerations and some experimental data, it could be inferred that short-term exposure to static fields above 5T may produce significant detrimental effects on health."**
> [World Health Organization, 1987]

Even though intense magnetic fields are dangerous to human health (with electrocution being an extreme case), health effects do not directly translate into mood swings. Furthermore, geomagnetic fields fluctuate at intensities far less (by a factor of 100 to 200) than household electrical fields that we encounter every day. In short, substantial evidence exists showing a link between fluctuations in the Earth's magnetic field and cycles in human behavior. The mystery revolves around the exact sequence of events responsible for the correlations.

Some may wonder about geomagnetism and its implications for mankind's free will. The *Unified Cycle Theory* states that physical cycles translate into fluctuations in human behavior. Even though mankind possesses free will and the ability to reason, when it comes to important monetary decisions, people typically follow trends and their emotions rather than using sound reasoning. This does not imply that all people blindly follow the crowd. It only means the majority do. I've arrived at this conclusion after years of reading, talking to brokers, and listening to how others make their investment decisions.

Most investors use herd-instincts, tips, and hunches more than strict fundamental analysis. When buying stocks, very few people research a company to evaluate its book value. And when they do, they usually don't pay attention to which assets are actually tangible, and which ones are worthless, like goodwill. And most investors, even professionals, don't have a clue about assets and liabilities hidden from corporate balance sheets. After Enron's collapse, Congress passed the Sarbanes-Oxley Law to correct these transparency issues. Yet, in spite of the law, most large financial institutions found loopholes, and they continued hiding assets from their quarterly reports until the credit-crisis erupted in 2007.

As an example of irrational decision making, the recent housing bubble provides an excellent case study. Before purchasing their homes, very few people sat down to ask a very basic and fundamental question: "Can I comfortably afford the monthly payments on my new mortgage?" The answer was a resounding – No. Rather, home-buyers fell under the influence of peer pressure – as the housing bubble inflated. It made a lot of people jealous when their neighbors and friends bragged about the vast sums of money being made on paper as home prices rose. An ever increasing number of people came to believe that home prices would rise forever. And they felt that by purchasing a home (regardless of the price), it could be used as an ATM-machine anytime they found their finances short of cash.

How Physical Cycles Affect Humans

A research team lead by Samuel McClure of Princeton University identified the conflict that arises when difficult financial decisions arise, and how our brains frequently favor the irrational-emotional choice. This decision making process originates as a conflict between our emotional brain and our logical brain. In a sense, this revelation is old news. Throughout history, these internal conflicts have been described in many ways – the conflict between good and evil; listening to one's guardian angel rather than the devil; or using common sense rather than impulse. In a Science article entitled *Separate Neural Systems Value Immediate and Delayed Monetary Rewards*, the McClure team wrote:

> "When humans are offered the choice between rewards available at different points in time, the relative values of the options are discounted according to their expected delays until delivery. Using functional magnetic resonance imaging, we examined the neural correlates of time discounting while subjects made a series of choices between monetary reward options that varied by delay to delivery. We demonstrate that two separate systems are involved in such decisions. Parts of the limbic system associated with the mid-brain dopamine system, including para-limbic cortex, are preferentially activated by decisions involving immediately available rewards. In contrast, regions of the lateral prefrontal cortex and posterior parietal cortex are engaged uniformly by intertemporal choices irrespective of delay. Furthermore, the relative engagement of the two systems is directly associated with subjects' choices, with greater relative fronto-parietal activity when subjects choose longer term options." [McClure *et al.*, 2004]

In a Harvard University Gazette article entitled *Brain Takes Itself on over Immediate vs. Delayed Gratification*, Steve Bradt describes the McClure team results in terms we can all relate to:

> "You walk into a room and spy a plate of gooey doughnuts dripping with chocolate frosting. But wait: You were saving your sweets allotment for a party later today. If it feels like one part of your brain is battling another, it probably is, according to a newly published study.... Researchers at four universities found two areas of the brain that appear to compete for control over behavior when a person attempts to balance near-term rewards with long-term goals. The research involved imaging people's brains as they made choices between small but immediate rewards or larger awards that they would receive later. The study grew out of the emerging discipline of neuroeconomics, which investigates the mental processes that drive economic decision making.... The researchers examined a much-studied economic dilemma in which consumers behave impatiently today but prefer/plan to act patiently in the future. For example, many people who are offered the choice of $10 today or $11 tomorrow choose to receive the lesser amount immediately. But if given a choice between $10 in one year or $11 in a year and a day, people often choose the higher, delayed amount.... In classic economic theory, this choice is irrational because people are inconsistent in their treatment of the day long time delay. Until now, the

cause of this pattern was unclear, with some arguing that the brain has a single decision-making process with a built-in inconsistency, and others, including the authors of the Science paper, arguing that the pattern results from the competing influence of two brain areas. 'Our emotional brain has a hard time imagining the future, even though our logical brain clearly sees the future consequences of our current actions,' said Laibson, an economist in Harvard's Faculty of Arts and Sciences. 'Our emotional brain wants to max out the credit card, order dessert, and smoke a cigarette. Our logical brain knows we should save for retirement, go for a jog, and quit smoking. To understand why we feel internally conflicted, it will help to know how myopic and forward-looking brain systems value rewards and how these systems talk to one another,' Laibson said. ... The finding supports the growing view among economists that psychological factors other than pure reasoning often drive people's decisions. The work may also cast light on other forms of impulsive behavior as well as drug addiction."[Bradt, 2004]

At this point, a good basis for the *Unified Cycle Theory* has been established. However, later in the book, chapters of additional evidence strengthen the case substantially. For now, a quick review concludes this chapter.

In nature, a harmonic, physical cycle lives that affects just about everything in our universe. Within the series, the frequency of each parent cycle equals a period exactly three times greater than its child cycle. As each cycle lengthens, its magnitude increases. In human affairs, starting with the shortest frequencies, the EUWS cycles impact stock market movements, economic recessions, final panics, economic depressions, civilization patterns, and war occurrence.

All of the cycles that affect mankind can be linked either directly or indirectly to one of the four basic physical forces – gravity, electromagnetic fields, the weak nuclear force, or the strong nuclear force.

Fluctuations in magnetic fields affects the way humans feel. Presumably, magnetism influences the operation of our electrochemical nervous system, which influences our limbic systems, which controls our emotions. As a consequence, geomagnetic cycles translate into collective mood swings affecting a large portion of the human populations. These mood swings usually over-rule our logical brains in making investment choices, war decisions, and political picks.

Chapter 7
Classification of Cycles

The last few chapters focused on geomagnetic cycles and their affect on mankind. However, forces other than geomagnetism influence humans. The cycles mentioned so far only represent a subset of all naturally occurring oscillations.

This chapter takes the most significant cycles, and classifies them by groups. Classification allows for better organization in the deductive process at this book's conclusion. The cycles are classified by the physical force most closely associated with the source of the repetitive pattern. Once again, these forces include gravity, electromagnetism, and the two nuclear forces.

The Gravitational Force -- Gravity and Motion Cycles

Gravity and motion cycles heavily influence atmospheric and climatic conditions on Earth. But the list is short. Only three gravitational cycles affect Earth's weather in a significant way.

The 24-Hour Daily Cycle – As the Earth spins around its axis, it creates a 24-hour cycle of night and day. While associated with gravity from the Sun, this cycle primarily involves motion. Earth formed from circulating gases, and the spinning motion has persisted ever since. Once set in motion, a body continues in motion until some counter-acting mass interrupts it. With no such resistance in sight, Earth will continue spinning in its 24-hour cycle for a long time. As Earth rotates, alternating periods of warm, bright days and cool, dark nights cause people to make minor adjustments. The daily cycle can influence small decisions such as when to work, what clothes to wear, and whether to stay indoors or outdoors.

The 365.25-Day Annual Cycle – As the Earth rotates around the Sun, the seasons of the year change. It's the constant gravitational pull from the Sun that keeps Earth in its yearly orbit around the Sun (and keeps our planet from flying out into space). Along with the Sun's gravity, this cycle results from the tilt in Earth's axis. Because of the tilt, the Northern Hemisphere receives more hours of sunlight during the months of April through August – hence, the Northern Hemisphere enjoys spring and summer during these months. Conversely, the Southern Hemisphere receives more daylight from October through February, coinciding with their spring and summer. Mankind adapted to the annual weather cycle with improved clothing and permanent housing.

The 100,000-Year Ice-Age Cycle – The short-term nature of the daily and seasonal cycles makes everyone aware of them. However, the same cannot be said of the ice-age cycle. Like night and day, ice-ages occur at regular intervals. Similar to the way seasons change due to the Earth's axial-tilt as it rotates around the Sun, the currently accepted climate theory identifies four long-term cycles that modulate wide fluctuations in the Earth's temperature every 100,000 years. During this cycle, global temperatures usually change anywhere from 10°C to 16°C (equivalent to 18°F to 30°F). In 1941, Serbian mathematician and astronomer, Milutin Milankovitch, outlined a theory about how global climate varies according to gravitational cycles. However, the Milankovitch Theory went mostly unnoticed until credible information about ancient temperatures became available during the 1970s. Unfortunately, ice-age frequencies lack the precision of the daily and seasonal cycles. More than likely, the next ice-age has already started. If not, one should start within a few thousand years. Ice ages begin in a way similar to the arrival of autumn in the Northern Hemisphere. As August rolls around, no guarantee exists that cooler weather will begin; however, on average, temperatures gradually decline with each passing day until cooler temperate become noticeable by late-September or early-October. The Milankovitch Theory proposes that ice-age cycles result from variations in the following gravitational cycles.

- A 405,000-year cycle in the eccentricity of Earth's orbit.
- A 96,600-year cycle in the eccentricity of Earth's orbit, as Earth alternates between states of <u>nearly circular</u> to <u>slightly elliptical</u> orbits.
- A 41,000-year obliquity cycle resulting from variations in Earth's axial tilt from 22.1° to 24.5°.
- A 21,000-year precession cycle in the terrestrial axis which causes the equinox-date to shift throughout the period.

Climatologists generally accept the Milankovitch Theory as the source of ice-age cycles. However, several problems arise. And a small group of scientists now openly question several aspects of the theory. In fact, the *Unified Cycle Theory* solves many of these issues. A more detailed analysis comes later in the book.

Classification of Cycles

Chart 7A captures the essence of gravity and motion cycles. A characteristic of the seasonal cycle includes temperature peaks and valleys closely matching theoretical turnings points – year after year. This makes its frequency easy to calculate after a few cycles. Furthermore, rather than reversing abruptly, seasonal temperature extremes turn gradually over a period of two to three months. Essentially, gravitational cycles resemble sine waves. In fact, by averaging daily temperatures from any area of the world over a period of twenty years, the resulting chart yields a near-perfect sine wave with a 365.25-day frequency. The sine wave fingerprint of gravitational cycles helps distinguish them from cycles related to other physical forces.

The Electromagnetic Force – Earth's Geomagnetic Cycles

Several naturally occurring cycles modulate Earth's magnetic field. As far as documenting the cause of the various geomagnetic cycles, physicists have focused mostly on the dominant factor – the Sun. However, in addition to the Sun, magnetic influences also originate from the Milky Way and the Moon. The following sections review the seven major geomagnetic cycles. The review starts with the most dominant cycle and proceeds down the list – ending with the least influential oscillation.

The Schwabe Sunspot Cycle, approximately 11-year period – German astronomer Samuel Heinrich Schwabe discovered the 11-year sunspot cycle in 1843. This solar cycle directly impacts the geomagnetic field. The Geomagnetic AA index consists of two components.[Feynman, 1982] First, the Solar Activity Component results from solar flare eruptions. Second, the Interplanetary Component is associated with coronal holes and recurrent high-speed solar wind streams. It's interesting to note that the interplanetary component of the geomagnetic AA field precedes changes in the sunspot cycle, and it's the best known predictor of a sunspot maximum. This leading indicator seems odd. How is it possible for a development on Earth to predict future sunspot numbers? Does this imply that an external force influences both sunspots and geomagnetic cycles? Unfortunately, a satisfactory answer cannot be found.

The Gleissberg Sunspot Cycle, approximately 87.8-year period – Named after German astronomer Wolfgang Gleissberg, this longer term cycle regulates the amplitude of the 11-year sunspot cycle. Similar to other magnetic field frequencies, scientists often disagree on the length of the Gleissberg Cycle – with estimates ranging from 70 to 100 years. However, two University of Arizona geologists recently estimated the frequency of this cycle at 87.8-years.[Peristykh and Damon, 2003]

The Suess Sunspot Cycle, approximately 208-year period – Named after Austrian chemist/physicist Hans Suess, this cycle also affects conditions on Earth. The same geologists who determined the Gleissberg frequency estimated the Suess Cycle at 208 years.[Peristykh and Damon, 2003] In combination with the Gleissberg Cycle as well as other cycles, some climatologists suggest that the Suess Cycle helps regulate oscillations in warm and cool periods on Earth. For example, recent globally cool periods spanning the Oort Minimum (1010-1050), the Wolf Minimum (1280-1340), the Spoerer Minimum (1420-1530), the Maunder Minimum

(1645-1715) and the Dalton Minimum (1795-1825) were separated by an average of 195 years. In their abstract *Persistence of the Gleissberg 88-Year Solar Cycle Over the Last 12,000 Years: Evidence from Cosmogenic Isotopes,* Peristykh and Damon note of the controversy surrounding the role played by sunspot cycles and geomagnetism in regulating global climate:

> **"The importance of the Gleissberg cycle is increased also by its significance in studies of solar–terrestrial connections. It is not our intention to discuss in detail the role of the Gleissberg cycle in climate change. Prior to NASA's Solar Maximum Mission satellite instrumental (ACRIM/I) monitoring of solar irradiance ... it was popular to refer to the 'solar constant'. Suggestions of a role of the Sun in climate change were frequently not taken seriously or even ridiculed because of the excesses of the Schwabe cycle correlations sometimes referred to as 'cyclomania'. However, since the demonstration of a ca. 0.1% change in solar irradiance during the NASA mission, the role of the Sun has been taken seriously leading to many publications. That was essentially initiated by Eddy [1976] who suggested 'a possible relationship between the overall envelope of the curve of solar activity and terrestrial climate' found from comparison between sunspot numbers and climatic records."** [Peristykh and Damon, 2003]

Short-Term Sunspot Cycle, 27-day period – Like other sunspot cycles, fluctuations in this short-term cycle synchronize with the Earth's magnetic field. Unfortunately, this cycle's periods vary so widely that it becomes almost impossible to predict turning points. Sometimes the frequencies widen to 40 days. Other times they shrink to 15 days. In some years, such as 2007 (see **Chart 7B**), the 27-day oscillations stabilized somewhat. In other years, they virtually disappear. Its extraordinary variability makes this cycle difficult to analyze. Nonetheless, because of its intensity and its effect on geomagnetism, it likely plays a vital role in human behavior cycles. However, time constraints prevented adequate research. For now, its influence remains unknown.

The Semiannual Geomagnetic Cycle, 182.625-day period – Scientists only show limited interest in the semiannual geomagnetic oscillation. **Chart 7C** demonstrates its strength. It shows the annual average of Geomagnetic AA since 1868, with a 25-day average employed to smooth the data. The x-axis represents the number of days elapsed since the start of the year. A clear pattern of two cycles per year appears. The *Unified Cycle Theory* hypothesizes that this semi-annual oscillation serves as a primary influence for the seasonal pattern exhibited in financial markets. An early proponent of the seasonal cycle in the stock market, Yale Hirsch, published the *Stock Trader's Almanac* noting the tendency. His son, Jeffrey, now heads the publication.[Hirsch, 2008]. Periods of geomagnetic strength during the January-March and August-October periods correspond with an unusually large number of panics and crashes. Conversely, declining geomagnetic activity coincides with an absence of financial panics. This inverse relationship between geomagnetism and market cycles appears in other frequencies as well. Even though geomagnetism affects seasonal market behavior, longer-term cycles occasionally obliterate its influence. However, when other cycles synchronize with the seasonal cycle, predictive powers increase. For stock market traders, this cycle definitely deserves attention.

The Eclipse Cycle, approximately 173.31-day period – The eclipse cycle results from different angles between the plane of the Moon's orbit around the Earth and the plane of the Earth's orbit around the Sun. Normally, these three heavenly bodies reside in different planes. However, about once every 173 days, these bodies come into perfect alignment. When that happens an eclipse occurs during either the corresponding full moon (a lunar eclipse) or new moon (a solar eclipse). **Chart 7D** shows the average Geomagnetic AA tendency for all eclipse cycles since 1868. The x-axis represents the number of days since the first new moon before a solar eclipse. So, days 0, 29, 58, 87, 116, and 145 represent the days around the time of a new moon. Notice how Geomagnetic AA tends to drop around the time of a new moon. And notice how the geomagnetic AA level rises in between those intervals (at the time of a full moon). However, two exceptions appear. At day 0, at the time of the first new moon before a solar eclipse, geomagnetic AA tends to peak – which is the exact opposite of its normal behavior at the time of a new moon. And slightly after day 73 (around the time of a full moon), geomagnetic activity falls instead of rising. Similar oscillations occur in stock and commodity markets. Other than these two exceptions, the eclipse cycle mostly reflects geomagnetic changes related to the Lunar Cycle, which comes next.

The Lunar Cycle, 29.53-day period – The 29.53-day cycle in Geomagnetic AA originates from the Moon's magnetic affect on Earth. This happens because of the Moon's distortion of Earth's magneto-tail. Details about the magneto-tail come later in Chapter 34. **Chart 7E** shows the average Geomagnetic AA tendency for all lunar cycles since 1868. This tendency also appears in stock market fluctuations.

For periods of 300 years and less, the seven preceding cycles combine to produce most of the change in Geo AA. However, it should not be assumed that these seven cycles account for all variability in geomagnetism. Other, less dominate cycles surely play a role as well. But this book only concentrates on dominant cycles.

Unlike gravitational cycles, magnetic cycles tend to exhibit erratic behavior. Gravitational cycles consistently display smooth sine wave appearances. However, magnetic cycles show a different set of characteristics. First, instead of rising and falling slowly and steadily, magnetic cycles ascend and descend in stair-step fashion. Second, instead of turning smoothly, magnetic cycles often reverse suddenly, with spikes. Third, instead of showing consistent frequencies, the lengths of geomagnetic periods deviate quite substantially.

As this book progresses, note the characteristics of the various cycles. These characteristics help in classification of the various cycles, especially the mysterious EUWS cycles. To demonstrate shifting geomagnetic frequencies, the study now moves back to the Schwabe Sunspot cycle. **Table 7.1** analyzes the 35 complete sunspot cycles between 1616 and 2000. This analysis separates the cycles by century.

Table 7.1 – Schwabe Sunspot Cycle, Variations by Century.

Century	Avr. Frequency (years)	Longest Cycle (years)	Shortest Cycle (years)
17th	11.13	15	8
18th	11.00	17	8
19th	11.22	14	7
20th	10.56	12	9

Between 1616 and 2000, 35 complete sunspot cycles developed.[Eddy, 2003] and [NOAA, NGDC, 2008] That translates into an average period of 10.97 years. Yet, from century to century, the frequency varied substantially. During the 20th Century, the periodicity shrank to 10.56 years. And a century earlier, the average frequency reached 11.22 years.

Looking at the extremes for individual cycles, the 20th Century showed the greatest stability, with periods ranging from 9 to 12 years. However, during the 18th Century, oscillations varied wildly from 8 to 17 years.

Inconsistencies in naturally occurring magnetic cycles have wreaked havoc on analysts attempting to estimate their frequencies. Literature on the topic creates confusion about which estimate to trust. Different statistical methods yield different frequency estimates for the same dataset.

As you read on, keep these magnetic field tendencies in mind. The wide variations in estimated frequencies could be interpreted as a sign these cycles only live in the imaginations of their discoverers. However, a better interpretation exists. Substantial variability provides a strong indication that electromagnetism modulates the cycle. This is one example of how a characteristic of a cycle helps determine its source.

Next, the analysis moves to the eclipse cycle. Of all the geomagnetic cycles, the eclipse cycle may seem the strangest. Yet, in Mike Jackson's *Out, Damned Spot* article, he describes altered physical conditions in both humans and rats at the time of a solar eclipse:

> "One early experimental study [Keshavan et al., 1981] bears the horrifying/fascinating title *'Convulsive threshold in humans and rats and magnetic field changes: Observations during total solar eclipse.'* Psychiatric patients undergoing electro-convulsive therapy were administered shocks of progressively increasing strength until the convulsive threshold was reached, both on a control day of 'normal' magnetic field variation, and on the day of a solar eclipse, when solar ionizing radiation was significantly cut off by the interposition of the moon. Magnetometer readings showed a 19 nT transient variation due to the eclipse, and convulsive thresholds in both the human patients and in laboratory rats were reduced by statistically significant amounts

compared to the control-day thresholds. The control day was chosen to be 28 days later (1 lunar synodic period and very close to one solar rotation period), to minimize differences related to those periods, and the authors therefore concluded that the small 19 nT field variation was responsible for depressing the thresholds, through a mechanism that remained unexplained."[Jackson, 2003]

Jackson also notes how the stock market mirrored the sunspot cycle between 1996 and 2002:

"Correlation, however statistically significant, does not necessarily prove causation. For example, look at the time series plotted in the figure on page 10. [Not shown here.] Both data sets exhibit short-term ... variations that appear to be random, superimposed on a slow systematic trend, rising to a maximum in 2000 and then declining. Not surprisingly, one of these variables (#2) is the Wolf sunspot index, which peaked at 169 in July, 2000. Have you guessed what the other variable is? (*hint:* To buy or not to buy - that is the question.) It is the Dow Jones Industrial Average stock-market index. For the 88-month interval beginning in January, 1996, these indices exhibit a highly significant statistical correlation (R2=0.69, p<0.01), although for the previous several decades, no such correlation exists. Is it conceivable that some causative connection took hold in 1996?"[Jackson, 2003]

While Jackson failed to pinpoint the "causative connection," the *Unified Cycle Theory* provides a solution. The sunspot cycle, via geomagnetism, acts as one of many variables affecting human behavior. When the sunspot cycle syncs with other important cycles, the correlation between stocks and sunspots can appear quite strong. That happened during the 1990s. However, when synchronization disappears, the correlation with sunspot activity can fade just as quickly.

The examination now moves to the seasonal cycle. A research team led by P.T. Nastos (University of Athens, Greece) studied the effects of geomagnetism on people in Athens. In their abstract, entitled *Environmental Discomfort and Geomagnetic Field Influence on Psychological Mood in Athens*, the Nastos team found emotional swings corresponding to seasonal tendencies in geomagnetism. The Nastos team wrote:

"This analysis implies significant contribution of environmental variations, expressed by a discomfort index, in the aggravation of psychological symptoms like depression, sleep disturbances, anxiety, aggressiveness, etc. Moreover, geomagnetic field variations expressed by the international geomagnetic index ... manifest significant indications that they contribute to the aggravation of sleep disturbances. A clear seasonal variation, with a maximum around August and a minimum at the end of the year, appears in the environmental index, while a double oscillation with a period of about six months is obvious in the geomagnetic index. The same more or less seasonal variation was mirrored in most of the psychological symptoms we analyzed in the present study."[Nastos et al., 2006]

This study notes the dual nature of the seasonal cycle, the gravitational one with an annual frequency and the geomagnetic one occurring semi-annually. In particular, the sum of patients with panicked and depressed moods (the two sentiment factors associated with market panics) closely correlated with the Geomagnetic AA cycle. See **Chart 7F**. The x-axis represents the month of the year. Note that stocks typically endure their greatest weakness during September-October. As measured by Nastos *et al.*, those same months accounted for the two highest totals of panicked and depressed patients.

Also, compare **Chart 7F** with **Chart 10E**. The Chapter 10 chart shows the seasonal trend in the Dow Jones Industrial Average. The two charts exhibit similar patterns in both the timing of the turns and the magnitude of the moves.

The Nuclear Forces – EUWS Cycles

Up to this point, all known cycles relate closest to either gravitational or electromagnetic forces. Nothing can be found related to either the strong nuclear force or the weak nuclear force. Nonetheless, neither force should be completely discounted as a potential factor modulating these cycles.

That's especially true for the mysterious EUWS cycles. These cycles control a wide range of global oscillations. While the EUWS cycles exhibit many magnetic field characteristics, and some gravitational tendencies, fluctuations in the weak nuclear force may propagate these cycles. The weak nuclear force remains a candidate because it affects all types of particles and their properties – from a short range.

Based on recent WMAP results, NASA estimates the universe consists of matter and energy in these percentages – 4.6% ordinary baryonic matter, 23.2% dark matter, 71.7% dark energy, and 0.5% neutrinos.[Hinshaw *et al.*, 2008] Looking at this composition, dark matter and dark energy account for 95% of the matter and energy in our universe. It's easy to imagine that these dominant components play some role in human cycles. Unfortunately, physicists know very little about dark matter and dark energy. Without better understanding, insight regarding their role in cyclic behavior will remain unclear.

Even though they only comprise 0.5% of our universal content, neutrinos may play a role. Unlike dark matter and dark energy, physicists already know quite a bit about neutrinos. Recently coming under the operation of Pennsylvania University, a neutrino detector in Homestake's South Dakota Gold Mine has collected data since 1970.

Examining the Homestake data from 1970 through 1994, a research team led by Stanford University's Peter Sturrock discovered cycles in solar neutrinos.[Sturrock *et al.*, 1997] This development indirectly implies the involvement of the weak nuclear force in cycles. That's because interactions involving neutrinos generally occur via the weak force.

Classification of Cycles

However, until more conclusive evidence emerges regarding neutrinos and the nuclear forces, observable cyclic patterns on Earth can only be attributed to gravity and electromagnetic forces. To conclude this chapter, cycles that affect Earth can be separated into three main classes:

(a) Gravitational and motion cycles that affect global climate.
(b) Geomagnetic cycles originating from the Sun, Moon, and galactic center.
(c) The mysterious EUWS cycles, which display characteristics that are both magnetic and gravitational.

Now with the basic cycles both identified and classified, the next chapter shows how cycles operate in nature.

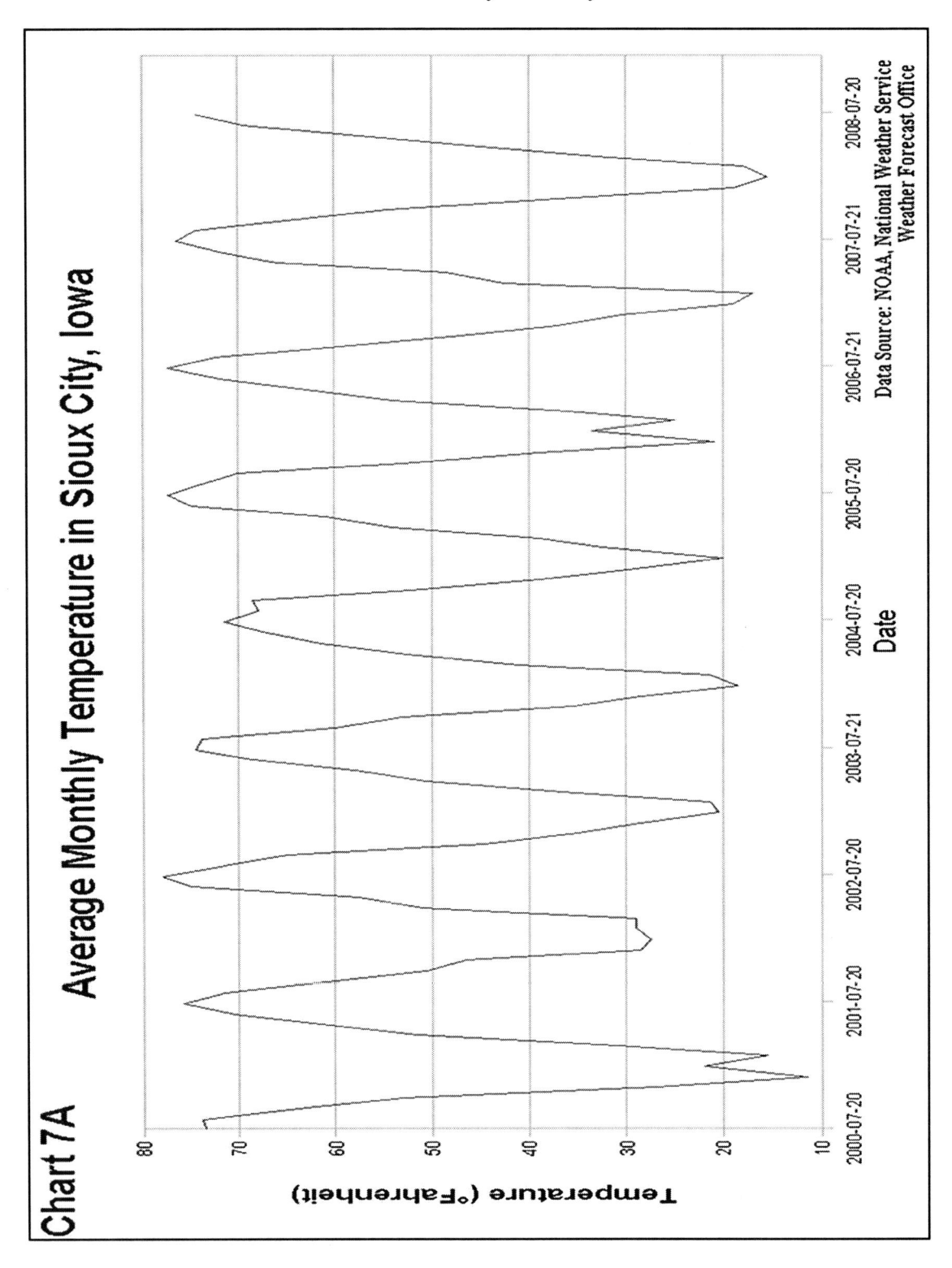

Chart 7A — Average Monthly Temperature in Sioux City, Iowa

Data Source: NOAA, National Weather Service Weather Forecast Office

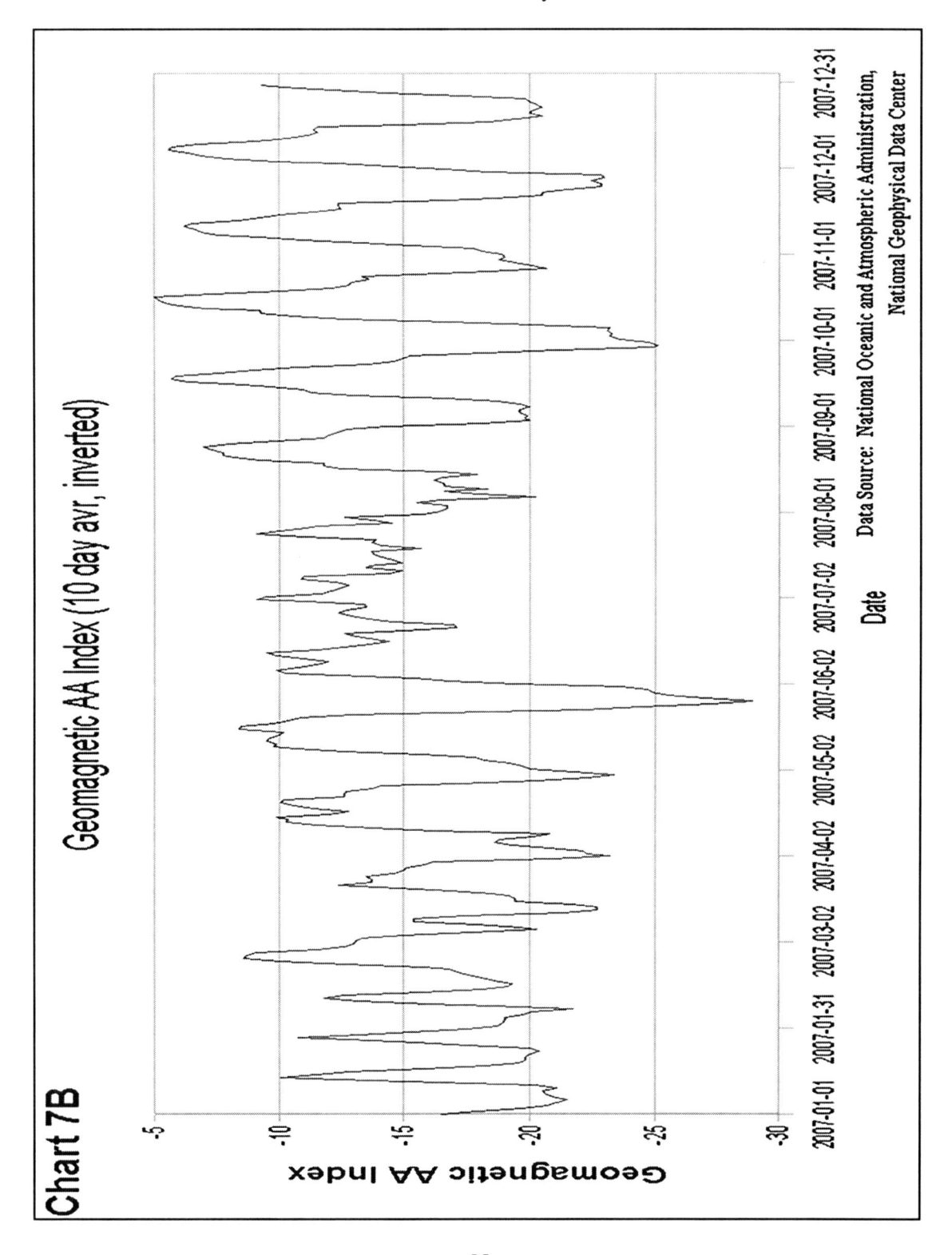

Classification of Cycles

Chart 7B

Geomagnetic AA Index (10 day avr, inverted)

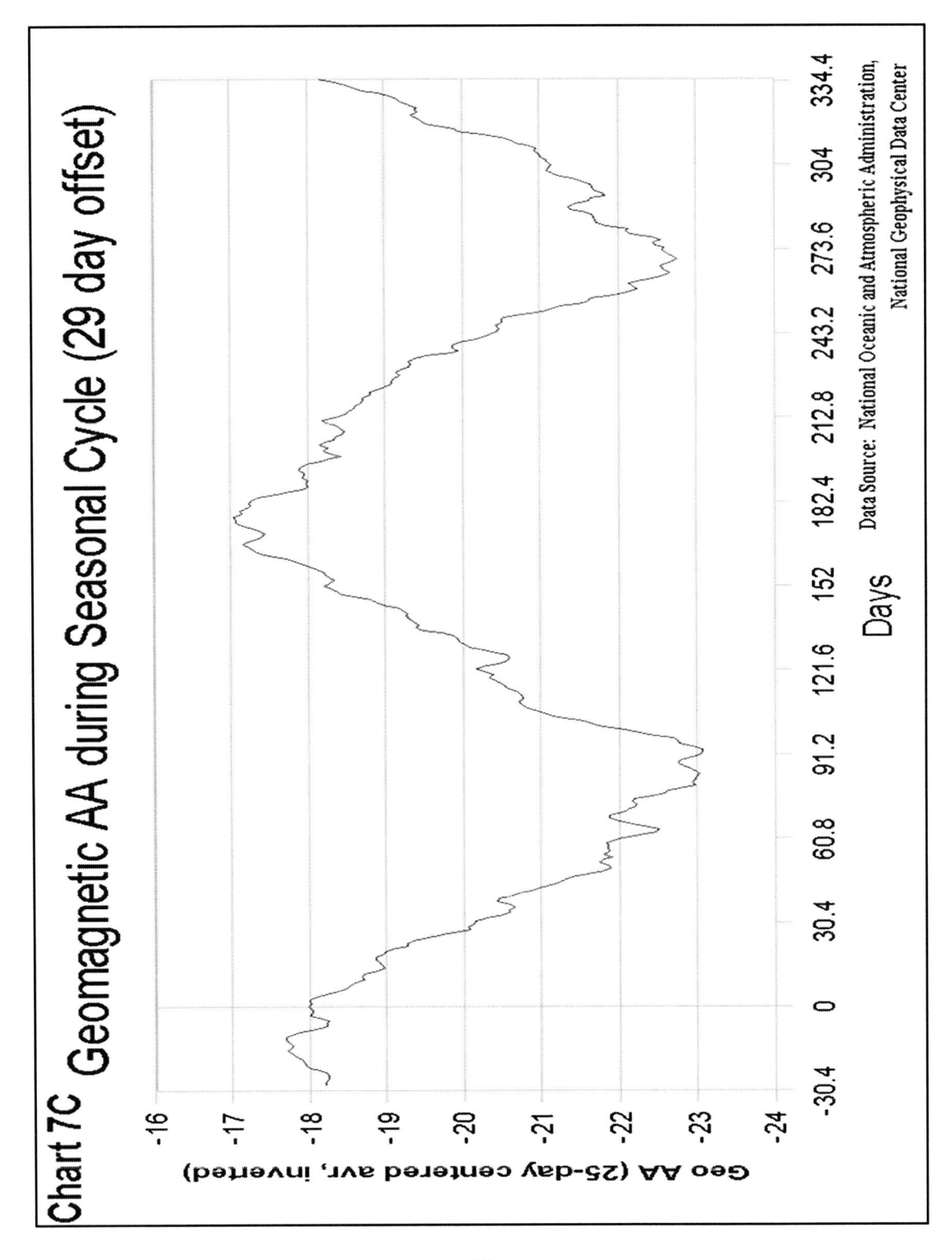

Chart 7C Geomagnetic AA during Seasonal Cycle (29 day offset)

Data Source: National Oceanic and Atmospheric Administration, National Geophysical Data Center

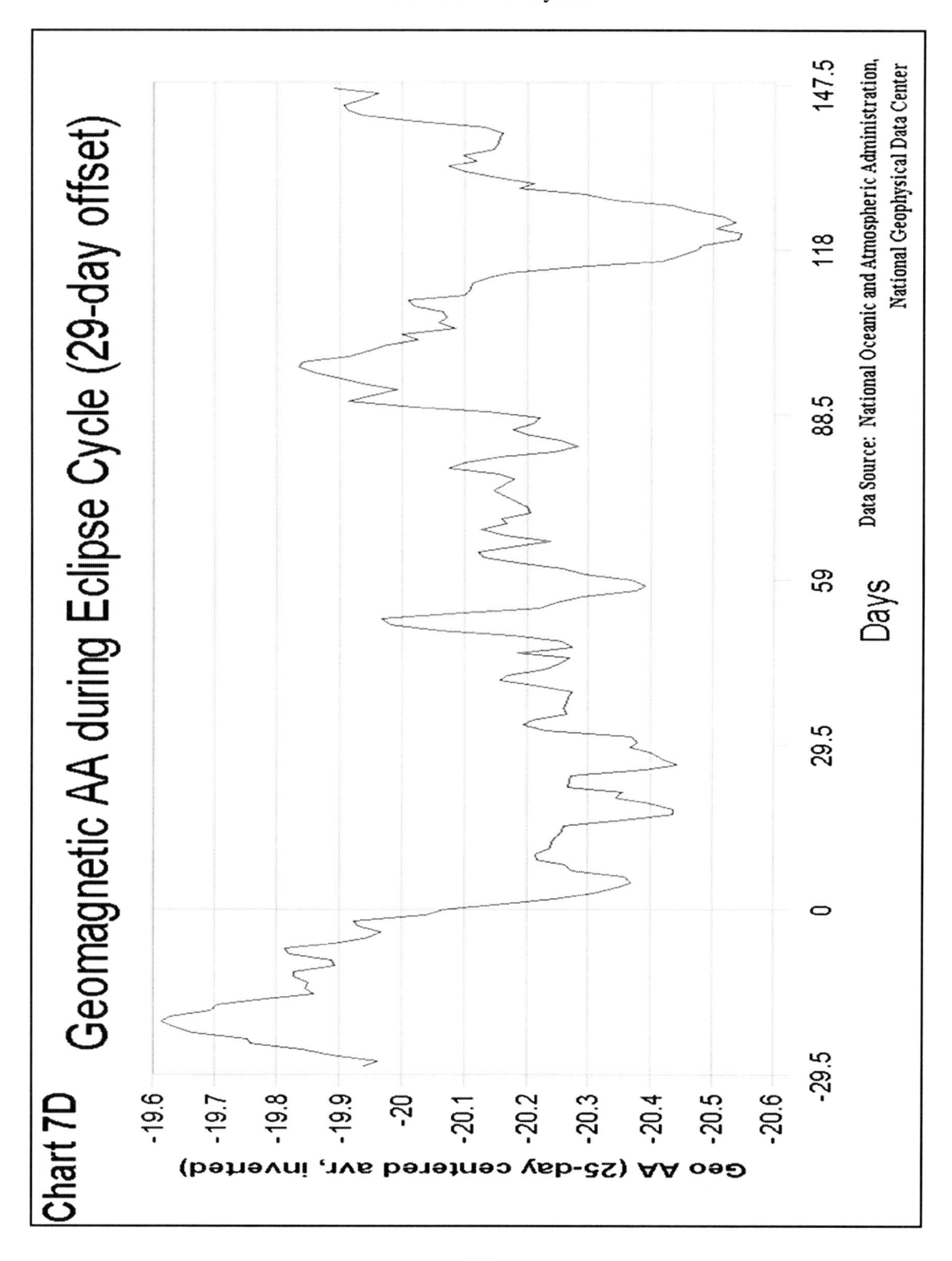

Classification of Cycles

Chart 7D: Geomagnetic AA during Eclipse Cycle (29-day offset)

Data Source: National Oceanic and Atmospheric Administration, National Geophysical Data Center

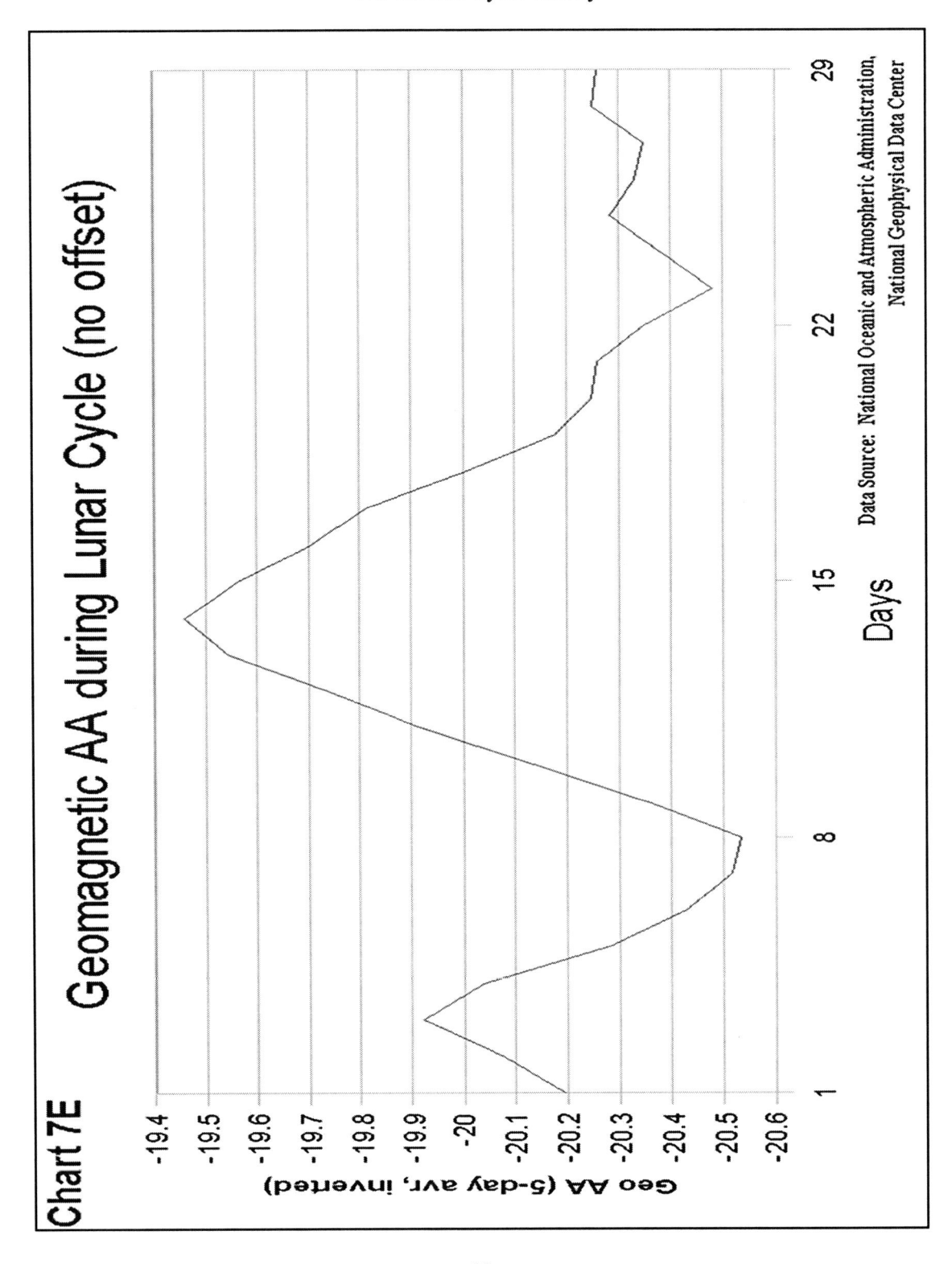

Chart 7E Geomagnetic AA during Lunar Cycle (no offset)

Data Source: National Oceanic and Atmospheric Administration, National Geophysical Data Center

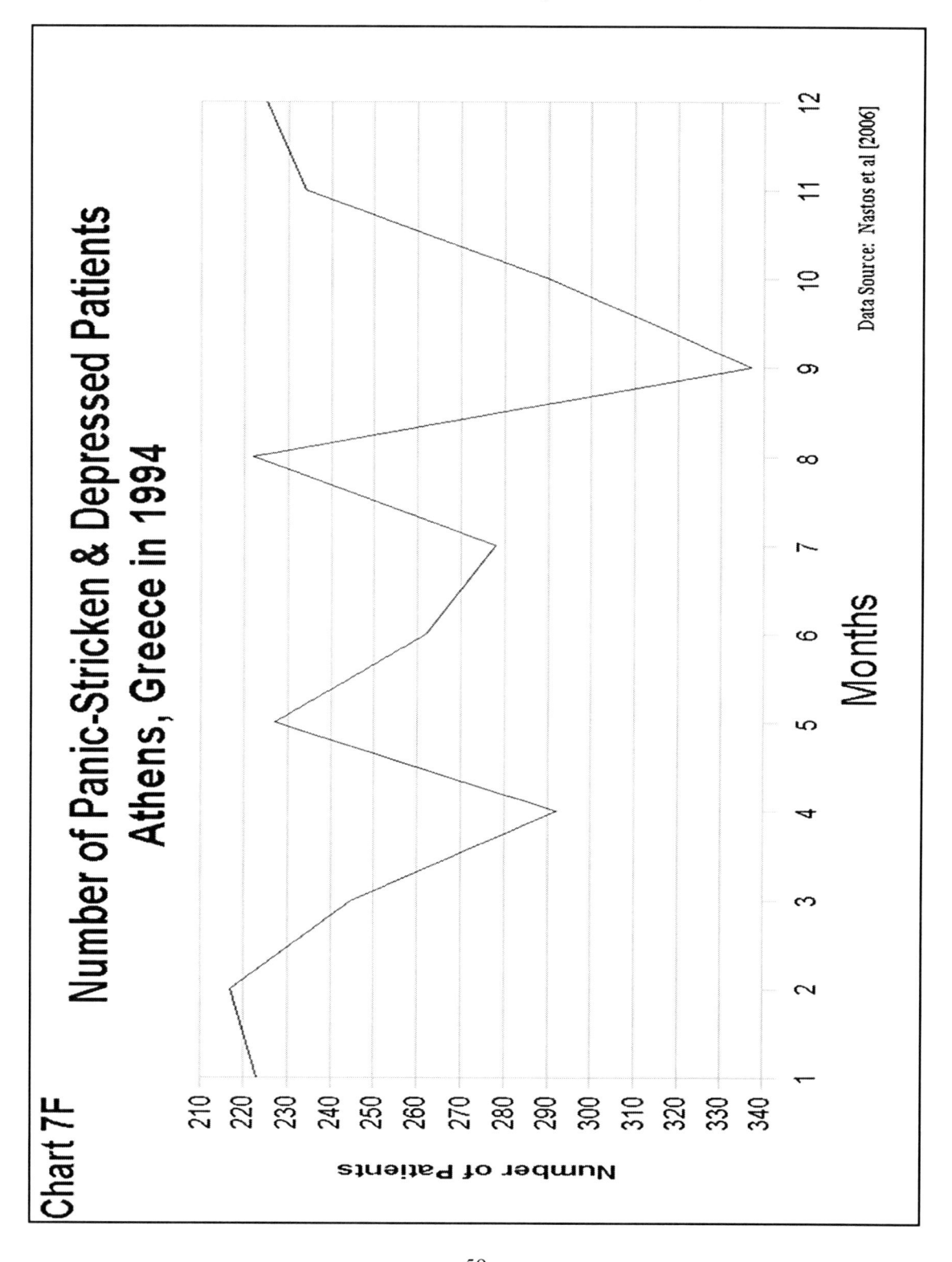

Classification of Cycles

Chart 7F

**Number of Panic-Stricken & Depressed Patients
Athens, Greece in 1994**

Data Source: Nastos et al [2006]

Chapter 8
The Turning Point Distribution Principle

This chapter shows how EUWS cycles reverse direction. All of the examples pertain to financial markets with frequencies less than 172 years. Even though these examples represent human behavior, the same distribution principles apply to EUWS cycles involving inorganic matter. In later chapters, EUWS cycles with frequencies of thousands, millions, and billions of years reinforce the patterns outlined in this chapter.

When it comes to explaining financial market movements, investors greatly overrate the impact derived from government planning, legislative policymaking, central banking decisions, and corporate news releases. At most, impact from these sources only lasts a few days. Eventually, long-term sentiment trends completely obscure reactions to breaking news.

Cyclic forces power market movements more consistently than any other factor. True, central banks can engage in tremendous inflationary policies. And market prices respond to these inflationary trends. But investors only respond to credit-inflation positively as long as cyclic forces remain in an up-phase. When EUWS cycles turn lower, sentiment turns negative, and investors respond differently to credit inflation. For example, in the early stages of credit inflation, equity markets usually boom. But when cyclic forces turn lower, investors often perceive inflationary policies as negative for stocks and bonds, but bullish for gold, silver, oil, and other commodities.

The Unified Cycle Theory

One portion of the *Unified Cycle Theory* addresses these inflection points – that critical point in time when market direction reverses. This chapter concentrates on market tops, when investor sentiment turns from positive to negative. At a market top, bulls greatly outnumber bears. But as the top approaches, the number of bullish investors jumping on the bandwagon slows. As with any type of market movement, tops form in a jerky fashion. The final rally can sometimes develop as an explosive blow-off that forces short-sellers to urgently liquidate bearish bets. Once the market forces bears to cover short-sales, the market moves lower. During long-term declines, weak rallies frequently interrupt the down-trend. These interruptions result from various short-term cycles running counter to the longer-term cycle. Once the shorter-term cycles move back in sync with the longer-term trend, the long-term trend resumes. This pattern of moving up and down in stair-step fashion describes the nature of markets. That's how they've always functioned, and probably always will.

The previous chapter identified three classifications of cycles – gravitational, geomagnetic, and EUWS. In this chapter, numerous examples show how geomagnetic and EUWS cycles interact to cause market fluctuations. Sometimes these cycles work together, and other times they oppose each other. Most of the time, some cycles point up, while others point down. But on rare occasions, the vast majority of the cycles align in harmony. When that happens, a major reversal almost always takes place.

Near important market tops, many cycles invariably reverse from higher to lower in close proximity. Because of the harmonic nature of the EUWS cycles, they demonstrate very precise theoretical turning points. However, that doesn't imply markets always top in accordance with theory. Actual tops develop less precisely. Yet, in spite of this uncertainty, a well defined list of alternative turning points emerges. The alternatives essentially encompass the first seven sub-cycles of the parent cycle. Sub-cycles close to a major theoretical turning point hold the highest probability of being achieved.

An array of reversal points, distributed sporadically around theoretical peaks and troughs, describes the turning point distribution principle (**TPD Principle**). In the remainder of this chapter, case-studies clarify this rule. These case studies focus on reversals associated with geomagnetic cycles and EUWS cycles. The 11-year sunspot cycle acts as the key geomagnetic cycle, while the 19-year panic cycle serves as the most important EUWS cycle.

In Chapter 7, Mike Jackson described a link between sunspot numbers and the Dow Jones Industrial Average starting in 1996. He noted a statistically significant correlation for seven years before the parallel movements ceased in 2003. Jackson concluded by wondering if some causative connection took hold in 1996.[Jackson, 2003] Actually, Jackson's example served as one of many similar cases. Over the past century, numerous market bubbles correlated with the sunspot cycle. This chapter details several of those cases.

As far as the causative connection Jackson sought, the *Unified Cycle Theory* provides two answers.

The Turning Point Distribution Principle

1) An extensive period of speculation and leverage must precede a sunspot maximum.
2) The sunspot cycle maximum must synchronize with either the 6.36-year or the 19-year cycle.

As time progresses, the correlation between markets and sunspot activity jumps from one market to another. To identify which market will become susceptible to Schwabe cycle dynamics, look for the market with the greatest run up, the wildest speculation, and the broadest participation. During the 1920s, USA stocks satisfied the criteria. Another smaller stock market speculation followed from 1932 to 1937. After the 1937 crash, speculators stayed away from equities until another mini-bubble peaked in 1968. During the 1970s, speculators concentrated on precious metals. During the 1980s, market mania shifted to Japanese stocks. Throughout the 1990s, a great speculation surrounded technology shares. In every one of these cases, the bubbles burst near the time of a sunspot maximum.

Sunspot cycle peaks can be roughly calculated thirty or more years in advance. And they become more precisely known two to three years prior to their tops. But the sunspot cycle doesn't act alone. If a sunspot maximum and a EUWS peak arrive together, then the probability of a reversal at the time of their synchronized peaks becomes exceedingly high. However, if their reversal dates fail to match, then the TPD Principle comes into play, and a wider range of possible reversal points emerge. Examples illustrate all of these points.

TPD Principle, Example 1 – 1929 Stock Market Crash

The bull market of the Roaring '20s reached a critical juncture in late-1928. At that time, a period of increasing sunspot activity reached a maximum. And by early 1929, the 11-year Schwabe cycle reversed course and headed downward. See **Chart 8A**. The 1929 downturn in sunspots became critical because, throughout history, an unusually large number of crashes immediately followed solar maxima. However, in 1929, stocks didn't immediately respond.

As 1929 began, two major cycles conflicted. An uptrend in the 19-year panic cycle offset bearish influence from declining solar activity. The 19-year EUWS cycle projected a theoretical peak on September 27, 1930. Because of this conflict, the entire spectrum between November 1928 and September 1930 became the primary window for a reversal of the bull market.

Of course, for a trader, a 2-year window may seem rather wide. To narrow down the expected reversal point further, the shorter-term cycles provided timing clues. In particular, the seasonal cycles and the eclipse cycles became somewhat synchronized to the downside during three different times within this window – October 1929, March 1930, and September 1930. As it turned out, the first of these potential turning points triggered the infamous '29 crash. By crashing in October 1929, the market split the difference between the sunspot maximum and theoretical peak of the 19-year EUWS cycle. In this way, the market settled the conflict.

Chart 8B provides a graphical representation of the various short-term cycles that came into

The Unified Cycle Theory

play in resolving the 1929 conflict. The markers at the top of the graph represent theoretical peaks of the various cycles, while the markers just above the x-axis represent theoretical bottoms for the cycles. A review of every cycle presented in **Chart 8B** follows.

a) 19-Year EUWS Peak – A theoretical peak of the panic cycle occurred on September 27, 1930.
b) Sunspot Maximum – Using a 1-year average, a sunspot maximum arrived during November 1928.
c) Seasonal Cycle – The span between early-August (the down-arrow) and late-October (the up-arrow) encompassed the time of the year when markets are weakest.
d) Eclipse Cycle – A theoretical peak of the eclipse cycle projected to October 3, 1929. (The actual eclipse cycle top came one month earlier.)
e) Full Moon – A full moon passed on October 18, 1929.
f) 86-Day EUWS Peak – A theoretical peak of the 86-day EUWS cycle occurred on October 20, 1929.

The TPD Principle involved the eclipse cycle as well. The eclipse cycle normally peaks on the first new moon before a solar eclipse. Following that reversal point, it takes six weeks for sentiment to shift from euphoria to panic. Then, on the first full moon after a solar eclipse, a panic-phase begins. A panic-phase usually lasts two weeks – ending at the time of the next new moon.

But in 1929, the entire eclipse pattern described above shifted one new-moon earlier. At that time, the top of the eclipse cycle arrived two new moons before the November 1 solar eclipse. This shift represented an eclipse-cycle with the 2nd highest probability of occurrence.

To summarize the 1929 crash, the TPD Principle came into play in four different ways.

- The market deviated from normal by topping 1 year after a sunspot cycle peak.
- The market deviated from normal by crashing 1 year ahead of the 19-year panic cycle peak.
- The market achieved the most likely seasonal scenario by crashing during September-October.
- The market achieved the 2nd most likely eclipse cycle scenario by topping two months prior to the November 1 solar eclipse.

That's all it took to turn the tide. In 1929, following the October 18 full moon and the 86-day EUWS cycle top on October 20, the final elements of hope quickly faded away. After steady losses for a few days, on October 23 terrified shareholders sold heavily, with the Dow losing 6.4%. On October 24, share prices fell another 2.1%. On Black Monday, October 28, the Dow crashed another 12.9%. Then the following day, Black Tuesday, the Dow crashed another 11.8%.

In 1929, the various deviations permitted a <u>best fit compromise</u>. With these compromises, each cycle came as close as possible to fulfilling their normal theoretical patterns. In essence, compromise best describes how the TPD Principle operates when cycles conflict – which happens more often than not.

TPD Principle, Example 2 – 1980 Silver Market Crash

During the 1970s, the precious metal markets enjoyed one of the greatest bull markets of all time. And similar to the majority of other great speculative bubbles, the end came at the time of a sunspot maximum. To illustrate how the precious metal markets reversed, silver serves as the poster child. The steady increase in sunspot numbers came to an abrupt halt by early-1980, and then leveled off for a few months. The slowdown in sunspot momentum coincided with the all-time high in the silver market. Then, after a brief pause, sunspot numbers increased marginally until reaching their eventual maximum later that year. See **Chart 8C** for sunspot activity from 1977 through 1984.

But the sunspot cycle operated with numerous other cycles in modulating the reversal. In markets, at least 15 different cycles interact to produce fluctuations. Many of these cycles counteract each other. With so many cycles involved, deviations frequently occur. **Chart 8D** shows the large number of cycles connected to the 1980 silver market crash. A review of every cycle presented in **Chart 8D** follows.

   a) Sunspot Maximum – Using a 1-year average, a sunspot maximum arrived during June 1980. During January 1980, a significant slowdown in sunspot activity preceded the maximum.
   b) Seasonal Cycle – The span between early-January (the down-arrow) and late-March (the up-arrow) covered the 2nd weakest time of the year for markets.
   c) Eclipse Cycle – A theoretical peak of the eclipse cycle came on January 17, 1980.
   d) 28.7-Day EUWS Peak – A theoretical peak of the 28.7-day cycle occurred on January 20, 1980.
   e) Full Moon – A full moon passed on March 2, 1980.
   f) 258-Day EUWS Peak – A theoretical peak of the 258-day cycle occurred on March 18, 1980.

This particular crash provided a case where market cycles matched theory – almost to perfection. By mid-January 1980, several important cycles aligned themselves in harmony. The actual peak of the bull market came on January 21, 1980. That transpired two trading days after the eclipse cycle turning point of January 17, 1980, and one day after a theoretical peak of the 28.7-day EUWS cycle.

Security traders predominantly know about seasonal weakness during the August-October period. Indeed, the autumn period contains almost all panics and crashes for USA equities. However, based on seasonal fluctuations in the Geomagnetic AA index, the *Unified Cycle Theory* identifies the January-March period as another time-frame vulnerable to market weakness. By including panics in foreign stocks and commodities (not just USA securities), the January-March period coincides with the second highest number of crashes – not too far behind the August-October period. Furthermore, it's virtually impossible to find a significant panic outside of these two seasonal windows.

The great silver speculation reached a crescendo by early-1980. As the new-year arrived, the sunspot maximum, the seasonal peak, the eclipse cycle peak on January 17, and the 28.7-day cycle peak on January 20 all combined to push the silver market to a climax at $50 per ounce on January 20. On the very next day, on January 21, the bottom fell out of the market. After the reversal, prices fell as fast as they had risen days earlier. The first wave of selling took silver prices down to $33 by mid-February, at the time of the next new moon.

Then a rally ensued. For the next two weeks, silver prices rallied – until the next full moon. In 1980, silver market changes mirrored the 6-week topping pattern traced out during the 1929 stock market crash. In fact, this pattern appears so frequently at market tops that it cannot develop by mere coincidence. In addition, the silver market top closely followed eclipse cycle and lunar cycle patterns.

Hence, on March 2, 1980, as the 6-week topping pattern completed its formation, and the full moon passed, the silver market entered the panic phase of its crash. The final blow came when the 28.7-day, 86-day, and 258-day cycles all turned lower on March 18. By the end of March 1980, the silver market had lost nearly 80% of its peak price – collapsing from $50 to $10 in a little more than eight weeks. The defining moment of the silver implosion came when the largest speculators in the market, the Hunt Brothers of Texas, fell into financial ruin. Eventually they declared bankruptcy because of lawsuits resulting from the crash. Prior to the crash, one of the brothers – Bunker Hunt, was listed among the wealthiest people in the world.[Tuccille, 2004]

TPD Principle, Example 3 – 1987 Stock Market Crash

The 1987 stock market crash provides another example of the workings of the TPD Principle. The same cycles entangled in examples 1 and 2 also produced the September-October 1987 crash. Once again, the sunspot cycle and the 19-year panic cycle acted as the primary modulators. Similar to 1929, a conflict arose between their turning points.

During late-1987, a sunspot minimum occurred at the time of a panic cycle peak. See **Chart 8E**. At that time, various formulas projected the next sunspot maximum for 1990. The 1987 scenario contrasted sharply with the one in 1929. Remember, in 1929, the market resolved a minor separation in their peaks by splitting the difference.

But the 1987 conflict arose with one cycle at a minimum while the other achieved a maximum. The TPD Principle allows for all possibilities. However, certain outcomes possess higher probabilities than others. Market panics rarely occur during the advancing phase of the sunspot cycle. The theoretical peak of the 19-year cycle arrived on December 27, 1987, slightly after a sunspot minimum. In combination, these circumstances implied that the stock market must either immediately reverse direction in 1987, or wait until 1990. In the United States, the panic occurred immediately – near a sunspot minimum.

Major crashes usually occur within a year of theoretical peaks of the 19-year panic cycle (a 5% leeway on either side of the theoretical peak). In 1987, the Dow Industrials topped about four months prior to the theoretical peak. That amounted to a 1.7% deviation from the projected top – well within tolerance.

Similar to other crashes, the 19-year panic cycle warned of a 1987 crash, while shorter-term cycles pinpointed the exact timing of the reversal. In pinpointing the timing, the seasonal cycle, the eclipse cycle, and the lunar cycle all played pivotal roles. See **Chart 8F**. A review of every cycle presented in **Chart 8F** follows.

   a) 19-Year EUWS Peak – A theoretical peak of the panic cycle occurred on December 27, 1987.
   b) Sunspot Minimum – Using a 1-year average, a sunspot minimum arrived in early-1987.
   c) Seasonal Cycle – The span between early-August (the down-arrow) and late-October (the up-arrow) encompassed the time of the year when stock market panics cluster.
   d) Eclipse Cycle – A theoretical peak of the eclipse cycle projected to August 24, 1987.
   e) Full Moon – A full moon passed on October 6, 1987.
   f) 86-Day EUWS Peak – A theoretical peak of the 86-day cycle occurred on October 2, 1987.

In 1987, a 19-year panic cycle peak provided a warning that something big was about to happen. However, the precise timing came from the seasonal cycle (which turned lower in early-August) and the eclipse cycle (with a theoretical peak on August 24). The stock market topped on the next day, August 25! After that, stock prices fell sharply for the next month.

The final blows came when an 86-day EUWS peak occurred on October 2 and a full moon passed on October 6. The full moon on October 6 coincided with the completion of a six week topping pattern. Two weeks later, on October 19, 1987, Wall Street suffered the largest one-day decline in its 200+ year history.

Except for the fact that it coincided with a sunspot minimum, the 1987 crash furnished another example of a panic developing exactly as theory predicted. In summary, it developed within four months of a 19-year panic cycle peak; it began three weeks after the autumn seasonal cycle turned lower; the Dow's final top came just one day after an eclipse cycle peak; a 6-week topping pattern formed ahead of the panic; the panic-phase of the crash started at the time of a full moon; the panic-phase started just four days after an 86-day cycle peak; and the panic-phase of the crash lasted two weeks, ending at the time of a new moon.

The Unified Cycle Theory

TPD Principle, Example 4 – 1990 Japanese Market Crash

The 1990 Japanese stock market crash exhibited most of the same cyclical characteristics as the previous three examples. Also like the others, it had its own quirks. The Japanese crash linked itself closely to the 1987 USA crash, with the 19-year cycle acting as the link.

However, the bull market in Tokyo climaxed one 2.12-year sub-cycle after the optimal time. (One 19.08-year EUWS cycle consists of nine 2.12-year cycles.) The TPD Principle identifies the 2.12-year sub-cycles as high probability alternative turning points. Considering that a sunspot maximum occurred at the same time as the 1990 theoretical peak of the 2.12-year cycle, the time around January 1990 emerged as a strong candidate for the Japanese bubble to burst.

And that's what happened. The 2.12-year separation from the 19-year theoretical peak served as the major difference between the 1987 New York crash and the 1990 Tokyo crash. Except for this one significant difference, almost all other cycles behaved the same.

See **Chart 8G**. It shows the sunspot maximum during late-1989. The increased solar activity coincided with the blow-off stage of the Japanese bubble. Soon after the sunspot maximum, the panic-phase of the crash arrived. See **Chart 8H**. A review of every cycle presented in **Chart 8H** follows.

 a) 19-Year EUWS Peak – A theoretical peak of the panic cycle occurred on December 27, 1987.
 b) 2.12-Year EUWS Peak – A theoretical peak occurred on February 8, 1990.
 c) Sunspot Maximum – Using a 1-year average, a sunspot maximum arrived during January 1990.
 d) Seasonal Cycle – The span between early-January (the down-arrow) and late-March (the up-arrow) covered the 2nd weakest time of the year for markets.
 e) Eclipse Cycle – A theoretical peak of the eclipse cycle occurred on December 28, 1989.
 f) Full Moon – A full moon passed on February 9, 1990.

Looking at all of the clues provided by the *Unified Cycle Theory*, the 1990 Japanese crash fits the mold quite well. At its height, on the last trading day of December 1989, the sunspot cycle, the seasonal cycle and the eclipse cycle all turned lower in unison. With this initial set of cycles reversing direction together, the Nikkei 225 Index reversed direction and headed lower starting with the first trading day of January 1990. Then, with the arrival of the late-January new moon, a counter-trend rally began. During early-February, Japanese stocks completed a 6-week topping pattern. That coincided with the arrival of a full moon and a 2.12-year EUWS theoretical peak. Completion of the 6-week topping pattern drove the final stake in the heart of the speculation. Panic followed. However, instead of a normal 2-week panic, the turmoil lasted 6 weeks – finally subsiding during late-March 1990.

The Japanese stock market crash of 1990 can be best summarized in this way: Two deviations

The Turning Point Distribution Principle

from the normal pattern occurred. First, the crash came one 2.12-year cycle after the theoretical peak of the 19-year panic cycle. Second, the panic-phase of the crash lasted six weeks instead of two. These two deviations allowed the Nikkei to crash in a way that made it fit better with the host of other cycles involved.

These examples complete discussion of the TPD Principle. The following rules summarize the Turning Point Distribution Principle:

- All dates reside on the list of possible turning points for a market.
- The theoretical turning points of the geomagnetic cycles and the EUWS cycles hold the highest probabilities of being achieved – within a small margin for error.
- In most instances, the various geomagnetic and EUWS cycles work in conflict. Some point higher, while others point lower.
- The longer-term cycles determine major market trends.
- Around the time of major inflection points, the major cycles yield their dominance to the short-term cycles. Short-term cycles play the major role in timing reversals.
- Market tops usually deviate from their theoretical peaks in a way that fits the collective pattern for all cycles involved – both long-term and short-term.
- Deviations from theoretical norms appear non-randomly. They center on <u>second most likely scenarios</u>, which primarily equate to theoretical turning points of competing cycles.

The remainder of this book concentrates on specific cycles. Nonetheless, conflicting influence from competing cycles can never be ignored. In that vein, this book rehashes the TPD Principle repeatedly in coming chapters.

Chart 8A — Int'l Sunspot Numbers (1 Year Avr.)

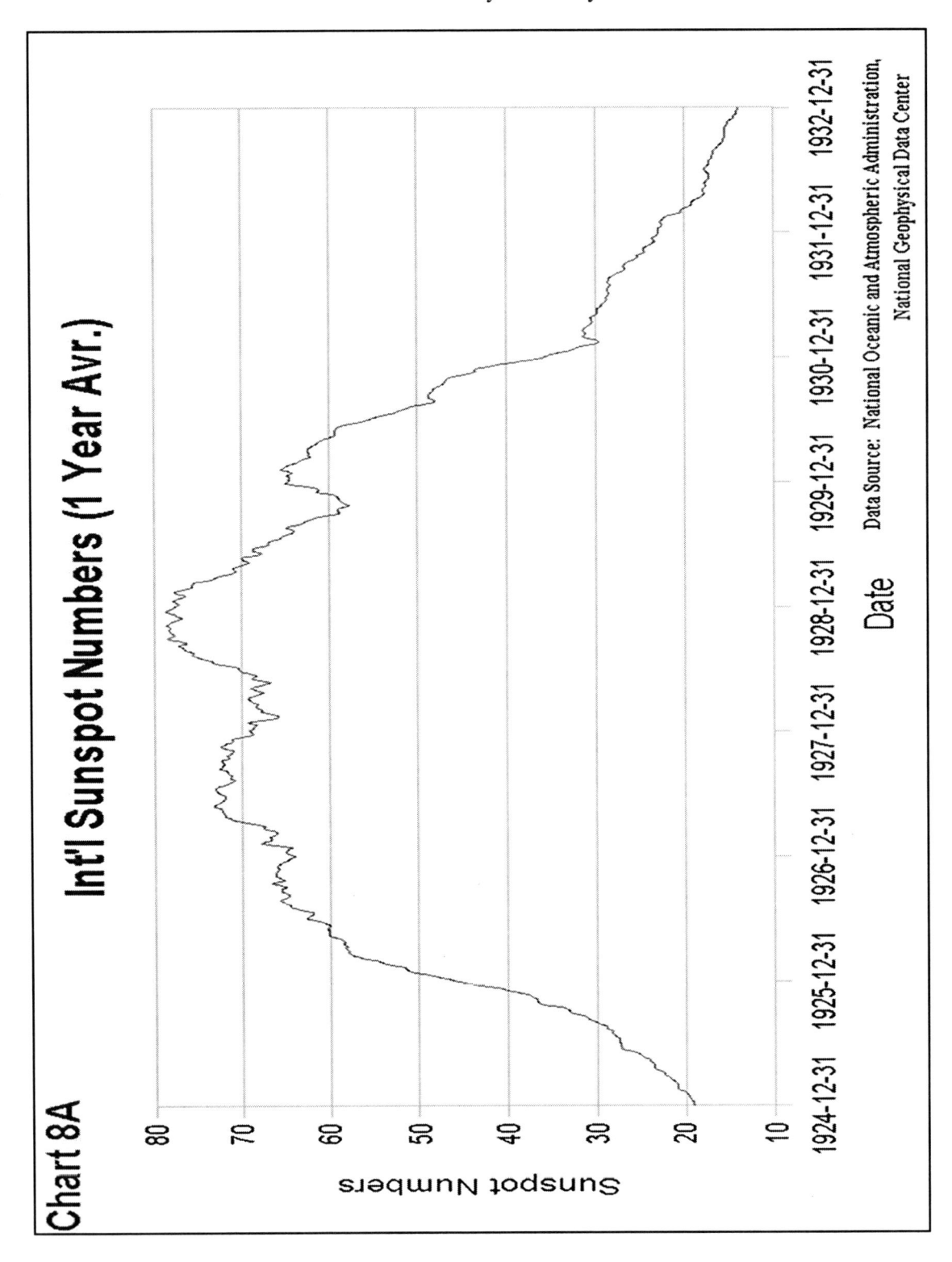

Data Source: National Oceanic and Atmospheric Administration, National Geophysical Data Center

The Turning Point Distribution Principle

Chart 8B
Dow Jones Industrial Average

19.1 Year High

86 Day

86 Day

Dow Industrials

1929-07-26　1929-08-23　1929-09-21　1929-10-20　1929-11-17　1929-12-16

Date

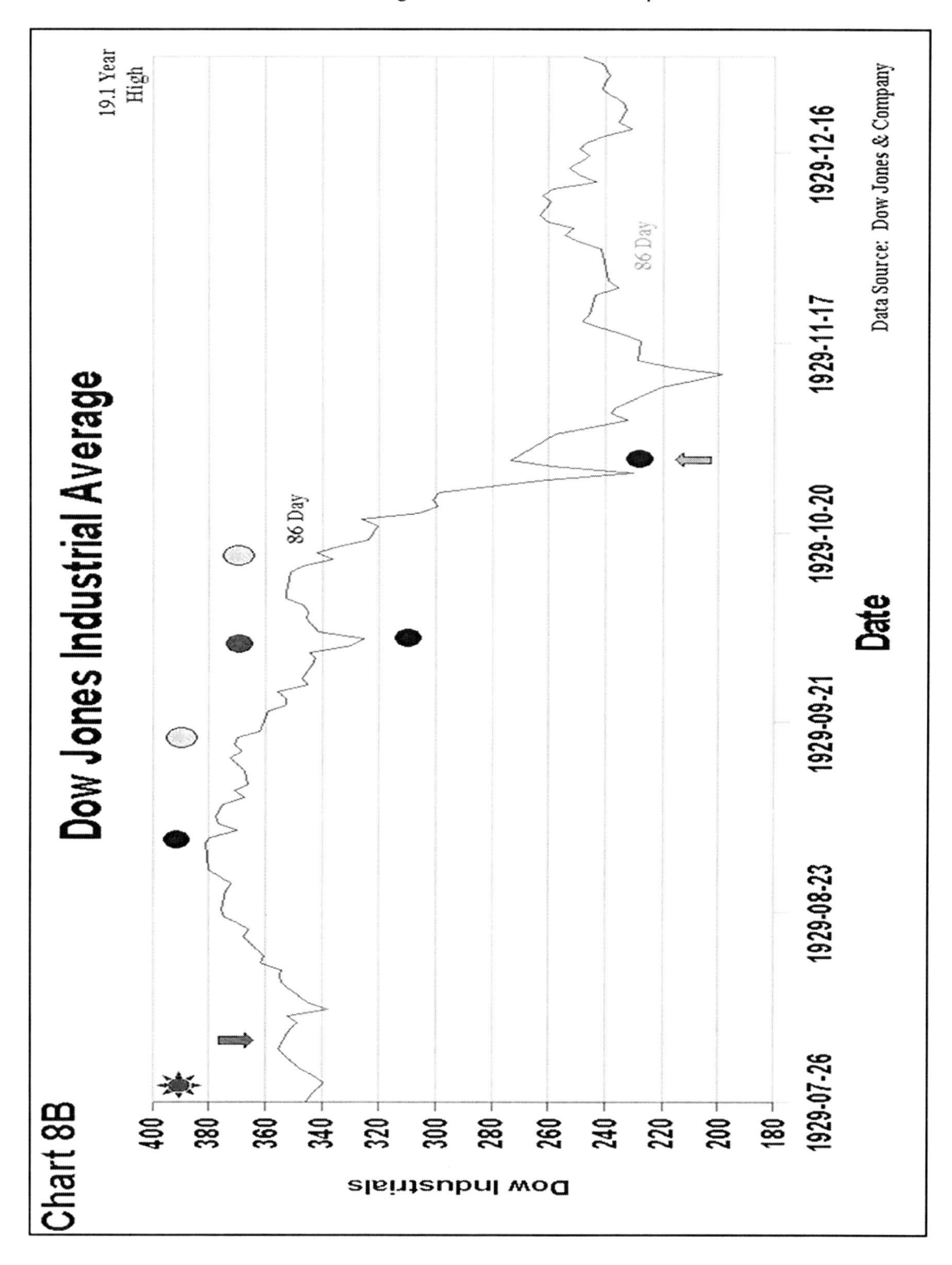

Data Source: Dow Jones & Company

The Unified Cycle Theory

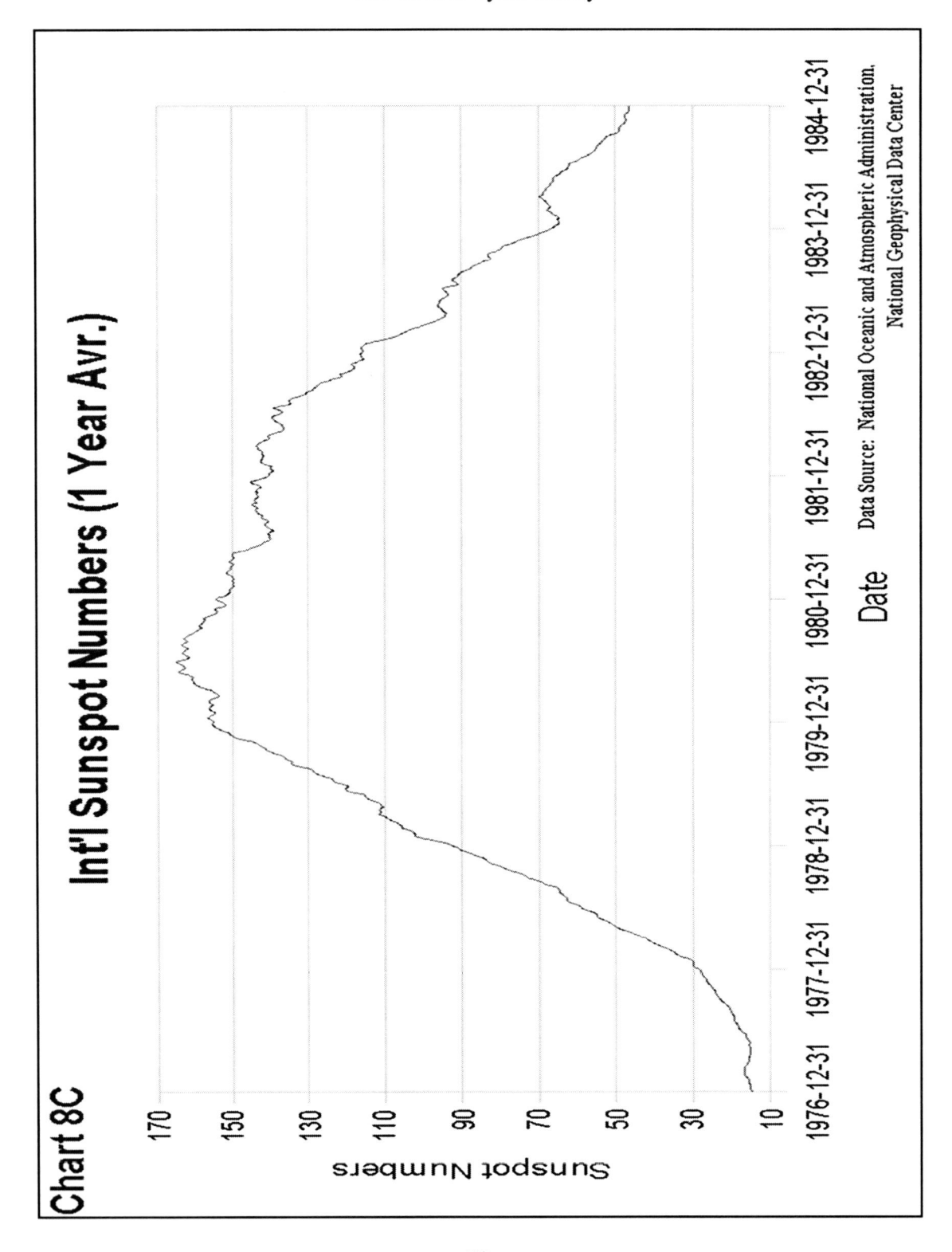

Chart 8C — Int'l Sunspot Numbers (1 Year Avr.)

Data Source: National Oceanic and Atmospheric Administration, National Geophysical Data Center

The Turning Point Distribution Principle

Chart 8D

Spot Silver Price

- 28.7 Day
- 36.9 Week
- 12.3 Week
- 28.7 Day

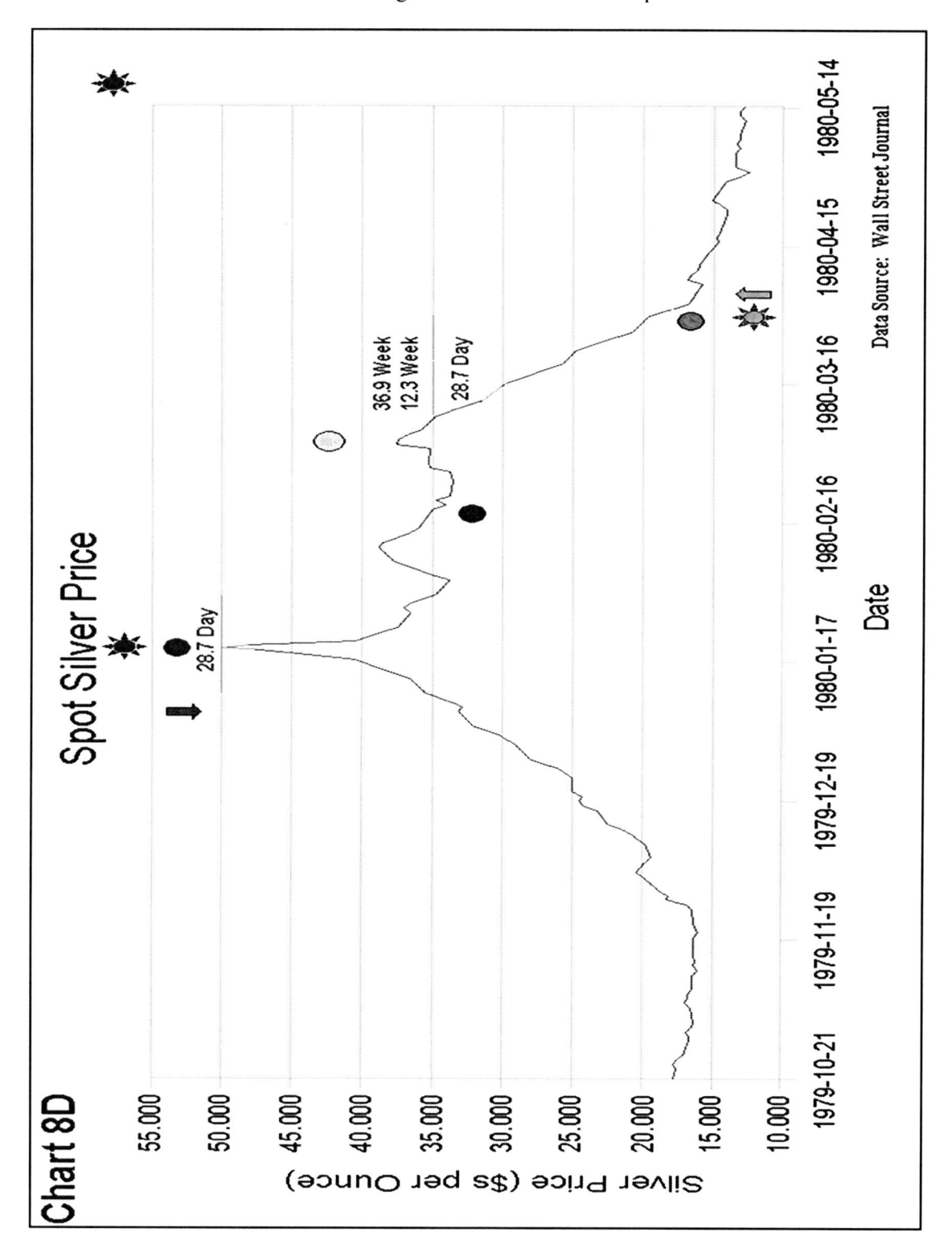

Data Source: Wall Street Journal

Chart 8E

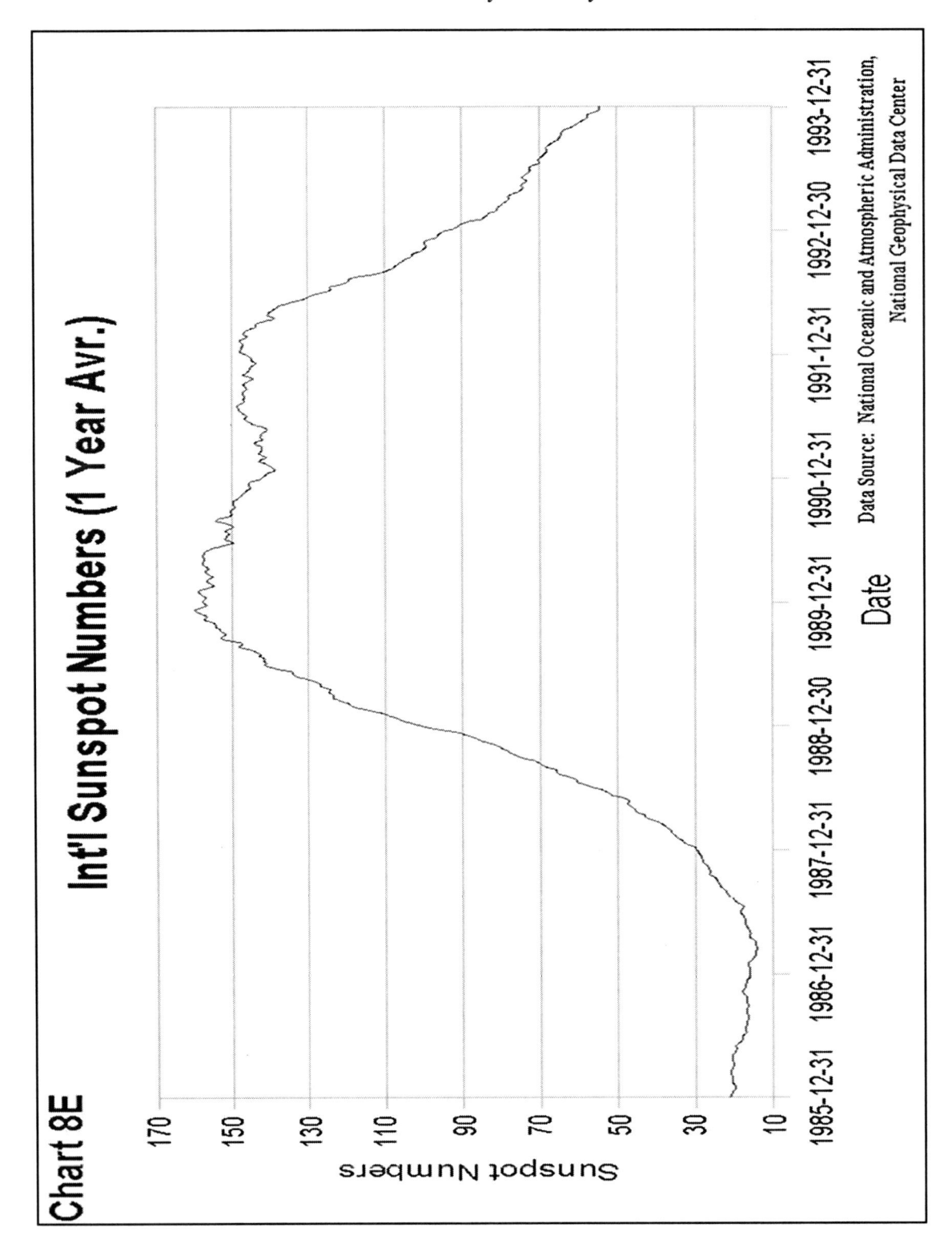

Int'l Sunspot Numbers (1 Year Avr.)

Data Source: National Oceanic and Atmospheric Administration, National Geophysical Data Center

The Turning Point Distribution Principle

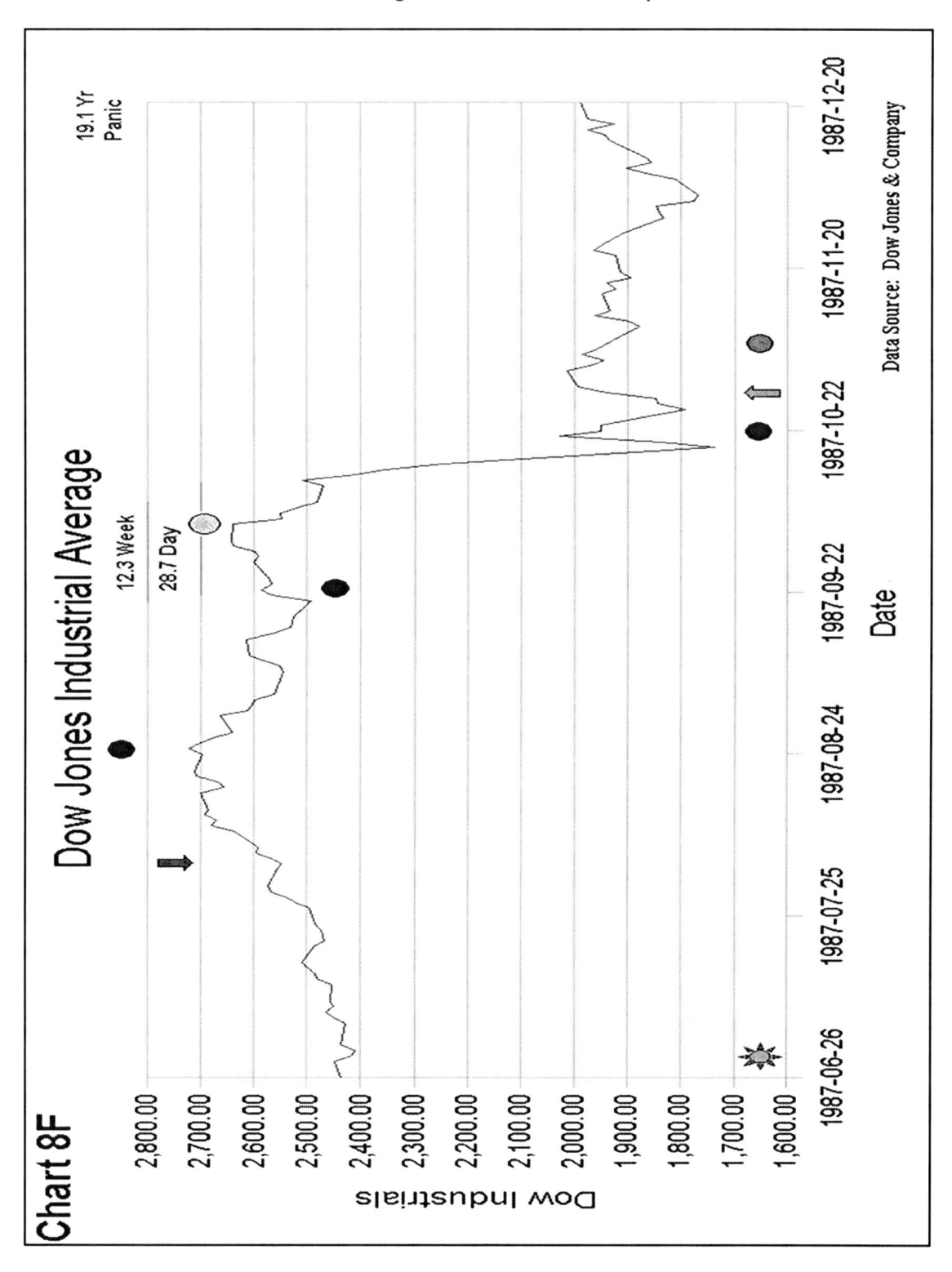

Chart 8G

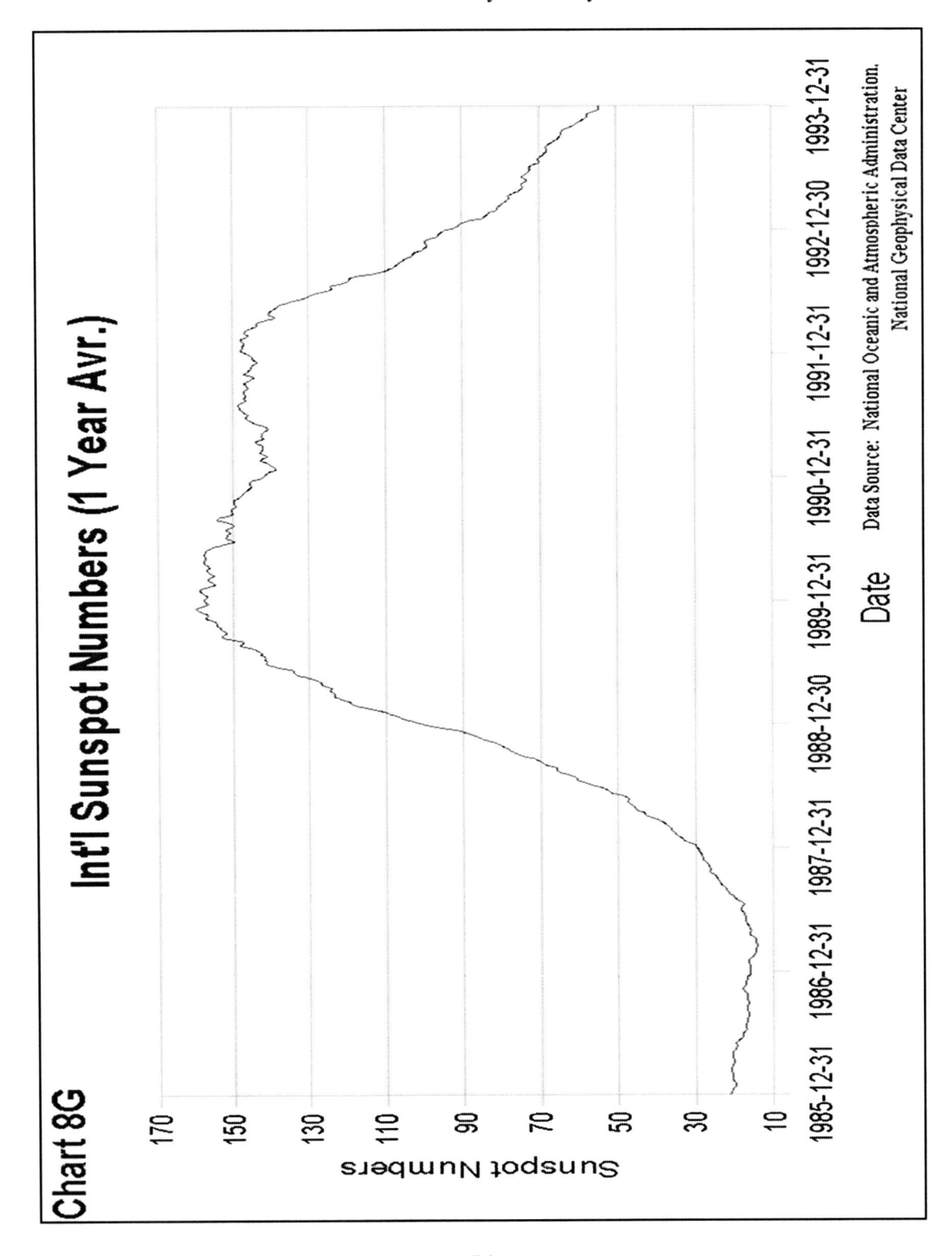

Int'l Sunspot Numbers (1 Year Avr.)

Data Source: National Oceanic and Atmospheric Administration. National Geophysical Data Center

The Turning Point Distribution Principle

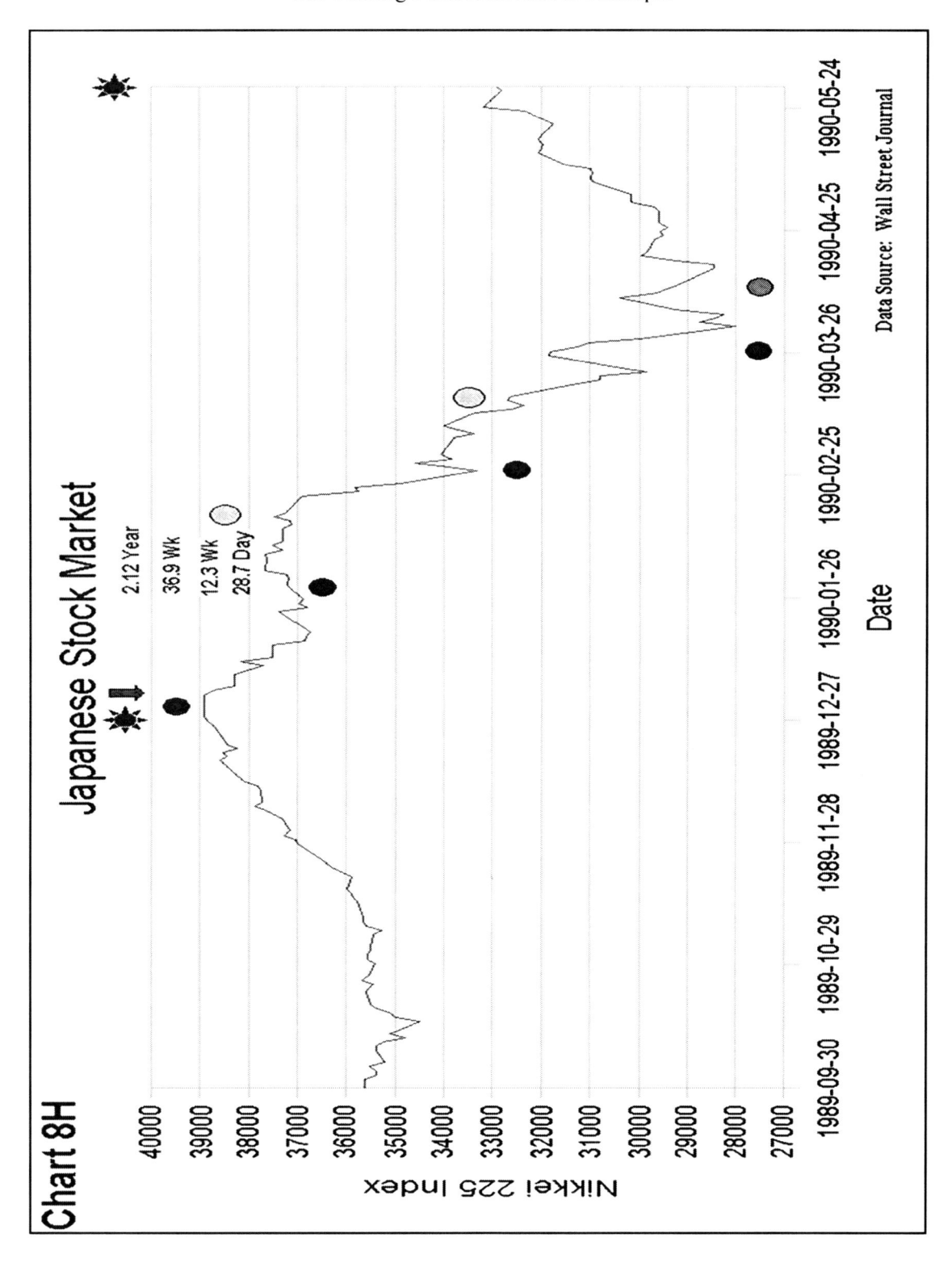

Chapter 9
Financial Leverage & Cycles

During the early 1980s, while the gold market received considerable attention from inflation hedgers, it came to my attention that gold prices fluctuated in phase with the lunar cycle. Previously, the claim that sunspot cycles affected market swings introduced me to the idea of celestial influences. Now, the idea that the moon also affected markets intrigued me further. Engulfed with curiosity, a quick research effort confirmed the correlation. See **Chart 9A**.

Except for the distortions caused by the eclipse cycle, the gold market swings moved in near-perfect unison with changes in the lunar cycle. Gold prices rose until the time of a full moon, then prices trended lower until the next new moon. And this cycle continued repeating. By calculating the probabilities of random occurrence, it became clear that these oscillations were not the result of mere chance.

After watching in amazement as this cycle repeated for several more months, the correlation suddenly ceased by mid-1982. Why? No immediate answer could be found. Nonetheless, this long string of synchronized moves always stuck with me. Nearly twenty-five years later, a plausible explanation finally surfaced.

Most of the physical cycles discussed in previous chapters repeat with moderate precision. They are predictable and observable. Plus, reasonable theories explain their oscillations. However, the same cannot be said about their influence on mankind. The assumption that external forces influenced human behavior in a consistent manner proved incorrect. And this incorrect assumption caused years of fruitless research. Eventually, it became apparent that a single physical force can result in two completely different reactions – depending on the responsiveness of the population at the time the force strikes. In other words, for human beings, cycle visibility fluctuates.

Of course, the last chapter already showed how competing cycles can temporarily obscure or cancel out the effects of their rivals. Finally, with this breakthrough in cycle visibility, it became clear why frequencies in human behavior were so difficult to determine. The combination of these two complicating factors – <u>competing cycles</u> plus <u>oscillating visibility</u> – made the already difficult task of studying human activity cycles all the more complex.

In formulating the *Unified Cycle Theory*, important breakthroughs included the ideas (1) that the Sun, the Moon, and eclipse cycles all influenced mankind through the same geomagnetic mechanism, (2) that our electrochemical nervous systems acted as the receptors for these cycles, and (3) that geomagnetic oscillations affected human emotions.

With these assumptions in place, the pieces started falling into place. The idea that collective moods swings influenced market behavior came next. At tops, people first become fearful, and then stocks fall in response. At bottoms, people first become confident, and then stocks rise after the fact. In short, collective emotional swings act as the stock market's engine. Along these lines, when markets become intensely emotional (especially when introducing fear into the equation), financial cycles become more visible than during tranquil times.

Drilling in further, the use of leverage emerges as an important causative factor in cycle visibility. To understand how leverage becomes involved, it's necessary to digress. Specifically, what's different about a bull market versus a bear market? See **Chart 9B**. It represents a continuous series of stock market leverage – using Broker Loans before the 1940s, and NYSE Margin Debt after that.

The last great liquidation of stock market leverage took place in the twenty years following the 1929 crash. Since then, market leverage has trended substantially higher, with only brief interruptions during relatively mild bear markets. During a bull market, increases in margin debt represent potential selling during the next downturn. In essence, leverage acts as a cloud hanging over the market. During a sharp downturn, margin calls can force speculators out of the market if they fail to immediately provide additional money to cover collateral deficiencies.

And that's the major difference between a bull market and a bear market. In a bull market, speculators feeling the effects of rising confidence can leisurely accumulate leveraged positions at their convenience. In a steep bear market, speculators overcome with fear are often forced to sell immediately as collateral vanishes. During a downturn, forced liquidations intensify the strains on markets already highly emotional to begin with.

Leisurely accumulation during a bull market, versus **emotional, forced selling during a bear market** accounts for a good portion of the difference needed to affect cycle visibility. As it relates to financial markets, alternating visibility based on leverage enhances the predictive powers of the *Unified Cycle Theory*. Additional examples enhance the case.

From the time the bull market peaked in 1929 to the subsequent bear market low in 1932, the stock market followed a bumpy path down. But these oscillations weren't random. Instead, they coincided almost perfectly with the six eclipse cycles encountered within that period. See **Chart 9C**. The 1929-32 bear market is unique because it's the longest bear market in the history of the United States – since daily market quotes became available. (The 1835-42 bear market lasted twice as long. However, infrequent quotes made research associated with geomagnetic cycles difficult.)

The eclipse cycles between 1929 and 1932 were so precise that a short-seller could have made more money by only shorting stocks during the down-phases of the eclipse cycles. Conversely, a bullish trader could have made money by timing purchases and sales with the start and end of the six eclipse cycles during that time.

As **Chart 9B** already showed, the period following the 1929 crash coincided with massive deleveraging of stock ownership. The Great Depression encompassed a time of tremendous fear. Forced liquidations persisted month after month. Even many strong holders, with very little leverage, found their positions weakened substantially as the depression progressed. The angst became so great that during his inaugural speech in 1933, Franklin Roosevelt tried to sooth the populace with the words: **"We have nothing to fear but fear itself."**

However, by the time Roosevelt took office in 1933 the marketplace had already forced a great deleveraging. The bulk of the unpleasant liquidation had already occurred. With debt reduced to somewhat manageable levels, further deflation became unlikely. Without significant leverage, the stock market moved independently of the EUWS cycles during much of the 1940s and early-1950s. See **Chart 9D**.

The EUWS cycles didn't completely disappear. However, the deviations from the major theoretical turning points became extreme. For example, the theory projected a peak in the 57-year depression cycle during 1949. Instead of a depression, one of the greatest bull markets of all time began! This example shows the importance of financial leverage in making cycles visible. As financial leverage decreases, the odds of depression occurring shrinks proportionally.

As the 1950s progressed, the attitude toward debt slowly changed. Margin loans and other forms of debt began working their way back onto the balance sheets of American consumers and businesses. By the 1960s and 1970s, people began ignoring the lessons from the 1930s. Debt became a problem again. With debt overextended, the EUWS oscillations became increasingly visible and more predictable.

The Unified Cycle Theory

Before examining developments since 1970, a step back to the 19th Century and the early part of the 20th Century helps put recent events into better perspective. The 172-year EUWS cycle peaked in 1835. After an 86-year down-phase, the cycle hit bottom in 1921. That theoretical low coincided closely with the end of World War I and the deflationary depression of 1920-21.

At that trough, the 172-year cycle projected a generally prosperous period for the next 86 years – up until 2007. And looking at the other major cycles in this period, the 57-year depression cycle projected a low for 1978. Hence, looking at the big picture, the EUWS cycles projected an especially prosperous period from 1979 to 2007. Indeed, during the entire 25-year span from 1982 to 2007, USA economic growth continued virtually non-stop.

Only two minor recessions in 1990 and 2001 interrupted the non-stop growth. Throughout the history of our country, no other period came close to rivaling this 25-year growth spurt. Though rare, this 25-year period of prosperity fit almost perfectly with trends projected by the 172-year and 57-year cycles. But danger lurked. See **Chart 9E**.

On January 24, 2007, all major EUWS cycles below 500 years in length turned lower in unison. That theoretical peak coincided exactly with the eruption of the sub-prime crisis. More importantly, with unprecedented debt expansion between 1970 and 2007, bankers and other lenders created a financial bubble unparalleled in the history of mankind. Nothing previously came remotely close to rivaling the leverage heaped upon every major sector of the USA economy. With this enormous debt expansion, the oscillations projected by the *Unified Cycle Theory* became increasingly visible and precise.

From the perspective of the *Unified Cycle Theory*, the long-term bull market from 1975 to 2007 fit the theory almost perfectly. However, the theory projects the mirror image of that for the next 28 years. With boatloads of leverage still embedded in the markets, a period of financial chaos should reach unprecedented proportions. As the turmoil now engulfing the USA financial system intensifies, the downward spiral it creates coincides perfectly with projections from the 172-year cycle.

With terror projected to reach a crescendo far in excess of the fear experienced during the 1930s, *Unified Cycle Theory* oscillations will likely remain visible for years to come. Similar to previous panics, government officials now work at calming markets and preventing deleveraging from following its natural course. However, unlike successful interventions from 1971 through 2006, government interventions have repeatedly failed since 2007. Long-term cyclic forces now point sharply lower. By working against deleveraging, at best, government agencies might delay the process. However, government intervention cannot stop these cycles. Delay tactics only create market distortions in the short-term and more pain in the long-term.

Financial Leverage & Cycles

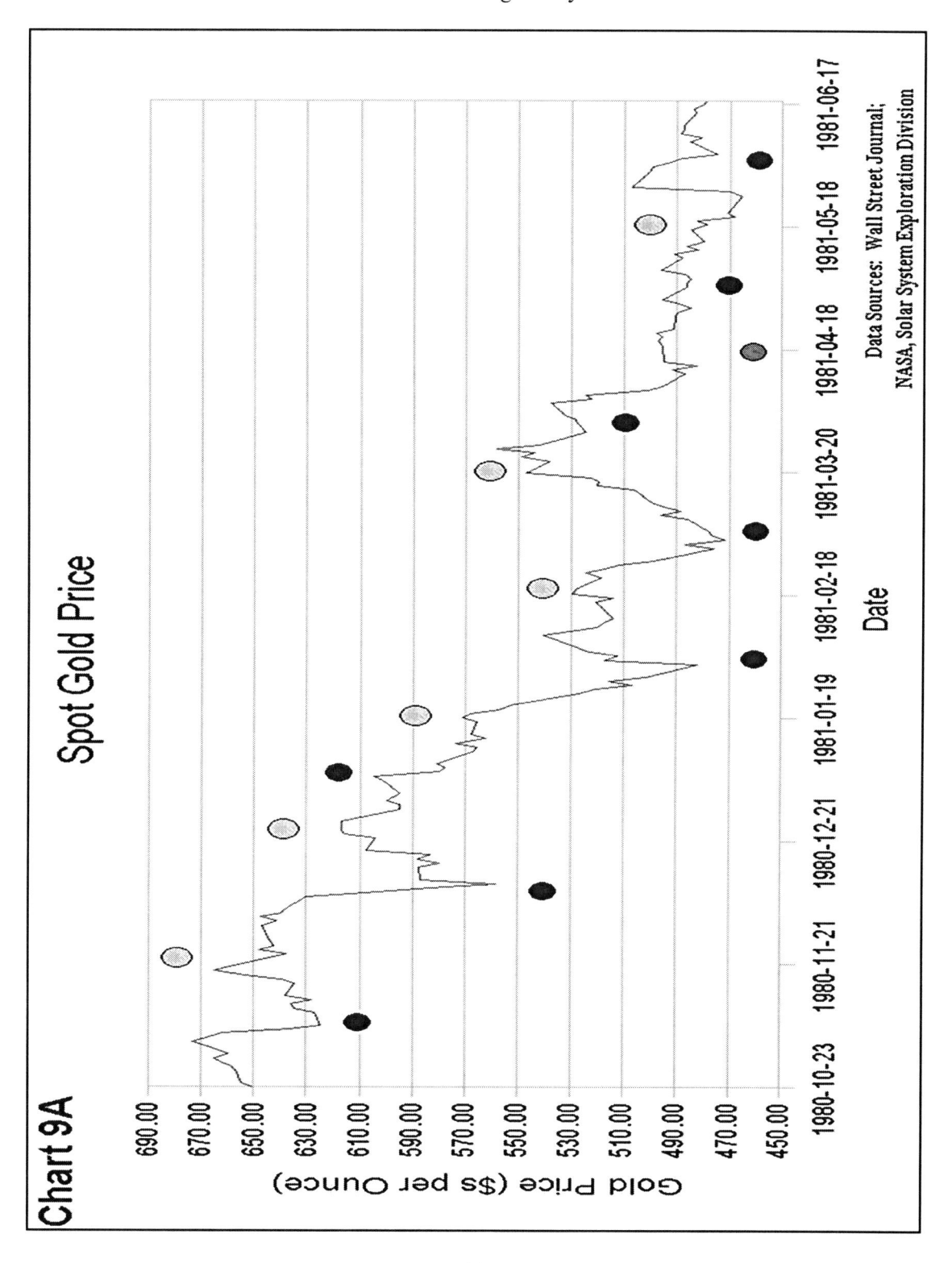

Chart 9A — Spot Gold Price

Gold Price ($ per Ounce) vs Date (1980-10-23 to 1981-06-17)

Data Sources: Wall Street Journal; NASA, Solar System Exploration Division

Chart 9B

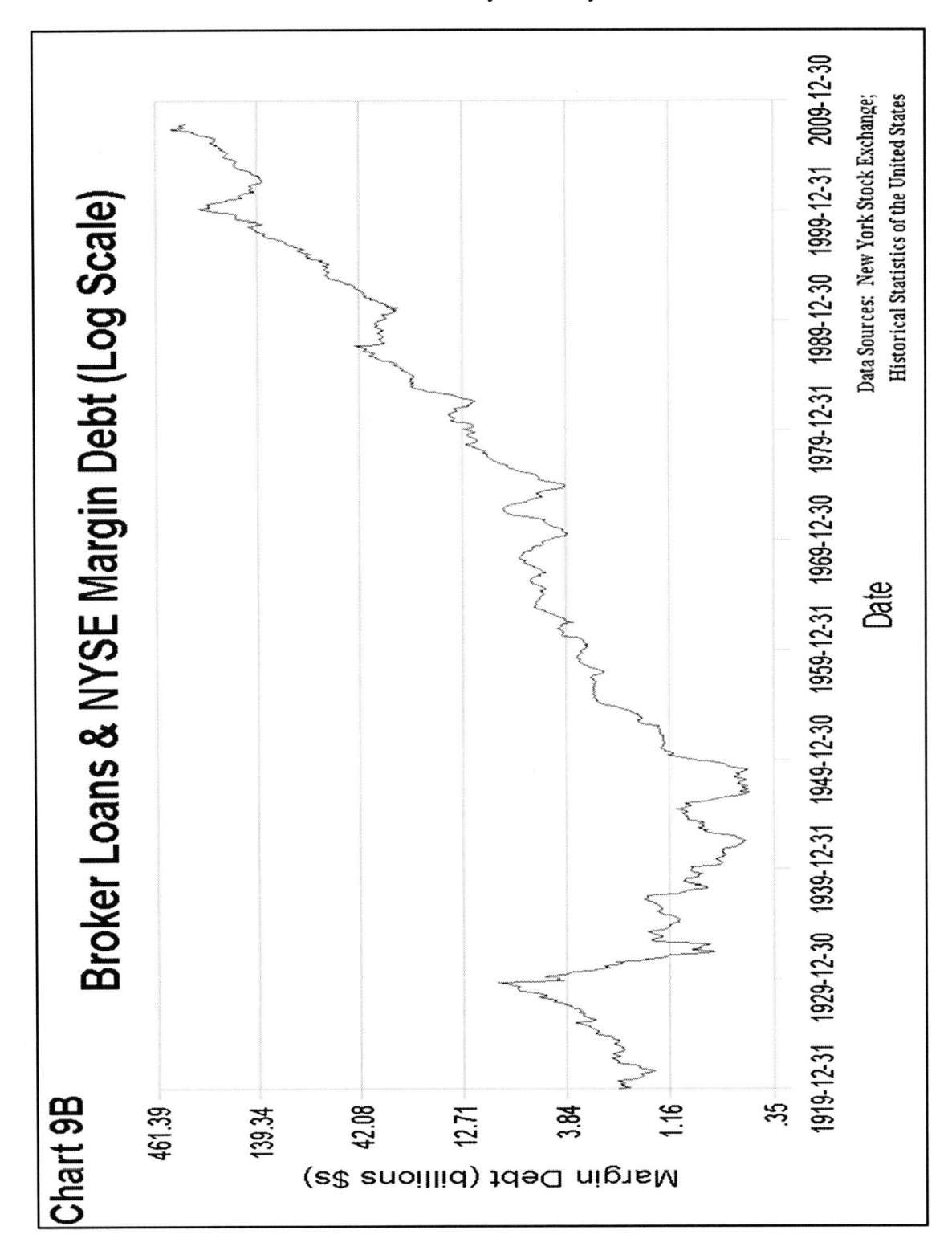

Broker Loans & NYSE Margin Debt (Log Scale)

Data Sources: New York Stock Exchange; Historical Statistics of the United States

Financial Leverage & Cycles

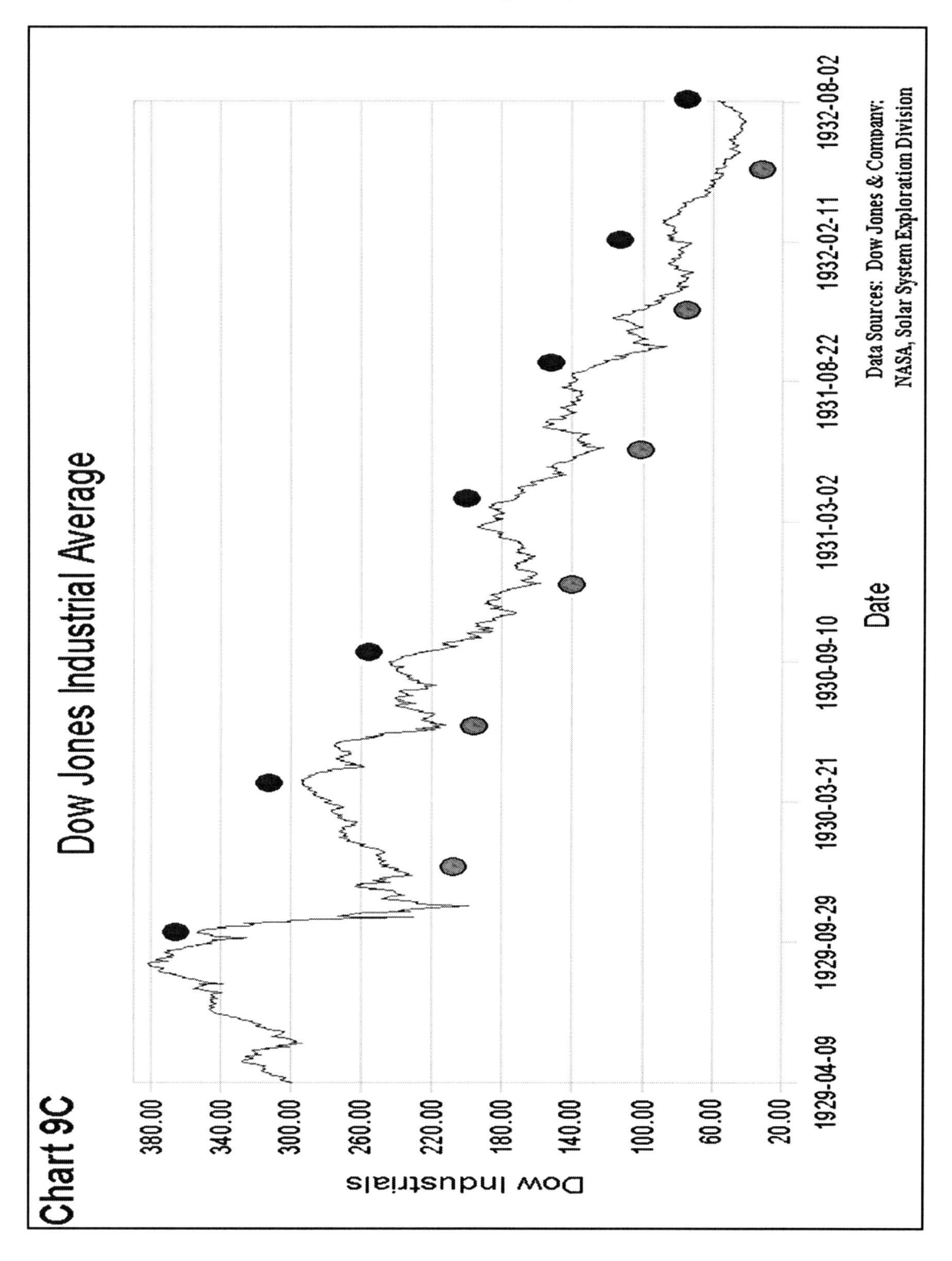

Chart 9C
Dow Jones Industrial Average

Data Sources: Dow Jones & Company; NASA, Solar System Exploration Division

85

Chart 9D Dow Jones Industrial Average

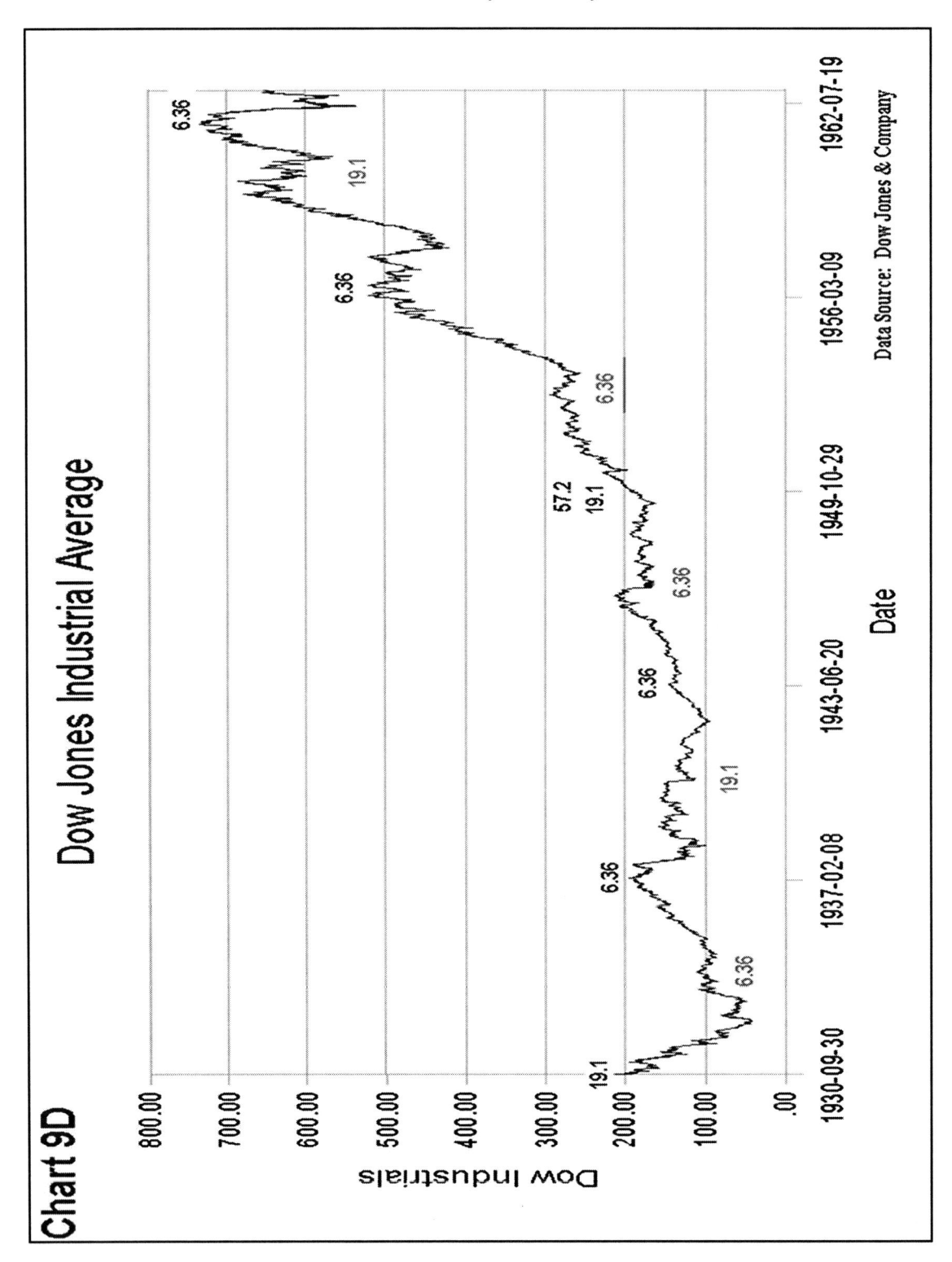

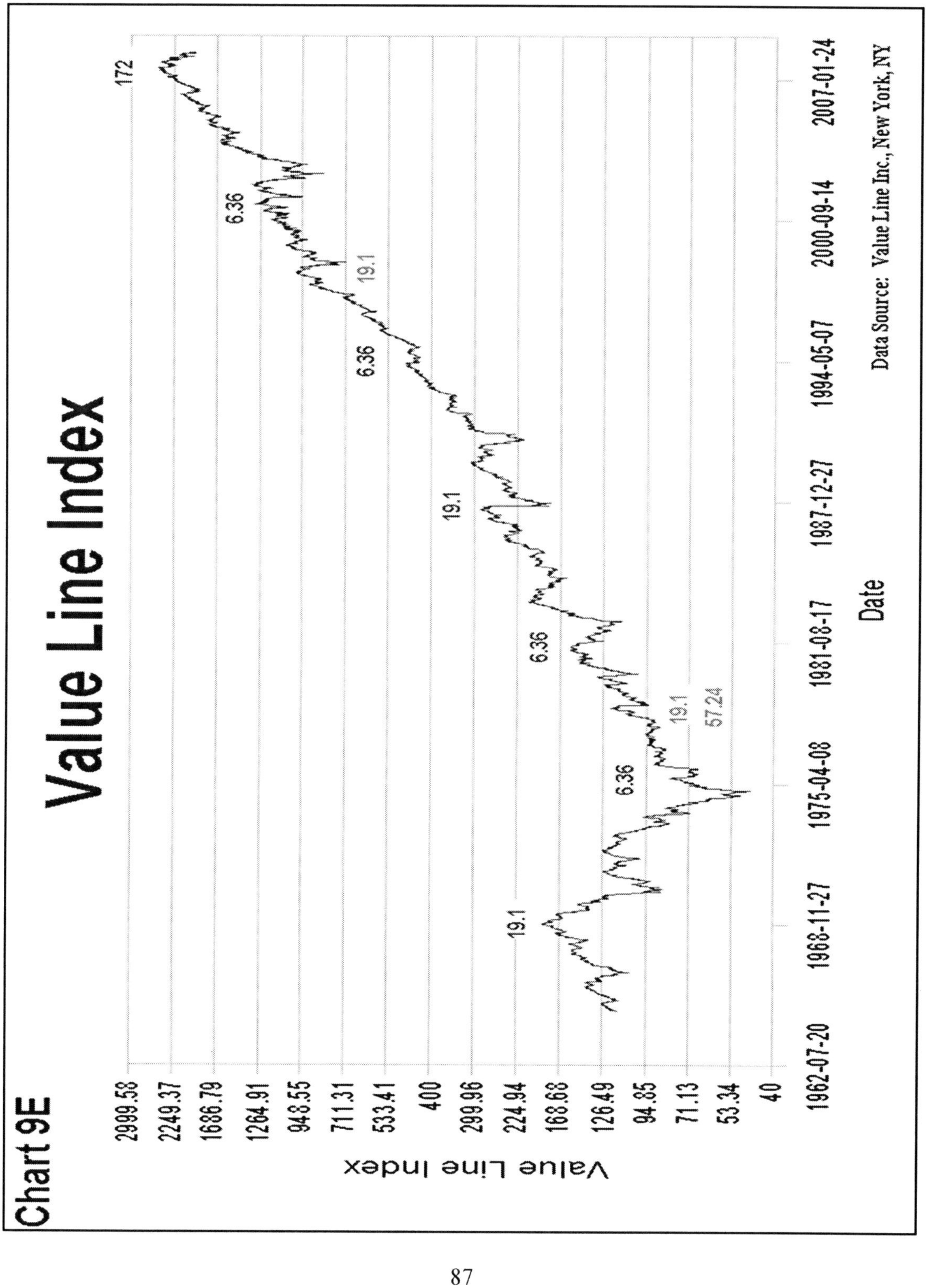
Chart 9E
Value Line Index
Value Line Index
2999.58
2249.37
1686.79
1264.91
948.55
711.31
533.41
400
299.96
224.94
168.68
126.49
94.85
71.13
53.34
40
1962-07-20
1968-11-27
1975-04-08
1981-08-17
1987-12-27
1994-05-07
2000-09-14
2007-01-24
Date
19.1
6.36
19.1
57.24
6.36
19.1
6.36
19.1
6.36
172
Data Source: Value Line Inc., New York, NY

Chapter 10
Electromagnetic Cycles

In Chapter 5, cycles related to fluctuations in Earth's magnetic field were identified and classified separately from other cycles. Chapter 5 contained several examples. However, those examples came mixed with other cycles. In this chapter, the analysis concentrates on individual magnetic cycles and their reliability.

Earth's short-term magnetic cycles result from the constantly changing positions of the Sun, Earth, and Moon, along with the effects of solar eruptions. The strength of each cycle corresponds to its frequency. (This review excludes the Suess and Gleissberg cycles.) The Schwabe sunspot cycle acts as the strongest modulator of geomagnetism. That's followed by the 27-day solar cycle, the seasonal cycle, and the eclipse cycle. At the end of the spectrum, the lunar cycle behaves with the shortest frequency and the weakest amplitude.

To validate the *Unified Cycle Theory*, the five geomagnetic cycles (sunspot, 27-day solar, seasonal, eclipse, and lunar) must be identifiable over long periods of time. Furthermore, they must influence the markets in a similar manner.

Simple tests involving the lengthy histories of the Dow Jones Industrial Average and the Geomagnetic AA Index strongly support the theory. Using the Dow's 12-stock predecessor as a proxy, both data series cover the years from 1885 to 2007. Before running the tests, a moving average was employed to de-trend the Dow. Then both data sets were spliced into buckets corresponding to the day of the particular cycle.

For example, for the lunar cycle, the day of a new moon represented day 1. On every succeeding day, the bucket index incremented by one. When the next new moon arrived, the bucket index was set back to 1. After sorting both datasets in this manner, daily values for the Dow and Geo AA were accumulated in the appropriate buckets, and then averaged. The study excluded the 27-day solar cycle because of difficulties related to the huge variation in its frequency.

After sorting and averaging the data, all four of these cycles correlated inversely to changes in the Geomagnetic AA Index. For all of the cycles, declining Geomagnetic AA caused investors to feel more confident, while rising Geomagnetic AA created fear. If they had not correlated, or correlated in different ways, this part of the theory would have been destroyed, placing everything back to square one. By statistically confirming that each geomagnetic cycle affects human behavior in an identical manner greatly boosted the credibility of the *Unified Cycle Theory*. The charts derived from this analysis come later in the chapter.

The study also revealed that geomagnetism operates with a lead time. The 11-year geomagnetic cycle led market changes by 2.75 years. The seasonal and eclipse cycles both led market trends by about one month. However, the geomagnetic-lunar cycle showed no noticeable lead time – probably due to the shortness of its frequency.

Actually, a delay may also exist in the geomagnetic-lunar cycle. However, the delay would be masked if it equals a month – since the lunar cycle renews itself every 29.53-days. The one-month delay could also explain why eclipse cycles tend to peak on the first new moon before a solar eclipse, instead of two new moons before a solar eclipse. Logically, the geomagnetic interference extreme should reach its greatest when the Sun, Earth, and Moon all reside in the same plane, and its weakest ½-cycle away from the plane.

These observed lead times suggest that changes in the Earth's magnetic field act like a battery in affecting humans. In exerting its influence, charge builds and dissipates slowly, while the geomagnetic power source fluctuates in the interim. At this time, the *Unified Cycle Theory* lacks a good explanation for this delay. It simply notes that geomagnetic fluctuations act as a leading indicator for market changes – leaving the lead-time characteristic as one of its unsolved mysteries.

Before reviewing case studies, a few reminders are necessary. First, the geomagnetic cycles operate alongside the EUWS cycles, with the EUWS cycles acting slightly more dominant. Thus, expected market changes don't always develop in line with Geomagnetic AA fluctuations. To identify conflicts, the charts include important EUWS cycles, when appropriate. When EUWS and geomagnetic cycles interfere, the resulting market moves tended to be sideways and choppy. Conversely, when geomagnetic cycles turn in unison with EUWS cycles, more intense, directional moves generally follow.

The review starts with the 11-year cycle, and it covers all stock market cycles between 1900 and 2007.

11-Year Geomagnetic Cycle

As previously noted, the Schwabe sunspot cycle repeats approximately every 11 years. Solar flares associated with sunspots affect the solar wind, which contributes to Geomagnetic AA oscillations. However, it needs to be reinforced that even though the sunspot cycle and the Geomagnetic AA cycle both fluctuate in 11-year cycles, they don't correlate perfectly throughout each cycle.

Sunspot cycles tend to rise quickly to a maximum, and then dissipate slowly to a minimum. But geomagnetic cycles tend to spend equal amounts of time advancing and declining. Most importantly, once adjusted for the lead time, markets tend to correlate closer to geomagnetic cycles than they do to sunspot cycles. Although, one portion of the sunspot cycle, that point in time when sunspots reach a maximum, tends to act as a reliable indicator of market tops.

Finally, in filtering out the case studies for this chapter, markets with the most leverage became the automatic choices. Using the leverage-visibility rule outlined in Chapter 9, the analysis in this chapter becomes more meaningful.

Study **Charts 10A1 and 10A2**. They show sunspot activity and Geo AA during the past century. Notice how Geomagnetic AA peaks tend to precede sunspot peaks by about 2.75 years. Also notice how sunspot numbers tend to rise rapidly during an up-phase, and then gradually dissipate over several years. Over the past century, stock market movements have done the opposite. While sunspot peaks identify market tops quite well, they're otherwise worthless as an indicator. In contrast, after adjustment for a 2.75-year lead time, the Geomagnetic AA Index follows market trends closely on the way up as well as on the way down. The charts in this chapter invert Geo AA data so they correlate with stock market activity. Additionally, the starting and ending dates for Geo AA charts are moved ahead 2.75 years so they match with stock market oscillations.

At the start of the 20th Century, the Geomagnetic AA Index provided a better indicator of the stock market top on January 19, 1906 than any of the EUWS cycles or the sunspot peak. Taking into account, the 2.75-year lead time, the Geomagnetic AA downturn coincided almost exactly with the actual market peak. See **Charts 10B1 and 10B2**. A 6.36-year cycle peak preceded the market's 1906 downturn by about 9 months. During June 1907, once the sunspot cycle peaked and the theoretical peak of the 2.12-year cycle arrived, the decline turned into a panic. By October of that year, the panic climaxed when the legendary financier J.P. Morgan came to the rescue and injected large sums of money into the stock market.

During late 1916, the stock market peaked in concert with the Geomagnetic AA Index. See **Charts 10C1 and 10C2**. However, a secondary peak in 1918 correlated with peaks in the sunspot cycle and the 2.12-year cycle. A more broadly-based average, constructed by Alfred Cowles of the Cowles Commission, showed that the 1916 stock market peak exceeded the one in 1918. Once the 1918 cycles turned lower, the economy collapsed into a deflationary depression that ended in 1920. This depressionary low coincided with the bottom of the 172-year EUWS cycle. The low also closely coincided with an important non-cyclic event. Six years earlier, on December 23, 1913, Congress passed the Federal Reserve Act – thus recreating a central bank. In 1920, in the midst of the unfolding deflation, the Federal Reserve came to the rescue. However, in the process, the Fed revealed its dark side. The first massive credit bubble in the history of the United States inflated as the Fed provided easy credit to fuel soaring stock prices throughout the Roaring '20s.

Based on the Geomagnetic AA Index, the bull market of the 1920s should have ended a year-and-a-half earlier than it did. See **Charts 10D1 and 10D2**. In fact, the advance-decline line market-proxy showed that stocks topped in 1928 – closely matching the Geo AA downturn. While the broad market declined, the tug-of-war with the upside of the 19-year EUWS cycle worked in a divergent manner, pulling the popular blue-chip stocks upward until September 1929. During October 1929, the stock market finally adjusted to declining geomagnetic levels – but only after the geomagnetic-seasonal and geomagnetic-eclipse cycles kicked in. During 1930, stocks rebounded partially. The rebound coincided with a counter-trend move in Geo AA. By late 1930, the 19-year EUWS cycle peak finally arrived. That happened about the same time as geomagnetic levels resumed their downtrend. With both the 19-year cycle and geomagnetic cycle finally moving lower together, the brunt of the bear market hit. This brutal phase continued until 1932.

In 1937, the stock market traced out a double-top before collapsing later that year. The first peak coincided almost exactly with the 6.36-year EUWS cycle. And the second peak occurred during a sunspot maximum. The 2nd crash of the Great Depression followed. The Geomagnetic AA Index peaked a few months after the second top. By late 1937, all three indicators were in sync to the downside. See **Charts 10E1 and 10E2**. A long bear market followed – lasting until 1942. For the stock market, 1942 coincided with a major low. From a cyclic standpoint, two important troughs sandwiched 1942. First, the EUWS registered a 19-year cycle trough during April 1940. And a 2.12-year trough arrived in May 1942 – almost exactly matching the stock market's low. A Geomagnetic AA minimum came a few months later. Hence, the 1937-42 bear market can be considered a near-perfect example of both the 19-year EUWS cycle and the Geo AA Index moving in sync from top to bottom – with the stock market following both cycles lower.

During the late 1950s and early 1960s, the 19-year cycle and the 11-year Geomagnetic AA Index were in direct conflict. Toward the end of 1958, the 19-year cycle bottomed at the very time the geomagnetic and sunspot cycles peaked. See **Charts 10F1 and 10F2**. This type of conflict often results in a choppy outcome, with the market failing to follow either cycle very closely. That's what happened in the 1950s and 1960s. By 1962, with Geo AA approaching a low, the stock market also hit bottom. As soon as geomagnetic activity came back into sync with the rising 19-year EUWS cycle, the Go-Go bull market of the 1960s came alive. The speculation that accompanied the 1960s bull market never reached the extremes of the 1920s.

However, it reached levels dangerous enough to show that traders once again ignored the ill-fated leveraging lessons from financial history.

In stark contrast to the conflicts from a decade earlier, by the late 1960s, geomagnetic activity and the 19-year panic cycle came back into perfect sync. The sunspot cycle, the Geomagnetic AA Index, and the 19-year EUWS cycle all peaked together in late-1968. At exactly the same time, the nastiest bear market since the Great Depression struck Wall Street. See **Charts 10G1 and 10G2**. Rather than using the Dow Industrials, the Value Line Index more accurately reflected what happened to the average stock from 1969 to 1975. During the first half of that bear market, the so-called "Nifty Fifty" growth stocks escaped the bear market. The thousands of companies making up the rest of the market got hammered. By the end of 1974, the average share had lost 75% of its 1968 peak price. Like the 1920s, the speculation of the 1960s ended badly. However, with the 57-year EUWS cycle approaching a trough in 1978, and with substantial deleveraging already taken place, the groundwork was laid for a new bull market.

Since 1968, every geomagnetic-sunspot cycle peak coincided with some type of major financial crash. From 1968 to 1974, secondary stocks succumbed to the cycle. During 1980, the geomagnetic-sunspot peak hit precious metals and other commodity markets. **Charts 10H1 and 10H2** show how the gold market correlated with Geo AA. The *Unified Cycle Theory* covers all types of markets, not only stocks. However, the sectors infected with speculation always emerge as the ones that correlate strongest with the various physical cycles. And precious metals fit the criteria in the late-1970s. The 19-year EUWS cycle bottomed in 1978 while the geomagnetic cycle troughed slightly ahead of that – during 1977. Speculation in the yellow metal started in the early 1970s, paused during 1975-76, and then returned stronger than ever from 1977 through early 1980. Then in early 1980, the gold bubble burst – coinciding with peaks in geomagnetism and sunspots. For the next two years, financial disaster struck gold bugs. By mid-1982, the gold market had lost 67% of its peak value.

Remember, all Geo AA charts in this chapter have the x-axis moved forward 2.75 years so their lead-time correlations appear clearer. During the 1980s, the great speculation in Japanese stocks represents the next in this sequence of geomagnetic-sunspot crashes. By the end of 1989, the bull reached a mature stage. As the first trading days of 1990 arrived, three important cycles combined to reverse the trend. Geo AA, sunspots, and the 2.12-year deviation of the 19-year EUWS cycle all peaked in early 1990. See **Charts 10I1 and 10I2**. After completing a 6-week topping pattern by mid-February, Japanese equities crashed for good. To this day, Japanese stocks have failed to recover losses resulting from the 1990 crash. In contrast to its lofty peak near 40,000 at the beginning of 1990, the Nikkei 225 Index traded in the vicinity of only 8,000 during November 2008.

In 2000, technology shares became the next casualty in the sequence of 11-year geomagnetic-sunspot crashes. Actually, the NASDAQ Index peaked in late-March 2000, while the sunspot cycle and the 6.36-year EUWS cycle hit their maxima during September of that year. And geomagnetic activity turned lower shortly afterward. See **Charts 10J1 and 10J2**. For most of the broader stock market averages, a double-top formed between March and September 2000. However, the relative strength in technology shares had already deteriorated substantially by the latter date. After the cycles turned lower in September, the partial rebound in NASDAQ ended abruptly. A relentless bear market followed. By late-2002, NASDAQ had lost nearly 80% of the peak value attained only three years earlier.

Seasonal Cycle

Two seasonal cycles exist. We're all familiar with the annual weather cycle. However, little is written about the semi-annual geomagnetic cycle. With their 2006 publication *Environmental Discomfort and Geomagnetic Field Influence on Psychological Mood in Athens*, Nastos *et al.* provided a welcome exception. Chapter 7 already introduced this cycle. Review **Chart 7F** again. Then compare it to **Charts 10K1 and 10K2**. They show similar seasonal tendencies.

The Dow's seasonal pattern mimics **Chart 7F** closer than it does the Geo AA **Chart 10K2**. This close correlation supports the hypothesis from Nastos *et al.* – that the seasonal behavior cycle result from a combination of both the annual weather cycle and the semi-annual geomagnetic cycle. Clearly, with two cycles per year, something other than axial-tilt causes the semi-annual geomagnetic cycle. A cause for this cycle cannot be found. However, I presume it originates from Earth's position relative to the Sun and the galactic center of the Milky Way.

Even though January-March seasonal weakness in the Dow appears minimal, an alternative set of data shows a different story. For USA equity markets, major panics during January-March appear conspicuously absent for the past 130 years. But that's not true for markets in general. The *Unified Cycle Theory* encompasses all markets worldwide. By including all global equity and commodity markets, the occurrence of January-March crashes increases dramatically.

The August-October period contained the 1720 South Sea Bubble crash, the 1873 stock market panic, the 1929 stock market crash, the 1987 stock market crash, the 1997 Southeast Asian equity market crash, and recently the 2008 global equity market crash.

However, the January-March period shows an equal number of infamous crashes. These include the implosion of the 1637 Tulip-mania, the 1720 Mississippi Bubble crash, the 1980 precious metal market crash, the 1990 Japanese stock market crash, and arguably the 2000 NASDAQ crash.

During 2000, NASDAQ briefly diverged from the rest of the market. This divergent pattern mimicked similar behavior in other major crashes. A more detailed inspection reveals how these divergences keep speculators bullish even as internal market conditions deteriorate. After the Dow Industrials topped in January 2000, speculators rushed into NASDAQ stocks as a safe haven, falsely believing that technology fundamentals were too strong to allow a significant decline in share prices. But NASDAQ provided the same "fools trap" as other divergent indexes had during the early phases of other crashes. Viewed from this perspective, the 2000 crash can be considered a January to early-April crash that mostly fit inside the expected seasonal weak period.

Now, the study shifts a step lower – to major declines that qualify as panics, but not crashes. In the domestic equity markets, the following panics essentially transpired during the August-October period: the 1857 panic from July 30 to October 12; the 1869 panic from July 27 to September 29 (associated with the Fisk-Gould gold crisis); the 1873 panic from August 2 to September 20 (resulting in closure of NYSE trading from September 21-29); the 1937 mini-crash from August 14 to October 23; and the 1998 LTCM crisis from July 17 to October 5. That's five panics in the August-October period.

Next, three panics started in the January-March period and ended in the August-October period. Those panics include 1907 from January 7 to March 25 and July 8 to November 15; the 1920 mini-depression from January 3 to February 25 and July 8 to November 19; and the start of the 1973-74 bear market from January 11 to March 23 in 1973 and the end from July 24 to October 4 in 1974.

Finally, three large bear markets spanned the entire calendar year, and cannot be classified as seasonal. Those three instances include the 1835-42 depression; the 1892 depression; and the 1968-70 bear market.

Summarizing this survey of panics and crashes by the time of the year they occurred, the tally appears as follows:

August-October	14	(46.7%)
January-March	13	(43.3%)
All other periods	3	(10.0%)

Once again, the seasonal tendencies of panics and crashes closely mimic the patterns shown in **Charts 7F, 10K1, and 10K2**. In this particular study of 30 panics and crashes, the pattern appeared even stronger than it did in the charts (which cover all time periods). As a side note, this seasonal study also reflects the importance of speculation in making cycles more visible. During tranquil times, markets often follow patterns other than the ones in **Charts 10K1 and 10K2**. However, when speculative fever heats up, markets increasingly revert to tracking the geomagnetic-seasonal cycle. In summary, the components of <u>annual temperature change</u> plus <u>semi-annual geomagnetic oscillation</u> combine to cause the seasonal market cycle.

Eclipse Cycle

On average, the Sun, Earth, and Moon align themselves on the same plane every 173.31 days to create the eclipse cycle. To verify this geomagnetic cycle's influences on markets, all eclipse cycles between 1885 and 2007 were analyzed for the Dow Jones Industrial Average and Geomagnetic AA. This time-frame included 257 eclipse cycles. That represents an extensive sample – covering more than twice as many cycles as the seasonal study. Review **Charts 10L1 and 10L2**.

Chapter 6 already noted physiological changes associated with a solar eclipse.[Keshavan *et al.*, 1981] However, the study in this chapter – computed from 257 eclipse cycles, depicted in **Charts 10L1 and 10L2** – qualifies as a more comprehensive study. Because the geomagnetic cycle acts as a leading indicator, a 29-day offset was used when comparing it to the Dow Industrials. Both of the chart patterns closely match the observations already presented earlier in this book.

Specifically, markets tend to peak on the first new moon prior to a solar eclipse. They tend to fall for one month after that, until the next new moon on day-29.5. Then markets rally until the next full moon. And that's followed by another decline until the next new moon on day-59. The market then hits bottom near the mid-point of the cycle around day-88.5. For individual markets, the time-span for a bottom generally ranges from the two weeks before the new moon at day-88.5 to two weeks after.

Lunar Cycle

It takes the Moon approximately 29.53 days to rotate around the Earth – using the Sun as the reference point for the cycle. By averaging data for the 1,508 lunar cycles since 1885, similar patterns emerge for both Geo AA and the Dow Industrials. Study **Charts 10M1 and 10M2**. The lunar cycle repeats in a simple way. Markets tend to rise ahead of a full moon. And conversely, they tend to fall after a full moon – continuing the decline until the next new moon. The only exceptions to this rule come (a) at the top of an eclipse cycle – at the time of a new moon, and (b) at the bottom of the eclipse cycle – at the time of a full moon.

27-Day Solar Cycle

The frequencies for the 29.53-day lunar cycle and the 27-day solar component of Geomagnetic AA come close to matching. For that reason, when they move in sync, attributing cause-and-effect becomes more difficult. Also disturbing, the frequency of the 27-day solar cycle varies erratically. While not impossible, this erratic behavior makes studying it more challenging. At this time, its impact on markets has not been assessed. However, based on the other geomagnetic cycles, an impact probably exists. That's especially true since the strength of the 27-day cycle approximates, or even exceeds, the power of the seasonal cycle.

In a sense, the only thing that should matter is the net effect from the sum of all components of Geomagnetic AA. Some may argue that if a correlation exists between markets and geomagnetic activity, then the Geomagnetic AA Index itself (which acts as the sum of all of its components) should be sufficient in researching a relationship to market activity. This argument deserves consideration. For example, throughout 1987, a smoothed 60-day average of Geomagnetic AA, used in combination with a 30-day lead time, tracked stock market activity fairly closely. See **Charts 10N1 and 10N2**.

But the entire process behaves in a complex way. The Geomagnetic AA Index correlates to market cycles in two ways. First, the longer-term 11-year geomagnetic component leads market changes by 2.75-years. This 2.75-year lead-time correlation has already been reviewed extensively. Second, the short-term cycles lead market changes by one month. The 1987 example looks enticing as a short-term indicator; however, this one example could simple result from coincidence. An extensive short-term study has not been conducted, thus preventing a definitive conclusion.

However, one thing remains certain. Many cycles constantly interact with each other. They often oscillate in conflicting ways. Without understanding each individual sub-component, it becomes difficult to detail a good theory about the entity. Regretfully, nothing substantial can be added here about the effects of the 27-day cycle. However, this area definitely deserves future research.

Galactic Cosmic Rays and the Geomagnetic AA Index

Before concluding this discussion on geomagnetism, it's important to look at its association with galactic cosmic rays (GCRs). The reason... various researchers suggest that GCRs play a role in modulating climate cycles, species extinctions, and various geological events.[Perry, 2007] and [Shaviv, 2002] Details come later in the book.

During the 1930s, the Department of Terrestrial Magnetism of the Carnegie Institution of Washington established a worldwide network of GCR monitoring sites, supervised by Dr. S.E. Forbush. Nonetheless, over the years, the monitoring sites changed, making it impossible to construct a perfectly continuous series.

NOAA's National Geophysical Data Center now collects and distributes GCR data. However, the NGDC has not made data from 1935 to 1953 available to the public. For this book, a composite GGR Index was manufactured by making appropriate adjustments to the data, and then averaging together the data from the GCR monitoring sites at Huancayo, Peru (1953 to 2006), Climax, Colorado (1953 to 2006), Kiel, Germany (1958 to present), Calgary, Canada (1964 to present), and Beijing, China (1993 to present). **Charts 10O and 10P** show how the GCR Index correlates with Geomagnetic AA, and **Charts 10O and 10Q** depict how the Index acts as a leading indicator for sunspot numbers.

Table 10.1 provides information about turning points for cycles in Geo AA, GCRs, and sunspots. Summarizing the results, on average, GCR peaks lead Geo AA peaks by seven months. However, for 5 of the 6 cycles, GCR tops concurred with Geo AA. The sole exception came in 1977, when the GCR peak arrived 41 months prior to the Geo AA top. Turning to the sunspot cycle, on average, GCR peaks arrived 36 months prior to sunspot maxima – ranging from 31 to 43 months ahead of the sunspot cycle.

For troughs, GCRs acted as a distinct leading indicator for both Geo AA and sunspots – extending the lead time far beyond that displayed at peaks. GCR valleys averaged 23 months ahead of Geo AA low points, with only 2 cycles concurrent. For sunspots, GCR valleys preceded sunspot minima by 68 months, on average.

Table 10.1 – Comparison of Turning Points: GCRs, Geotic AA, and Sunspots.

GCR Max.	GeoAA (Inv)	Sunspot Max	GCR Min.	Geo AA (Inv)	Sunspot Min
May 1955	Oct 1955	Sept 1958	Aug 1958	Jan 1961	Dec 1964
Nov 1965	Oct 1965	June 1969	Sept 1969	Jan 1975	Dec 1976
(June 1974) May 1977	(July 1977) Oct 1980	June 1980	May 1983	May 1983	Mar 1987
May 1987	July 1987	Dec 1989	Aug 1990	(Mar-Aug 90) Mar 1992	Nov 1996
Mar 1998	Feb 1998	Oct 2000	Nov 2003	Jan 2004	May 2008
Mar 2008	Jan 2008				

In conclusion, many cycles exist in nature. This book cannot cover them all. Hence it only includes cycles that influence mankind and nature the most. Nonetheless, for any excluded cycles, the basic *Unified Cycle Theory* concepts remain intact. Different classifications of cycles exist with varying impacts. Gravitational cycles affect our planet differently than electromagnetic cycles. And the mysterious EUWS cycles create a third set of oscillations. Combined, they generate endless fluctuations in our environment. Separated and identified, they enhance our ability to predict future events on Earth – thus providing us with adequate warnings concerning future environmental changes.

This chapter examined cycles associated with Earth's magnetic field. Among other things, geomagnetism affects human behavior. Especially noteworthy, geomagnetic cycles correspond to oscillations in equity and commodity markets – adjusted for slight lead times.

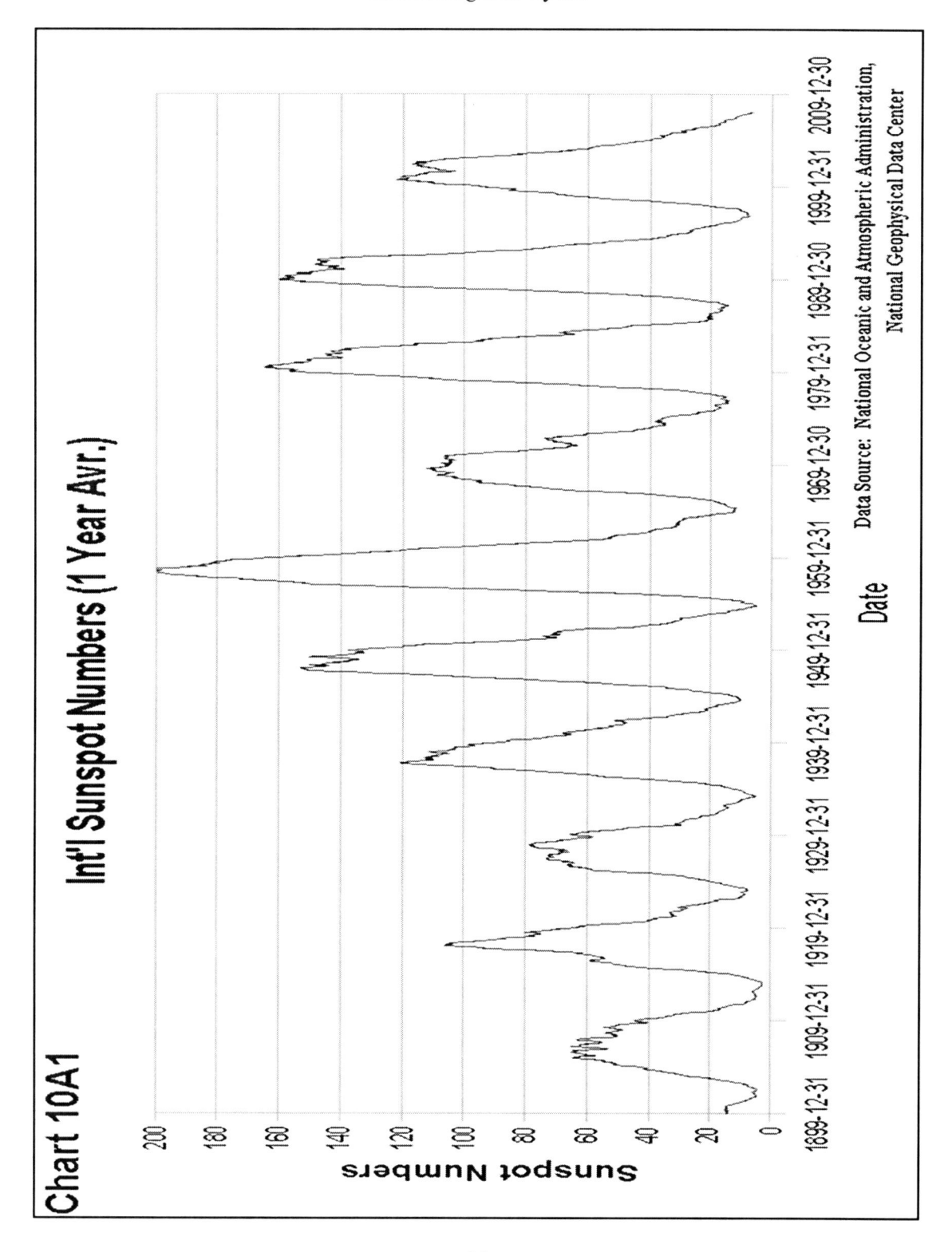

Electromagnetic Cycles

Chart 10A1 — Int'l Sunspot Numbers (1 Year Avr.)

Data Source: National Oceanic and Atmospheric Administration, National Geophysical Data Center

Chart 10A2

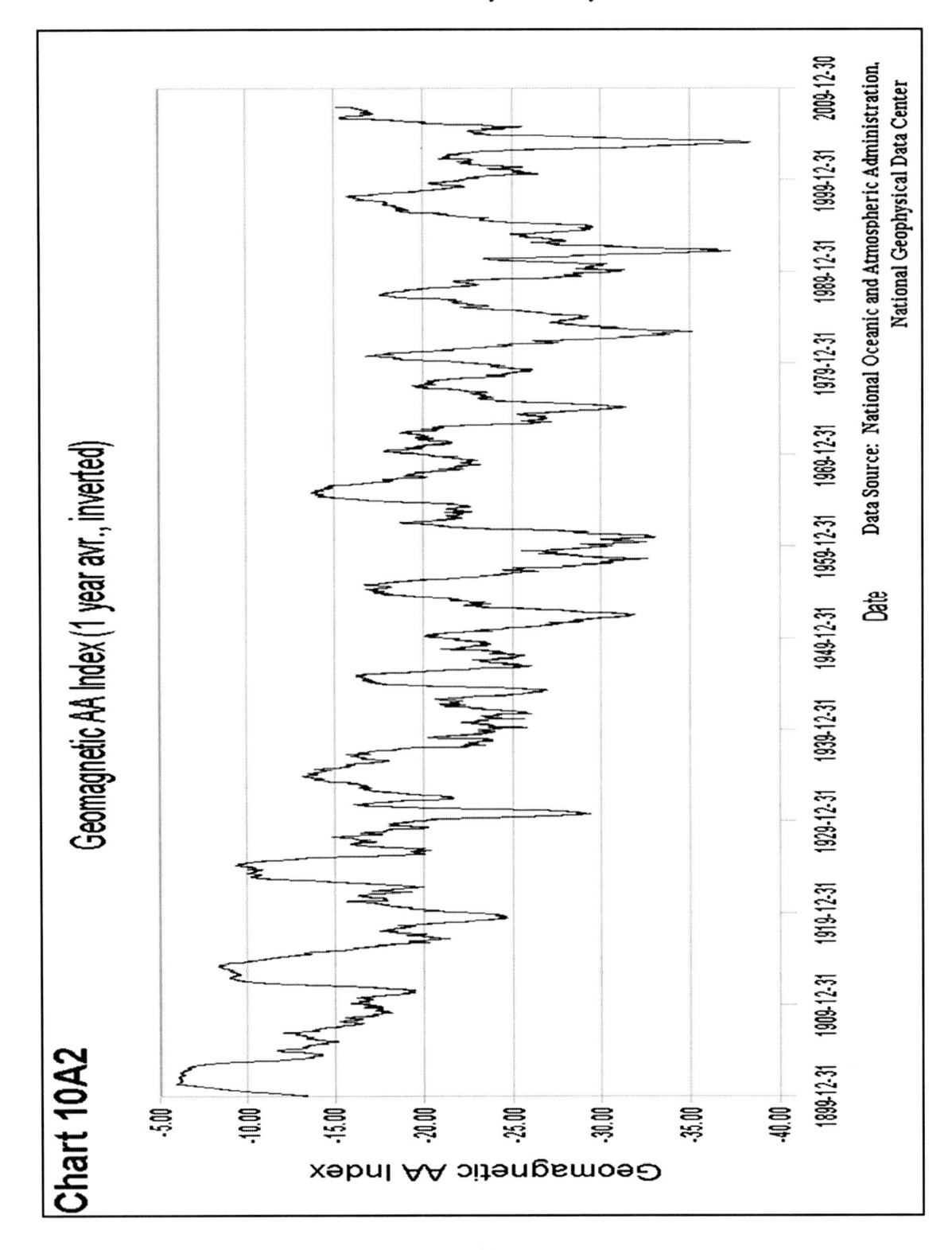

Geomagnetic AA Index (1 year avr., inverted)

Data Source: National Oceanic and Atmospheric Administration, National Geophysical Data Center

Chart 10B1
Dow Jones Industrial Average

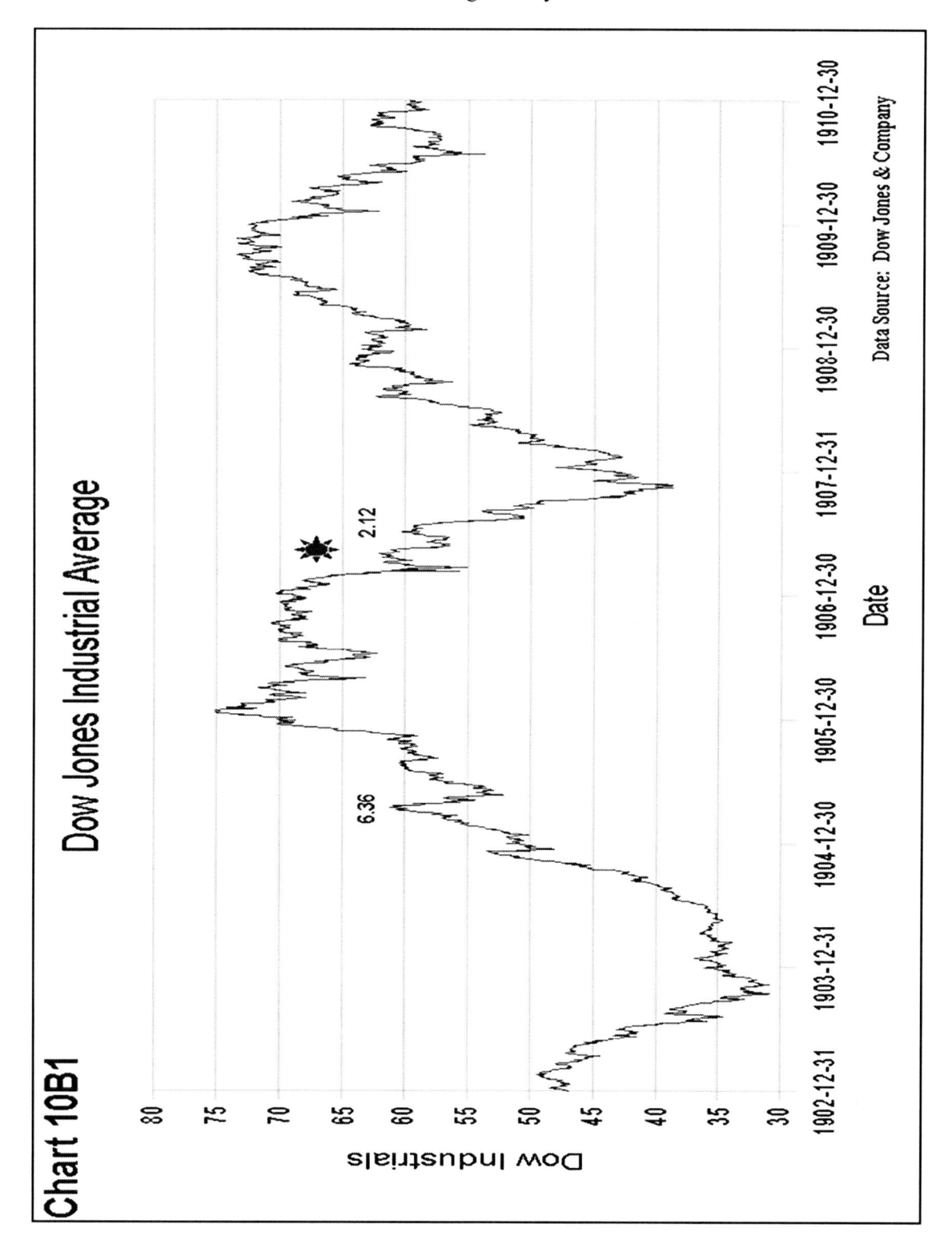

6.36

2.12

Date

Data Source: Dow Jones & Company

Chart 10B2

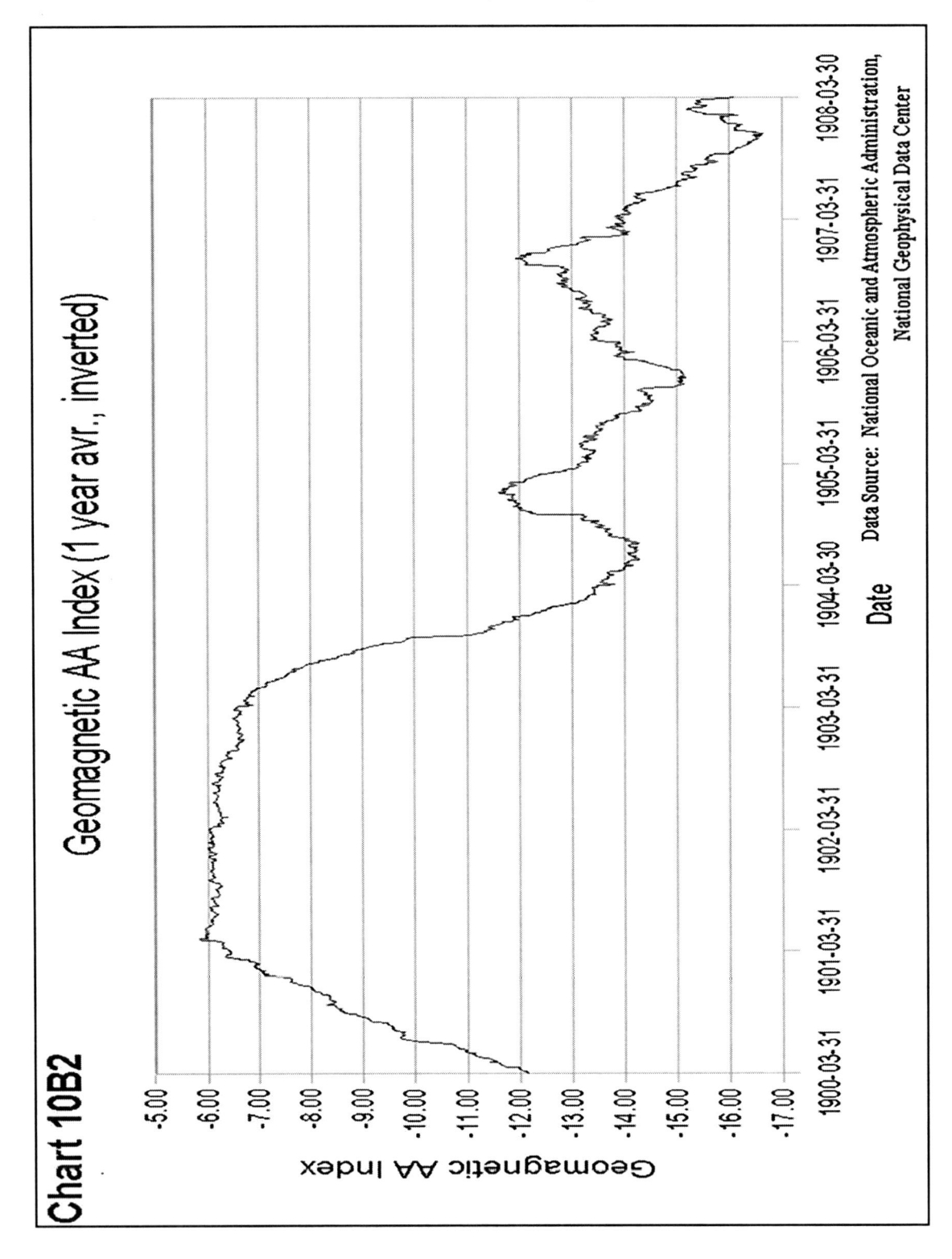

Geomagnetic AA Index (1 year avr., inverted)

Data Source: National Oceanic and Atmospheric Administration, National Geophysical Data Center

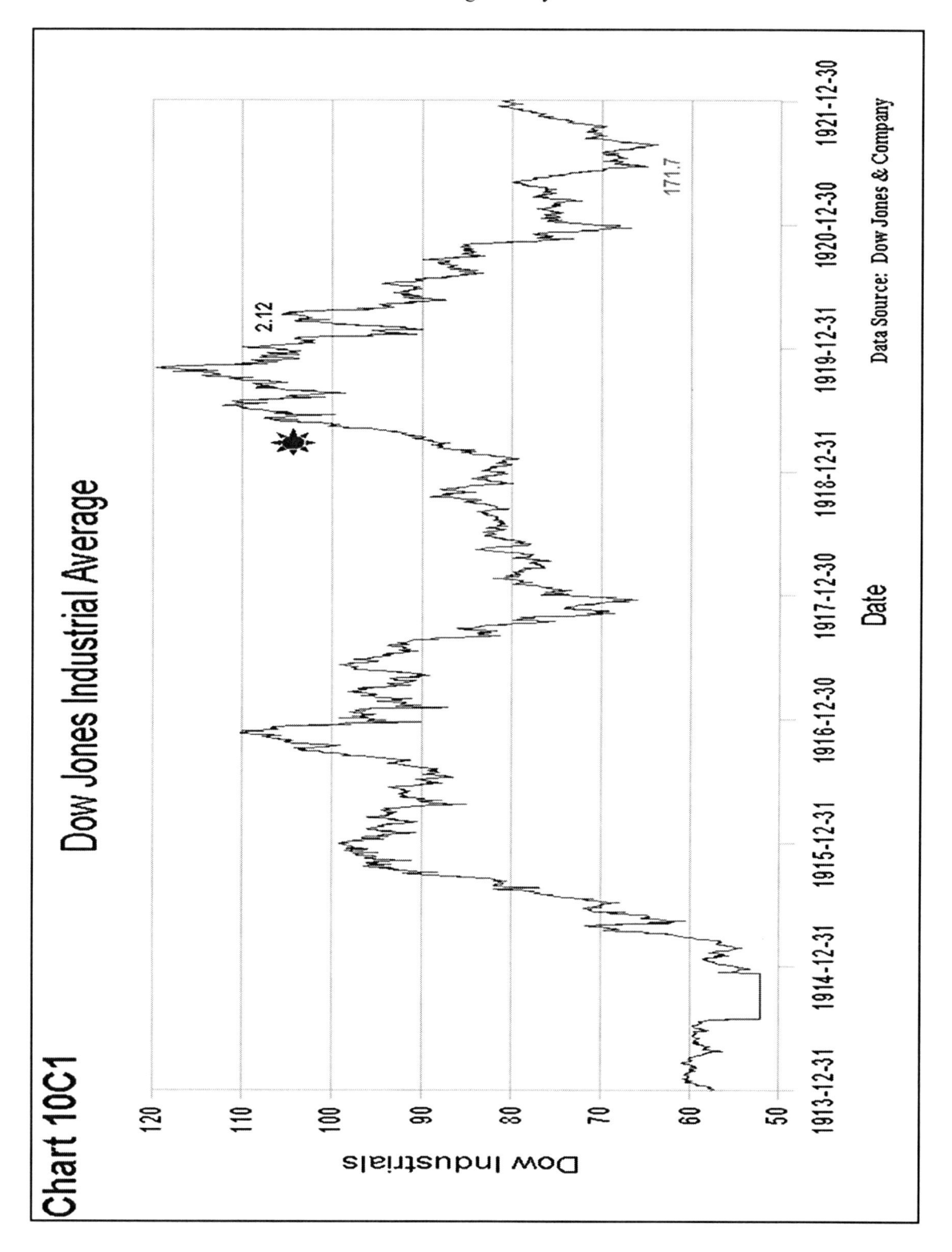

Electromagnetic Cycles

Chart 10C1

Dow Jones Industrial Average

Data Source: Dow Jones & Company

Chart 10C2

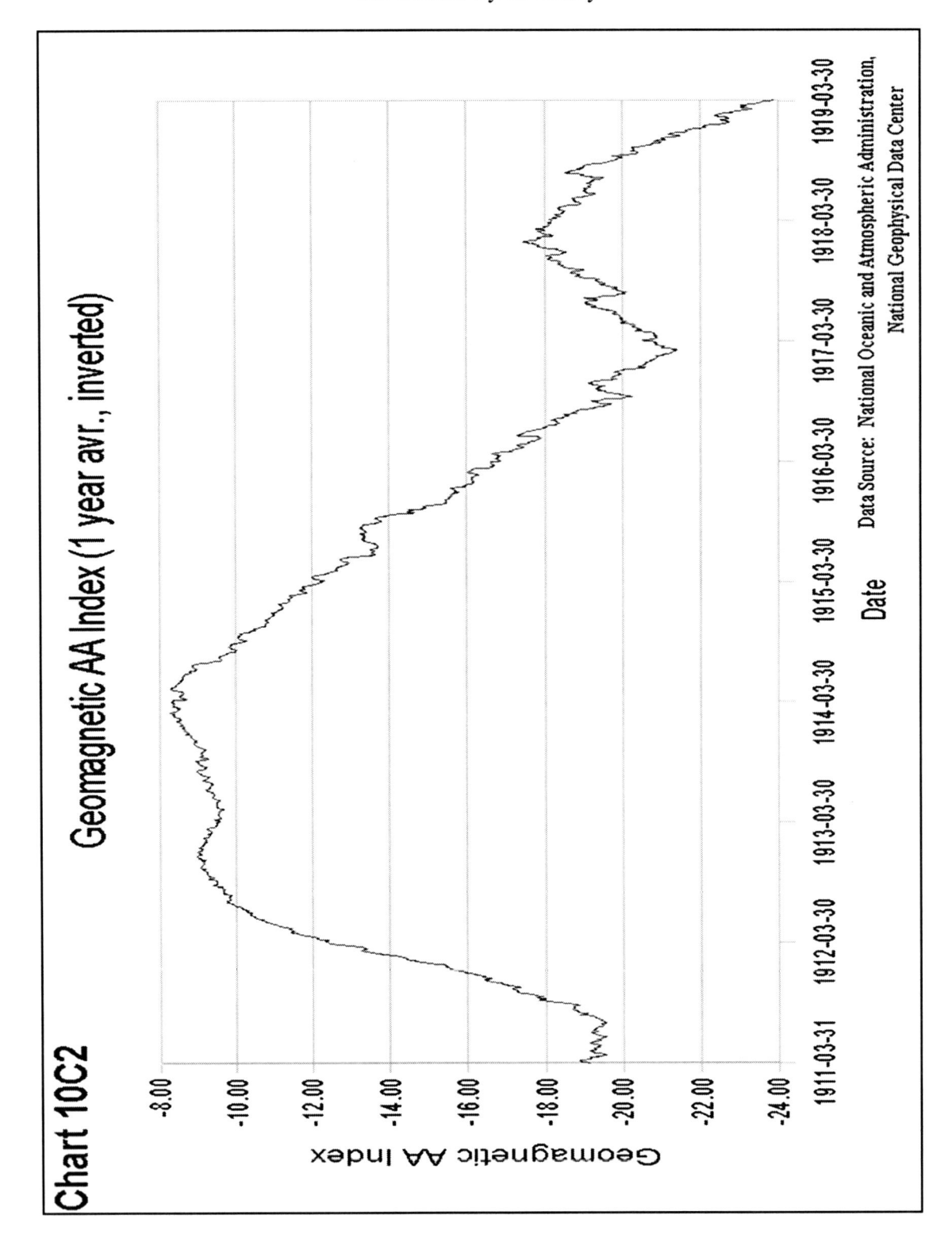

Geomagnetic AA Index (1 year avr., inverted)

Data Source: National Oceanic and Atmospheric Administration, National Geophysical Data Center

Chart 10D1
Dow Jones Industrial Average

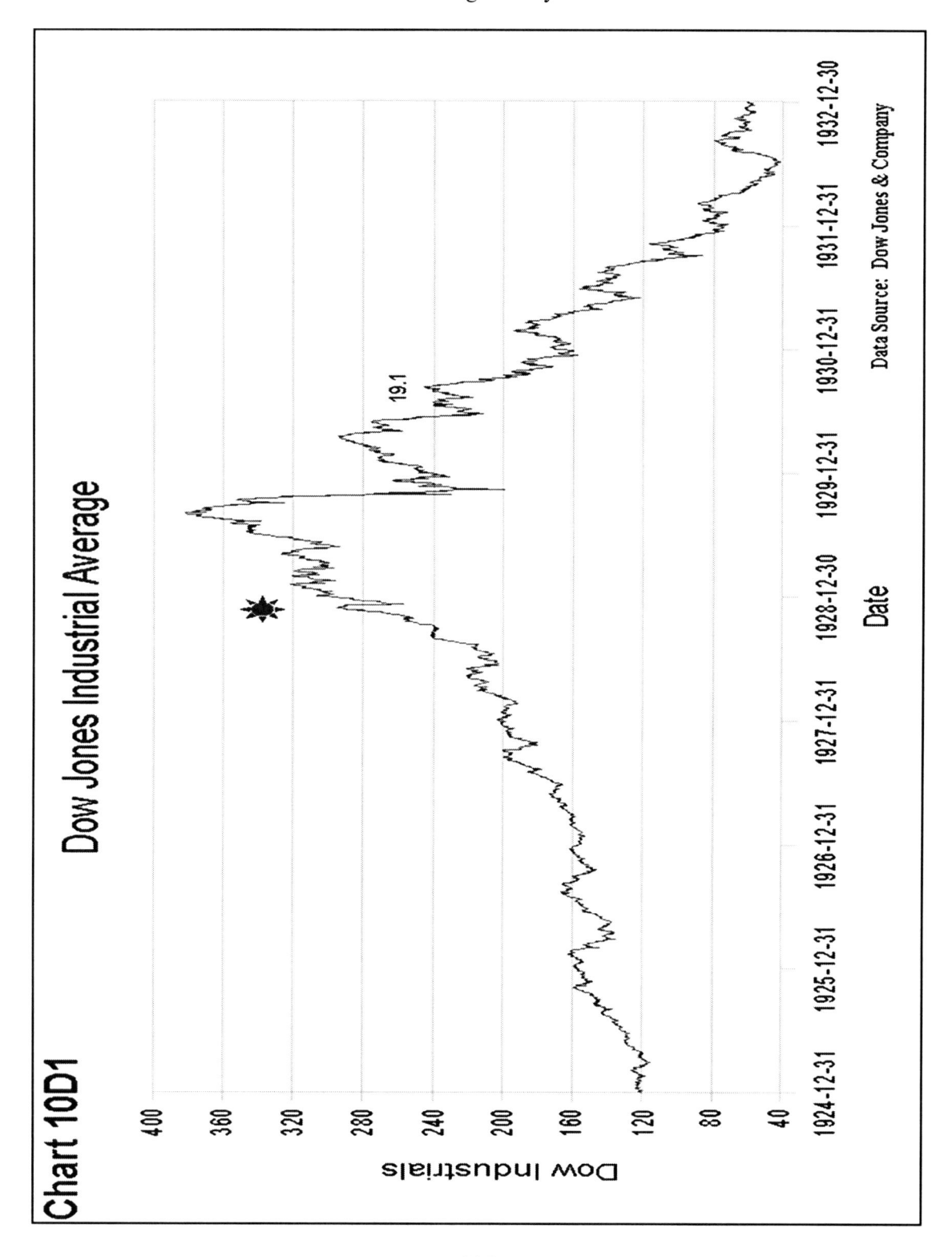

19.1

1924-12-31 — 1925-12-31 — 1926-12-31 — 1927-12-31 — 1928-12-30 — 1929-12-31 — 1930-12-31 — 1931-12-31 — 1932-12-30

Date

Dow Industrials

Data Source: Dow Jones & Company

Chart 10D2

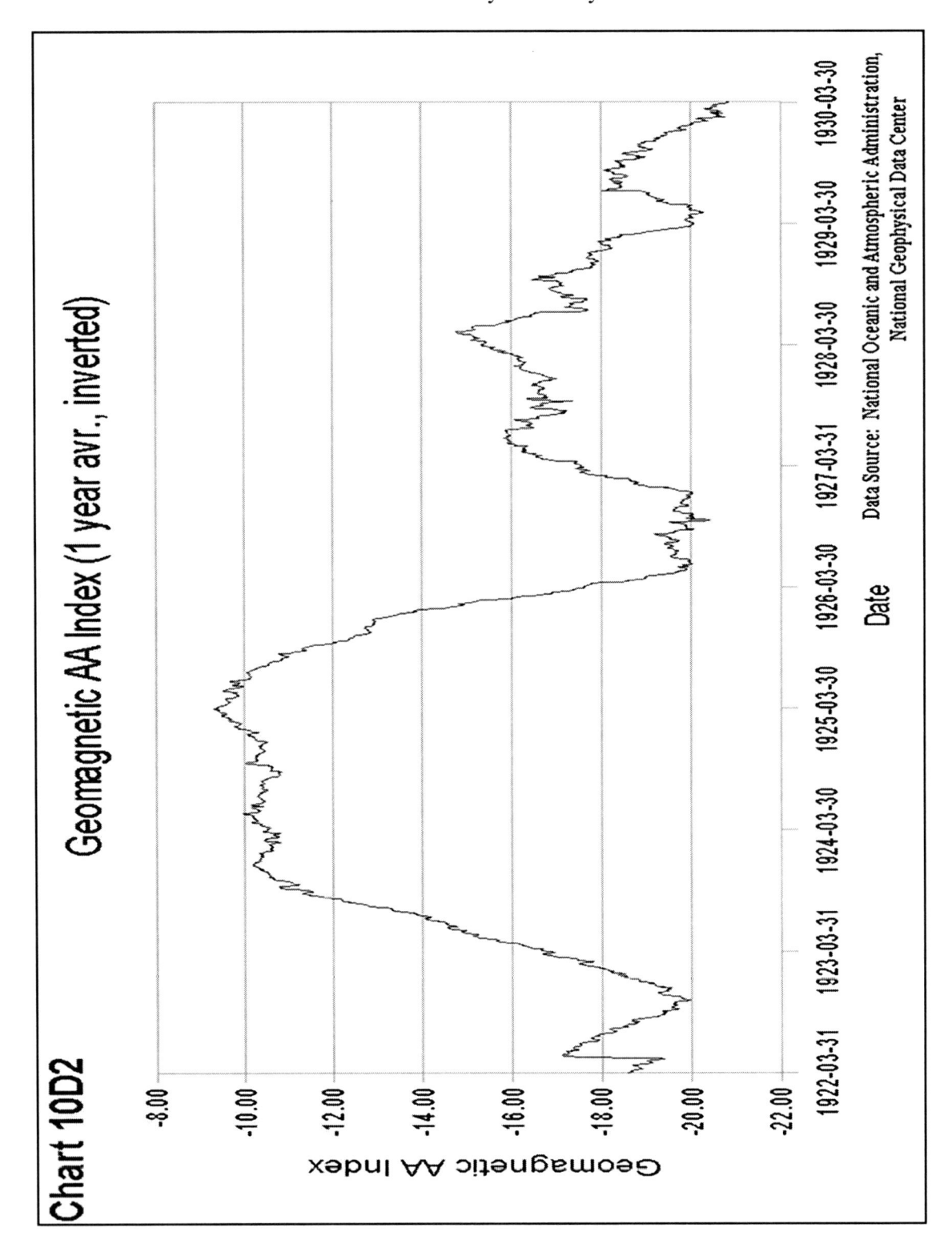

Geomagnetic AA Index (1 year avr., inverted)

Data Source: National Oceanic and Atmospheric Administration, National Geophysical Data Center

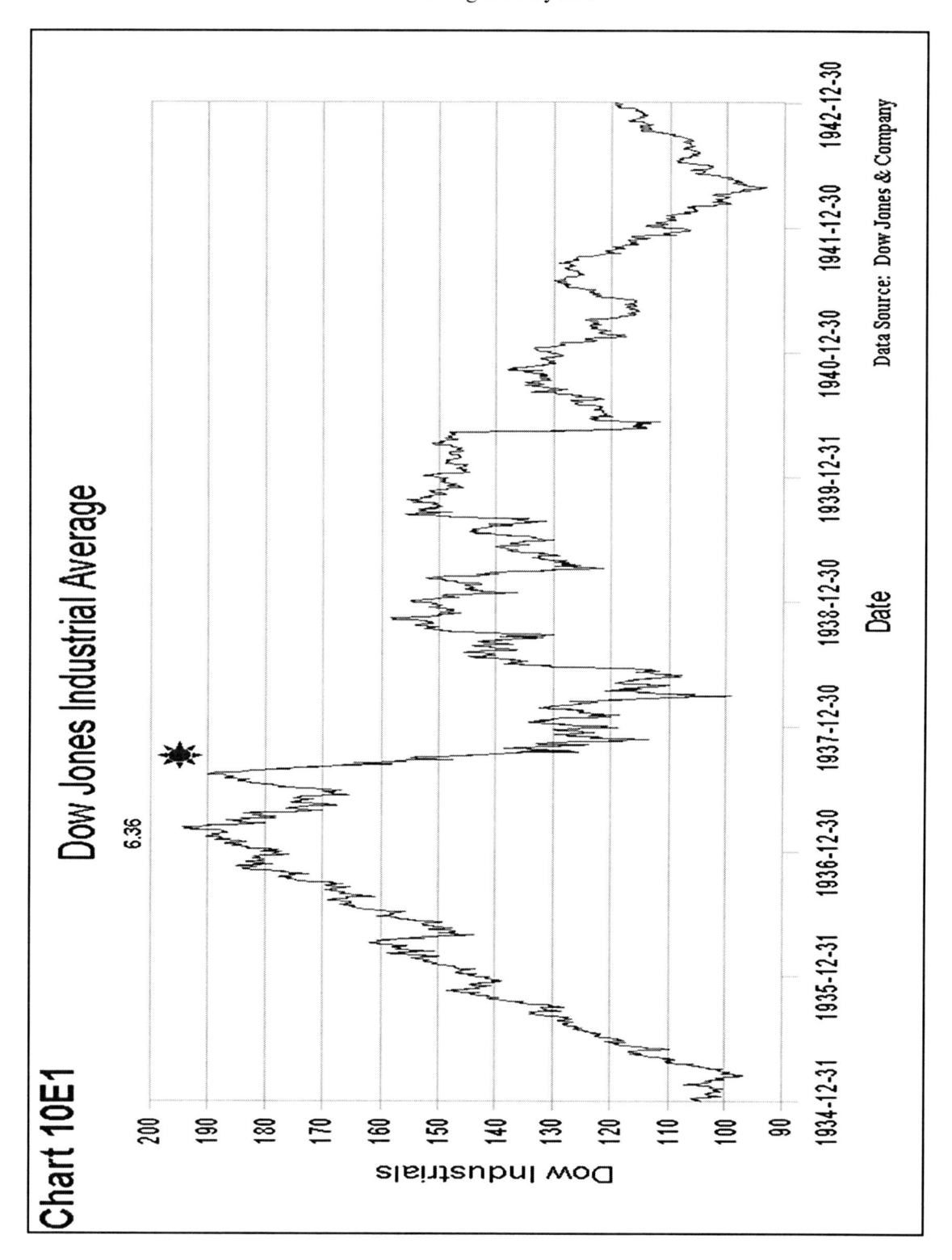

Electromagnetic Cycles

Chart 10E1

Dow Jones Industrial Average

6.36

Data Source: Dow Jones & Company

Chart 10E2

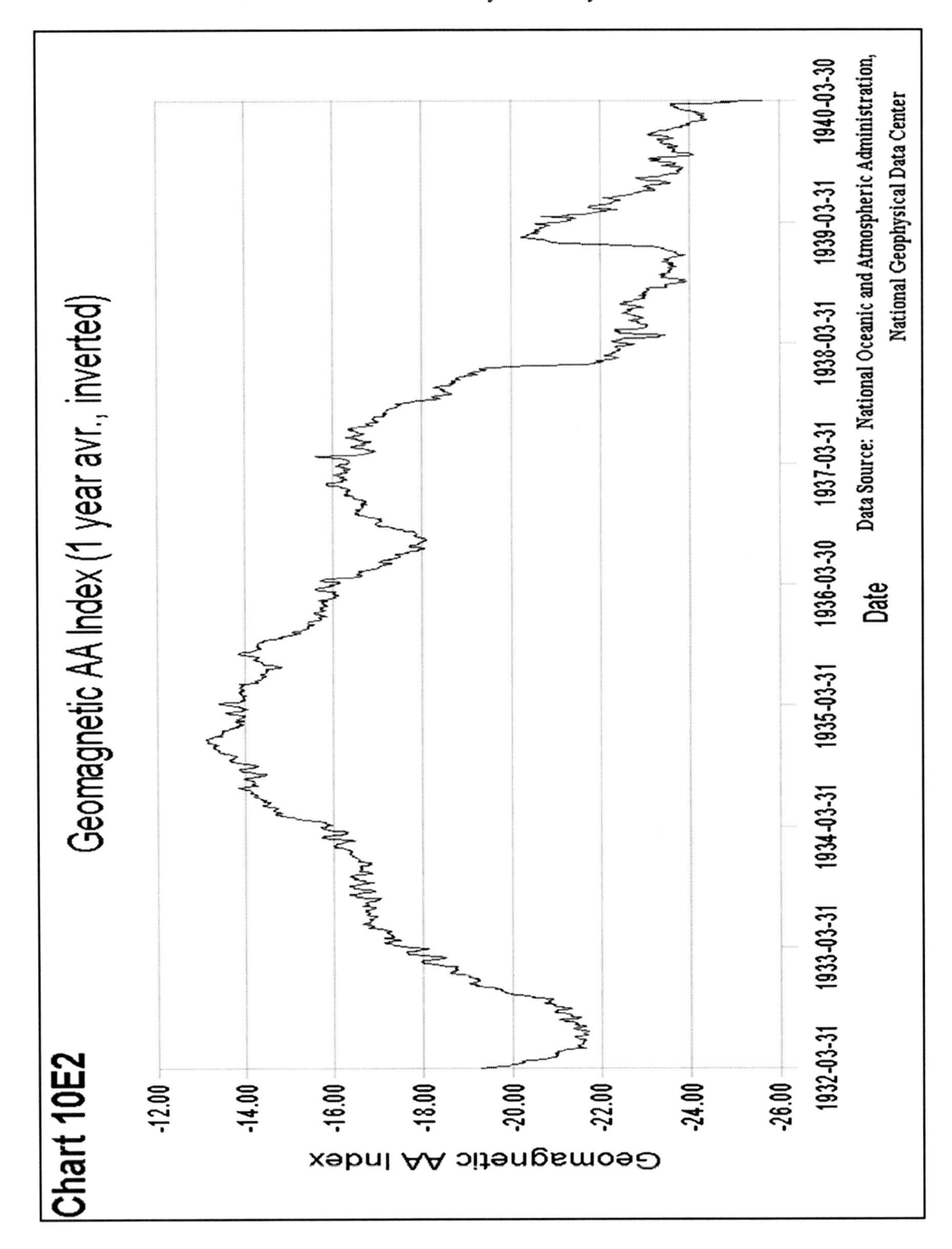

Geomagnetic AA Index (1 year avr., inverted)

Data Source: National Oceanic and Atmospheric Administration, National Geophysical Data Center

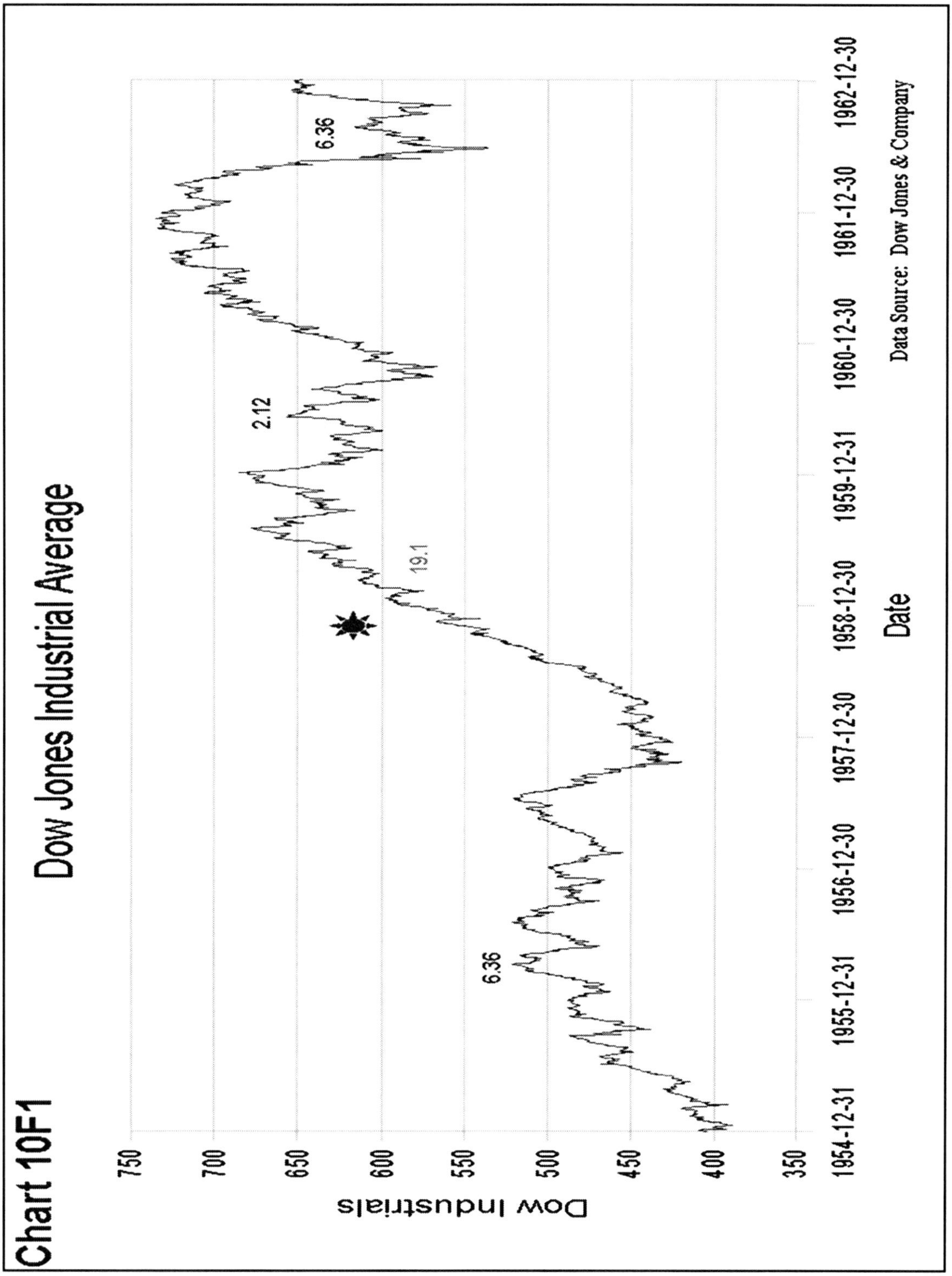
Chart 10F1
Dow Jones Industrial Average
Dow Industrials
750
700
650
600
550
500
450
400
350
1954-12-31
1955-12-31
1956-12-30
1957-12-30
1958-12-30
1959-12-31
1960-12-30
1961-12-30
1962-12-30
Date
6.36
19.1
2.12
6.36
Data Source: Dow Jones & Company

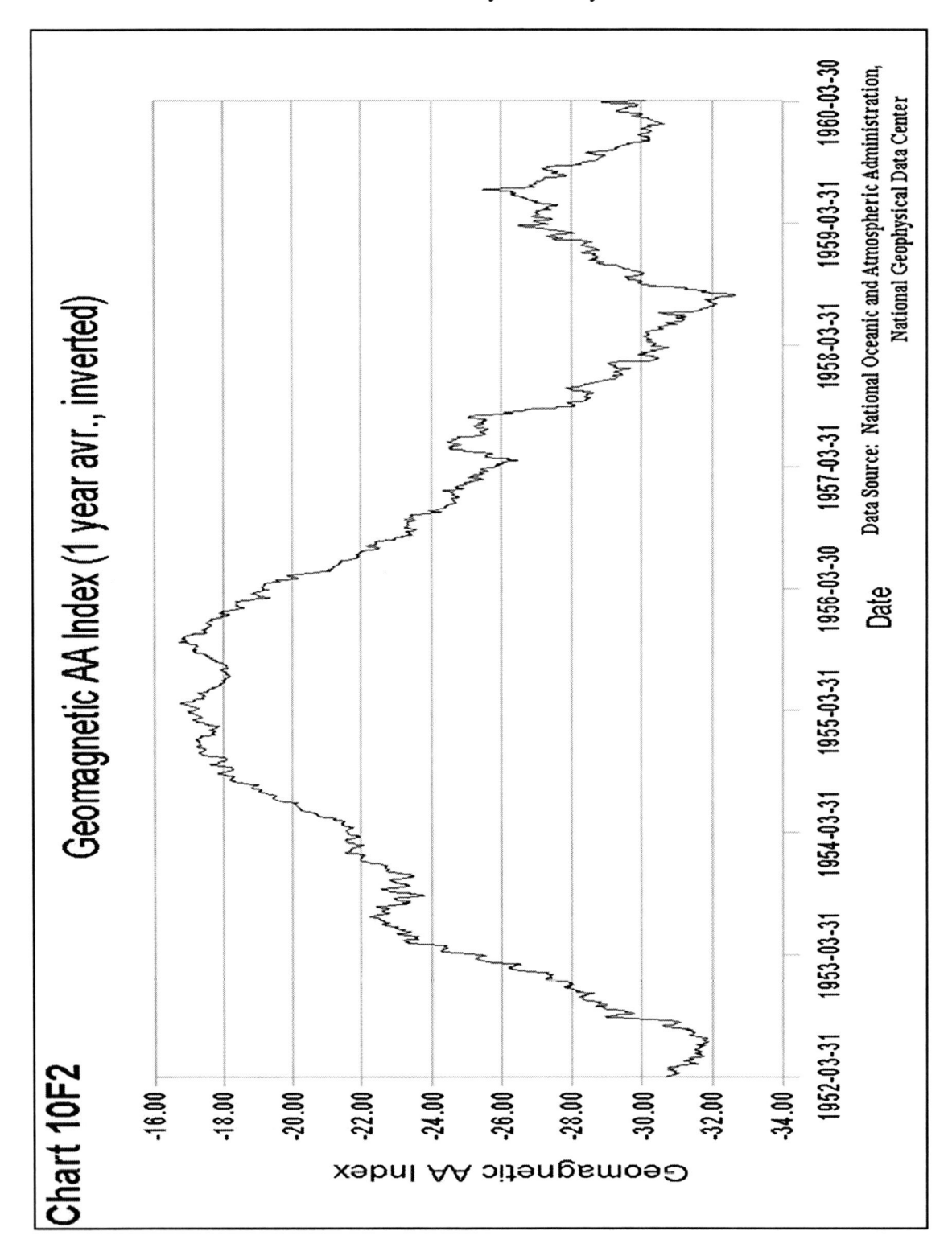

Chart 10F2: Geomagnetic AA Index (1 year avr., inverted)

Data Source: National Oceanic and Atmospheric Administration, National Geophysical Data Center

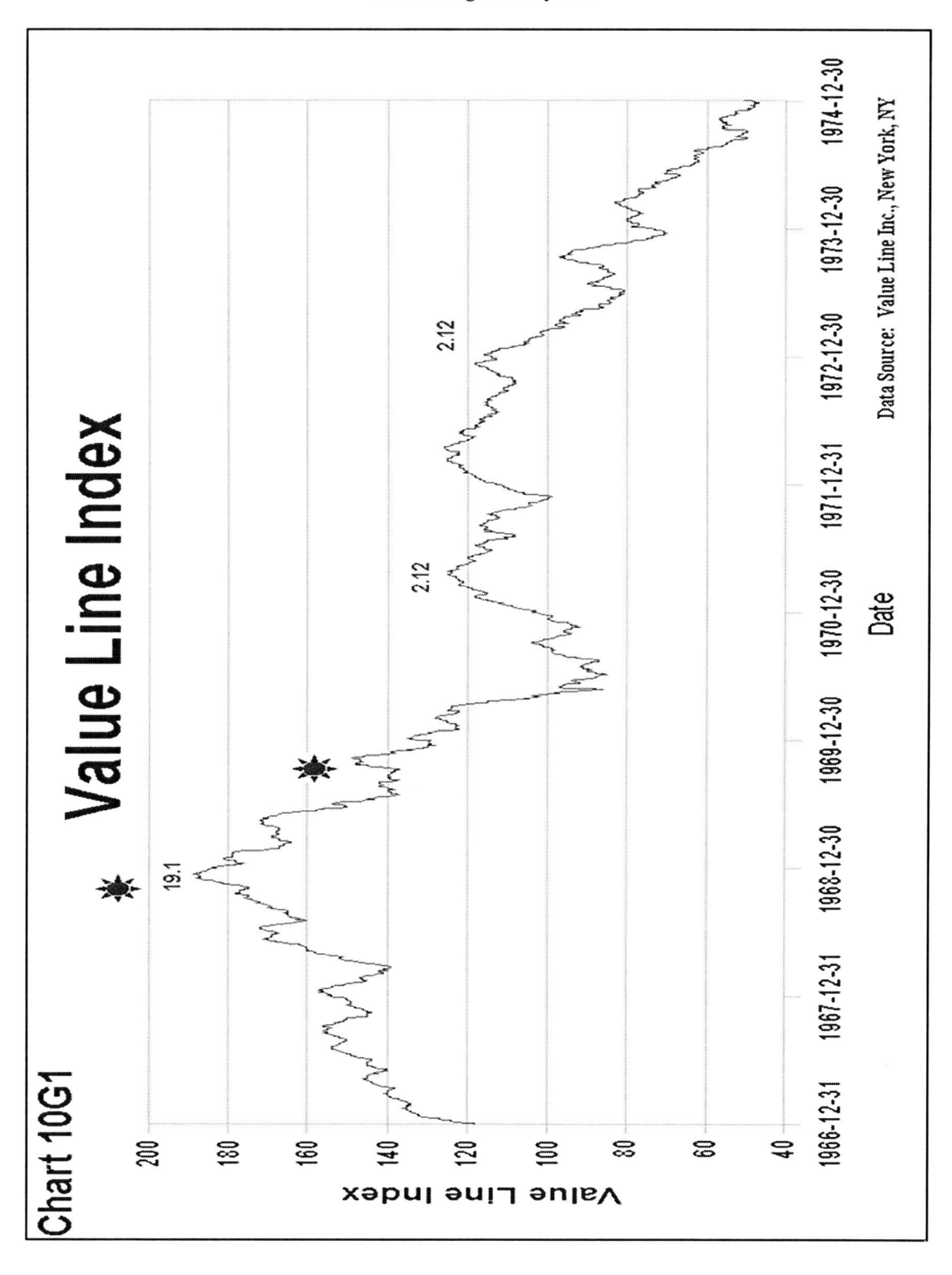

Electromagnetic Cycles

Chart 10G1

Value Line Index

Chart 10G2

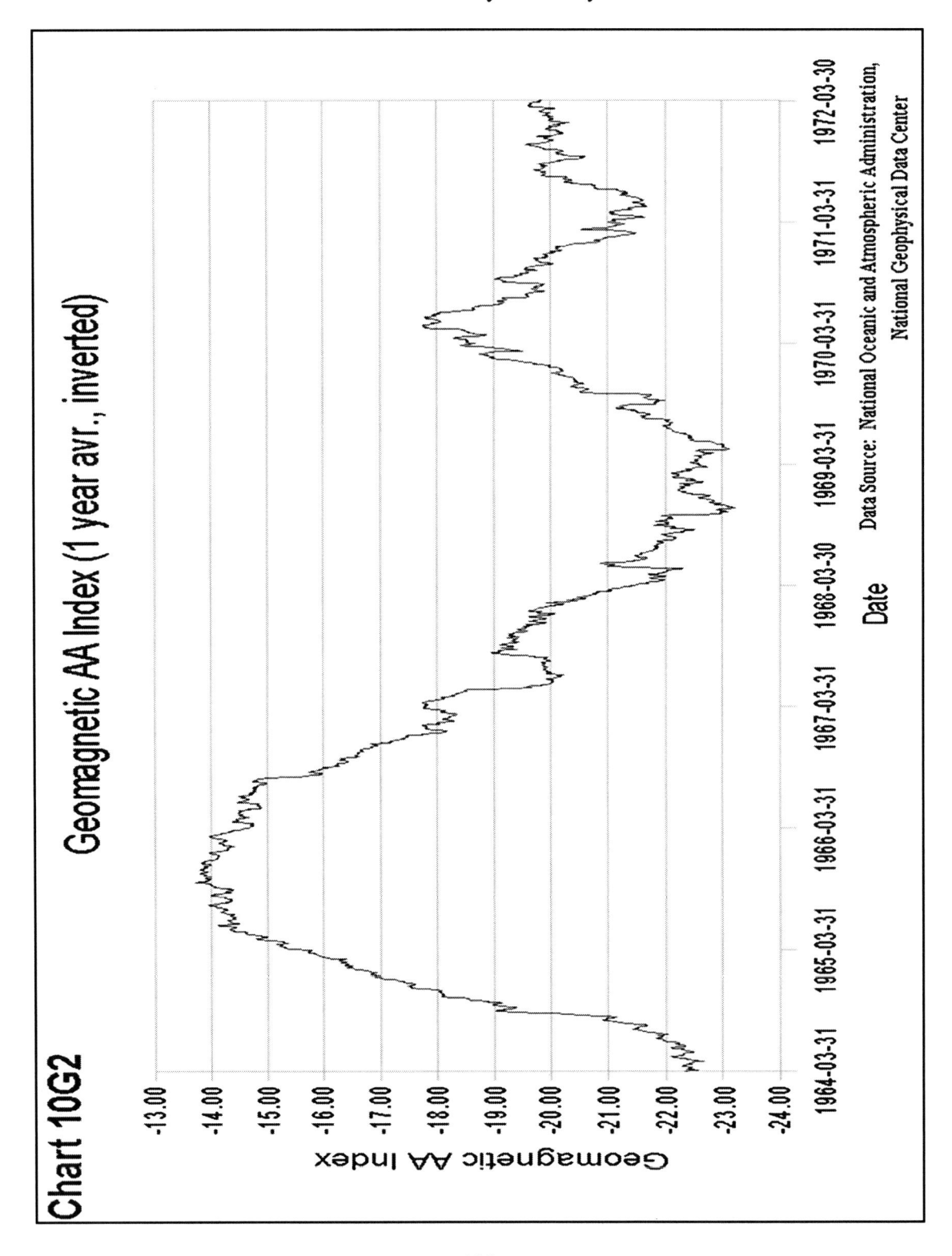

Geomagnetic AA Index (1 year avr., inverted)

Data Source: National Oceanic and Atmospheric Administration, National Geophysical Data Center

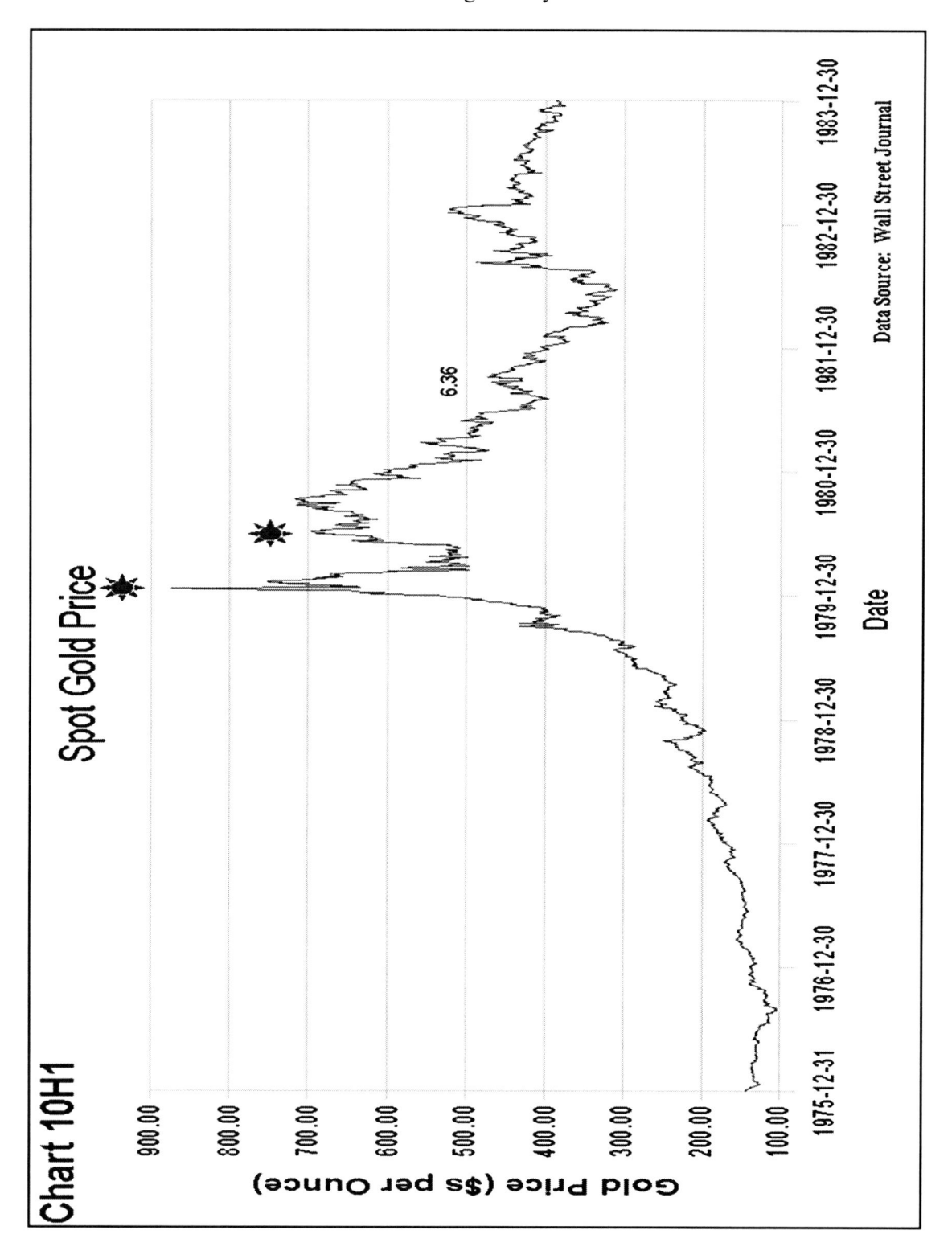

Electromagnetic Cycles

Chart 10H1 — Spot Gold Price

Gold Price ($ per Ounce) vs Date

6.36

Data Source: Wall Street Journal

Chart 10H2

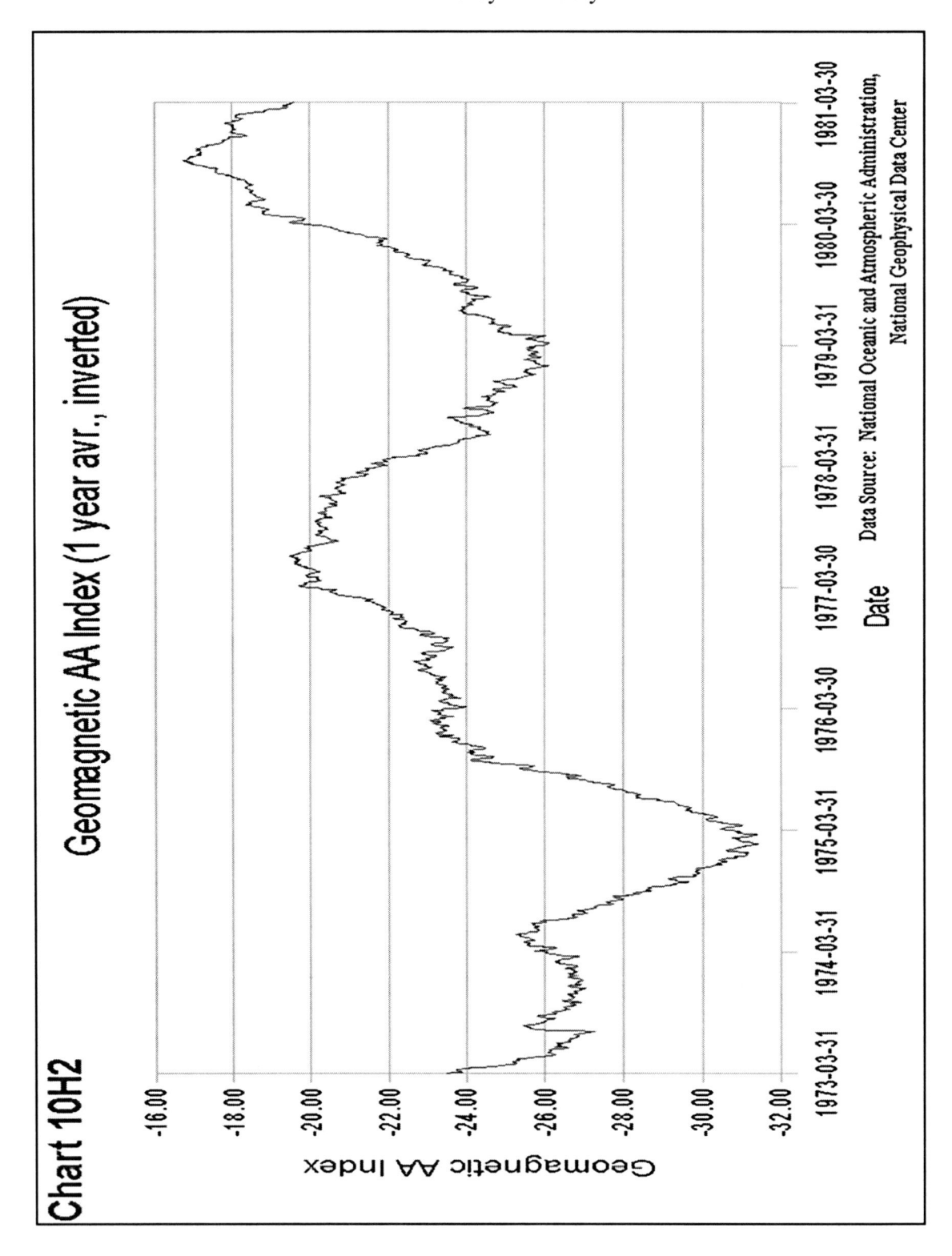

Geomagnetic AA Index (1 year avr., inverted)

Data Source: National Oceanic and Atmospheric Administration, National Geophysical Data Center

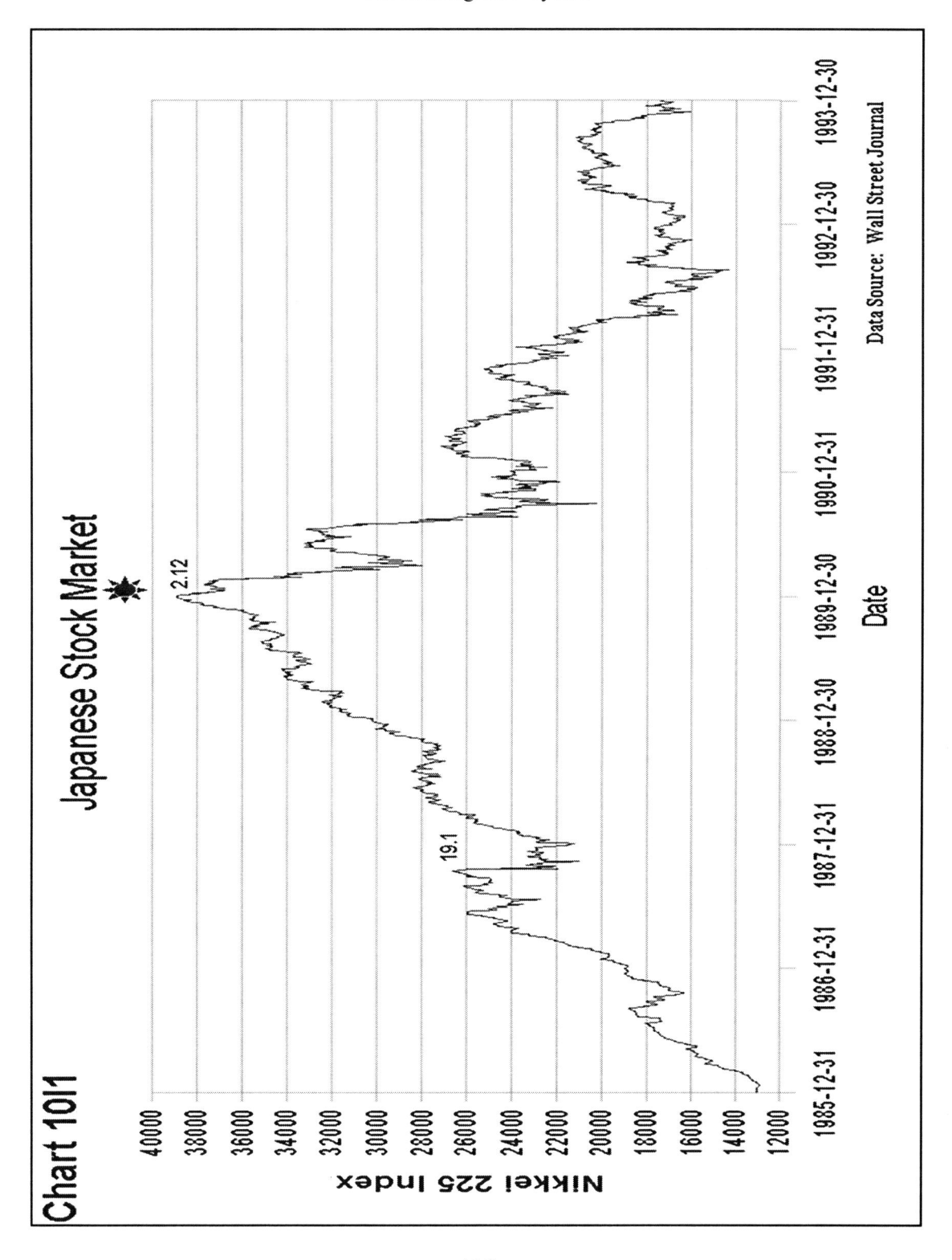

Electromagnetic Cycles

Chart 10I1
Japanese Stock Market

Nikkei 225 Index vs Date (1985-12-31 to 1993-12-30)

Data Source: Wall Street Journal

Chart 10|2

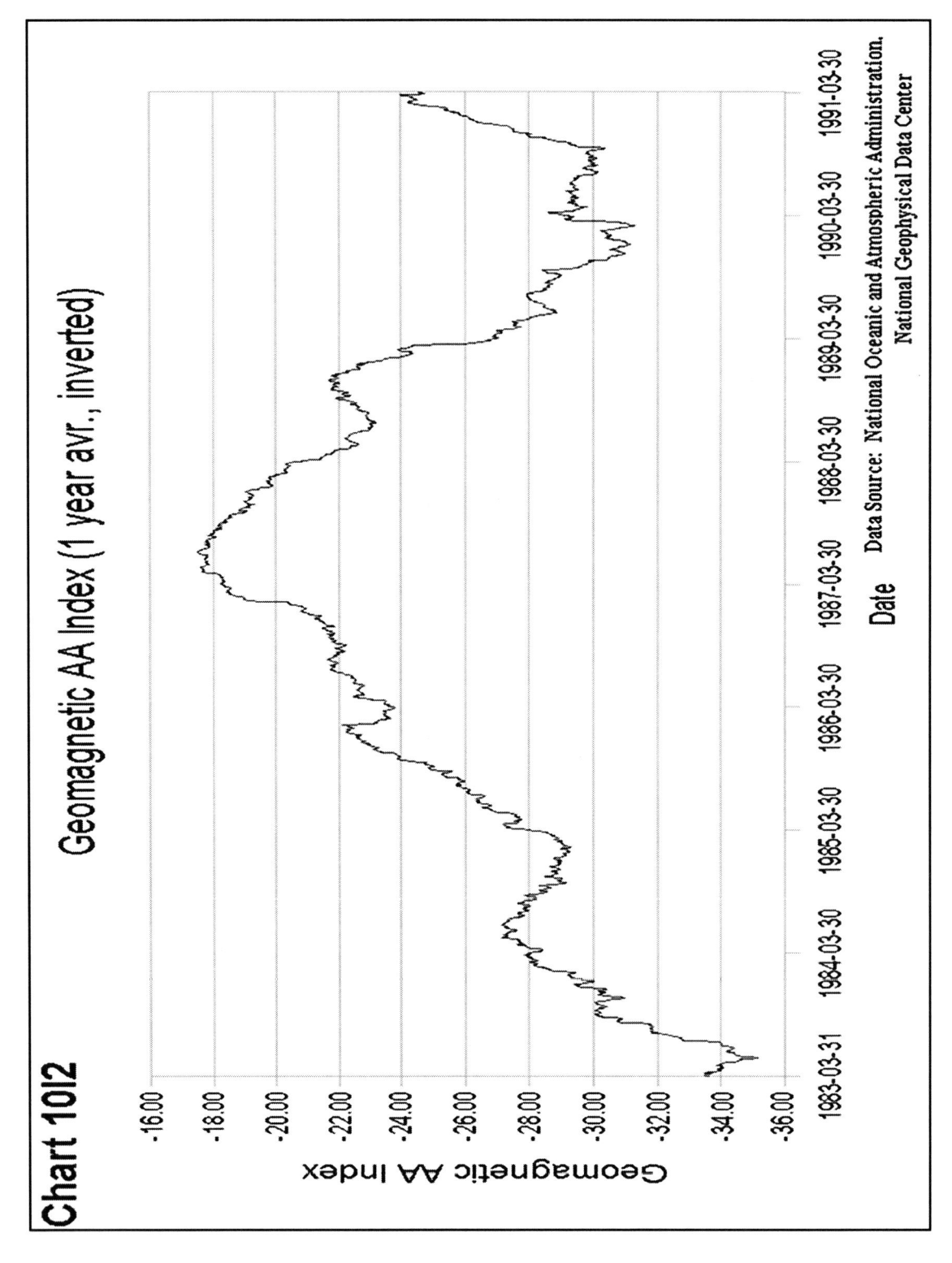

Geomagnetic AA Index (1 year avr., inverted)

Date

Data Source: National Oceanic and Atmospheric Administration, National Geophysical Data Center

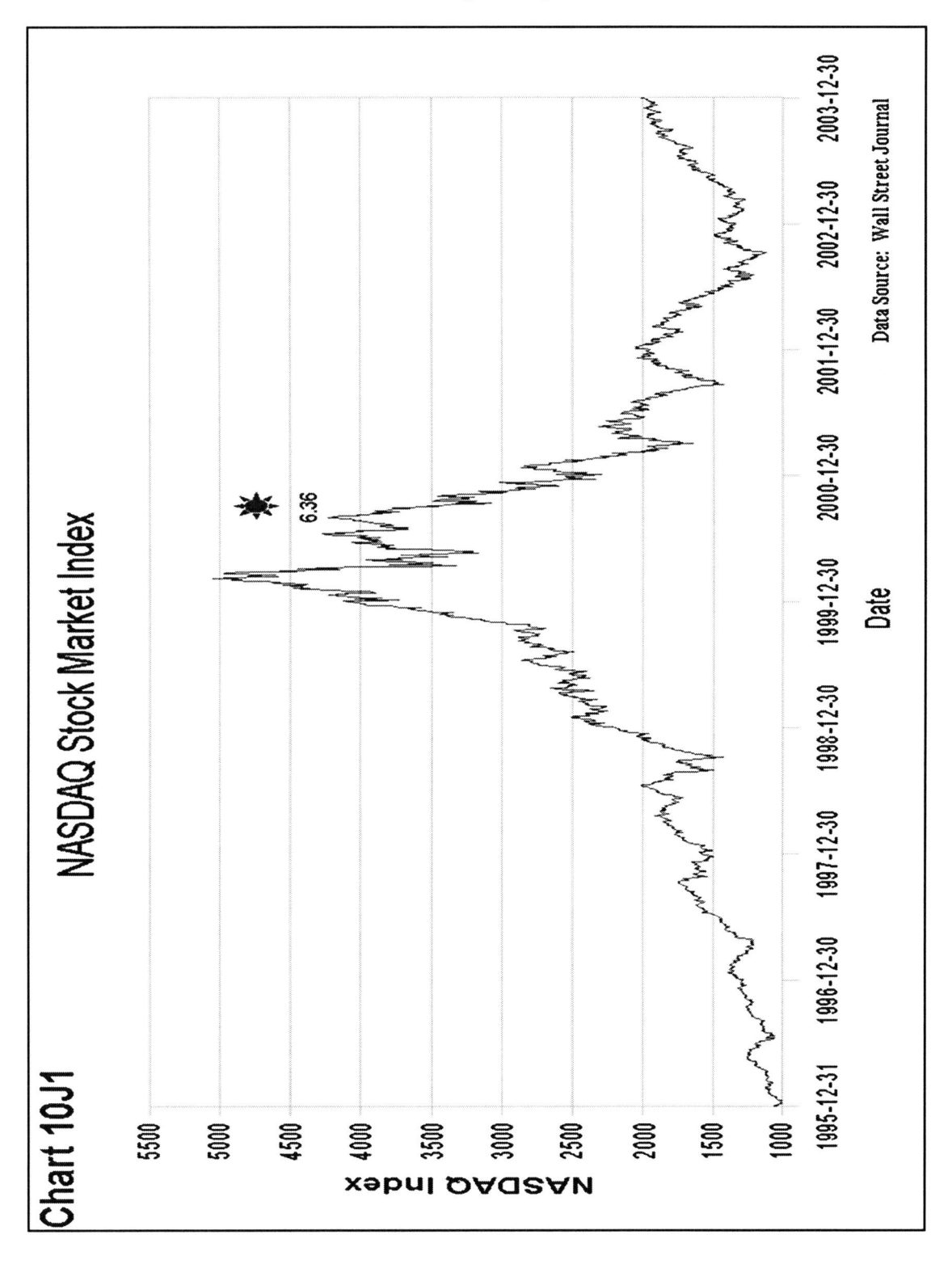

Electromagnetic Cycles

Chart 10J1

NASDAQ Stock Market Index

6.36

Data Source: Wall Street Journal

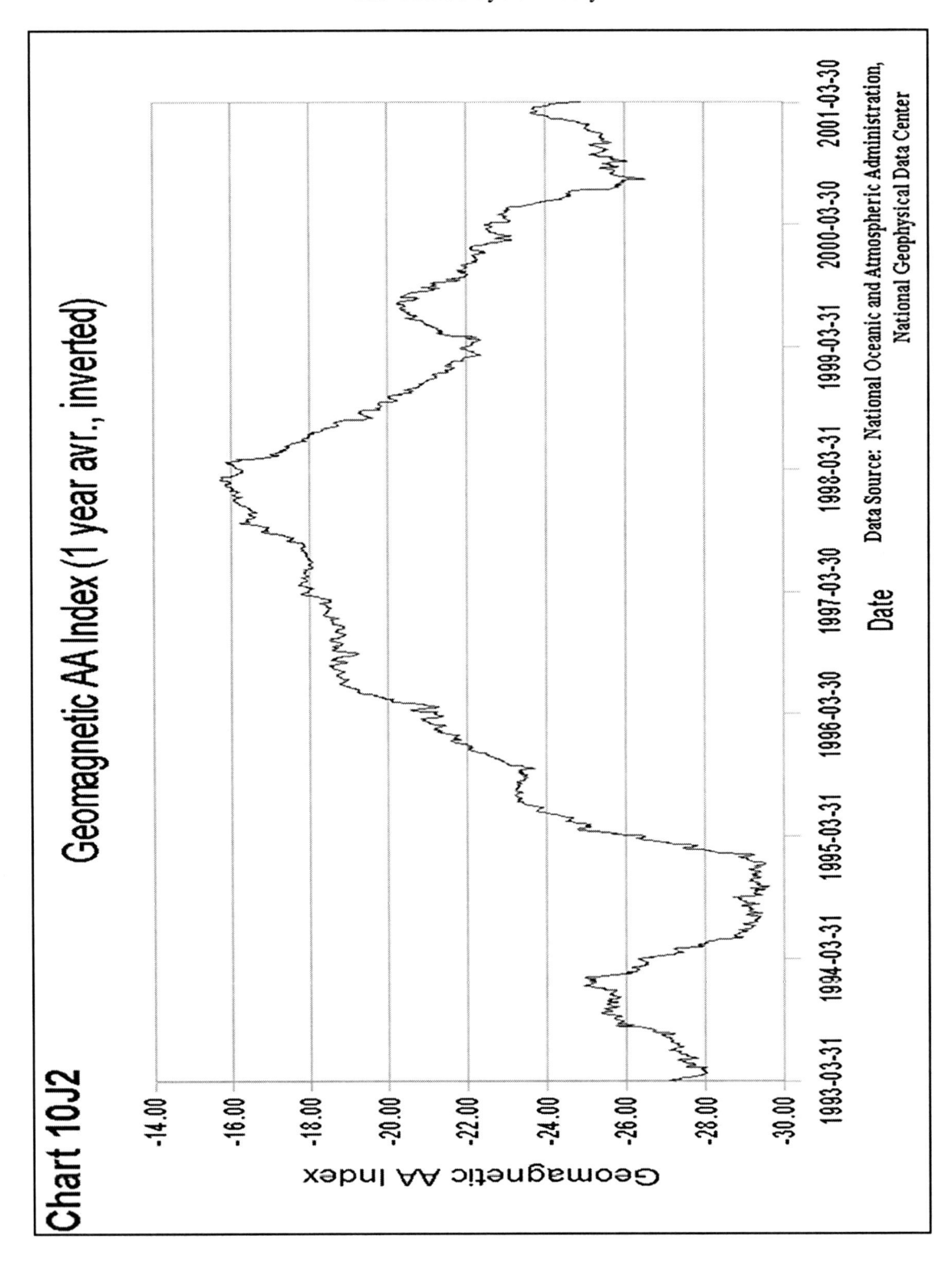

Chart 10J2: Geomagnetic AA Index (1 year avr., inverted)

Data Source: National Oceanic and Atmospheric Administration, National Geophysical Data Center

Chart 10K1

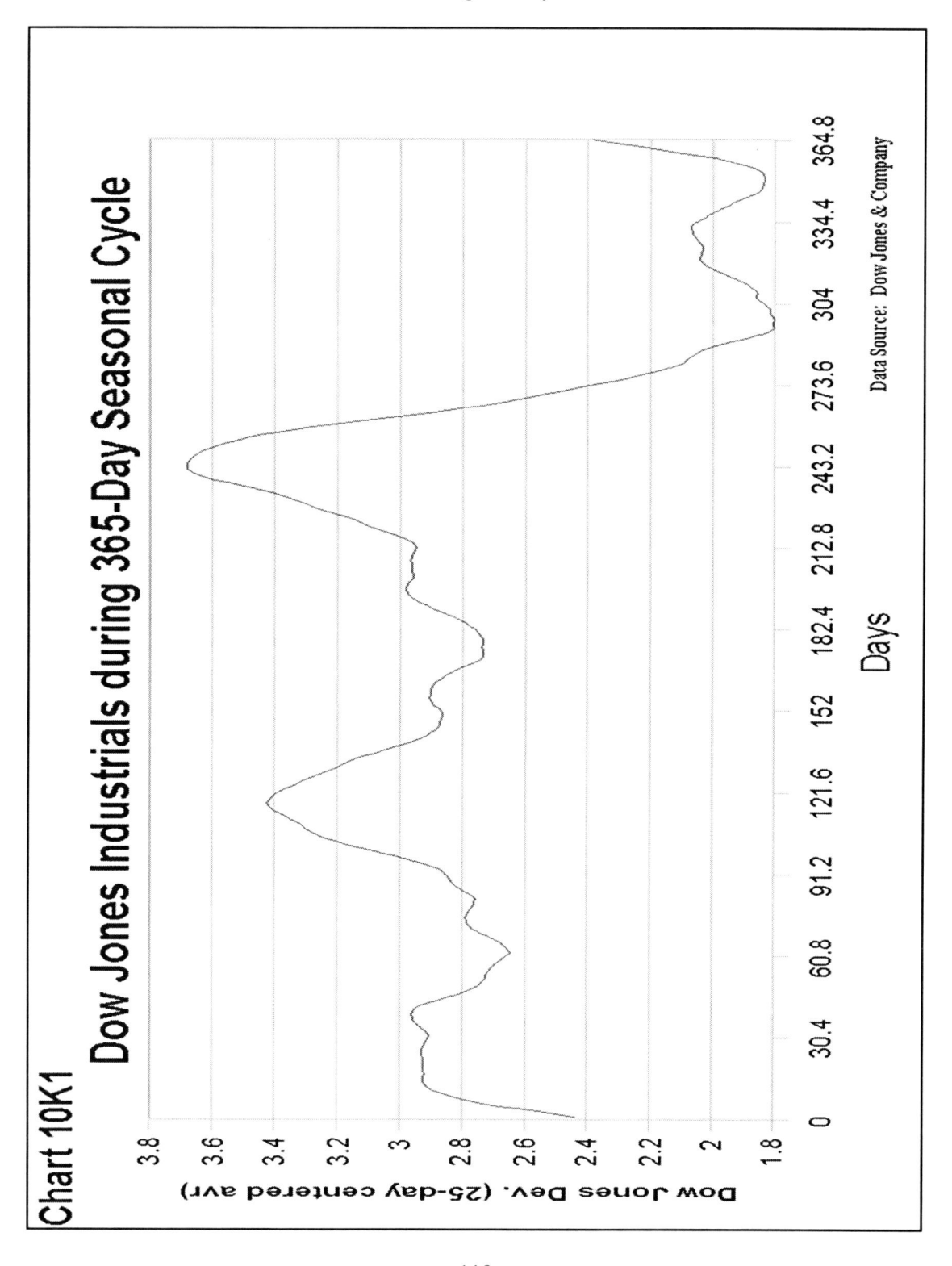

Dow Jones Industrials during 365-Day Seasonal Cycle

Data Source: Dow Jones & Company

Chart 10K2

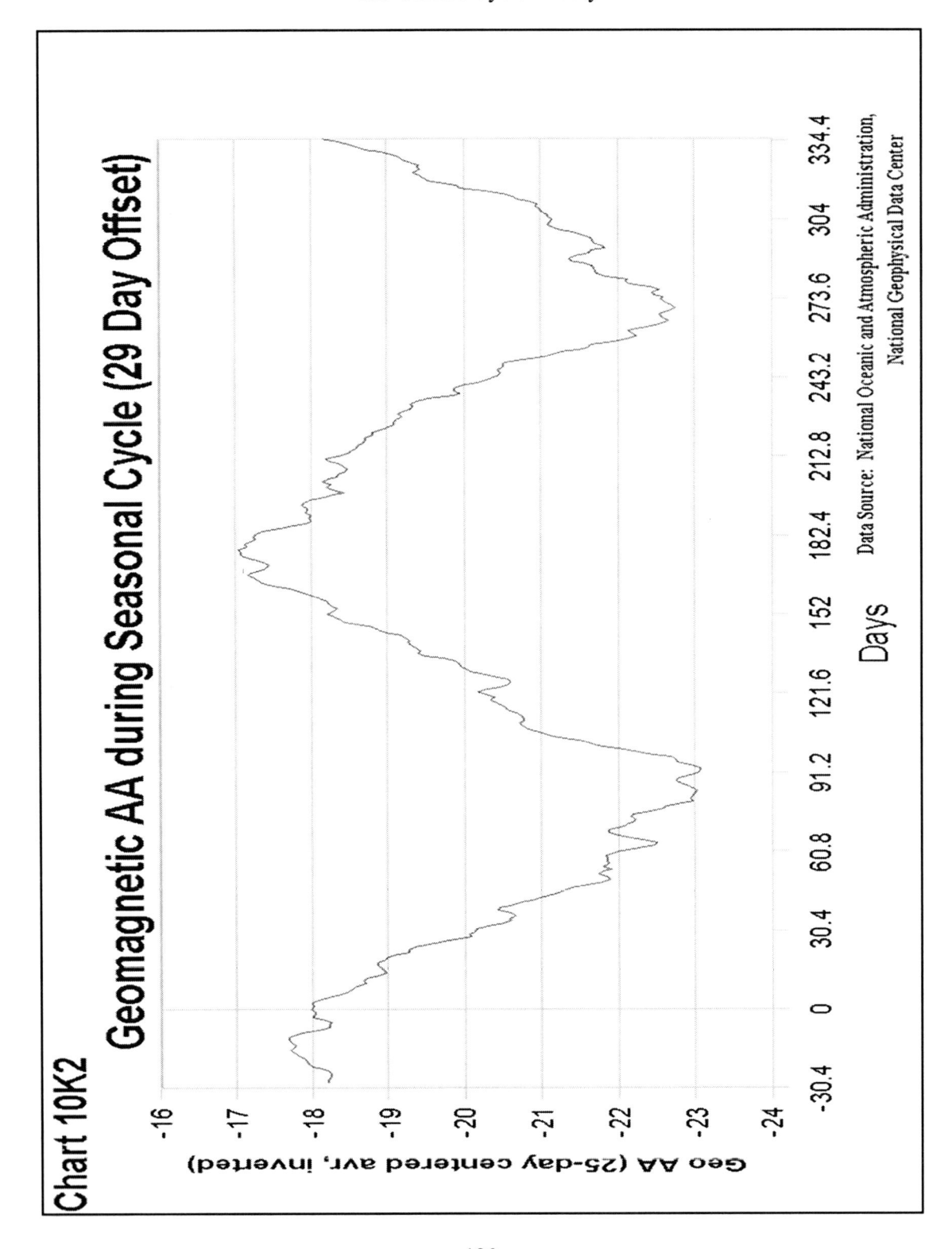

Geomagnetic AA during Seasonal Cycle (29 Day Offset)

Y-axis: Geo AA (25-day centered avr, inverted)
X-axis: Days

Data Source: National Oceanic and Atmospheric Administration, National Geophysical Data Center

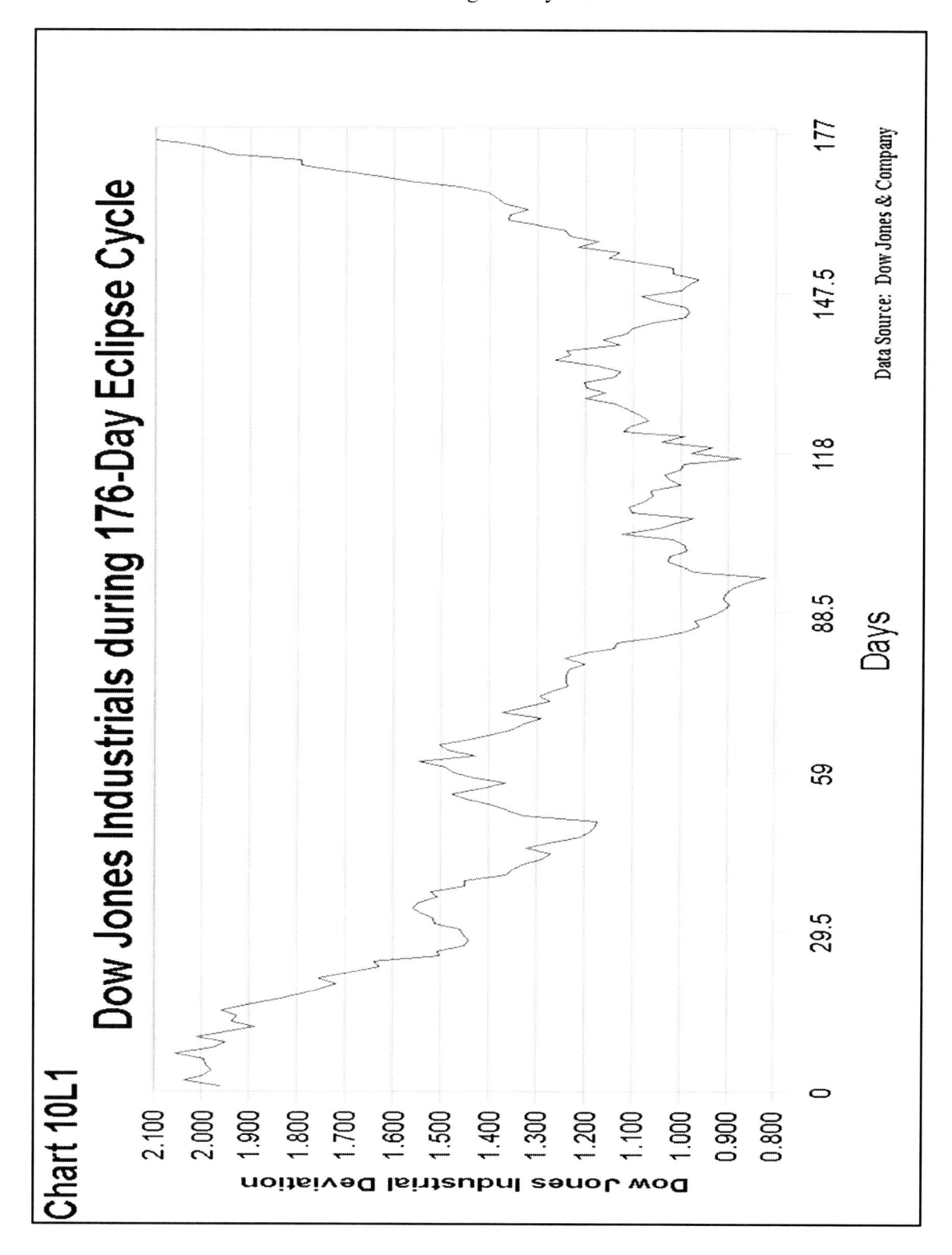

Electromagnetic Cycles

Chart 10L1

Dow Jones Industrials during 176-Day Eclipse Cycle

Dow Jones Industrial Deviation vs *Days*

Data Source: Dow Jones & Company

121

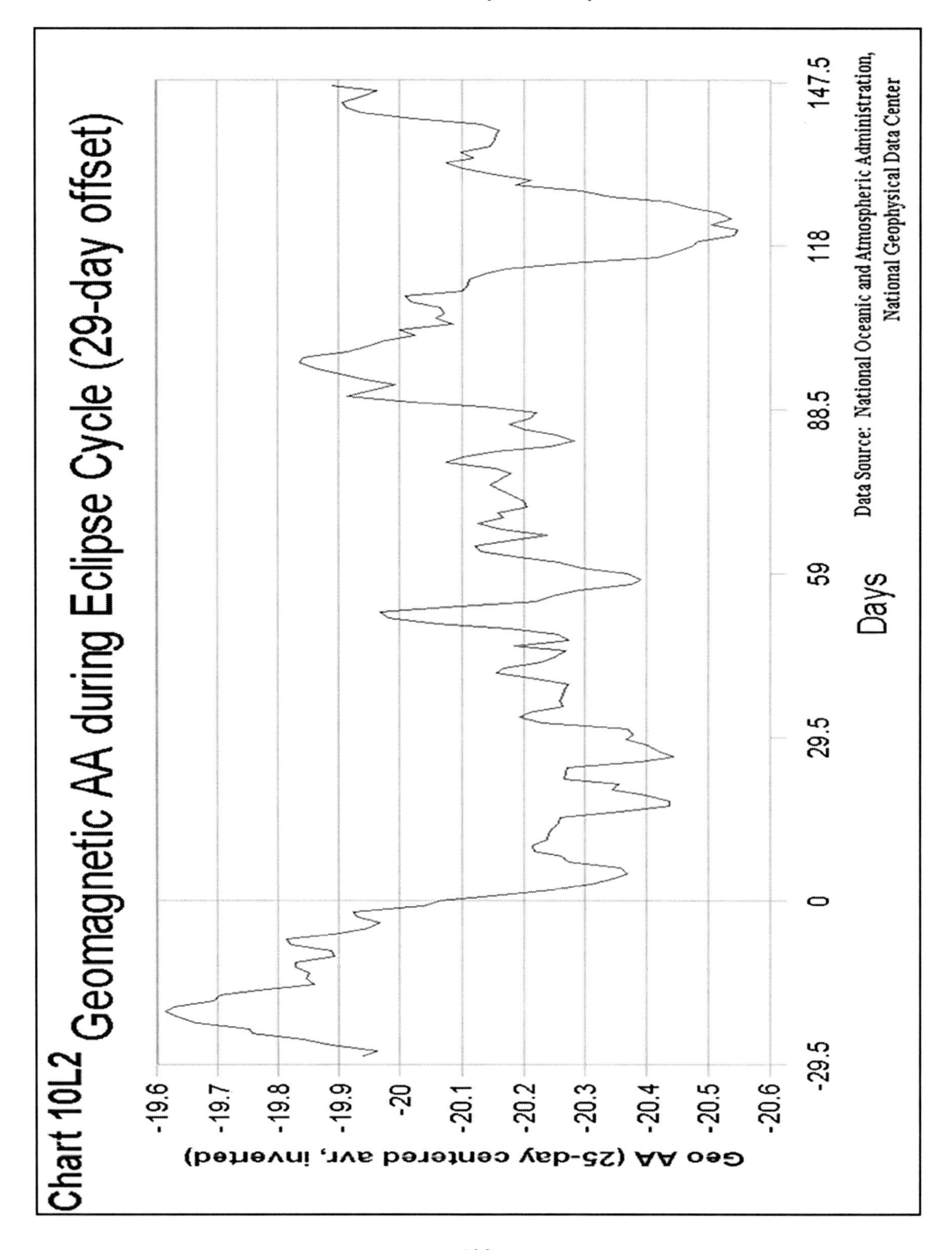

Chart 10L2 Geomagnetic AA during Eclipse Cycle (29-day offset)

Data Source: National Oceanic and Atmospheric Administration, National Geophysical Data Center

Chart 10M1

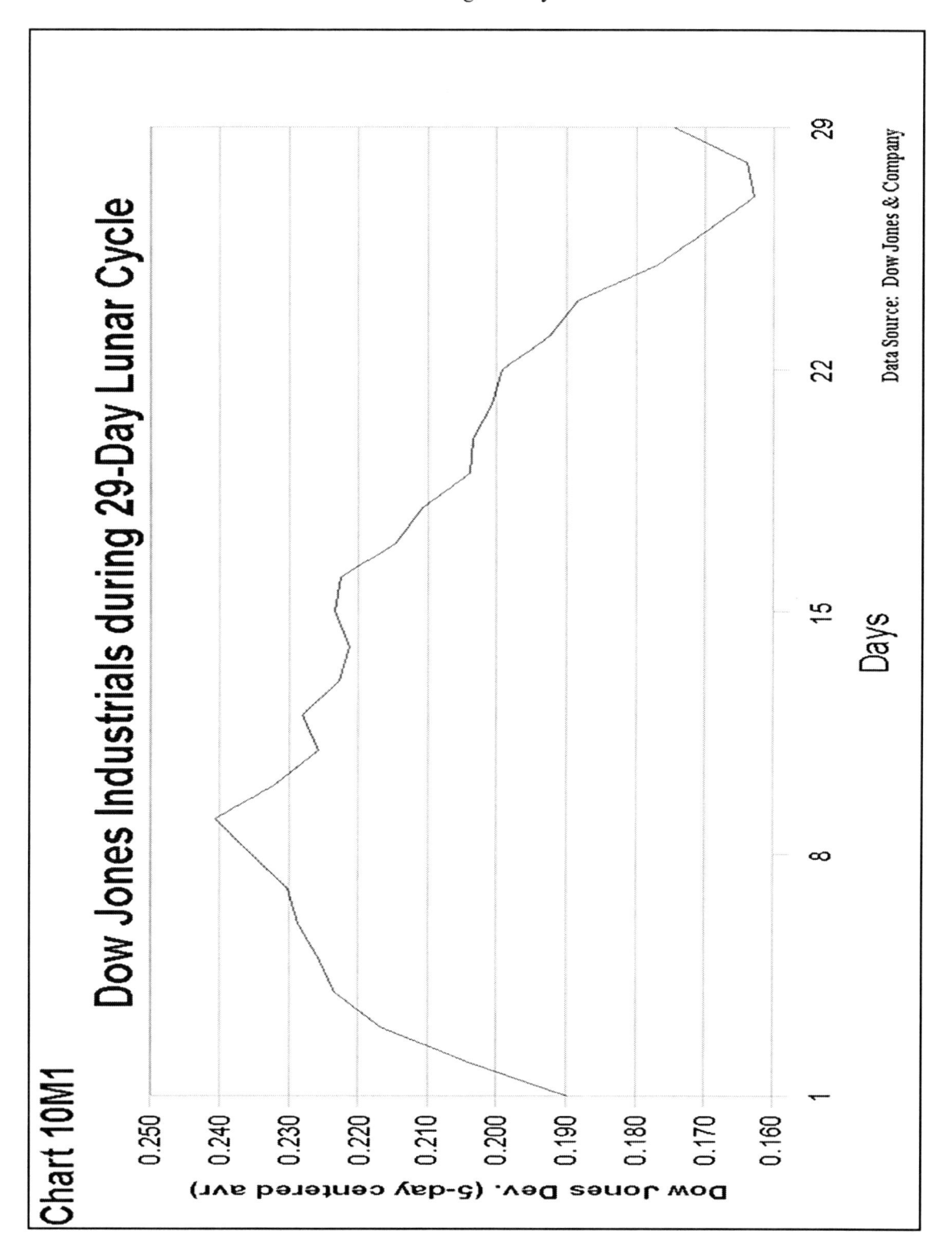

Dow Jones Industrials during 29-Day Lunar Cycle

Dow Jones Dev. (5-day centered avr)

Days

Data Source: Dow Jones & Company

Chart 10M2

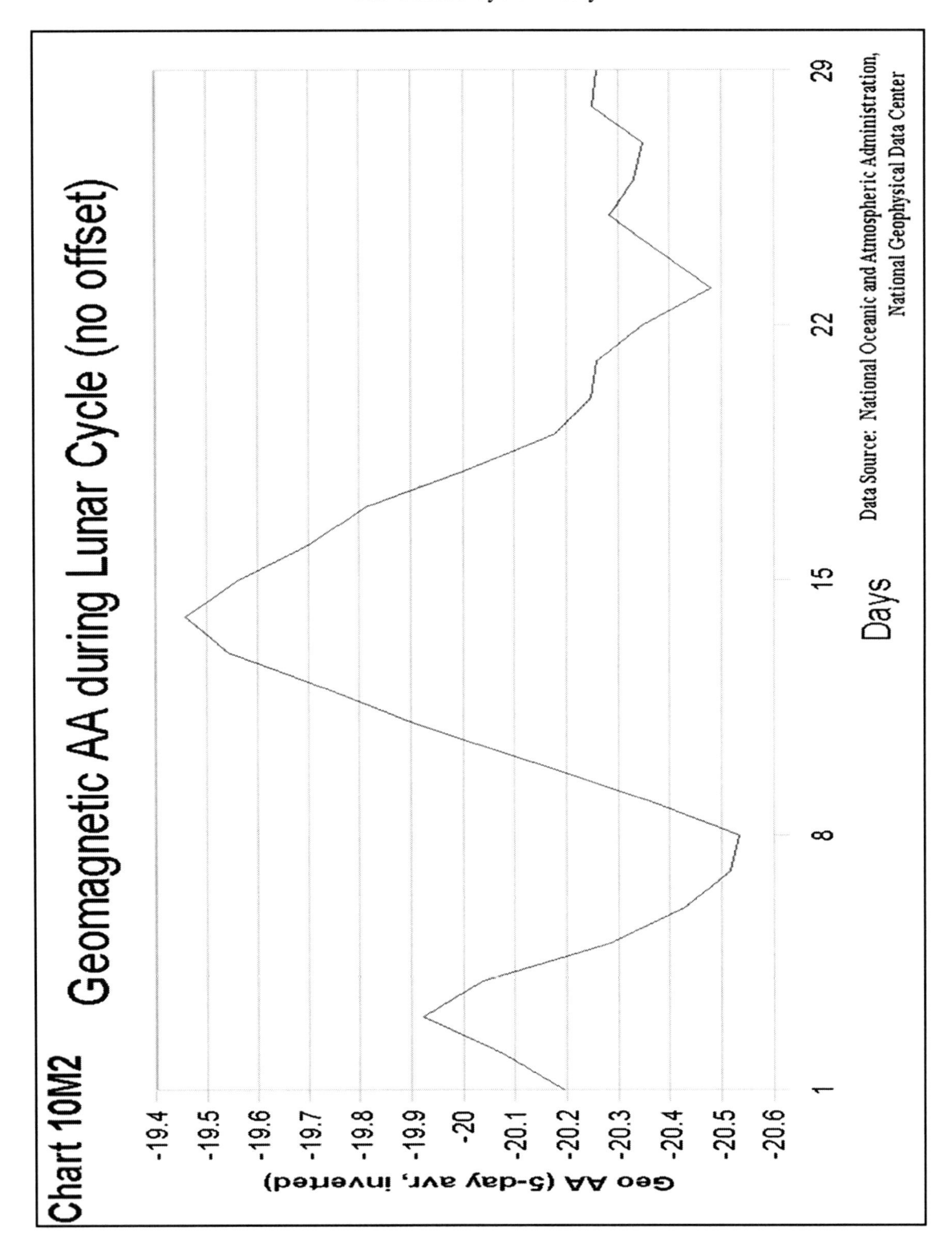

Geomagnetic AA during Lunar Cycle (no offset)

Data Source: National Oceanic and Atmospheric Administration, National Geophysical Data Center

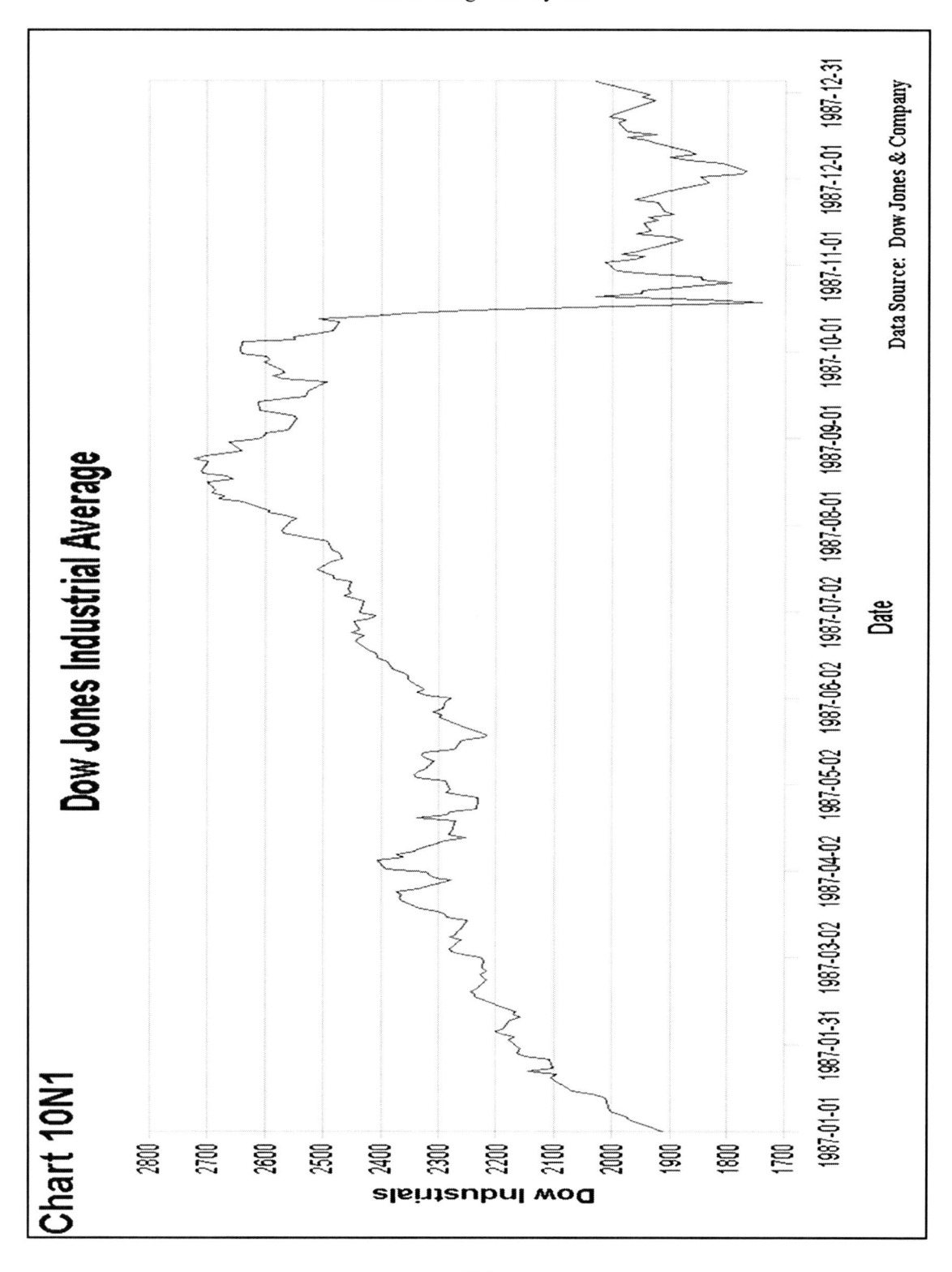

Electromagnetic Cycles

Chart 10N1

Dow Jones Industrial Average

Data Source: Dow Jones & Company

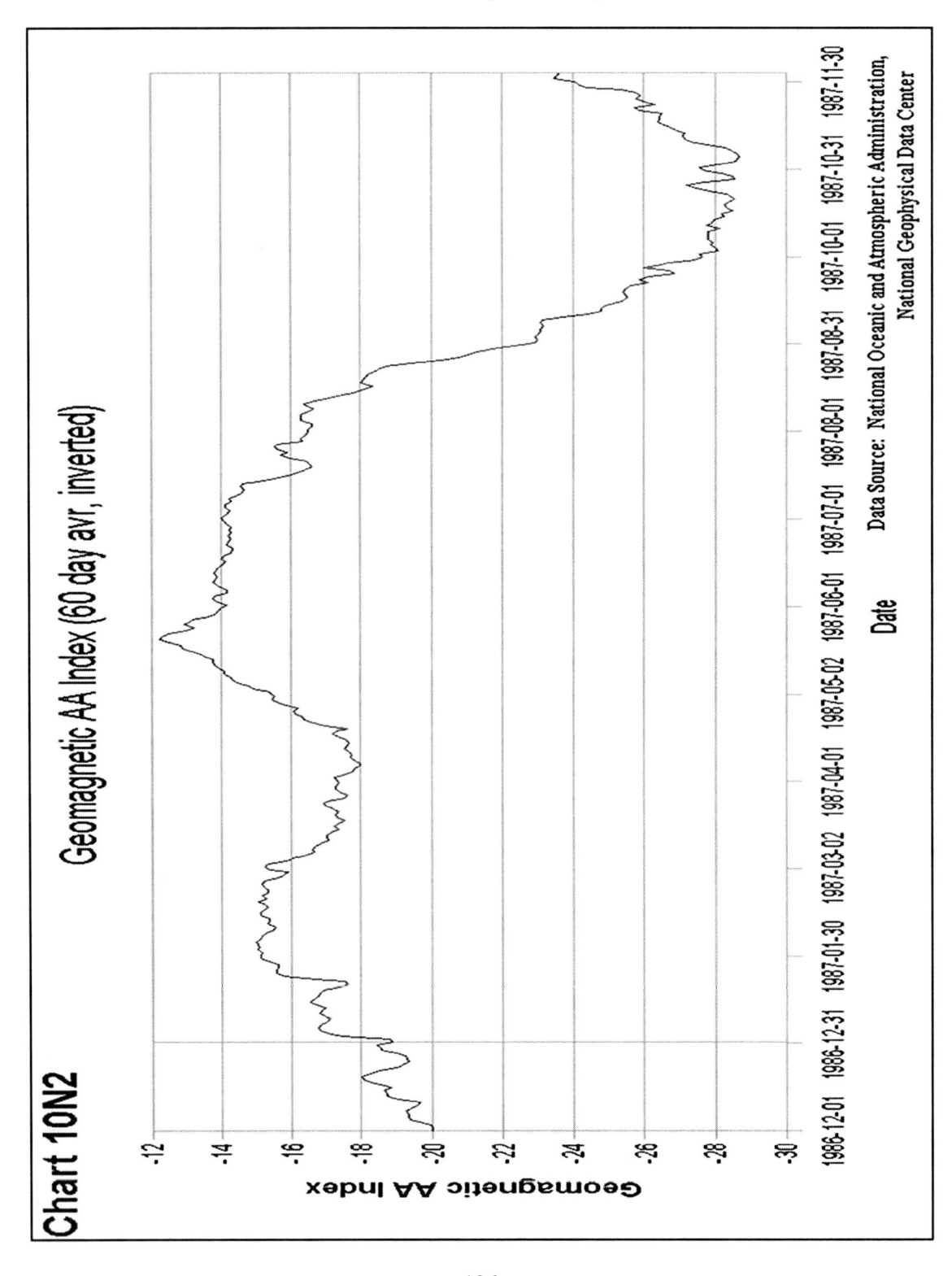

Chart 10N2 — Geomagnetic AA Index (60 day avr, inverted)

Data Source: National Oceanic and Atmospheric Administration, National Geophysical Data Center

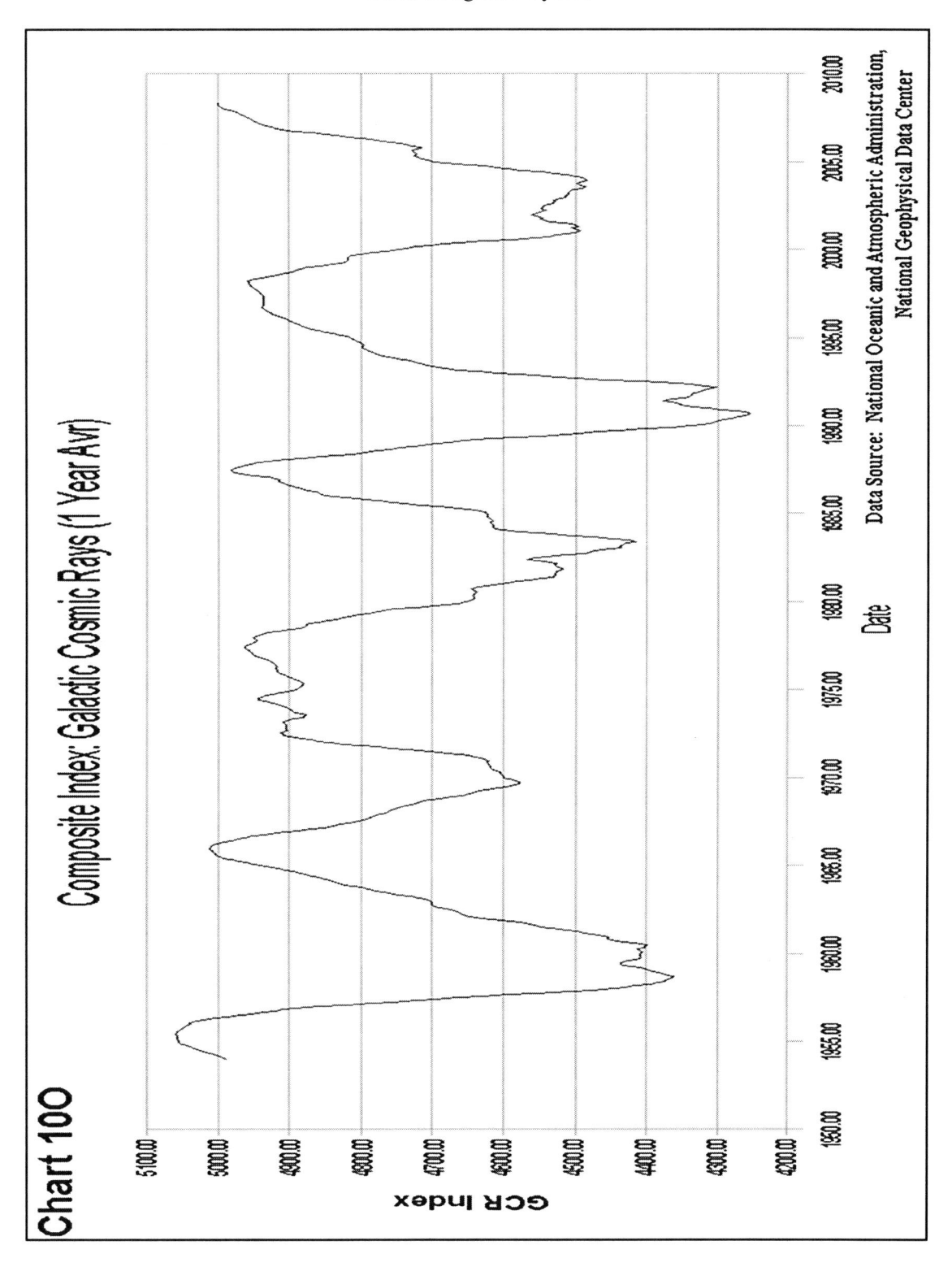

Electromagnetic Cycles

Chart 100

Composite Index: Galactic Cosmic Rays (1 Year Avr)

Data Source: National Oceanic and Atmospheric Administration, National Geophysical Data Center

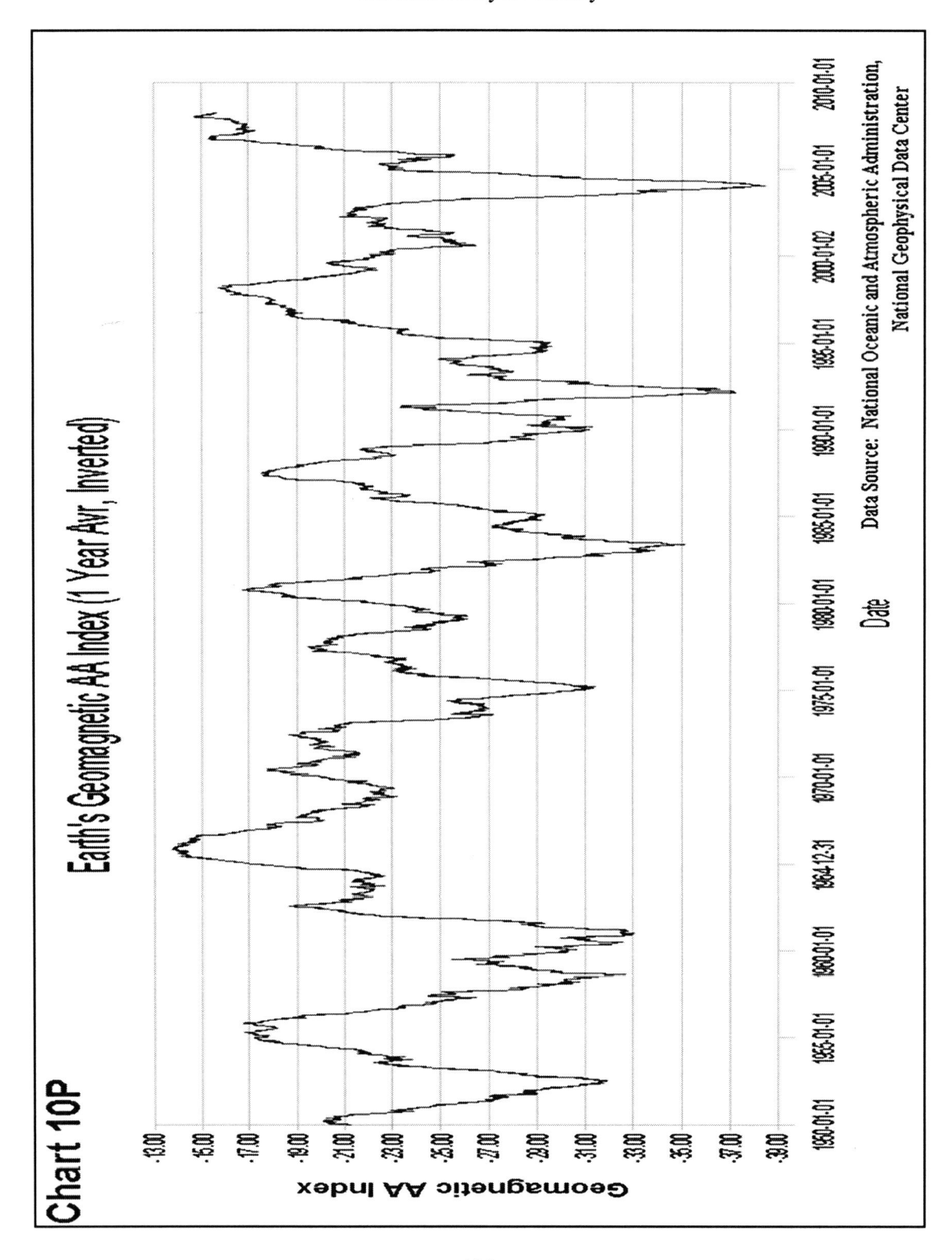

The Unified Cycle Theory

Chart 10P — Earth's Geomagnetic AA Index (1 Year Avr, Inverted)

Data Source: National Oceanic and Atmospheric Administration, National Geophysical Data Center

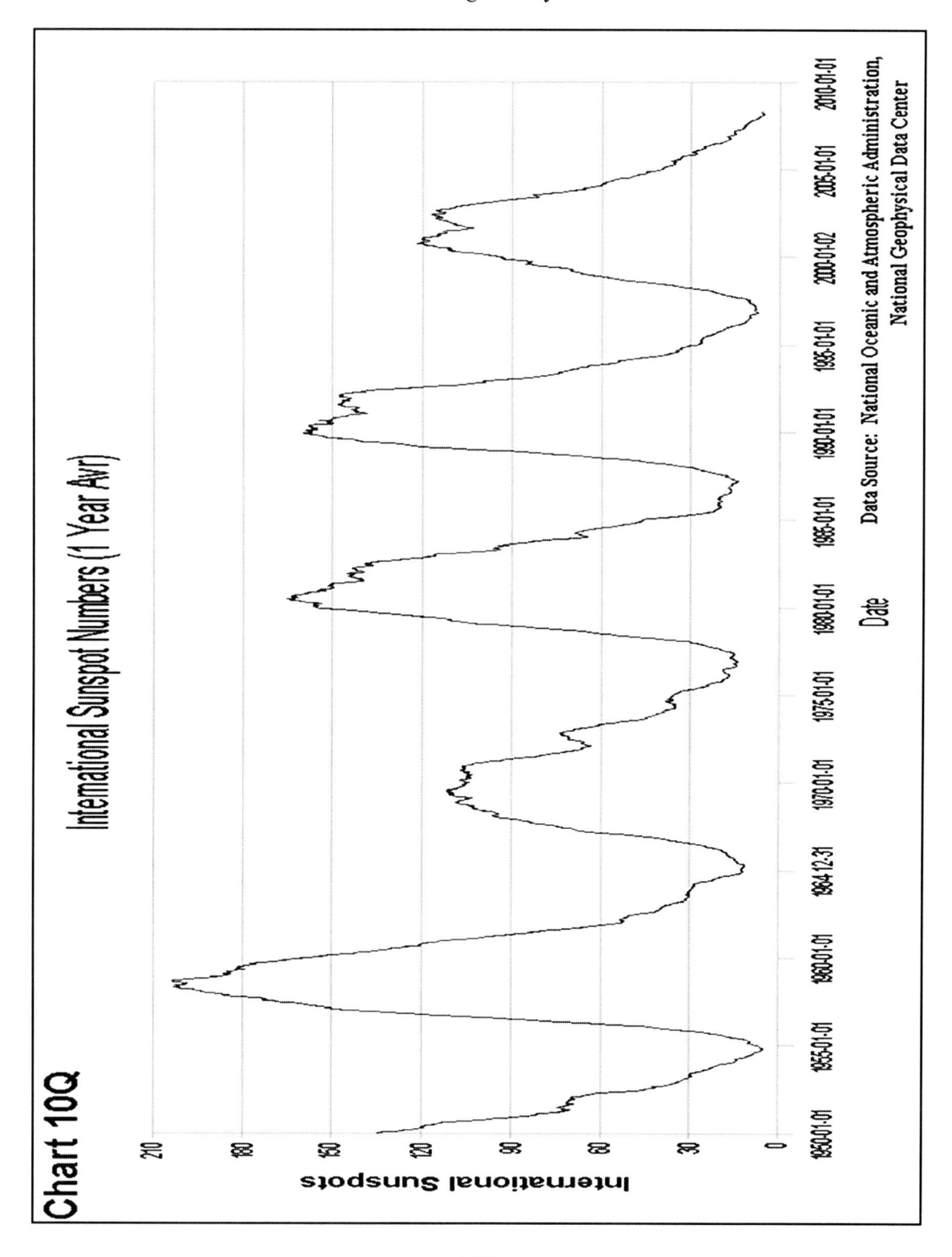

Electromagnetic Cycles

Chart 10Q — International Sunspot Numbers (1 Year Avr)

Data Source: National Oceanic and Atmospheric Administration, National Geophysical Data Center

Chapter 11
Gravitational & Motion Cycles

This chapter reviews gravitational cycles resulting from the interaction of the Sun, Earth, planets, and Moon. These cycles primarily affect global climate. Chapter 5 already reviewed the two gravity and motion cycles you are most familiar with – the daily cycle and the annual cycle.

As part of the annual weather cycle, seasonal rain and temperature changes affect the growing season, which alters agricultural supplies, which influences food prices. During years with adequate rain, agricultural prices generally hit their annual low during harvest. Within the course of a year, agricultural price oscillate based on available supplies, while demand stays relatively constant. In addition to commodities, seasonal temperatures also influence human behavior.[Nastos *et al.*, 2006]

The daily and annual cycles are so much a part of our lives that there's no need to review them further. Instead, this chapter concentrates on the unfamiliar gravitational cycles – those associated with the periodicity of ice-ages.

100,000-Year Ice-Age Cycle

Since the 1970s, climatologists have generally explained ice-age cycles with the Milankovitch Theory. This theory states that ice-age cycles result from changes in Earth's precession, obliquity, and eccentricity. In an article entitled *Gravitational Explanation of the Glacial Periods and Calculation of the Precession of Mercury's Orbit*, physicist Dr. Mariano Gonzalez Ambou explains how Milankovitch cycles contribute to ice-ages:

> **"It is known that due to the small gravitational attraction of the Moon and the other planets, the orbital parameters of the Earth change with time. The inclination of the Earth's axis varies between 22.1 and 24.5 degrees with a period of some 40,000 years. The eccentricity of the orbit varies between 0.005 and 0.006 approximately, with a period of some 100,000 years. The terrestrial axis takes some 26,000 years to describe a complete precession circumference.... Spectral analysis ... shows a certain number of fundamental frequencies. The frequency of greatest intensity corresponds to a cycle of some 100,000 years, then there are three notable cycles, one of 40,000 and two others of 24,000 and 19,000 years each. The climatic variation of the last three frequencies may be explained from the variations in the inclination of the Earth's axis which produce a variation in average sunshine in the high latitudes of the northern hemisphere (where there is more continental area) in the different seasons.... What continues to be an enigma is the main cycle of 100,000 years in the glacial periods. It is thought that this cycle must be linked to the variation of the eccentricity of the terrestrial orbit, but it is not known how."**[Ambou, 2008]

Ambou hints at one of the problems with the Milankovitch Theory – a good explanation of the 100,000-year glacial period. Within the current glacial cycle, we reside at the summit. The current annual global temperature averages 60°F. That's quite a bit warmer than the 53°F average recorded during the last stages of the ice-age 11,700 years ago. Yet, global temperatures now sit slightly below the 61°F maximum which prevailed as ancient Egyptian civilization emerged around 3,500 BC. By the time the Little Ice-Age ended 400 years ago during the Maunder Minimum, the average global temperature dipped to 58.5°F. Throughout history, similar temperature fluctuations have been the norm. In fact, climate swings far greater than those registered during the last 100,000 years developed quite often during ancient times.

Rocks, ocean sediments, and ice cores permanently captured evidence of these paleoclimate oscillations. But most of this evidence sat unnoticed until the development of new technologies allowed proper readings. Within the past thirty years, as technology advanced, the lingering evidence finally revealed its long cyclical history. Awareness of these cycles only became widely known as the 21st Century approached. This awareness sparked interest in two different theories.

First, support for the Milankovitch Theory mushroomed when the climate history closely matched its predictions.

Second, the history showed carbon-dioxide cycles closely matched temperature movements. This fact spawned theories that fluctuations in atmospheric CO_2 cause global temperature changes. And the corollary to this gained popular acceptance – industrial carbon emissions cause global warming.

However, both the Milankovitch Theory and the thesis that atmospheric carbon causes global warming have serious shortcomings. Both theories appear to be vastly over-rated. They both may play a role in temperature changes. However, their contributions appear to be minor when compared to the influence from EUWS cycles. The task becomes determining the precise percentage each factor contributes in modulating climate cycles.

In a paper entitled *Historical Carbon Dioxide Record From the Vostok Ice Core*, researchers from the Laboratoire de Glaciologie et de Géophysique de l'Environnement provide evidence that the current warming trend developed from naturally occurring cycles:

> "**There is a close correlation between Antarctic temperature and atmospheric concentrations of CO_2 (Barnola *et al.*, 1987). The extension of the Vostok CO_2 record shows that the main trends of CO_2 are similar for each glacial cycle.... According to Barnola *et al.* (1991) and Petit *et al.* (1999), these measurements indicate that, at the beginning of the deglaciations, the CO_2 increase either was in phase or lagged by less than ~1000 years with respect to the Antarctic temperature, whereas it clearly lagged behind the temperature at the onset of the glaciations.**"[Barnola *et al.*, 2003]

This revealing observation shows that atmospheric carbon-dioxide levels lag global temperature changes instead of causing them! If CO_2 levels caused temperature changes they should lead, not lag. It's often dangerous to assume cause and effect when correlations exist. And that appears to have happened with the theory that carbon-dioxide changes cause global-warming. The Vostok history implies that mankind's fossil fuel usage simply adds to atmospheric CO_2 content – with little impact on temperature.

The truth cannot be denied. Our planet has warmed more than 1°F since the start of the industrial revolution. But this 1°F jump may have resulted from a different force. Based on current evidence, it appears that most, if not all, of the increase resulted from naturally occurring EUWS cycles. And that's where attention should focus. In fact, the *Unified Cycle Theory* predicts that mankind needs to worry more about global cooling than warming.

Ice ages have an unusual characteristic of gradually cooling for 90,000 years followed by abrupt warming for the next 10,000 years. With 11,700 years having passed since the end of the last ice-age, another cooling period seems imminent. Even though a new ice-age may be in the process of starting, there's no need for alarm. While ice-age cooling phases last 9 times longer than warming phases, temperatures drop slowly during the cooling phase. Hence, glaciers won't descend upon heavily populated Northern Hemisphere continents anytime soon.

Similar to global warming during the past 100 years, climatologists probably over-rate gravity's affect on global temperatures. Reality seems to lie closer to a shared contribution between gravitational cycles and EUWS cycles. In addition to Ambou's concern about explaining its 100,000-year periodicity, Milankovitch Theory lacks in other areas. These Milankovitch issues follow.

Milankovitch Problem 1 – A Limited History with Insufficient Understanding

Geologists remain more skeptical of the Milankovitch Theory than astronomers and physicists. Currently, only 5 million years of history have been studied. And for this limited history, the correlations are less than desirable. Going beyond the last 5 million years, astronomers still don't fully understand longer-term orbital patterns. In an abstract entitled *Successive Refinements in Long-Term Integrations of Planetary Orbits*, a research team led by UCLA geophysicist Ferenc Varadi wrote:

> "The long-term evolution of the orbits of the major planets is an important problem for astronomy, as well as the geosciences. Numerical simulations of planetary orbits have become fairly common in the past decades, but their main goal so far has been the understanding of orbital dynamics and not necessarily very high accuracy. It does not particularly matter for an astronomer whether Earth's orbital eccentricity was low or high at a given time, as long as the reasons for its variations are well understood. For the geosciences, however, the actual numbers are more important since geological records of isotope ratios are routinely compared to and interpreted using orbital variations... The exact connections between Earth's orbital variations, its climate evolution, and its geological record are still a matter of debate... Paleoclimatologists need therefore more accurate orbital data, covering longer time intervals, as they try to unravel the complex, evolving interactions between Earth's orbital parameters and climate... Despite considerable progress in the last two decades, we still do not know the details of the evolution of planetary orbits on the time scale of tens of millions of years."[Varadi *et al.*, 2002]

Milankovitch Problem 2 – 41,800 Year Frequency and Obliquity

A research team led by Anna-Maria Nador of the Geological Institute of Hungary studied lake sediments from the Pannonian Basin in Hungary. They detailed their findings in an abstract entitled *Milankovitch-Scale Multi-Proxy Records from Fluvial Sediments of the last 2.6 Ma, Pannonian Basin, Hungary*. In the V-1 borehole from the Pannonian Basin, a 41,800 year cycle emerged from 2.58 to .99 Ma – covering 39 cycles over 1.6 million years. [Nador *et al.*, 2002]

These oscillations almost exactly match the EUWS frequency of 41,728 years. And they are somewhat removed from Milankovitch's obliquity frequency – generally stated at approximately 40,000 to 41,000 years. The Pannonian Basin data suggests that the 41,728-year EUWS cycle plays a dominant role in climate cycles.

Milankovitch Problem 3 – Small Eccentricity vs. Large Climate Changes

John Wilkins, a physicist at Ohio State University, described another Milankovitch problem in *Companion to Energy*:

> "The 100,000 year cycle should not cause the variations observed if the couplings are linear, since the eccentricity should merely in that case change the size of the precession effects. The eccentricity changes are also rather small, as the orbit of Earth is remarkably circular even at its greatest eccentricity."[Wilkins, 2008]

A number of climatologists have been perplexed by this problem. The *Unified Cycle Theory* provides an explanation – showing how the 125-kyr EUWS frequency fits climate cycles better than eccentricity cycles. This explanation comes in a later chapter.

Milankovitch Problem 4 – Disappearing Cycles

Wilkins continued his critique of the Milankovitch Theory in *Companion to Energy* with this observation:

> "A five million year record from Lake Baikal of silica-bearing sediments using multiple proxies for climate shows evidence of the 41,000 year cycle throughout the record, but particularly strongly 1.8 to 0.8 million years BP. The 23,000 year cycle has been strongest during the past 400,000 years. The data also show that the 100,000 year cycle has been strong only during the last 800,000 years. The record also showed major cooling episodes around 1.7 and 2.7 million years BP.... An analysis of a ten million year record found similar results.... Similarly, a 65 million year record shows that the Milankovitch cycles' effects are quite varied—the 100,000 year cycle is very pronounced sometimes, and not large at others."[Wilkins, 2008]

Ever since climatologists first observed the association between Milankovitch and temperature cycles, they found it mystifying how these cycles fade in and out. However, the *Unified Cycle Theory* solves the disappearing cycle conundrum. Two cycles with slightly different frequencies but equivalent amplitudes produce alternating periods of visibility. When the two cycles oscillate in sync, the combined amplitudes make the resulting cycle highly visible. However, when the cycles move out of sync, their amplitudes tend to cancel, causing the cycle to disappear. In a later chapter, examples show how this happens.

Milankovitch Problem 5 – Cycles Not Apparent in All Datasets

In *Companion to Energy*, Wilkins expressed an additional concern with the Milankovitch Theory:

> "Examination of data from a lake in Hungary from 2.6 to 3.05 million years BP do not show the Milankovitch timing at all, but rather shorter cycles. It is unclear how these 'sub-Milankovitch' cycles could be accommodated in the model."[Wilkins, 2008]

The *Unified Cycle Theory* also solves this problem. The EUWS frequencies of 515, 1545, 4636, and 13909 years equate to frequencies misidentified as sub-Milankovitch cycles. These shorter-term EUWS cycles correlate closely to known climate cycles.

Milankovitch Problem 6 – Asian Monsoon Patterns

Wilkins continued his critique of the Milankovitch Theory in *Companion to Energy* with this observation:

> "The phase shift found in a study of the Asian monsoon, and the fact that there are different phase changes for the precession and obliquity data, seems to indicate there is a problem in our understanding of the Milankovitch mechanism. The physics of the situation has no easy explanation. This is an area ripe for more understanding."[Wilkins, 2008]

Essentially, Milankovitch problems 5 and 6 are nearly identical. Asian monsoon patterns correlate closely to EUWS cycles below the 21,000-year precession threshold. Examples come later in the book.

Milankovitch Problem 7 – 41,667 Year Frequency and Obliquity

In *Origin of the 100 Kyr Glacial Cycle: Eccentricity or Orbital Inclination?*, a pair of physicists, Richard Muller (California Berkeley) and Gordon MacDonald (California San Diego), note that spectral analysis of delta Oxygen-18 data over the last 800 Kyr shows a peak at 0.024. This translates into a frequency of 41,667 years, making it much closer to the 41,728 year EUWS cycle than it does to the 41,000 year obliquity cycle.[Muller and MacDonald, 2002]

Milankovitch Problem 8 – The Stage-11 Problem

In *Origin of the 100 Kyr Glacial Cycle: Eccentricity or Orbital Inclination*, Muller and MacDonald also mention other issues:

> "Three strong peaks are present in the eccentricity spectrum: near 0.0025 cycles per Kyr (400 Kyr period), near 0.008 cpkyr (125 Kyr period) and near 0.0105 cpkyr (95 Kyr period). The disagreement between the spectra of the climate and the spectrum of the eccentricity is evident. The absence of the 400 Kyr peak in the climate data has long been recognized (for a review see Imbrie *et al.*, ref 8), and

models have been devised to suppress that peak. The cancellation that takes place between the cycles near 400 Kyr has been referred to by Imbrie *et al*.... as the 'Stage-11 Problem', since no cancellation is seen in the isotopic data."[Muller and MacDonald, 2002]

Predictions from the *Unified Cycle Theory* match the observations from Muller and MacDonald. The EUWS includes cycles at 125.2, 375.6, and 1127-kyrs. All three correspond to climate cycles. Hence, even though the absence of a 400-kyr climate cycle disrupts the Milankovitch Theory, it fits quite well with the *Unified Cycle Theory*.

Milankovitch Problem 9 – Temperature Maximum Precedes Insolation Peak

Muller teamed up with fellow Berkeley physicist, Daniel Karner, who identified additional Milankovitch issues. In *A Causality Problem for Milankovitch*, the Berkeley duo noted that the insolation peak at 125 Ka doesn't match the temperature maximum at 135 Ka at Devil's Hole. The 10,000 year deviation seems quite damaging:

> "In January 1999, Tezer Esat and collaborators published new results of measurements of coral terraces of the Huon Peninsula in Papua New Guinea... They showed that sea level rose and peaked around 135 ka, to a point close to ... present levels. At the Fall meeting of the American Geophysical Union in San Francisco, Gideon Henderson and Niall Slowey reported new results from U-Th dates of aragonite-rich sediments from the slopes of the Bahamas... They had three samples spanning Termination II, and they arrived at an age of 135 ±2.5 Ka for the termination. At the same meeting, Christina Gallup and collaborators presented new U-Th and U-Pa ages of fossil corals in Barbados... They too found that most of the rise in sea level took place prior to the expected insolation warming. They report that the sea level had risen to within 18 meters of the present level by 135.8 ±0.8 Ka."[Karner and Muller, 2000]

Prior to the last ice-age, temperatures at Devil's Hole approached their maximum around 135 Ka. That happened at the time of an insolation minimum. The minimum implies that global temperatures should have been cool around the time of 135 ka, not warm. The insolation maximum followed 10,000 years after that, occurring around 125 Ka. During those 10,000 years, insolation increased sharply; yet global temperatures remained flat.

Where insolation failed, the *Unified Cycle Theory* shined. Theoretical peaks of the 41.7, 125.2, and 375.6-kyr cycles occurred at 128.6 Ka. The period from 150 Ka to 128.6 Ka covered the final portion of the up-phase for these 3 cycles. By 135 Ka, the up-phase for all three cycles was nearly complete – essentially matching the temperature data from Esat, Henderson, Slowey, and Gallup. The climate record from Devil's Hole closely follows the *Unified Cycle Theory* prediction, with temperatures rising from 155 Ka to 125 Ka.

Milankovitch Problem 10 – Variable Frequencies, Dotted with Rapid Changes

Another problem appears that climatologists failed to mention. It seems unlikely that climate cycles in the range of 10 thousand to 1 million years could originate from gravitational sources. Gravity and motion cycles imply smooth climate transitions – similar to the gradual transitions for daily and annual temperature cycles.

Instead, these long-term climatic oscillations display characteristics that resemble the Schwabe sunspot cycle. They show periodicity, however, they fluctuate in highly erratic patterns. One cycle can easily span a period twice as long as the previous cycle. Furthermore, these fluctuations tend to spike. If these cycles originated solely from gravity and motion, they should exhibit smooth, sine-wave patterns. But they don't. Their erratic behavior implies they possess electromagnetic properties. Gravity and motion cycles cannot come close to explaining the erratic characteristics of ice-age fluctuations.

Milankovitch Problem 11 – A Catch All for Every Climate Cycle

In their conclusion of *A Causality Problem for Milankovitch*, Karner and Muller bluntly jab the theory:

> "**And the Milankovitch theory has other difficulties, such as the 'Stage-11 problem', that have been addressed with other ad hoc mechanisms. If we allow every discordant measurement to have its own explanation, we do not have a theory that can make predictions, and that means it really isn't a theory at all.... Even if we have no simple answer, at least we can conclude that the Milankovitch theory must not be a procrustean bed into which every observation will be forced. Different aspects of climate may have different driving forces – and some may even be unrelated to insolation... There is already evidence that different proxies in the same core can be measuring completely different aspects of climate. Precession, for example, is dominant in the atmospheric oxygen signal in the Vostok ice core, and simultaneously very small in the temperature proxy for the same core. In the sea floor, eccentricity can be strong in one proxy, and yet virtually absent in another. Now that we have lost the simple Milankovitch picture, we must look at the data again, as if for the first time, regard climate to be multidimensional, and be open to new ideas unbiased by our prior theoretical prejudices.**"[Karner and Muller, 2000]

The conclusion from Karner and Muller nicely leads into the next discovery. A European team consisting of Bridget Wade from the University of Edinburgh and Heiko Palike from Stockholm University discovered a 1.2 million year cycle in the Antarctic ice sheet. They attribute it to Milankovitch dynamics. In their abstract, entitled *Oligocene Climate Dynamics*, Wade and Palike wrote:

> "**Our data allow a detailed examination of Oligocene paleoceanography, the evolution of the early cryosphere, and the influence of orbital forcing on glacioeustatic sea level variations. Spectral analysis reveals power and coherency ... with an additional strong imprint of the eccentricity and 1.2 Myr obliquity amplitude cycle, driving ice sheet oscillations in the Southern Hemisphere.**"[Wade & Palike, 2004]

However, instead of eccentricity and obliquity, Wade and Palike probably observed the 1.13 million year EUWS cycle. The preceding quote highlights how quickly researchers attribute any cycle in the range of 10 thousand years to 1 million years to Milankovitch dynamics. This example also brings to light an interesting coincidence. The four largest Milankovitch cycles roughly separate themselves by a factor of 3 – corresponding closely to EUWS frequencies. And the smallest Milankovitch cycle corresponds to the 13.9-kyr EUWS by a harmonic of 1.5. See **Table 11.1**.

Table 11.1 – Frequency Comparison: Milankovitch Cycles vs. EUWS Cycles.

Milankovitch Freq.	EUWS Freq.	Comments
1,200,000	1,126,655	Wade & Palike - eccentricity and obliquity.
400,000	375,552	400-kyr cycle in eccentricity, but not in climate.
100,000	125,184	Attributed to eccentricity.
41,000	41,728	Attributed to obliquity.
21,000	13,909	Precession. 21-kyr cycle ~ 1.5 x EUWS.

The comparison brings some important questions to the forefront: Did this matchup simply happen by accident? Or does an unknown factor force the semi-correlation? At this point, no attempt will be made to answer these questions. However, these semi-correlations provide food for thought and potential clues about the nature of our universe. After presenting more evidence, a later chapter reexamines the link between Milankovitch cycles and EUWS cycles.

As this chapter shows, gravity and motion cycles impact our environment. And these cycles involve quite a bit more than the simple rotation of Earth around the Sun. For example, ice-age cycles partially result from the combined gravitational influences of the Sun, Moon, and planets. However, to fully understand how these cycles impact climate, gravitational cycles must be isolated from climatic changes produced by the EUWS cycles. Both cycles appear to influence climate, but the EUWS cycles tend to be dominant.

Now, attention moves to the EUWS cycles. The EUWS cycles fill the gaps left by gravity and motion. The central theme of this book revolves around cycles that impact Earth the most. And the astonishing EUWS cycles fit those criteria better than any of the other cycles. This book dedicates the remaining chapters to these mysterious cycles.

Chapter 12
EUWS Cycles

The upcoming chapters contain the most interesting portions of this book. They examine the cycles associated with the Extra-Universal Wave Series (EUWS). While the force behind these cycles remains unknown, the characteristics associated with their oscillations allow an educated guess about their origin.

With the *Unified Cycle Theory's* core model in place, predicting additional frequencies became as easy as multiplying by three. In the beginning, the predictions seemed more like a game, rather than something destined to provide a long list of accurate cycles. But after finding supportive evidence for several of the predicted cycles, the nature of the search changed. Curiosity changed to dumbfoundedness as cycle after cycle appeared. Within six months, strong evidence appeared for every one of the EUWS cycles under 22.2 billion years. **Table 12.1** contains all of the predicted EUWS frequencies between 28 days and 50 trillion years. The question kept coming to mind: What does this sequence imply?

Table 12.1 – Theoretical EUWS Cycles.

Abbr. Freq.	Extended Freq.	Comments
28.7 days	28.67889 days	Predicted and found – cycles in stocks and solar neutrinos
86 days	86.03667 days	Predicted and found – stock market cycle
258 days	258.11 days	Predicted and found – stock market cycle
2.12 years	2.12 years	1st core cycle -- Bi-annual stock cycle and solar neutrinos
6.36 years	6.36 years	2nd core cycle -- Economic recession cycle
19.1 years	19.08 years	3rd core cycle -- Financial panic cycle
57 years	57.24 years	4th core cycle -- Economic depression cycle
172 years	171.72 years	5th core cycle -- Minor civilization cycle
515 years	515.16 years	6th core cycle -- Civilization cycle
1545 years	1.54548 Kyr	Predicted and found – dark-age cycle
4636 years	4.63644 Kyr	Predicted and found – climate cycle
13.9 Kyr	13.90932 Kyr	Predicted and found – climate cycle
41.7 Kyr	41.72796 Kyr	Predicted and found – climate cycle
125 Kyr	125.18388 Kyr	Predicted and found – climate cycle
376 Kyr	375.55164 Kyr	Predicted and found – climate cycle
1.13 Myr	1.126655 Myr	Predicted and found – climate cycle
3.38 Myr	3.379966 Myr	Predicted and found – climate cycle
10.1 Myr	10.139894 Myr	Predicted and found – climate cycle and geological periods
30.4 Myr	30.419683 Myr	Predicted and found – cycles in climate, geology, & evolution
91.3 Myr	91.259049 Myr	Predicted and found – cycles in climate, geology, & evolution
274 Myr	273.777146 Myr	Predicted and found – cycles in climate, geology, & evolution
821 Myr	821.331437 Myr	Predicted and found – cycles in climate, geology, & evolution
2.46 Gyr	2.463994 Gyr	Predicted and found – cycles in climate, geology, & evolution
7.39 Gyr	7.391983 Gyr	Predicted and found – cycles in climate and geology
22.2 Gyr	22.175949 Gyr	Predicted and found – cycles in climate and geology
66.5 Gyr	66.527846 Gyr	Predicted
200 Gyr	199.583539 Gyr	Predicted
599 Gyr	598.750617 Gyr	Predicted
1.8 Tyr	1.796252 Tyr	Predicted
5.4 Tyr	5.388756 Tyr	Predicted
16.2 Tyr	16.166267 Tyr	Predicted
48.5 Tyr	48.498800 Tyr	Predicted

EUWS Cycles

Knowing the actual cause of the EUWS cycles isn't absolutely imperative. Nevertheless, finding the source will greatly enhance the aesthetics of the *Unified Cycle Theory*. As **Table 12.1** shows, the EUWS cycles cover a wide range of events. However, the list is actually more extensive. The advancing phases of these cycles correlate with the following events.

Astronomy related events (Giga-year timescale)
- reduced asteroid impacts
- possible decrease in the star formation rate
- possible decline in the formation of universes (reductions in 'Big Bangs')
- increased universal temperatures (as temperature increases, universal expansion accelerates)

Climate related events (Mega-year and kilo-year timescales)
- increased CH_4 levels
- increased CO_2 levels
- increased Be-10 levels
- increased atmospheric Carbon-14
- reduced delta-Oxygen-18 sediment levels
- increased global temperatures
- increased rainfall

Biological related events (Mega-year and kilo-year timescales)
- expansion in gene diversity
- increased number of species

Economic and social events for humans (cycles less than 2000 years)
- emergence from dark-ages
- expansion in civilizations
- increased cooperation among people
- increased confidence
- increased prices in commodity markets
- increased prices in security markets

During the declining phases of EUWS cycles, everything listed above happens in reverse. As examples, for large timescales, global temperatures decrease, carbon-dioxide levels fall, and species become increasingly extinct. For shorter timescales, civilizations disappear, cooperation among people decreases, and market prices fall.

It's also important to emphasize that most EUWS frequencies below 40 million years represent cycles found by others. Within the last century, various scholars already discovered all of the main components of the EUWS cycles related to human activities. The monumental breakthrough with the *Unified Cycle Theory* isn't that these cycles exist. Rather, the breakthrough lies with the fact that these cycles act together as a cohesive group. And because of this, they share a common origin. Knowing that these cycles share a common origin immediately invalidates every existing theory regarding their source.

Whatever the source, any theory explaining the origin of the EUWS cycles must account for every frequency from 28.7 days all the way to 22.2 billion years. None of the existing theories come close to doing that. A number of researchers have proposed that cycles in the million to billion year range originate from galactic cosmic rays coming from our Milky Way galaxy. At one time, I leaned toward that same idea.

However, the sheer size of the largest, verifiable EUWS cycle completely eliminates our galaxy as a possible source. The most recent trough of the 22.2-gyr cycle points back to the Big Bang. Hence, it's possible that the Big Bang itself touched off this series of cycles. However, for that to be true, the trough of the 22.2-gyr cycle must coincide <u>exactly</u> with the Big Bang. In this particular case, close doesn't count.

The latest estimate of the age of the universe rests at 13.73 ±0.12 Ga.[Hinshaw *et al.*, 2008] The estimated 22.2-gyr trough comes at 13.82 Ga. The difference equals a 91.3-myr sub-cycle defined by the TPD Principle. It's possible that estimates of the age of the universe and the 22.2-gyr frequency will be revised so their troughs match. However, even if that happens, it still seems unlikely that a Big Bang could modulate itself.

If the Big Bang failed to set off this sequence of cycles, then only one other possible explanation exists. And that possibility lies outside of our known universe! Hence, the EUWS cycles serve as the first credible evidence that something lies beyond our observable universe.

Looking at the totality of the EUWS cycles, the *Unified Cycle Theory* predicts the possibility of cycles longer than 22.2 billion years that modulate repeated big bangs. If the EUWS cycles set off the Big Bang, then these cycles certainly control big bangs in other universes as well. This idea isn't new. A few physicists have already become uncomfortable with the singularity of the Big Bang.

University of California Davis physicist Andreas Albrecht, one of the early developers of Alan Guth's cosmic inflation theory, now has second thoughts about inflation. Albrecht, along with another leading theoretical physicist, Joao Magueijo, co-authored an abstract entitled *A Time Varying Speed of Light as a Solution to Cosmological Puzzles*. In their paper, they outline a theory that the speed of light varies over time.

> **"Cosmologists have long been dissatisfied with the 'Standard Big Bang' (SBB) model of the Universe. This is not due to any conflict between the big bang theory and observations, but because of the limited scope offered by the SBB to explain certain striking features of the Universe. From the SBB perspective the homogeneity, isotropy, and 'flatness' of the Universe, and the primordial seeds of galaxies and other structure are all features which are 'built in' from the**

beginning as initial conditions. Cosmologists would like to explain these features as being the result of calculable physical processes. A great attraction of the Inflationary Cosmologies is that they address these issues by showing on the basis of concrete calculations that a wide variety of initial conditions evolve, during a period of cosmic inflation, to reflect the homogeneity, isotropy, flatness and perturbation spectrum that we observe today."[Albrecht and Magueijo, 1999]

While admitting that their theory hasn't been worked out to the same detail as cosmic inflation, Albrecht and Magueijo did show how their theory addresses the flatness issue, the cosmological constant, and the entropy problems.

Next, the pair took their theory a step further. In a transcript entitled Einstein's *Biggest Blunder,* Albrecht and Magueijo propose that a variable speed of light paves the way for multiple big bangs. Specifically, Magueijo stated:

"And you might think this is the end of the Universe, but of course in the picture of this theory it's just creating the conditions for another Big Bang to happen again — a sudden drop in the speed of light, another sudden discharge of all this energy into another Big Bang. So it is possible that actually our Big Bang is just one of many, one of many yet to come, and one of many which there were in the past already — maybe the Universe is just this sequence of Big Bangs all the time."[Cal Davis Cosmology, 2008]

But the cosmic inflation revolt doesn't end with Albrecht and Magueijo. Princeton University physicist, Paul Steinhardt, another early developer of Guth's inflation model, now openly questions its validity. Christian Science Monitor writer P.N. Spotts interviewed Steinhardt in an article entitled *The Big Bang (One More Time)*:

"He and Cambridge University physicist Neil Turok have unveiled a model in which the universe has no beginning or end, but replenishes itself in a cycle of expansion and contraction. Each expansion is triggered by its own big bang.... They hold that much of their model works well in a four-dimensional universe of height, depth, width, and space-time. They add that it finds its true home in the 9 to 10 dimensions of string theory, which tries to explain how the four forces of nature emerged from one unified force early on. One variation, known as M theory, holds that the universe consists of two parallel sheets, or membranes. The two membranes are separated by a 'fifth' dimension a tiny fraction of a centimeter wide. Steinhardt and Turok's calculations describe the membranes meeting in a slap, triggering the big bang. On the membrane humans inhabit, the bang yields the particles, energy, and forces familiar to scientists. The other contains 'we know not what,' Steinhardt says. The duo posits the second one may be home to 'dark matter.' Over trillions of years, the membranes expand, growing darker, colder, and less dense, until the logic of Steinhardt's equations brings them back together in another cosmic slap. The membranes

resume expanding, even as they drift apart, only to repeat the cycle. Steinhardt says this model yields all the features of the inflationary model, without inflation."[Spotts, 2002]

While Steinhardt presents interesting ideas, multiple big bangs don't necessarily have to be so complex that they involve membranes. The basis for a simpler big bang cycle centers on ideas discussed by University of Maryland geoscientist, Charles Breiterman. Breiterman raises a relevant question about whether the Earth is an open or closed system. In his abstract, entitled *Considering the Earth as an Open System*, Breiterman wrote:

"This review article synthesizes research from several disciplines by conceiving of the Earth as an open system significantly influenced by inputs of matter from outer space. It carefully and critically surveys the topics of mass extinctions, interplanetary dust/micrometeorites, delivery of complex pre-biotic organic molecules from space, ice-ages, small comets and panspermia. The evidence, if verified by continuing investigations, shows that the Earth was an open system just after planet formation. Since then, the Earth has been significantly influenced by inputs of extraterrestrial matter, but the inputs are not critical to the sustained functioning of the system."[Breiterman, 2004]

Opinions vary on this subject. My own view sits close to Breiterman's, but not exactly the same. Reality probably lies closer to Earth acting as a hybrid system with some elements open and others closed. Furthermore, these ratios vary with time. In other words, the openness of Earth's ecosystem oscillates – just like everything else in our universe.

Taking this one step further, the huge 22.2-gyr length of the largest observed EUWS frequency suggests that a multiverse exists as an open system. Our own observable universe functions as a small part of the multiverse. It does so in conjunction with a large set of other universes. Each of these individual universes, similar in operation to our own but varying in size, could be considered hybrid systems. Each universe essentially sustains itself; however, sometimes universes exchange elements. And the openness of each universe varies with time. While each individual universe behaves as a self-contained unit, nothing in the structure of the multiverse prevents individual universes from interacting with each other.

This universal interaction would mimic the behavior of galaxies in our universe. Galaxies generally stay separated, but every once in a while, they crash and form a new, larger galaxy. Additionally, the magnetic field surrounding galaxies tends to contain matter and energy inside its field, however, some leakage occurs. In an identical manner, universes could merge and share matter with each other. In fact, under this model, a universe could simply be viewed as a super-large galaxy. Albeit, during a contraction phase, the black hole at the center of a universe attains such a large mass that new properties appear. These new properties then permit a universe to completely implode – resulting in endless cycles of big crunches followed by big bangs.

This simple universal model basically operates on size. Each collection of matter in the multiverse displays different properties depending on size. For example, a small collection of galactic gas collapses to form a planet. A larger mass compresses to form a star similar to our Sun. A collection of matter greater than 10 solar masses forms a supernova, which eventually collapses, then explodes. More than 50 solar masses may result in supernova that collapse into black holes without subsequent explosions. In each case, a combination of mass and elemental composition determines the properties of the mass-collection.

Something similar could happen when collections of mass reach a threshold well beyond the size of a galaxy. With attainment of that threshold, a galaxy suddenly changes into a universe. In an open multiverse, things never balance within each individual universe. For example, the amount of matter never equals anti-matter and the number of electrons fail to equal positrons. Inside an open multiverse, any combination of matter and energy is possible for each constituent universe.

Of course, this is all conjecture. However, the conjecture fits the evidence supplied in this book. If a series of waves originates from outside our universe, it implies the existence of other universes inside a multiverse. And no logical reason emerges requiring these individual universes to remain separated from each other.

Evidence of an oscillating universe isn't limited to the EUWS cycles themselves. In winning the 2006 Nobel Prize for physics, John Mather (coordinator of the COBE satellite program launched by NASA in 1989) and George Smoot (an astrophysicist at the University of California at Berkeley) may have uncovered key evidence of EUWS cycles making their mark in remote sections of our universe. In its news release entitled *U.S. Duo Win Physics Nobel for backing up Big Bang*, Reuters noted their measurements of the microwave background radiation:

> **"[They] showed temperature variations in background radiation in space, in the range of a hundred-thousandth of a degree, that offered clues as to how galaxies, stars and planets were formed as matter coalesced.... The team also found 'ripples' in space, or small variations in the microwave background that provided new clues about galaxy and star formation and why matter had been concentrated in a specific place rather than spreading out."** [Lannin & Edmonds, 2006]

These ripples in space could easily be related to the EUWS cycles. Throughout all timescales, the EUWS cycles alternate between "hot and cold" phases that produce expansions and contractions. These expansions and contractions seem to disperse matter unevenly throughout our universe. Hence, alternating EUWS phases probably trigger the ripples observed by Mather and Smoot.

Actually, these ripples could be explained by viewing gravity in a non-traditional way. The EUWS cycles mesh well with the perplexing gravitational distortions attributed to dark matter and dark energy.

Isaac Newton initially developed theories of gravity in 1687. In 1905, Einstein's theory of General Relativity refined gravitational theory – stating that the speed of gravity equals the speed of light. However, Einstein was never fully decided on whether his equation should include dark energy. Astronomers following Fred Zwicky speculated about additional sources to Einstein's equation in the form of non-light emitting material, called dark matter.

Physicists have never observed either dark matter or dark energy. Theorists contrived these components to account for unexplained gravitational effects. Dark matter accounts for the missing matter that holds galaxies together. In the opposite way, dark energy accounts for the missing force that pushes the universe apart.

A pair of physicists from the UK, HongSheng Zhao of the University of St Andrews and Baojiu Li of Cambridge University, propose that dark matter and dark energy are actually the same entity in their abstract entitled *Dark Fluid: Towards a Unification of Empirical Theories of Galaxy Rotation, Inflation and Dark Energy*:

> **"Gravity, the earliest and the weakest of the known forces, has never been very settled.... Empirical theories of Dark Matter like MOND gravity and of Dark Energy like f(R) gravity were motivated by astronomical data. But could these theories be branches rooted from a more general hence natural framework? Here we propose the natural Lagrangian of such a framework based on simple dimensional analysis and co-variant symmetry requirements, and explore various outcomes in a top-down fashion. Our framework preserves the co-variant formulation of GR, but allows the expanding physical metric be bent by a single new species of Dark Fluid flowing in space-time.... The Dark Fluid framework naturally branches into a continuous spectrum of theories with Dark Energy and Dark Matter effects."** [Zhao & Li, 2008]

The pair replaces the terms dark energy and dark matter with a new force that explains both. They call this new force <u>dark fluid</u>. A press release from the University of St. Andrews explains several key properties of the dark fluid theorized by Zhao and Li:

> **"Dr HongSheng Zhao, of the University's School of Physics and Astronomy, has shown that the puzzling dark matter and its counterpart dark energy may be more closely linked than was previously thought... In Dr Zhao's model, dark energy and dark matter are simply different manifestations of the same thing, which he has considered as a `dark fluid". On the scale of galaxies, this dark fluid behaves like matter and on the scale of the Universe overall as dark energy, driving the expansion of the Universe. Importantly, his model, unlike some similar work, is detailed enough to produce the same 3:1 ratio of dark energy to dark matter as is predicted by cosmologists.... However, the Universe might be absent of dark-matter particles... The findings of Dr Zhao are also compatible with an interpretation of the dark component as a modification of the law of gravity rather than particles or energy. Dr Zhao concluded. 'No matter what dark**

matter and dark energy are, these two phenomena are likely not independent of each other.'"[St. Andrews Press, 2008]

Looking closer at the dark fluid properties theorized by Zhao and Li, and comparing them to the properties exhibited by the EUWS cycles, a surprisingly close match emerges. That's especially true for the alternative interpretation from Zhao and Li – that dark fluid actually represents a modification of the law of gravity. And the 3:1 ratio of dark energy to dark matter equals the way the EUWS cycles sub-divide.

As it relates to the EUWS cycles, the rapid expansion of our universe could represent effects from the still-expanding 66.5-gyr EUWS cycle – which physicists attribute to dark energy. At the same time, the contracting phases of EUWS cycles smaller than 22.2-gyr could provide the forces holdings galaxies together – which physicists attribute to dark matter.

If the EUWS cycles cause the behavior attributed to dark energy and dark matter, then the EUWS cycles must modulate cycles in gravity. This emerges as a distinct possibility because, in spite of Newton's well defined laws over 300 years ago, gravity remains a somewhat mysterious force. For example, a gravitational wave has never been observed, and physicists still debate Einstein's assessment that the speed of gravity equals the speed of light.

In a narrative entitled *The Speed of Gravity - What the Experiments Say*, physicist Tom Van Flandern from the University of Maryland's Army Research Lab explains the variety of problems associated with gravity operating at the speed of light:

> **"Standard experimental techniques exist to determine the propagation speed of forces. When we apply these techniques to gravity, they all yield propagation speeds too great to measure, substantially faster than light speed. This is because gravity, in contrast to light, has no detectable aberration or propagation delay for its action, even for cases (such as binary pulsars) where sources of gravity accelerate significantly during the light time from source to target. By contrast, the finite propagation speed of light causes radiation pressure forces to have a non-radial component causing orbits to decay ...; but gravity has no counterpart force proportional to v/c to first order. General relativity (GR) explains these features by suggesting that gravitation (unlike electromagnetic forces) is a pure geometric effect of curved space-time, not a force of nature that propagates. Gravitational radiation, which surely does propagate at light speed but is a fifth order effect in v/c, is too small to play a role in explaining this difference in behavior between gravity and ordinary forces of nature. Problems with the causality principle also exist for GR in this connection, such as explaining how the external fields between binary black holes manage to continually update without benefit of communication with the masses hidden behind event horizons. These causality problems would be solved without any change to the mathematical formalism of GR, but only to its interpretation, if gravity is once again taken to be a propagating force of nature in flat space-time**

with the propagation speed indicated by observational evidence and experiments: not less than [20 trillion times the speed of light!] Such a change of perspective requires no change in the assumed character of gravitational radiation or its light speed propagation. Although faster-than-light force propagation speeds do violate Einstein special relativity (SR), they are in accord with Lorentzian relativity, which has never been experimentally distinguished from SR-at least, not in favor of SR. Indeed, far from upsetting much of current physics, the main changes induced by this new perspective are beneficial to areas where physics has been struggling, such as explaining experimental evidence for non-locality in quantum physics, the dark matter issue in cosmology, and the possible unification of forces. Recognition of a faster-than-light speed propagation of gravity, as indicated by all existing experimental evidence, may be the key to taking conventional physics to the next plateau." [Van Flandern, 2008]

In addition to fitting better with universal observations, Van Flandern's estimate of the speed of gravity fits well with characteristics of the EUWS cycles. If EUWS cycles originate from outside our universe, and if they modulate gravity, they must do so at incredibly fast speeds. By doing so at speeds in excess of 20 trillion times the speed of light, gravitational influences from one side of our universe would reach the other side almost instantaneously! Hence, if the EUWS cycles modulate gravity, then their effects spread throughout our universe in unison. In fact, their effects may spread throughout much of the multiverse with only relatively minor delays.

If EUWS cycles act as the instigator to star formation, and they do so concurrently throughout our universe, then the EUWS force must travel at super-fast speeds. If EUWS cycles travel at speeds equal to Van Flandern's estimate of the speed of gravity, then that would fit the criteria for concurrent SFRs. However, if EUWS forces traveled at the speed of light as Einstein suggested, then the SFR in one area of our universe should vary substantially from the SFR in another area.

The preceding pages deal with the EUWS cycles in the giga-year range, and higher. These giga-year cycles provide the basis for a model of the formation of our universe, stars, and planets. However, at the other end of the spectrum, a set of relatively short-term EUWS cycles produce different outcomes. Among other things, these relatively short-term cycles correspond to fluctuations in human behavior.

Some of these human activity cycles were identified more than a century ago. But the original estimates lacked the precision that can now be attained with modern, more extensive databases. Nevertheless, these early pioneers deserve credit and recognition for their breakthrough efforts in the study of cycles. That recognition follows.

In 1862, Clement Juglar, a French statistician, wrote *Des Crises Commerciales et de leur Retour Périodique en France, en Angleterre et aux États-Unis* introducing the idea that the economy operates in an environment of recurring crises separated by about 10 years, but ranging anywhere from 7 to 11 years. However, what Juglar really observed was a mixture of three different cycles – the 19-year EUWS cycle, the shorter-term 6.36-year EUWS cycle, and interference from the 11-year Schwabe sunspot cycle. Juglar played an important role by introducing the world to the concept of economic cycles.

In 1923, Joseph Kitchin published an article in the Harvard University Press entitled *Cycles and Trends in Economic Factors*. The article described the existence of a 40-month cycle resulting from a study of USA and UK statistics between 1890 and 1922. Unfortunately, a study covering only 32 years lacks the longevity needed to firmly establish a human action cycle. Since then, the Kitchin cycle has been mostly discredited. Nonetheless, because of his stature, Kitchin's article heightened interest in economic cycles in the United States.

Even though they both made mistakes, the work of Juglar and Kitchin sparked needed interest in business cycles. Their work paved the way for more meaningful discoveries. In retrospect, the errors made by Kitchin and Juglar are quite understandable. For periods less than 50 years, the EUWS cycles compete with geomagnetic cycles for dominance. When these two types of cycles combine, the resulting sequence becomes quite distorted. Thus, analyzing these shorter-term time-series becomes exceedingly difficult.

In fact, the shorter-term EUWS cycles are more difficult to identify than the ones with frequencies in excess of 1 million years. Normally, one would expect the opposite. However, for frequencies under a million years, cycles in gravity, motion, solar activity, and geomagnetism all interfere with the EUWS cycles. But on the really long-term timescales, the EUWS cycles stand virtually alone, without interference. Hence, don't be surprised by the crystal-clear appearance of giga-year cycles in the next few chapters. But then, they become slightly more obscure while stepping down the ladder toward the shorter-term cycles.

For Kitchen and Juglar, these competing cycles caused too many analytical problems. Actually, the same held true for just about every other analyst who followed them. However, as cycle lengths increased, especially for periods greater than nineteen years, distortions from the geomagnetic cycles were less problematic. Because of this, the merits of past discoveries held up much better for cycles with frequencies greater than nineteen years. The *Unified Cycle Theory* merely aided in making small refinements to the earlier estimates. The Kuznets cycle illustrates this.

Simon Kuznets, a professor at the Wharton School of the University of Pennsylvania, won the 1971 Nobel Prize in Economics for his work in detailing economic activity during the 19th and first-half of the 20th Century. Kuznets also discovered a cycle in investment activity spanning between 15 and 20 years – called the Kuznets Infrastructural Investment Cycle. A few years later, Edward Dewey refined the length of this cycle to 18.33 years, calling it a real estate investment cycle. This book refines its frequency further, to 19.08 years. Its defining characteristic comes near its theoretical peaks, with a long sequence of panics and crashes repeatedly hitting the marketplace.

The Unified Cycle Theory

Next, as mentioned in Chapter 2, Russian economist Nikolai Kondratieff discovered a cycle spanning 50 to 60 years, often called the Long Wave or the Kondratieff Wave. During 1925, in his paper *The Major Economic Cycles*, Kondratieff wrote about a long wave in commodity prices and economic activity. Associated with this, Edward Dewey discovered a 57-year cycle in wars. This book refers to the 57-year frequency as an economic depression cycle. Its length has been refined to 57.24 years.

Raymond H. Wheeler, a former professor at the University of Kansas, deserves credit for the discovery of both the 170-year cycle and the 510-year civilization cycle. Wheeler started studying economic cycles in the 1930s. The book *Climate: The Key to Understanding Business Cycles* best outlines Wheeler's findings.[Zahorchak, 1983] Similar to other cycles, the original discovery provided a good estimate based on the data available at the time. Wheeler did many things well. However, some of his conclusions were slightly wrong. Here's the list of Wheeler's discoveries and mistakes.

1) Essentially, Wheeler correctly estimated the length of the civilization cycle at 510 years. However, based on more extensive data, and working within the confines of the *Unified Cycle Theory*, the new estimate equals 515.16 years.

2) Wheeler believed civilization power alternated between the East and the West. This conclusion has no logical basis. The 515-year cycle affects all nations, all cultures, and all civilizations, all at the same time. As each successive civilization cycle reaches a peak, weak nations implode. At the same time, strong nations may stagnate or weaken, yet they survive. What Wheeler observed as an oscillation in power between East and West was simply an oscillation in the financial and economic strength of various nations at the top of the civilization cycle. At civilization cycle peaks, powerful nations tend to become over-confident, over-leveraged, over-taxed, and over-consuming. Over-confidence tends to make people less cautious, and it allows excessive unproductive activity to seep into the structure of all sectors of the economy. These underlying weaknesses eventually lead to civilization collapse. But it doesn't always happen this way. Some empires remain financially strong during cyclic up-phases. By avoiding excesses and staying financially solvent, some empires actually survive through multiple civilization cycles.

3) Wheeler correctly concluded that civilization cycles develop from physical cycles in nature. However, he incorrectly concluded that climate caused these cycles. Weather cycles do fluctuate closely in tandem with the 515-year and 172-year EUWS cycles. In spite of this, climate changes act as a coincident indicator. Weather cycles do not closely correspond to EUWS cycles under 19 years. Whatever causes these fluctuations in human behavior must be consistent for all frequencies. And climate change fails to offer that consistency.

4) Wheeler identified civilization cycle peaks at 570 BC, 60 BC, 450, 960, 1470, and 1980. He missed the actual peaks by roughly 170 years. The actual peaks came at 912

BC, 397 BC, 118, 633, 1148, 1663, and 2178. Within the scope of the *Unified Cycle Theory*, the mistake was understandable. A 170-year error on a 515-year cycle approximately equals the length of its first sub-cycle. Furthermore, true peaks for the 515-year cycle can only be derived from all types of data, including sunspot data and climate data – not just civilization data. Using these criteria, the *Unified Cycle Theory* turning points fit better than Wheeler's.

In spite of his mistakes, Wheeler made important contributions to understanding cycles. He was a great thinker and innovator. He did many things right. Now, with additional data, better information on physical cycles, and a better theory to explain these patterns, Wheeler's original estimates can now be improved.

The 1545.48-year dark-age cycle became the first non-core cycle discovered from the original six EUWS cycles. Alvin Toffler, an American writer and futurist, first discovered a civilization cycle spanning 1000 to 2000 years. Toffler outlines his civilization ideas in *The Third Wave*.[Toffler, 1980]

More recently, Dr. Sing C. Chew, a professor of Sociology at California State University-Humboldt, wrote about a civilization cycle ranging from 1000 to 1700 years. In his book *The Recurring Dark Ages: Ecological Stress, Climate Changes, and System Transformation*, Chew refers to these oscillations as a dark-age cycle.[Chew, 2006] This book sticks with Chew's dark-age terminology.

Chew suggests that dark-ages result from man-made climate changes – due to deforestation. While Chew's guess about the cause may be wrong, he provides excellent documentation about the timing of dark-age periods. His dark-ages closely match troughs in the 1545-year EUWS cycle.

Actually, Chew's errors in determining the cause can be applied to the discoverer of every other EUWS cycle. All of these analysts produced excellent work regarding the timing of cycles, but they errored while making stabs at the causes.

Except for the next two chapters, which cover five cycles, the remainder of this book dedicates one full chapter to each EUWS cycle. The evidence used to validate these cycles varies. For the longer cycles, evidence includes star formation rates, global climate indicators, cosmic ray flux, and geological dating. Looking at the different types of signals, temperature oscillations act as the characteristic most commonly shared by all the EUWS cycles. Temperature changes can be seen in every EUWS cycle between 57.24-years and 22.2-gyr. For cycles with frequencies below 1600 years, a combination of general history and commodity prices provide the bulk of the evidence. For the shortest cycles, daily security quotes during the last 100 years provide the database.

Chapter 13
Giga-Year EUWS Cycles

The *Unified Cycle Theory* predicts the existence of cycles with extremely long periods. In this chapter, attention centers on cycles with periods in excess of 7 billion years. Even though the theory predicts these cycles, understandably, supporting evidence appears in a limited fashion. Yet, strongly supportive evidence actually exists.

Chart 13A shows how extra-universal temperatures may have fluctuated during the last 600 billion years, and then it projects how these temperatures might behave 600 billion years into the future.

This model combines theoretical cycles with periods of 7.39, 22.2, 66.5, 200, and 599-gyr into one composite series. The arrow in the middle, at the bottom, represents the Big Bang. According to the theory, the optimal time for the Big Bang occurred at 13.82 Ga – during an extremely cold period in the multiverse.

This theoretical date for the Big Bang coincides closely with the latest figure from NASA. During February 2008, WMAP placed the age of the universe at 13.73 ±0.12 Ga.[Hinshaw *et al.*, 2008] The deviation approximately equals one 91.3-myr sub-cycle, using the TPD Principle. For the 66.5, 200, and 599-gyr cycles, the theoretical low at 13.82 Ga must be classified as speculative. However, the same cannot be said of the 7.39 and 22.2-gyr cycles.

22.2 Billion Year EUWS Cycle

The 22.2-gyr cycle has only encountered two theoretical turning points since the formation of our universe. Its trough came at 13.820 Ga. And its only observable peak came at 2.732 Ga. And these two theoretical turning points coincide closely with the most dramatic event in the history of our universe (the Big Bang) and the most dramatic event in Earth's history (the 2.7 Ga Event). The unusual activity surrounding 2.7 Ga prompted New Mexico Tech geologist Kent Condie to write an abstract entitled *What on Earth Happened 2.7 Billion Years Ago?*[Condie, 2003]

Throughout the entire family of EUWS cycles, they exhibit the same characteristics from parent to child. Near theoretical troughs, temperatures plummet, matter clumps together, and organisms die. Near theoretical peaks, the exact opposite happens – temperatures rise, matter explodes, and life sprouts.

These descriptions match what happened at both ends of the spectrum with the 22.2 Ga cycle. As the 22.2-gyr cycle approached an ultimate cold phase at its theoretical trough at 13.82 Ga, the ultimate clumping of matter occurred. The extreme cold produced the final death spiral – ending with a Big Crunch. Near its theoretical low, the 22.2-gyr cycle reversed, and the ultimate explosion occurred – the Big Bang. And our universe began its current expansion.

Well, at least until 2.7 Ga. During Earth's brief 4.54 billion year history, its most dramatic change occurred at the time of the 2.7 Ga Event. And these events happened in the reverse order of the Big Bang. Near the theoretical top of the 22.2-gyr cycle, one final upward spurt in temperatures occurred. In a sense the Earth exploded with great geological upheaval, and life on Earth encountered its first major evolutionary spurt. In a sense, the 2.7 Ga Event exceeded the Cambrian Explosion in jump-starting life on Earth.

Then, at the height of this tremendous period of activity, the 22.2 Ga cycle reversed. The down-phase of this cycle began about 2.7 billion years ago, which means it will take another 8.4 billion years before it hits bottom. When that happens, extreme cold will prevail. But don't worry. The Earth will no longer exist by the time the 22.2-gyr cycle hits bottom. Before then, the Sun will have used up most of its nuclear energy and expanded to a size that scorches Earth!

As the 22.2-gyr cycle reversed at 2.70 Ga, global temperature plummeted. A research team, led by geologist Robert Kopp of the California Institute of Technology, believes that the first Snowball Earth formed 400 million years after the 2.70 Ga Event – with ice sheets possibly approaching the equator around 2.3 Ga.[Kopp *et al*., 2005]

In additional to a frigid Earth, after the 2.70 Ga peak, the great explosion in life reversed abruptly.[Ding *et al*., 2006]. And the record-breaking geological upheavals ceased.[Condie, 2008] While not nearly as dramatic as the reversal of the Big Crunch turning into a Big Bang, the 2.70 Ga Event behaved as the exact opposite of the formation of our universe. Clearly, the

22.2-gyr cycle reached its peak at 2.70 Ga.

Using a measuring stick of 11.088-gyr (½ cycle), the 22.2-gyr cycle only missed its theoretical trough by roughly one 91.3-myr sub-cycle, or 0.8%. And the 22.2-gyr cycle only missed its theoretical peak by roughly one 30.4-myr sub-cycle, or 0.3%. Of course, errors in estimates for both <u>the age of the events</u> and <u>the frequencies of the EUWS cycles</u> mean the true deviations could differ somewhat. Nonetheless, throughout this book, the ages of geological and climatic events match EUWS theoretical turning points so consistently that it's hard to image a large degree of inaccuracy in any of the estimates.

Marked by a ½-cycle with the Big Bang at the trough and the 2.70 Ga Event at the peak, the 22.2-gyr cycle represents the start of a long string of cycles that left their imprint on Earth's geology, ocean sediments, ice cores, and biological diversity. And the EUWS cycles continue to leave their imprint to this date. The impact can be seen in cycles as short as 28.7 days in the stock market.

The amplitude indicated by the ½-cycle of the 22.2-gyr frequency exceeds all other EUWS cycles. In spite of its unsurpassed power, the 7.39-gyr cycle comes somewhat close in rivaling the strength of its parent. And that's the nature of all cycles in this series.

Each child cycle derives its frequency from its parent by a factor of 1/3. And the amplitude of each child cycle drops by a small degree from its parent. However, the drop in magnitude isn't so large that it becomes obscure. In fact, sometimes a child cycle's oscillation appears large enough to confuse it with its parent. But knowledge of the theoretical turning points helps eliminate confusion. With two oscillations in between, every third turning point matches its parent.

These simple rules pre-identify each and every EUWS cycle. Their theoretical turning points are etched in stone. (See the **Appendix** for a complete list of frequencies and turning points for all EUWS cycles.) This book starts with the 22.2-gyr frequency. Then it progresses by zooming in closer to study each successive EUWS sub-cycle.

All of the rules that apply for the 22.2-gyr cycle apply to the 28.7-day cycle – and every cycle in between. The twenty chapters that follow may seem like overkill. Yet, the manner in which the EUWS cycles sub-divide seems incredible and unimaginable. Hence, to validate such an incredible sequence, it's preferable to present too much evidence rather than not enough.

7.39 Billion Year EUWS Cycle

At any major EUWS trough, the *Unified Cycle Theory* predicts a <u>cold period</u> in which matter clumps together. For the giga-year cycles, this means stars should form at high rates during EUWS valleys. Conversely, smaller numbers of stars should form near EUWS peaks. Unfortunately, when it comes to star formation, the *Unified Cycle Theory* performs below expectations. Nonetheless, rather than throw out the evidence, this book presents start formation rates (SFR) along with their unexpected deviations.

According to the most recent data from the GAIA project sponsored by the European Space Agency, SFR does exhibit a cyclical pattern. See **Chart 13B**. However, instead of changing direction every 2.46-gyr, SFR fluctuates around every other 2.46-gyr benchmark. Low SFR occurred at EUWS theoretical peaks at 10.12, 5.20, and 0.27 Ga – exactly as the *Unified Cycle Theory* predicts. However, star formation patterns skipped beats between these theoretical peaks.

This unexpected behavior probably means some other factor interferes with the EUWS influence on SFR. In fact, a group of University of Washington astronomers led by G.S. Stinson studied SFR extensively within small galaxies. In an abstract entitled *Breathing in Low Mass Galaxies: A Study of Episodic Star Formation*, Stinson et al. wrote:

> **"Given their susceptibility to galactic 'weather', the cool baryonic content of dwarf galaxies should fluctuate, not just from galaxy to galaxy, but within an individual galaxy as a function of time. The expected temporal variations in the amount of cool gas should be accompanied by variations in the star formation rate (SFR). These variations in the SFR can be seen within the Local Group, where dwarf galaxies are sufficiently well resolved that their past SFRs can be directly inferred from their color-magnitude diagrams. As expected, past star formation in dwarf galaxies appears to be complex, with many systems exhibiting multiple episodes of star formation.... With these simplified but self-consistent models, we show that star formation in low mass galaxies can undergo a natural 'breathing' mode, where episodes of star formation trigger gas heating that temporarily drives gas out of the cool phase and into a hot halo. Subsequent gas cooling and gas accretion then allows gas to settle back into the halo, and star formation begins again. This episodic mode of star formation is superimposed on the net infall of gas into the galaxy, leading to galaxies with both episodic star formation and significant intermediate age populations, in line with observations."**[Stinson *et al.*, 2007]

In addition to density, mass, and time, the 7.39-gyr and 2.46-gyr EUWS cycles appear to play an important role in SFR. Stinson's team notes that star formation requires relatively cool matter within the host galaxy. This tantalizing fact implies that the EUWS cycles should play a role. However, based on data from the European Space Agency, the SFR frequency equals two 2.46-gyr cycles – falling in between the 7.39-gyr cycle and the 2.46-gyr cycle.

In the four dwarf galaxies analyzed by Stinson *et al.*, their data showed a different SFR for each galaxy. However, outside of that primary frequency, all four galaxies showed higher SFR between 0.6 Ga and 1.6 Ga (a 2.46-gyr EUWS cold period) and lower SFR around 2.732 Ga (a 2.46-gyr EUWS hot spot).

Study **Chart 13B** again. It shows how the GAIA SFR Index coincides with every other 2.46-gyr cycle. The chart shows SFR inverted, because that's how star formation should correlate with the EUWS. Low SFRs occurred near the EUWS hot spots of 10.12 Ga, 5.20 Ga, and 0.27 Ga – exactly as the *Unified Cycle Theory* predicts. However, SFR ignored the hot spots in between – at 12.50 Ga, 7.66 Ga, and 2.73 Ga.

Stronger evidence supporting the 7.39-gyr cycle comes from something that didn't happen – The Great Unconformity. A geological unconformity occurs when an age-gap appears in the deposit of sediment in Earth's crust. An unconformity creates an absence of sediment between two rock masses with distinctly different ages. An unconformity can result in two possible ways:

a) Erosion sweeps away sediment, leaving an absence of geological evidence for a period of time.
b) Earth's ecology fails to deposit sediment for a period of time.

The largest of the unconformities goes by the name The Great Unconformity. It can be seen in the geology in many parts of the world. At Frenchman Mountain in Nevada, The Great Unconformity left a 1.15-gyr gap ranging in time from 0.55 Ga to 1.70 Ga.[Rowland, 2008]

Chart 13C shows how The Great Unconformity and other major geological events meshed with theoretical turning points of major EUWS cycles. Tick marks on the x-axis of **Chart 13C** represent theoretical peaks of the 274-myr cycle. A verbal description of **Chart 13C's** visuals follows:

3) **2.70 Ga Event** – 2.73 Ga corresponded to theoretical peaks of the 22.2-gyr and 7.39-gyr cycles. That coincided with the 2.70 Ga Event, which unleashed record-breaking tectonic activity, along with significant global warming.

4) **1.64 Ga, Great Unconformity Begins** – The beginning of The Great Unconformity coincided with the theoretical peak of the 274-myr cycle at 1.64 Ga. According to the model presented in **Chart 13C**, this point marked the beginning of the coldest period in Earth's 4.54 billion year history. This theoretical cold period resulted from the combined synchronization downturns of the 22.2-gyr, 7.39-gyr, 2.46-gyr, 821-myr, and 274-myr cycles at 1.64 Ga.

5) **1.50 Ga, Probable Cold Period** – 1.50 Ga corresponded to the theoretical trough of the last 2.46-gyr cycle. This period probably corresponded to extremely cold conditions on Earth; however, The Great Unconformity washed away much of the evidence. The *Unified Cycle Theory* predicts bitter cold for this time, but scientists have not yet substantially confirmed this prediction.

6) **0.70 Ga, Snowball Earth Period** – 0.70 Ga corresponded to the theoretical trough of the last 821-myr cycle. Even though the 2.46-gyr cycle was in an advancing stage at that time, the 22.2-gyr and 7.39-gyr cycles were in declining phases – as they will continue to be for about another billion years. By analyzing the EUWS

cycles in combination, the declining phases of the 22.2-gyr and 7.39-gyr cycles offset the positive influence from the advancing phase of the 2.46-gyr cycle. The net result produced a theoretical cold period that roughly matched the theoretical cold period at 1.50 Ga. However, in this particular case, the scientific community has substantially confirmed that a Snowball Earth episode occurred around 0.70 Ga.

7) **0.54 Ga, Great Unconformity Ends** – At 0.54 Ga, global warming corresponded with a theoretical peak of the 274-myr cycle and the continued up-phase of the 2.74-gyr cycle. At this time, global temperatures rose enough to melt Snowball Earth's ice shelf, and spawn the Cambrian Explosion.

8) **0.27 Ga, 2.46-Gyr Cycle Peak** – 0.27 Ga coincided with a theoretical peak of the 2.46-gyr cycle. This also precisely corresponded with the warmest temperatures on Earth during the last billion years! Since then; however, the synchronized downturns of the 22.2-gyr, 7.39-gyr, and 2.46-gyr cycles project bitter cold in another 140 million years. The coming climate drop should surpass the coldness encountered during the Snowball Earth episode at 0.70 Ga.

A geological unconformity represents the exact opposite of a tectonic upheaval. Tectonic upheavals occur at major EUWS peaks with global warmth, leaving substantial geological trails. Conversely, unconformities occur at major EUWS troughs along with extremely cold climates, leaving limited geological trails.

Once the effects of EUWS cycles are understood, the reasons they produce unconformities become clear. Their characteristics exactly match the reasons unconformities appear.

a) **Erosion sweeps away sediment**. Ice sheets erode Earth's crust faster and more extensively than either rain or wind. The Great Unconformity coincided with the coldest period in Earth's history. It only seems logical that ice erosion wiped away massive amounts of the geological record during this time.

b) **Earth's ecology fails to deposit sediment for a period of time**. Reduced tectonic activity occurs near EUWS troughs. With limited tectonic activity taking place, less sediment becomes available to produce a good, permanent geological record.

Almost certainly, reduced tectonic activity and extensive ice erosion combine to produce unconformities near major EUWS troughs. As a general rule, larger EUWS frequencies produce greater unconformities; and conversely, smaller EUWS frequencies produce minimal unconformities. In coming chapters, this tendency can be seen in gaps in the geological timescale. Because of the size of the EUWS cycles covered in this chapter, data only exists for a period less than one full cycle. Nonetheless, for the data that can be found, theoretical long-term EUWS movements closely match with extraordinary events that transpired since our universe formed. These major events include:

- The Big Bang at 13.73 Ga closely corresponded to theoretical troughs of the 22.2-gyr and 7.39-gyr cycles.
- The 2.7 Ga Event closely matched theoretical peaks of the 22.2-gyr and 7.39-gyr cycles.
- The Great Unconformity between 1.7 Ga and 0.55 Ga closely coincided with a combination of EUWS cycles; however, downturns in the 22.2-gyr and 7.39-gyr cycles primarily modulated the unconformity.
- Star Formation Rates correspond to a secondary EUWS frequency at 4.92-gyr – half way in between the 7.39-gyr and 2.46-gyr cycles. The role played by EUWS cycles in SFR remains intriguing, but debatable.

As the EUWS frequencies become shorter, thus providing more complete cycles since Earth's formation, available evidence proliferates. And with this increased availability, the evidence turns quite compelling.

Chart 13A — 599, 200, 66.5, 22.2, & 7.4 Gyr Cycles

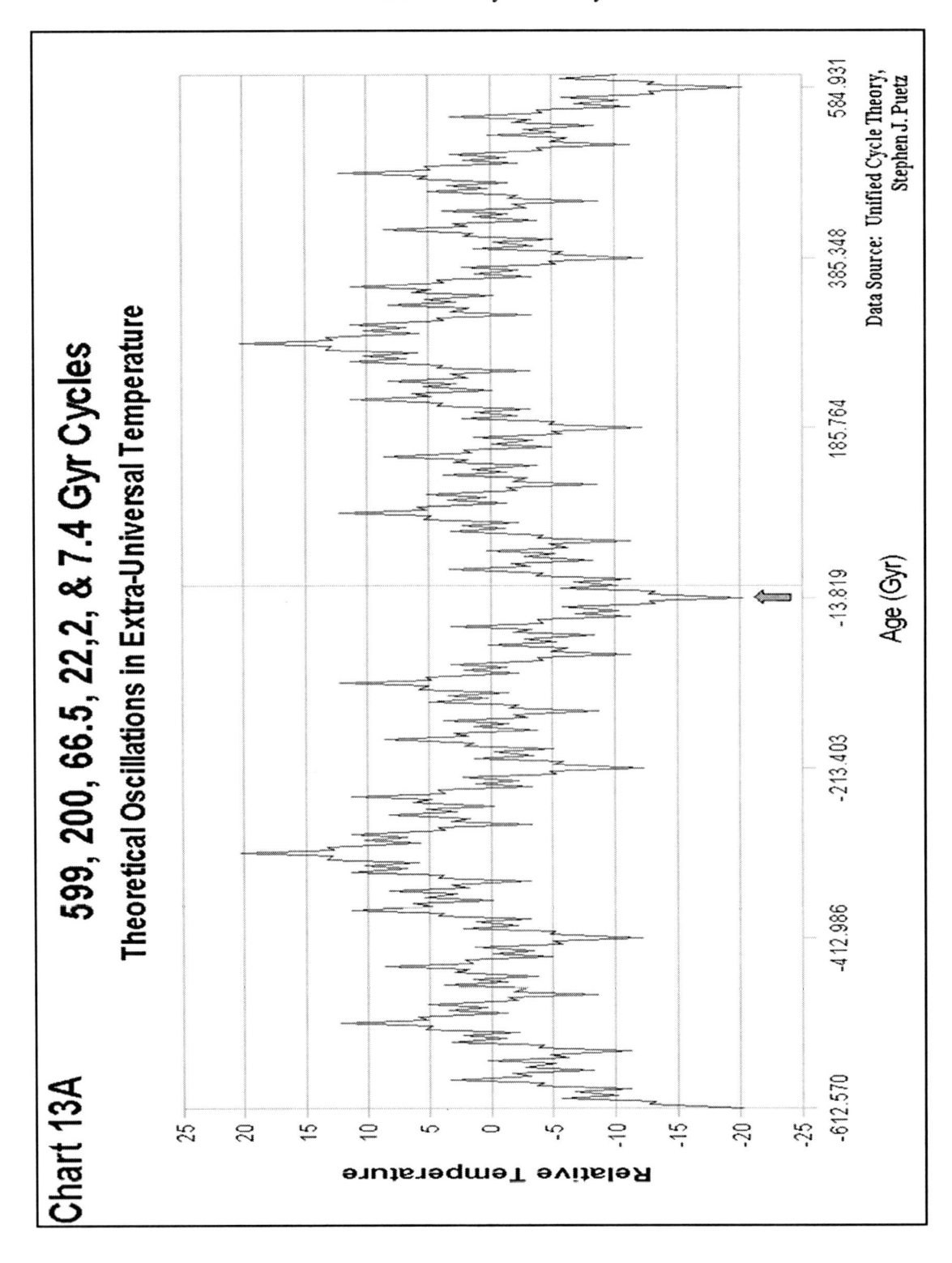

Theoretical Oscillations in Extra-Universal Temperature

Data Source: Unified Cycle Theory, Stephen J. Puetz

Giga-Year EUWS Cycles

Chart 13B

Star Formation Index

Star Form Rate (M/Gyr) vs *Gyr*

Data Source: European Space Agency, GAIA Science Advisory Group

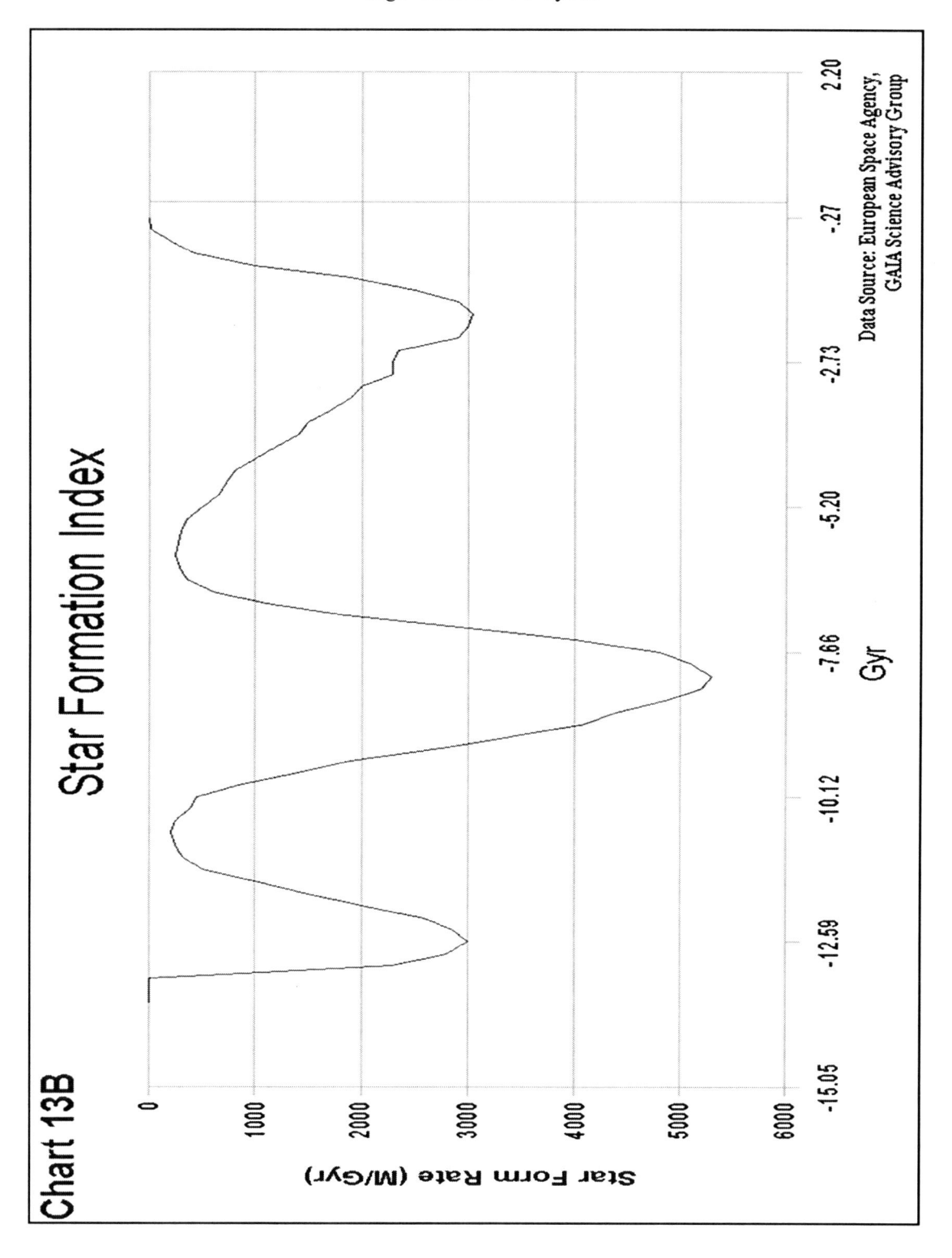

Chart 13C .030, .091, .274, .821, 2.46, 7.39, & 22.2 Gyr Cycles

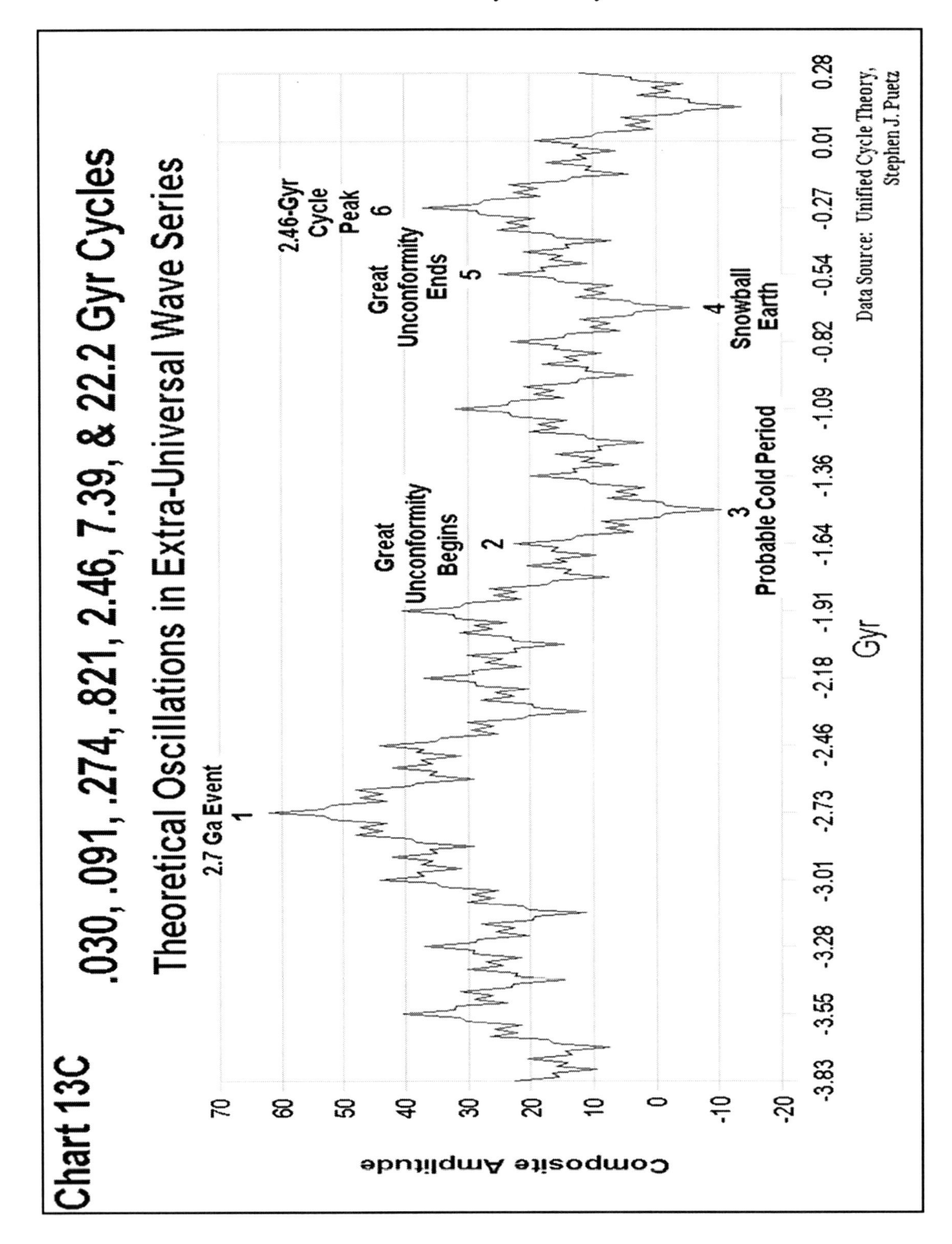

Theoretical Oscillations in Extra-Universal Wave Series

Data Source: Unified Cycle Theory, Stephen J. Puetz

Labeled points:
1. 2.7 Ga Event
2. Great Unconformity Begins
3. Probable Cold Period
4. Snowball Earth
5. Great Unconformity Ends
6. 2.46-Gyr Cycle Peak

X-axis: Gyr
Y-axis: Composite Amplitude

Chapter 14
2.46 Gyr, 821 Myr, and 274 Myr EUWS Cycles

Evidence supporting the EUWS cycles with frequencies of 274-myr, 821-myr, and 2.46-gyr comes from the same sources. Because of that fact, this chapter reviews the three cycles as a group. Since Earth formed, only two 2.46-gyr peaks have occurred – at 0.267 Ga and at 2.732 Ga. The list below shows all 821-myr and 274-myr peaks since Earth first developed. (All times are in billions of years before present.)

821-myr Peaks	**0.267**	**1.089**	**1.911**	**2.732**	**3.553**	**4.375**
274-myr Peaks	0.542	1.363	2.184	3.006	3.827	
274-myr Peaks	0.815	1.637	2.458	3.279	4.101	

The first set of evidence comes from the geological timescale. This timescale separates ages based on geological changes. Timescale classifications include eons, eras, and periods. Geological eons roughly correspond to EUWS peaks of the 2.46-gyr and 821-myr cycles – with two eons starting one 274-myr sub-cycle away from theoretical peaks of 2.46-gyr cycles, and another eon starting one 274-myr sub-cycle away from a theoretical peak of a 821-myr cycle. The Hadean eon began when the Earth formed, which coincided with a cool period (a EUWS trough). See **Table 14.1**.

Table 14.1 – Geological Eons vs. 2.46-Gyr and 821-Myr Theoretical Peaks.

Eons	Timescale	EUWS Peaks	Comment
Phanerozoic	.542 Ga to date	0.542 Ga	.542 occurred 1 274-myr sub-cycle before the 2.46-gyr peak of 0.268 Ga
Proterozoic	2.5 to .542 Ga	2.458 Ga	2.458 occurred 1 274-myr sub-cycle after the 2.46-gyr peak of 2.732 Ga
Archean	3.8 to 2.5 Ga	3.827 Ga	3.827 occurred 1 274-myr sub-cycle before the 821-myr peak of 3.553 Ga
Hadean	4.54 to 3.8 Ga	4.512 (trough)	Earth formed close to a 274-myr minimum at 4.512 Ga

Table 14.2 demonstrates how geological eras correspond to theoretical 274-myr peaks. By assuming these two datasets behave independently, the close correlation between geological eras and 274-myr peaks seems quite amazing. With only 13 geological eras and only 16 EUWS theoretical peaks during the last 4.15 billion years, 11 of the dates match fairly closely to form 274-myr era-pairs. In addition, one pair matched perfectly at 0.542 Ga. If the two datasets were completely independent, a perfect match of a 274-myr cycle (rounded to the nearest Myr) with 16 tries would only happen about 5.8% of the time. The odds of these two sequences being random become even more remote considering 11 pairs matched within 0.090-gyr.

Table 14.2 – Geological Eras vs. 274-Myr Peaks.

Eras	Timescale	EUWS Max	Comment
Cenozoic	.065 Ga to date	.006 Gyr AP	Closely matched a theoretical 274-myr peak
Mesozoic	.251 to .065 Ga	.268 Ga	Closely matched a theoretical 274-myr peak
Paleozoic	.542 to .251 Ga	.542 Ga	Perfectly matched a theoretical 274-myr peak
Neoproterozoic	1.00 to .542 Ga	1.089 Ga	Closely matched a theoretical 274-myr peak
Mesoproterozoic	1.60 to 1.00 Ga	1.637 Ga	Closely matched a theoretical 274-myr peak
Paleoproterozic	2.50 to 1.60 Ga	2.458 Ga	Closely matched a theoretical 274-myr peak
Neoarchean	2.80 to 2.50 Ga	2.732 Ga	Closely matched a theoretical 274-myr peak
Mesoarchean	3.20 to 2.80 Ga	3.279 Ga	Closely matched a theoretical 274-myr peak
Paleoarchean	3.60 to 3.20 Ga	3.553 Ga	Closely matched a theoretical 274-myr peak
Eoarchean	3.80 to 3.60 Ga		(No corresponding 274-myr peak)
Lower Imbrian	3.85 to 3.80 Ga	3.827 Ga	Closely matched a theoretical 274-myr peak
Nectarian	3.92 to 3.85 Ga		(Matched a 91.3-myr peak at 3.918 Ga)
Basin Groups	4.15 to 3.92 Ga	4.101 Ga	Closely matched a theoretical 274-myr peak

2.46 Gyr, 821 Myr, and 274 Myr EUWS Cycles

The geological timescale provides intriguing match ups with EUWS peaks; however, the strongest evidence comes from hard geological data. The first set of data deals with Earth's temperature during its first 3 billion years. Throughout the 20th Century, scientists surmised Earth remained hot during the Archean and Hadean eons, and then gradually cooled after that. See **Chart 14A**.

This ancient data comes from University of Ottawa geologist Jan Veizer. After reviewing the data, it was segregated into 50 to 100 million year bins, averaged, and plotted. The result appears in **Chart 14A**. The Oxygen-18 climate proxy confirms the consensus opinion, hot temperatures prevailed during Earth's early history, and then they gradually declined after 2.7 Ga.

In addition to the long-term downtrend, the paleoclimate data shows definite cyclical patterns. The maxima in **Chart 14A** closely align with theoretical peaks of both the 821-myr cycle and the 274-myr cycle – with 821-myr peaks showing greater amplitudes.

In spite of this extensive Oxygen-18 history, with the turn of the 21st Century, a number of geologists began to chip away at the consensus. In an abstract entitled *Atmospheric Composition and Climate on the Early Earth*, Penn State geologists James Kasting and Tazewell Howard wrote:

> **"Oxygen isotope data from ancient sedimentary rocks appear to suggest that the early Earth was significantly warmer than today, with estimates of surface temperatures between 45 and 85°C. We argue, following others, that this interpretation is incorrect—the same data can be explained via a change in isotopic composition of seawater with time. These changes in the isotopic composition could result from an increase in mean depth of the mid-ocean ridges caused by a decrease in geothermal heat flow with time. All this implies that the early Earth was warm, not hot. A more temperate early Earth is also easier to reconcile with the long-term glacial record. However, what triggered these early glaciations is still under debate. The Paleoproterozoic glaciations at approximately 2.4 Ga were probably caused by the rise of atmospheric O_2 and a concomitant decrease in greenhouse warming by CH_4. Glaciation might have occurred in the Mid-Archean as well, at approximately 2.9 Ga, perhaps as a consequence of anti-greenhouse cooling by hydrocarbon haze. Both glaciations are linked to decreases in the magnitude of mass-independent sulfur isotope fractionation in ancient rocks. Studying both the oxygen and sulfur isotopic records has thus proved useful in probing the composition of the early atmosphere.... Major glaciations are known to have occurred in the Neoproterozoic at approximately 0.6 and 0.75 Ga and in the Paleoproterozoic at approximately 2.4 Ga. All three of these periods show evidence for continental scale glaciation at low paleolatitudes and are thought by some authors, including us, to represent 'Snowball Earth' episodes."** [Kasting and Howard, 2006]

The research by Kasting and Howard sheds some light on the early global environment. The suspected 2.4 Ga glacial period deserves special scrutiny. Prior to that, the 2.7 Ga Event coincided with global warming and great tectonic upheaval. At 2.7 Ga, the warm phase abruptly ended. 300 to 400 million years later, a blanket of ice covered Earth. A research team led by California Institute of Technology geologist Robert Kopp places the age of the first "Snowball Earth" at 2.3 Ga. In an abstract entitled *The Paleoproterozoic Snowball Earth: A Climate Disaster Triggered by the Evolution of Oxygenic Photosynthesis*, Kopp's team described the conditions:

> **"Herein we discuss an alternate hypothesis, one in which the evolution of cyanobacteria destroyed a methane greenhouse and thereby directly and rapidly triggered a planetary-scale glaciation, the 2.3–2.2 Ga Makganeyene Snowball Earth.... In our alternative scenario, an evolutionary accident, the genesis of oxygenic photosynthesis, triggered one of the world's worst climate disasters, the Paleoproterozoic snowball Earth. Intensive investigation of the time period of the Paleoproterozoic glaciations may reveal whether a novel biological trait is capable of radically altering the world and nearly bringing an end to life on Earth."** [Kopp *et al.*, 2005]

Within the last ten years, University of Wisconsin geologist John Valley also joined the exodus from the camp backing a hot Hadean eon. In an abstract entitled *A Cool Early Earth*, Valley nicely describes why he recently changed his mind about Earth's paleoclimate:

> **"Single crystals of zircon began to add new information about the early earth in the 1980s, when a few rare grains from the Jack Hills and Mount Narryer regions of Western Australia became the most ancient terrestrial material known at that time—the oldest dating back almost 4.3 billion years.... It was not until 1999 that technological advances allowed further study of the ancient zircon crystals from Western Australia— and challenged conventional wisdom about the earth's earliest history.... The zircons' 18O values ranged up to 7.4. We were stunned. What could these high oxygen isotope ratios mean? ... Those high-18O rocks, if buried and melted, form magma that retains the high value, which is then passed on to zircons during crystallization. Thus, liquid water and low temperatures are required on the surface of the earth to form zircons and magmas with high 18O; no other process is known to do so. Finding high oxygen isotope ratios in the Jack Hills zircons implied that liquid water must have existed on the surface of the earth at least 400 million years earlier than the oldest known sedimentary rocks, those at Isua, Greenland. If correct, entire oceans probably existed, making the earth's early climate more like a sauna than a Hadean fireball."** [Valley, 2005]

Valley, Kopp, Kasting, and Howard all present strong cases showing that pre-Cambrian Earth encountered much cooler temperatures than earlier thought. Using their information, and combining it with the Oxygen-18 data from Veizer (once again, see **Chart 14A**), I constructed my own estimate of global temperatures during the last 3.5 billion years.

Constructing this paleoclimate series involved these steps:

- Flattening the slope of the Oxygen-18 data from Veizer – to adjust for the isotopic composition changes suggested by Kasting and Howard.
- Retaining the cyclical turning points of Veizer's Oxygen-18 data because no reason emerges to cast doubt on the accuracy of the oscillations in the data.
- Skewing the data so the Snowball Earth episodes reflect the appropriate intensity.

The resulting paleoclimate construction appears in **Chart 14B** and **Chart 14C**. Interestingly, the cold climate associated with 2.9 Ga and the 2.3 Ga snowball Earth episode both coincided with troughs in Veizer's Oxygen-18 data. The only conflict comes from disagreement about the degree of coolness at these lows. In assigning hard numbers to these relative temperatures, evidence provided by Valley, Kopp, Kasting, and Howard was weighted heavily.

Chart 14B shows the 22.2-gyr peak at 2.73 Ga (four arrows); the 2.46-gyr peak at 0.268 Ga (double arrows); and the 821-gyr peaks in between (single arrows). **Chart 14C** shows the same paleoclimate construction, except it marks 821-gyr peaks with double arrows and 274-gyr peaks with single arrows. **Chart 14D** depicts a global temperature model – as predicted by major EUWS cycles.

The model predicted ice-age epochs at 3.14, 2.32, 1.50, and 0.68 Ga. Comparing the model to temperatures from **Chart 14B**, the only potential miss came at 1.5 Ga. However, even in that case, Veizer's Oxygen-18 data indicates a substantial decline in global temperatures from around 1.9 Ga to 1.5 Ga. Unfortunately, geological data fails to confirm a 1.5 Ga glacial period. But remember, The Great Unconformity wiped away much of the geological record between 1.70 Ga and .55 Ga. Additionally, the TPD Principle came into play at the 3.14 Ga theoretical trough. Instead of arriving at the theoretical trough, the expected frigid period arrived a little less than 274-million years late – at around 2.9 Ga.

The model correctly projected the 2.46-gyr peak at 0.27 Ga. Unlike pre-Cambrian estimates, global temperature trends throughout the Phanerozoic eon contain a relatively high degree of accuracy. Also matching the model, global temperatures declined steeply during the last 270 million years. The model projects the climatic downtrend will continue for another 140 million years – potentially ending with a Snowball Earth episode colder than the previous ones. For mankind, the upcoming challenge becomes figuring out how to keep Earth's atmosphere warm for the next 140 million years to prevent global freezing!

The fact that global temperature cycles closely coincide with the EUWS model provides one set of supportive data. However, the uncertainty surrounding pre-Cambrian temperatures casts some doubt on the validity of temperature comparisons. The same cannot be said of plate tectonic intervals. Kent Condie, a geochemist at New Mexico Institute of Mining and

Technology, describes the periodicity of tectonic events in an abstract entitled *Supercontinents and Superplume Events: Distinguishing Signals in the Geologic Record*:

> **"Although both supercontinent formation and superplume events occurred at 0.28, 1.9 and 2.7 Ga, global warming at these times indicates the dominance of superplume events in controlling climate.... The distribution of U/Pb zircon ages coupled with Nd isotopic data suggest two major peaks in juvenile crust production rate, one at 2.7 Ga and another at 1.9 Ga.... So what is different about the superplume events that may be associated with peaks in juvenile crust production at 1.9 and 2.7 Ga? These events, hereafter called *catastrophic superplume events*, must be triggered by some process other than plate shielding. They also differ from shielding superplume events in that they are short-lived, less than 100 Myr duration, in contrast to shielding events that last for more than 200 Myr. Because large volumes of continental crust are associated with these superplume events, they must be more intense and perhaps more widespread than shielding events."**[Condie, 2003]

Dates of the .28 and 2.7 Ga continental superplume events almost identically match theoretical peaks of the 2.46-gyr cycle at .27 and 2.73 Ga. In addition, the 1.9 Ga superplume event matches the 1.91 Ga theoretical peak of the 821-myr cycle. Great tectonic expansions take place during the final stages of these major EUWS peaks. It's almost as if Earth attempts to explode at these peaks. Even though actual explosions never took place, Earth threw around its continental shelves substantially during these major EUWS periods.

The initial breakthrough in determining Earth's crust formation came in 1994 from a pair of researchers associated with the Montana Bureau of Mines and Geology. In an abstract entitled *Progressive Growth of the Earth's Continental Crust and Depleted Mantle – Geochemical Constraints*, M.T. McCulloch and V.C. Bennett plotted the uneven rate of Earth's crustal formation. See **Chart 14E**.

In addition to perfectly tracking the 821-myr EUWS cycle, the crustal formation reached its "highest high" at the 2.732 Ga peak of the 22.2-gyr cycle!

In his abstract, Condie shows the same spacing as McCulloch and Bennett. Condie also shows that supercontinents traverse cycles of formation and breakup equal to the 821-myr periodicity. He notes crustal formations spaced at 821-myr peaks at 0.27, 1.09, 1.91, and 2.73 Ga – all corresponding to warm periods. In addition, Condie indicates continental breakups at 821-myr troughs at 0.68, 1.5, and 2.32 Ga – all corresponding to cold periods. Condie places a question-mark after the 1.5 Ga breakup, indicating less certainty than the other continental splits. Nonetheless, the 1.5 Ga timeframe does match the cold climate scenario from Veizer's delta Oxygen-18 data, as well as matching the prediction from the EUWS model.

But the evidence gets even better. The changes associated with these cyclic peaks include much more than increased global warmth and great tectonic upheavals. Life itself evolves tremendously during these times. Condie describes the evolutionary indicators:

> "The Late Archean peak in juvenile crust production rate is also centered at 2700 ±50 Ma, thus confirming a strong correlation between supercontinent formation and juvenile continental crust production.... [In addition] stromatolites, layered structures thought to be deposited by microbial mat communities, are widespread in the Proterozoic with a prominent peak in number of occurrences at 1.9–1.8 and 2.7 Ga. The 1.9–1.8 Ga maxima include number of stromatolite occurrences, diversity of stromatolites, and the number of occurrences of microdigitate stromatolites. In addition to stromatolites, there are maxima in the reported occurrences of microfossils, oncoids, and chemo fossils (biogenic chemical remains) at about 1.9 Ga."[Condie, 2003]

As a follow-up to his 2003 publication entitled *What on Earth Happened 2.7 Billion Years Ago*, Condie created a PowerPoint presentation for his students. The presentation gives a near perfect visual representation of all major EUWS cycles.[Condie, 2008] In particular, the following frames provide amazing visuals of the EUWS cycles:

- Frame 4 – **Juvenile Continental Crust Age Distribution**. Important crust formations occurred near the 821-myr peaks at 2.73, 1.91, 1.09, and .27 Ga.
- Frame 6 – **Deformed Granitoid Zircon Ages**. The record peak for zircon ages came at 2.7 Ga – the theoretical peak for the 2.46, 7.4, and 22.2-gyr cycles.
- Frame 11 – **Peaks in Crustal Preservation Rate at 2.7 Ga and 1.88 Ga** – Each occurred almost exactly one 30.4-myr sub-cycle after the theoretical peaks of the 821-myr cycle – a good TPD Principle example.
- Frame 13 – **Orogenic Gold Deposits**. Peak gold mineralization occurred around the 821-myr EUWS peaks at 2.73, 1.91, and .27 Ga.
- Frame 18 – **Stromatolite Occurrence per Unit Volume of Juvenile Crust**. Life on Earth experienced great evolutionary peaks near 821-myr peaks at 3.55, 2.73, and 1.91 Ga.
- Frame 19 – **Delta Carbon-13 levels**. Peaks in atmospheric carbon occurred near 821-myr peaks at 3.55, 2.73, and 1.91 Ga.
- Frame 23 – **2.7 Ga Global Mantle Plume Event**. The great upheaval at 2.7 Ga coincided with the theoretical peak of the 2.46, 7.4, and 22.2-gyr cycles.
- Frame 25 – **Komatiite Formation**. Peaks in komatiite formation occurred near 821-myr peaks at 3.55, 2.73, and 1.91 Ga.
- Frame 26 – **Flood Basalts Formed at Fortescue around 2.764 Ga and Ventersdorp around 2.725 Ga**. These gigantic volcanic floods happened around the time of the theoretical peak of the 22.2-gyr cycle at 2.732 Ga. This suggests that increased volcanic activity occurs in proportion to the amplitudes of EUWS cycles. Evidence of periodic massive volcanic eruptions indicates that Earth tries to explode at major EUWS peaks.
- Frame 46 – **13 Cycles between .268 Ga and 3.827 Ga**. This detailed graphic shows the evolution of depleted mantle clustered around each of the theoretical peaks of the 274-myr cycle – with heavier concentrations at 821-myr peaks.

The work by Condie goes a long way in validating the EUWS cycles in the giga-year range. His work provides a long list of closely related events. As a reminder, concentrate on the correlations, not the causes of the events. At this point in the analysis, gathering a wide range of data about turning points, amplitudes, and frequencies overrides all other concerns.

Chart 14F shows Condie's juvenile crust age-estimates, broken into 100-myr buckets, and presented as a percentage of Earth's total crust. He published his findings in the abstract *Episodic Continental Growth and Supercontinents: A Mantle Avalanche Connection?*[Condie, 1998] Once again, great mantle upheavals occurred near theoretical peaks of the 821-myr cycle.

Condie's observation that significant evolutionary changes took place around 2.7 Ga requires more thorough investigation. Geologists recognize the Cambrian Explosion (around 542 Ma) as the time when a wide variety of species first appeared on our planet. However, the first evolutionary explosion actually took place more than 2 billion years earlier. UK geologists Corey Archer and Derek Vance published *Coupled Fe and S Isotope Evidence for Archean Microbial Fe(III) and Sulfate Reduction* noting the abundance of microbial life on Earth around 2.7 Ga.[Archer and Vance, 2006]

The most powerful evidence of EUWS-related evolutionary cycles comes from a research team at the Academy of Sciences in Shanghai, China. Led by Guohui Ding, the group used the latest technology to date the origin of a large database of genes. Entitled *Insights into the Coupling of Duplication Events and Macroevolution from an Age Profile of Animal Transmembrane Gene Families*, the abstract's synopsis describes their process:

> "**The interplay of information-processing life and force-driven environment has characterized Earth's evolutionary history since its beginning some 4 billion years ago. The study of macroevolution has seen a growing appreciation of this interplay. Previously, a large-scale effort was mounted to collect and analyze the paleontological and geochemical data. In the meantime, more and more genomes have been sequenced. The growing molecular sequence database with these paleontological data will provide important opportunities to investigate this interplay. Using the transmembrane proteins of 12 genomes, Ding and his colleagues have devised a sophisticated pipeline to date 1,651 duplication events grouped into 786 gene families, and have mapped the distribution of duplication events to the profile of macroevolution. They showed that the oxidation events played a key role in the major transitions of this density trace, and that the pulse mass extinction time points in the Phanerozoic phase coincide with the local peaks of the age distribution. Through some mathematical transformation of the density trace of the transmembrane gene duplicates during the Phanerozoic phase, they reported a potential cycle similar to the cycle detected by paleontologists. They concluded that a dramatically changed environment affected the evolution of life and left some imprint in the molecular level that can be detected.**"[Ding *et al*, 2006]

Later in the abstract, Ding *et al.* noted the unusual activity around the 2.7-Gyr Event, and began their study from that point forward.

"The observed age distribution deviates remarkably from the baseline distribution at about 2.75 Gyr ago. Thus, we ignored the bins whose age was before 2.75 Gyr for further analysis."[Ding *et al*, 2006]

Life multiplied tremendously around 2.7 Ga; however, I disagree with Ding's decision to ignore bins before 2.75 Ga. Valuable information resides in all of his team's data. By employing moving averages and calculating rates-of-change, important evolutionary history appears in Ding's dataset from 3.55 Ga to present. By comparing its 110-myr moving average against its 330-myr moving average, the geometric rise in diversity transforms itself into a leveled (but oscillating) data series. Using this methodology, **Chart 14G** shows the resulting evolutionary cycles for the last 3.5 billion years.

In line with other major events at the same time, the highest rate of genetic change corresponded to the 22.2-gyr EUWS peak at 2.732 Ga (five arrows)! The second highest growth rate came just after the 2.46-gyr EUWS peak at .268 Ga (triple arrows). And the oscillations in between coincided with theoretical peaks of the 274-myr cycle (single arrows) and the 821-myr cycle (double arrows). These oscillations, along with additional evidence coming in later chapters, strongly indicate that EUWS cycles regulate life on Earth.

Throughout this book, the bulk of the evidence concentrates on <u>EUWS peaks</u>. However, for the remainder of this chapter attention moves to <u>EUWS troughs</u>. EUWS low points heavily correlate with cold periods on Earth. In an abstract entitled *Cosmic Ray Diffusion in the Dynamic Milky Way: Model, Measurement, and Terrestrial Effects*, physicist Nir Shaviv correlates ice-age epochs with meteorite-impact cycles. He believes cycles of meteorite strikes indicate passage through the Milky Way's spiral arms. In turn, he proposes that the spiral arm passage corresponds to increased cosmic ray exposure.[Shaviv, 2002]

Shaviv lists IAEs at 20, 160, 310, 446, 592, 747, 920 (questionable), 2300, and 2950 million years ago. He then shows that ancient iron meteorites crashed to Earth at sporadic intervals – especially concentrated around times of IAEs. He postulates the cyclical nature of meteorite impacts provides indirect evidence of cosmic rays impacting Earth at the same times.[Shaviv, 2002]

Chart 14H shows the histogram constructed by Shaviv.

The chart inverts meteorite impact numbers so they sync with EUWS cycles. Notice how the absence of meteorite impacts coincide with EUWS peaks. Conversely, heavy bombardment coincides with EUWS troughs. Actually, with the exception of the 1089-to-815 Ma cycle (when the correlation shifted to 1-to-1), two meteorite cycles appear for every 274-myr EUWS cycle. Shaviv established the frequency at 143-myr.

The Unified Cycle Theory

Ignoring Shaviv's cause and effect conclusions, and concentrating solely on his data, it moderately conforms to the <u>clumpiness of matter</u> rule that applies to all universal masses during EUWS troughs. Specifically, our universe appears to contract during EUWS troughs, forcing large portions of matter to compress and merge. This includes more meteorites getting jogged from their orbits and falling to Earth during EUWS troughs. Oscillations in the gravitational force associated with EUWS cycles could provide one possible explanation for the clumpiness behavior.

Whatever the reason, various forms of matter cover our universe **<u>unevenly</u>**. If our universe expanded evenly, it would consist of a huge cloud of gas continually shrinking in density. But that's not the case. Clumps of matter gather in pockets throughout our universe. Dispersed throughout the universe, these clumps appear as moons, planets, stars, galaxies, and galaxy-clusters. Each clump of matter causes dramatic changes in localized densities – due to the gravitational influence of its mass.

This book has already supplied numerous examples. The European Space Agency calculated that cool gases clump together to form stars in a cycle spanning roughly 4.92 billion years – twice the frequency of the 2.46-gyr EUWS cycle. With a similar deviation, the 274-myr EUWS cycles approximately equals twice the frequency of meteorite impacts, as calculated by Shaviv. On Earth, alternating periods of warm mantle explosions and cold mantle breakups take place in phase with the 821-myr EUWS cycle. And the Big Bang occurred extraordinarily close to the 22.2-gyr cycle trough at 13.82 Ga.

As different as each of these cycles appear at first glance, they all demonstrate a periodicity that share similar expansion/contraction characteristics. In studying these oscillations, and especially in developing the *Unified Cycle Theory*, it's important to differentiate among the various types of cycles by their motion. Earlier, cycles were classified by their originating force. Now, the classification parameter changes from originating-force to spin.

Three types of spin appear in our universal:

1) Normal Spin Cycles – These cycles reoccur in an endless manner. As soon as one cycle finishes, the next one begins – essentially replicating the frequency of the previous cycle. Examples of spin cycles include electrons revolving around the nucleus of an atom, a moon circling around a planet, planets rotating around a star, stars moving around a galactic center, and galaxies moving within a galaxy-cluster.

2) Implosion Death Spirals – Spin cycles can terminate when an external factor throws balanced attractive-repulsive forces in favor of attraction. When that happens, a death spiral begins. The circulating mass gets pulled into a much larger mass at the center of the spiral. Actually, this can be viewed as a death-birth cycle. As soon as the death spiral puts to rest the previous set of spins, a new larger mass forms with its own spin cycles. Examples of implosion cycles include the formation of black holes, stars, planets, moons, and asteroids – as they develop from either gases or continually

accumulate small amounts of matter.

3) Implosion-Explosion Cycles – In some cases, if the collection of matter reaches a large enough mass, the death spiral associated with an implosion can reverse. At the inflection point of the implosion, when gravity fails to provide the power needed for further contraction, various repelling forces overwhelm gravity. At that critical moment, an offsetting explosion develops. While implosions act as death spirals, subsequent explosions disperse matter throughout the universe. In this way, explosions give birth to new universal possibilities. Hence, implosion-explosion cycles serve as the ultimate recycling systems. Examples of implosion-explosion cycles include supernova implosions-explosions, and as proposed by the *Unified Cycle Theory*, big crunch-big bang cycles.

To summarize this chapter, EUWS fluctuations influence matter within the multiverse in a way that spawns all three classes of cycles. In doing so, EUWS cycles affect a variety of seemingly unrelated events including cycles in big crunches and big bangs, star formation rates, meteorite impacts, formation and breakup of Earth's continents, and life on Earth.

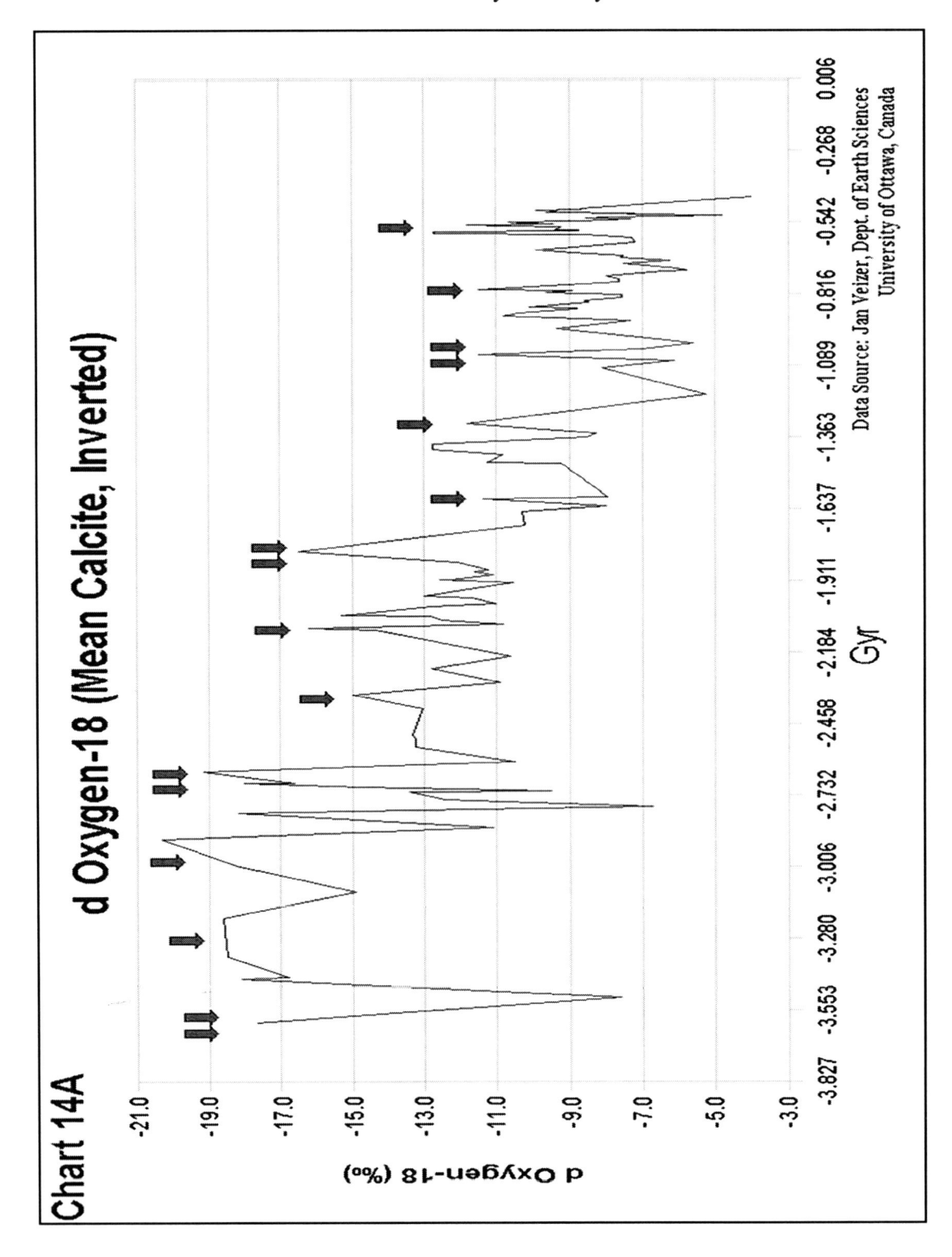

The Unified Cycle Theory

Chart 14A: d Oxygen-18 (Mean Calcite, Inverted)

Data Source: Jan Veizer, Dept. of Earth Sciences, University of Ottawa, Canada

2.46 Gyr, 821 Myr, and 274 Myr EUWS Cycles

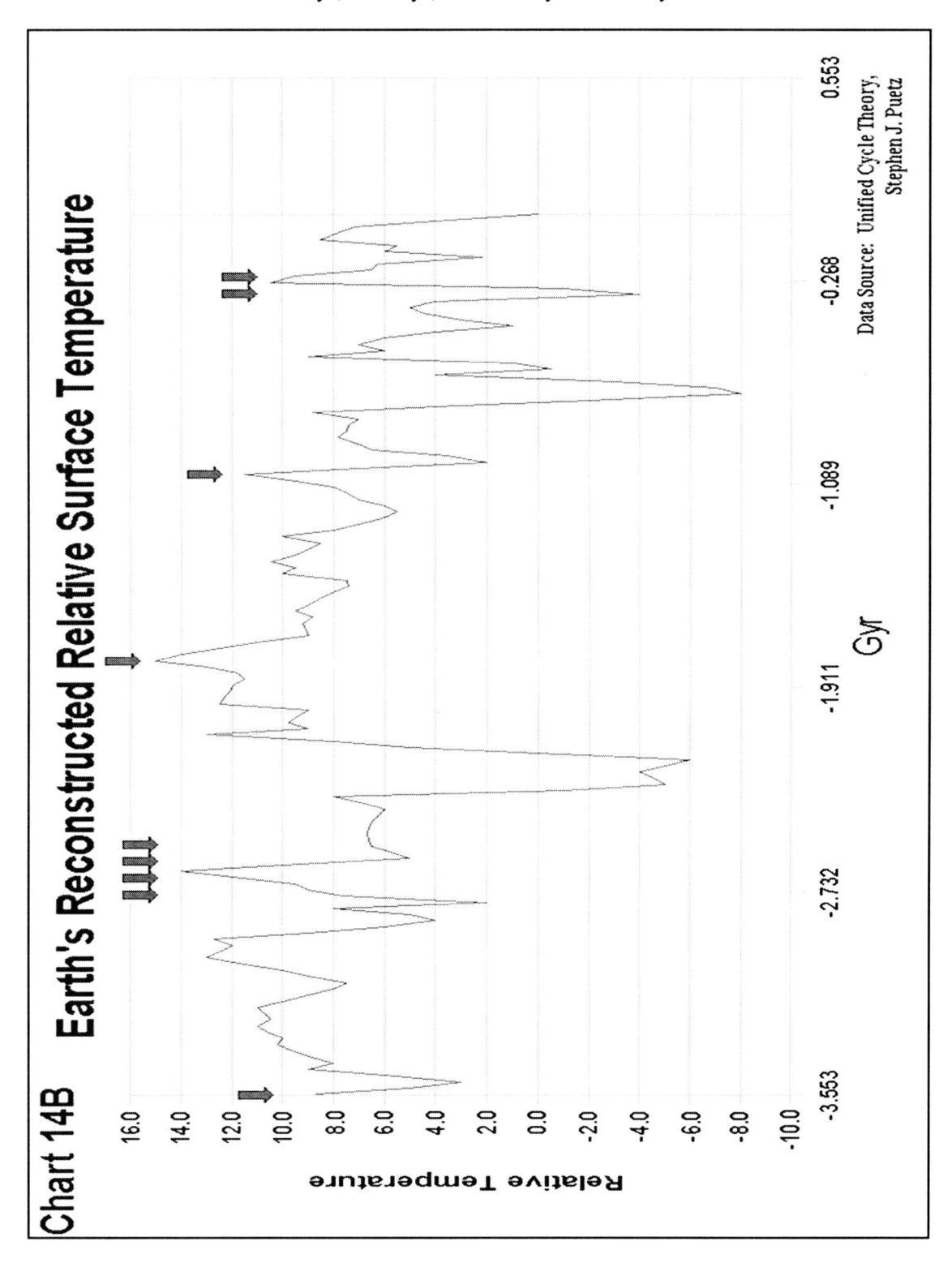

Chart 14B — Earth's Reconstructed Relative Surface Temperature

Data Source: Unified Cycle Theory, Stephen J. Puetz

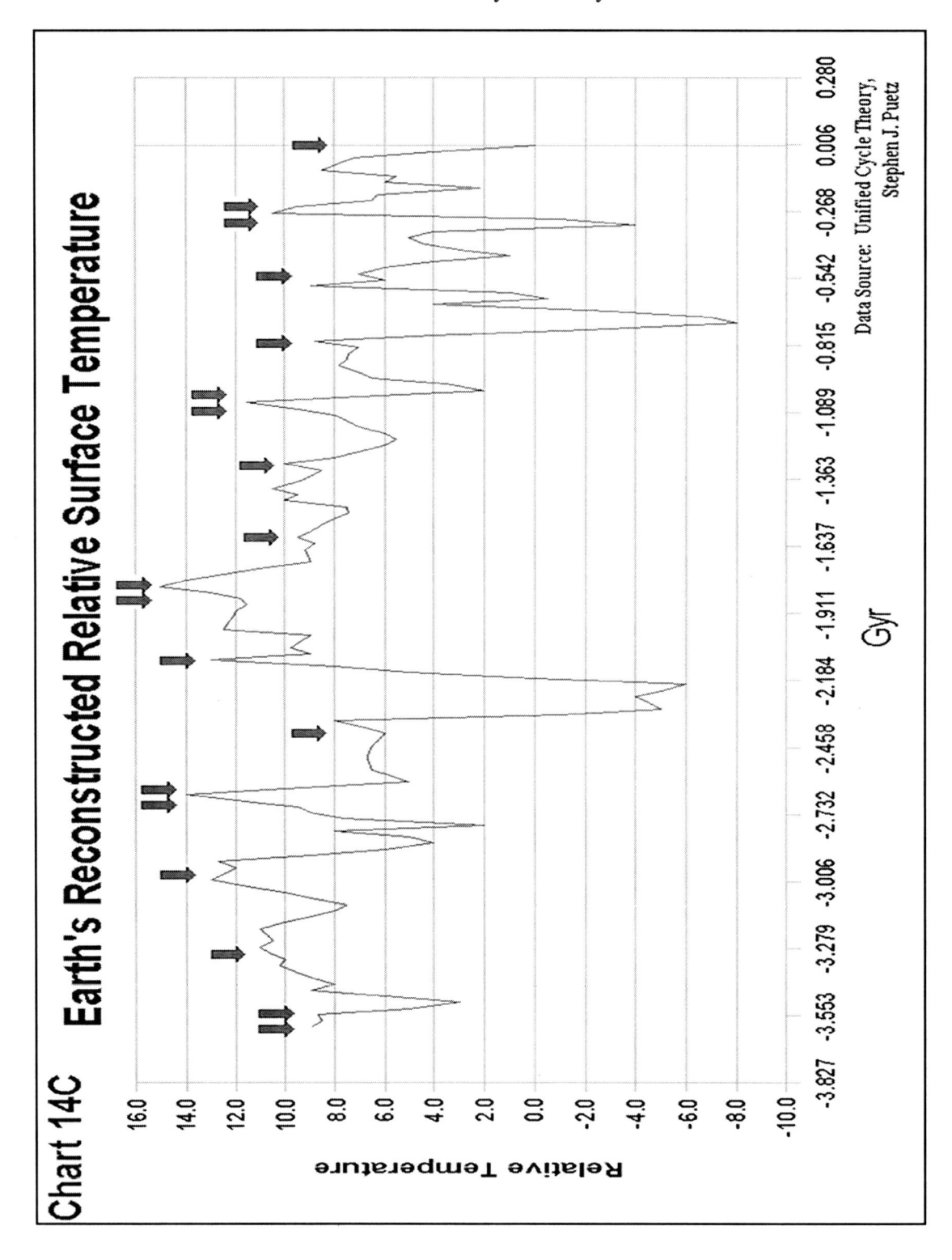

Chart 14C Earth's Reconstructed Relative Surface Temperature

2.46 Gyr, 821 Myr, and 274 Myr EUWS Cycles

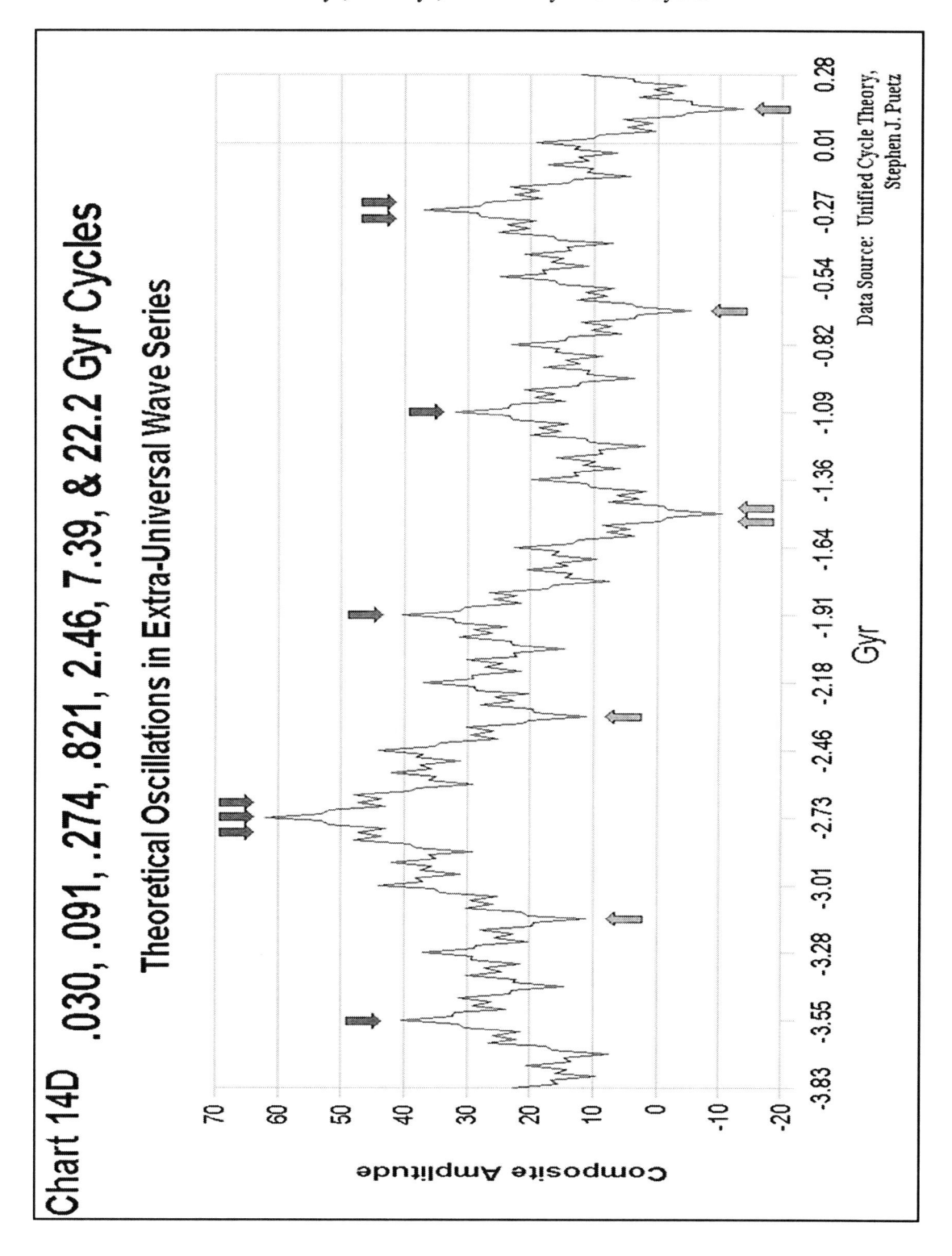

Chart 14D .030, .091, .274, .821, 2.46, 7.39, & 22.2 Gyr Cycles

Theoretical Oscillations in Extra-Universal Wave Series

Data Source: Unified Cycle Theory, Stephen J. Puetz

Chart 14E
Timetable of Earth's Crust Formation

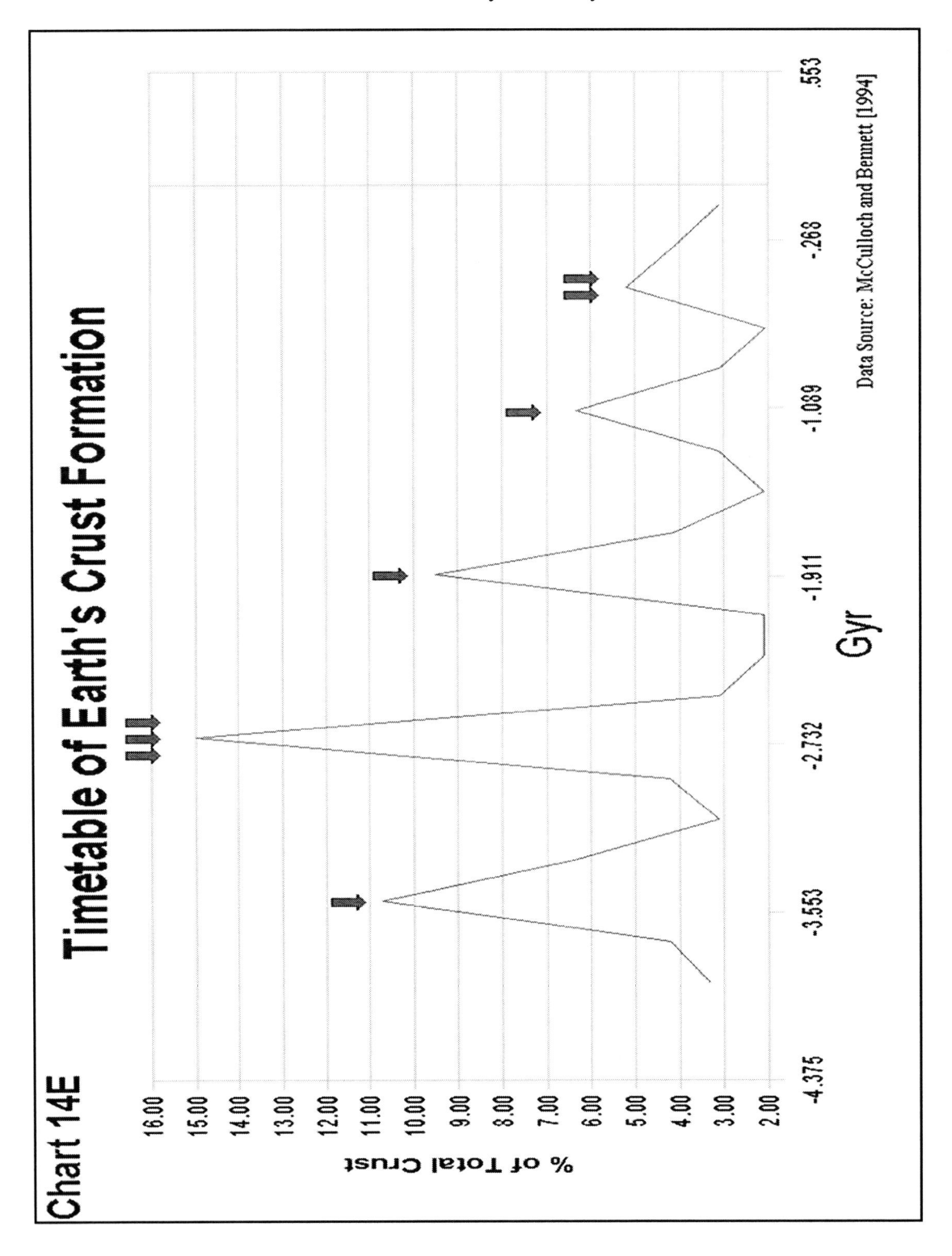

Data Source: McCulloch and Bennett [1994]

2.46 Gyr, 821 Myr, and 274 Myr EUWS Cycles

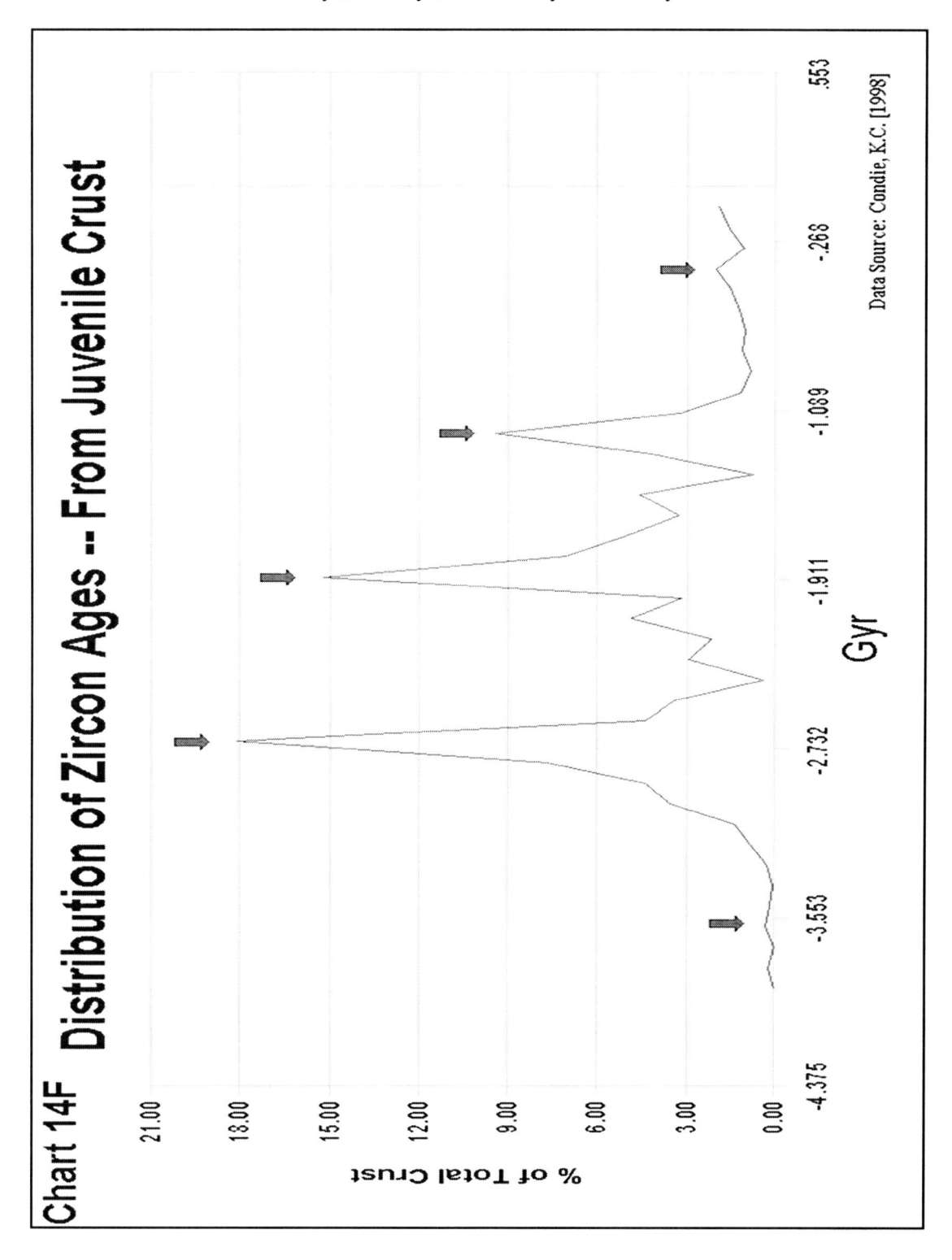

Chart 14F Distribution of Zircon Ages -- From Juvenile Crust

Data Source: Condie, K.C. [1998]

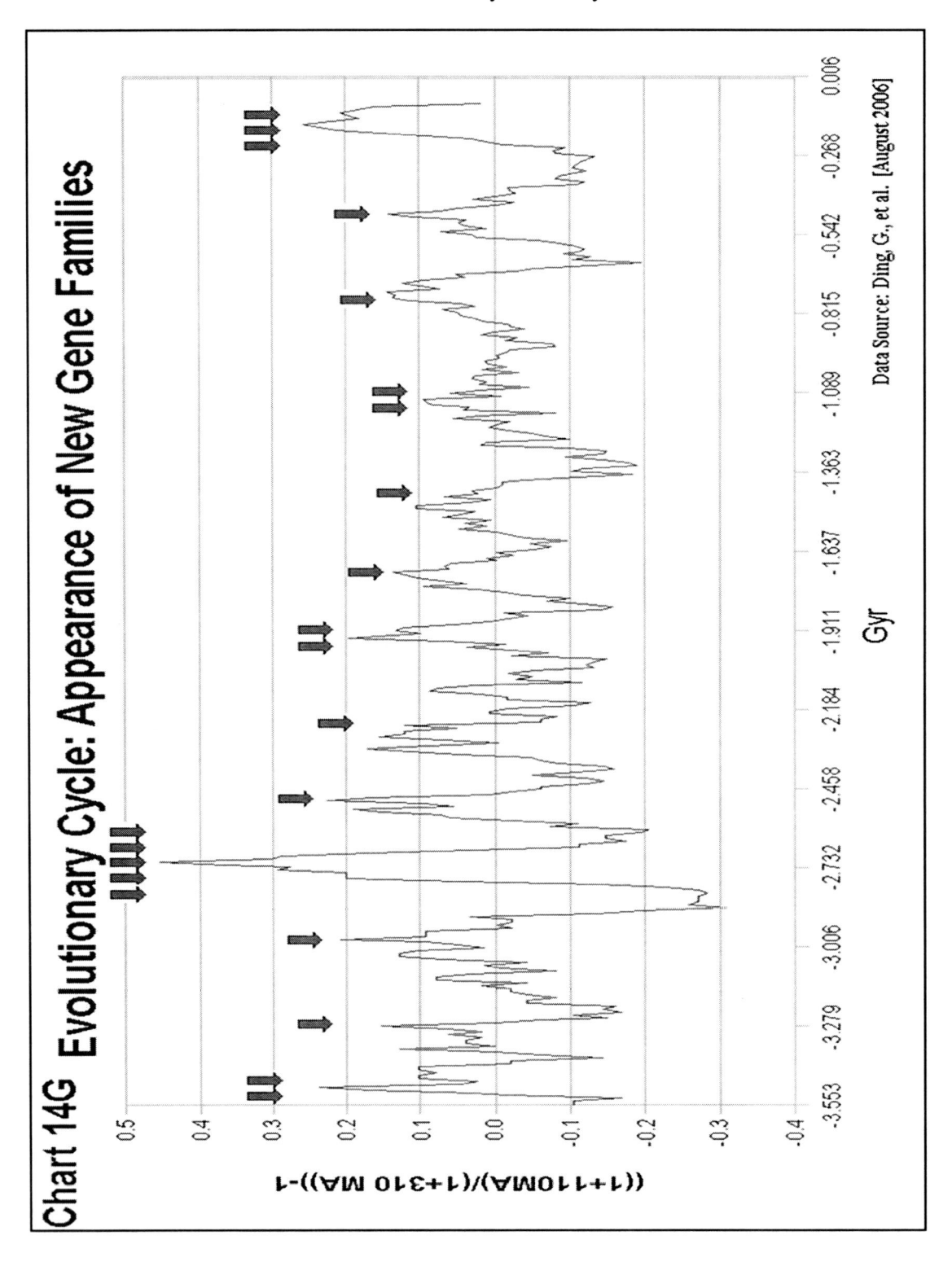

Chart 14G Evolutionary Cycle: Appearance of New Gene Families

Data Source: Ding, G., et al. [August 2006]

2.46 Gyr, 821 Myr, and 274 Myr EUWS Cycles

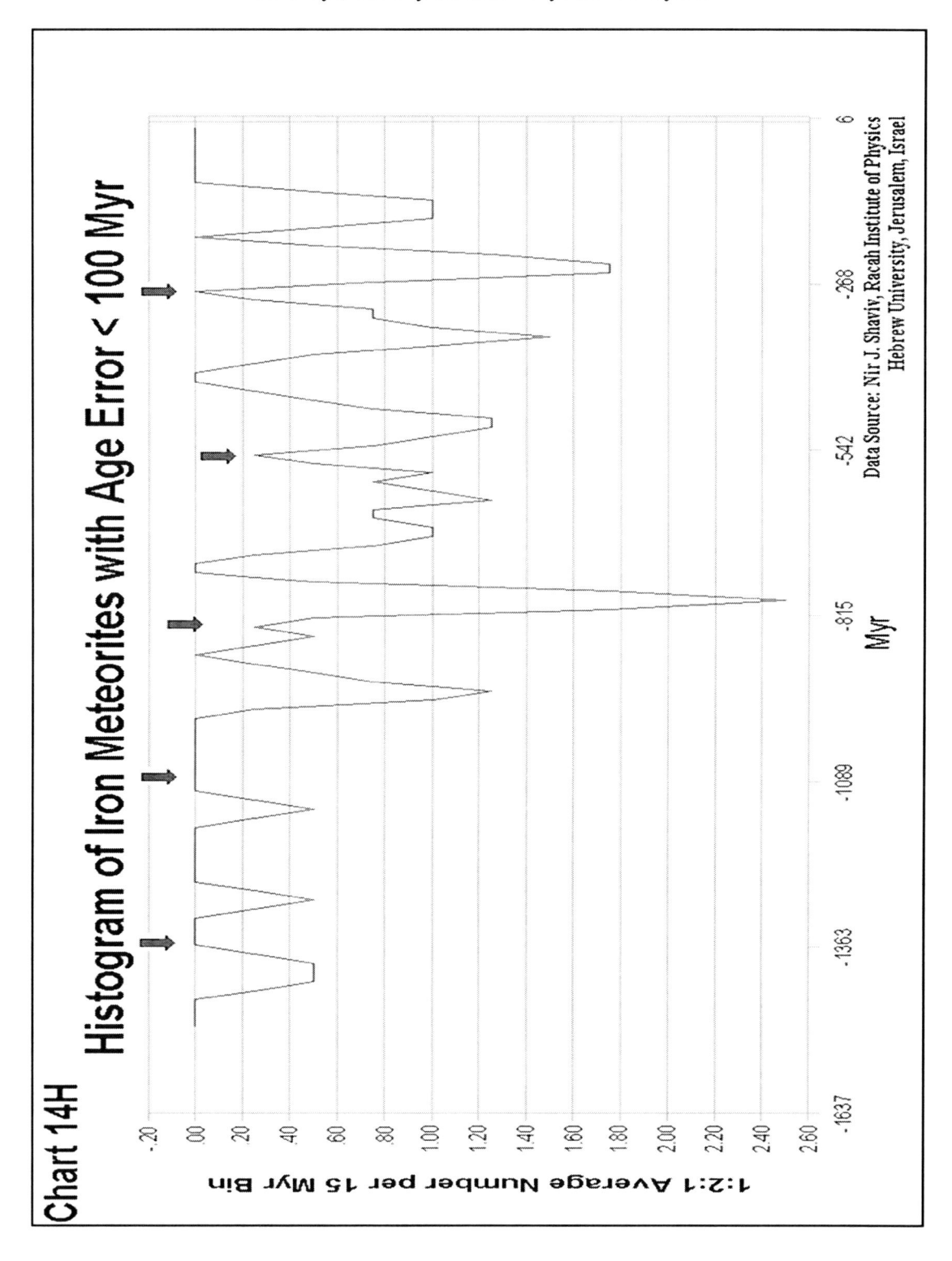

Chart 14H — Histogram of Iron Meteorites with Age Error < 100 Myr

Data Source: Nir J. Shaviv, Racah Institute of Physics, Hebrew University, Jerusalem, Israel

Chapter 15
91.3 Million Year EUWS Cycle

By zooming in further on the EUWS cycles, the 91.3 million year cycle appears next. The most recent peaks of the 91.3-myr cycle came at 85.38, 176.64, 267.90, 359.16, 450.41, and 541.67 Ma. Similar to its parent cycles (and in combination with the 30.4-myr cycle), theoretical peaks of the 91.3-myr cycle closely correspond to the geological timescale. **Table 15.1** shows the correlation. Column 3 marks every third theoretical peak in bold, followed by a double-asterisk – representing 91.3-myr theoretical peaks.

The last 630 million years covered 13 geological periods, 21 peaks of the 30.4-myr cycle, and 7 peaks of the 91.3-myr cycle. For all practical purposes, all 13 of the geological periods matched with a 30.4-myr theoretical peak. The only noteworthy miss came with the 17 million year deviation at the start of the Triassic period.

For the 30.4-myr cycle, 4-of-21 theoretical peaks came within 0.72-myr of actual geological period breakpoints. The chance of an independent event randomly coming within 720 thousand years of a 30.4-myr cycle equals about 5%. Using the probability mass function, the chance of getting 4 near-direct hits in 21 trials, with the probability of a success equaling 5%, comes to about 1.5 chances in a hundred.

Table 15.1 – Geological Periods vs. 91.3-Myr and 30.4-Myr Theoretical Peaks.

Periods	Timescale (Ma)	EUWS Max	Comment
Neogene	23.03 to present	24.54 Ma	Only 1.5 million years from a perfect match.
Paleogene	65.5 to 23.03	54.96 Ma	One 10.14-myr sub-cycle before 54.96 peak.
		85.38 **	91.3-myr peak. No matching period.
		115.80 Ma	No matching period.
Cretaceous	145.5 to 65.5	146.22 Ma	Only .72 million years from a perfect match.
		176.64 **	91.3-myr peak. No matching period.
Jurassic	199.6 to 145.5	207.06 Ma	Missed by two 3.38-myr sub-cycles.
		237.48 Ma	No matching period.
Triassic	251.0 to 199.6	**267.90 ****	Substantial 17-myr deviation.
Permian	299.0 to 251.0	298.32 Ma	Only .68 million years from a perfect match.
Pennsylvanian	318.1 to 299.0	328.74 Ma	One 10.14-myr sub-cycle after 328.74 peak.
Mississippian	359.2 to 318.1	**359.16 ****	Almost a direct hit for the 91.3-myr cycle.
		389.58 Ma	No matching period.
Devonian	416.0 to 359.2	420.0 Ma	Missed by only a 3.38-myr sub-cycle.
Silurian	443.7 to 416.0	**450.41 ****	Missed by two 3.38-myr sub-cycles.
Ordovician	488.3 to 443.7	480.83 Ma	Missed by two 3.38-myr sub-cycles.
		511.25 Ma	No matching period.
Cambrian	542.0 to 488.3	**541.67 ****	Almost a direct hit for the 91.3-myr cycle.
		572.09 Ma	No matching period.
		602.51 Ma	No matching period.
Ediacarian	630.0 to 542.0	**632.93 ****	Missed by only a 3.38-myr sub-cycle.

Turning to the 91.3-myr cycle, 2-of-7 theoretical peaks came within 0.33-myr of actual geological breakpoints. The chances of that randomly happening one time equals only 0.7%. Using the probability mass function, the chances of getting 2 near-direct hits in 7 trials, with the probability of a success equaling 0.7%, equals 1 chance in a thousand.

Probability of hitting four 30.4-myr peaks randomly = $(21!/4!(21-4)!) * (.05^4 * .95^{17}) = 1.5\%$
Probability of hitting two 91.3-myr peaks randomly = $(7!/2!(7-2)!) * (.007^2 * .993^5) = 0.1\%$

Both scenarios indicate a non-random relationship exists between geological timescales and the EUWS cycles. While the geological timescale provides moderately supportive evidence, records from ancient zircons, evolutionary cycles, and global temperatures buttress the significance of the 91.3-myr cycle much further.

Global temperature oscillations during the Phanerozoic eon closely correspond to the 91.3-myr EUWS cycle.[Frakes *et al.*, 1992] See **Chart 15A**. During the past 542 million years, Phanerozoic temperatures closely matched predictions from the EUWS model depicted in **Chart 15B**. The two notable exceptions amounted to 30.4-myr TPD Principle deviations. The projected 404 Ma low arrived about 30.4 million years early, and the projected low at 40 Ma came about 30.4 million years late.

Other than those two deviations, theory mostly matched recorded history. Especially impressive, the highest global temperature during the last billion years came precisely at the 2.46-gyr EUWS theoretical peak at 268 Ma. Once again, fact exactly matched theory. (Notice the 4 down arrows in **Chart 15A**.) Furthermore, the last four theoretical peaks of the 91.3-myr cycle closely coincided with four of the warmest periods on Earth during the last 400 million years!

Using the probability mass function, the chance of getting 4 hits within 10 million years prior to 4 successive 91.3-myr cycles, with the probability of a success equaling 11%, comes to about 1.5 chances in ten thousand.

Probability of hitting four successive 91.3-myr peaks randomly = $(4!/4!) * (.11^4 * .89^0) = 0.015\%$

This calculation indicates that the 91.3-myr cycle and global climate share an exceptionally close link. Looking into the future, the model projects a global warming period ending in 5.88 million years. Following the projected 5.88-myr AP peak, the model calls for a new Snowball Earth episode – as all EUWS cycles between 22.2-gyr and 91.3-myr join forces in falling into a valley of coldness.

In addition to oscillations in temperature, the list of EUWS correlations expands further to include carbon-dioxide cycles. Sprinkled throughout history, paleo-records show evidence of temperatures consistently rising and falling in conjunction with atmospheric carbon-dioxide and the EUWS cycles. In his abstract entitled *CO₂ and Climate: A Geologist's View*, Utrecht University (The Hague, Netherlands) geologist Harry Priem discusses the connection between carbon-dioxide levels and global temperature changes:

> **"How was the alternation between warmer and cooler periods during the last ice-ages related to the atmospheric CO_2 concentration? Air recovered from bubbles in polar ice cores reflects changes both in temperature and in atmospheric CO_2 concentration over the last 250,000 years. The fluctuations in CO_2 concentration appear to track those in temperature to a remarkable degree. However, the temperature changes are much greater than would be expected if changing atmospheric CO_2 content would be the major climate forcing factor. Moreover, closer analysis reveals that the fluctuations in CO_2 generally *lag behind* those in temperature. Only in a few cases the changes are in phase to**

within the time resolution of the ice core data – at least some hundreds of years. Never did the CO_2 changes precede temperature changes. This pattern can be explained by release of CO_2 from ocean water during warming and solution during cooling. A changing atmospheric CO_2 level was thus not the *cause* of the change in temperature, but still could have amplified the temperature change in a positive feedback."[Priem, 1997]

What Priem describes holds true for all timescales. A number of conditions change simultaneously when the EUWS cycles fluctuate. The list includes atmospheric carbon-dioxide concentrations. And the nature of the correlation suggests that carbon-dioxide levels do not provide the stimulus for global temperature changes. At this point, however, it's presumptuous to guess cause-and-effect. As usual, merely note that a correlation exists.

Even though Priem showed that atmospheric CO_2 fluctuations probably don't cause global temperature changes, his research lacks the longevity to identity mega-year cycles in carbon concentrations. Actually, a complex set of factors combine to form the carbon cycle. Throughout Earth's history, methane, carbon-dioxide, water, oxygen, plants, animals, volcanoes, and the oceans all interacted to contribute to the carbon cycle. Shortly after its formation, Earth's atmosphere contained heavy doses of methane (CH_4). During the past 2 billion years, the atmosphere shifted to carbon-dioxide and oxygen. This book will not get into the details of the carbon cycle. Simply note that global carbon fluctuations correspond to EUWS cycles.

In a research effort led by Miriam Katz of Rutgers University, indications of both the 30.4-myr and the 91.3-myr cycle appear in their abstract *Biological Overprint of the Geological Carbon Cycle*. In particular, sharp drops in buried organic carbon (covering time spans of 20 to 30 million years) followed theoretical peaks of the 91.3-myr cycle at 176.6 Ma and 85.4 Ma.[Katz *et al.*, 2005] The Katz study shows (1) organisms have trouble surviving after major theoretical peaks of the EUWS cycles, and (2) EUWS cycles correlate closely with the carbon cycle.

The 91.3-myr cycle also shows its impact on life. In addition to investigating issues related to Milankovitch cycles, Berkeley physicist Richard Muller emerged as a leading expert in the area of mass extinctions and evolutionary changes. Muller recently released his data in the publication *Diversity Data Parsed in 171 Different Ways*.[Muller, 2008] His data shows the 91.3-myr cycle played a major role in the evolution of the phylum porifera – which represents a biological classification of sponges. Sponges first appeared in large numbers at the time of the Cambrian Explosion – shortly after a 91.3-myr theoretical peak at 541.67 Ma. See **Chart 15C**.

After the initial Cambrian burst of life, many sponges soon became extinct. However, approximately every 91.3 million years, new types of sponges developed. A new up-cycle started about 35 million years ago, with the next peak due in 5.88 million years.

Muller's data also reveals the 91.3-myr cycle in the evolution of the phylum Brachiopoda.

91.3 Million Year EUWS Cycle

Phylum Brachiopoda refers to the biological classification for lamp shells. Lamp shells distinguish themselves as two-valved marine animals with an exterior similar to clams. Lamp shells slowly appeared after the Cambrian Explosion, and then reached their heyday near the time of the 2.46-gyr cycle peak, at 268 Ma. See **Chart 15D**. At the 268 Ma peak, the fate of lamp shells suddenly reversed. Approximately 10 million years later, the phylum nearly became extinct. Clearly, something quite dramatic happened to Earth's environment at that time.

After becoming nearly extinct, the number of lamp shell species recovered nicely by the time of the next 91.3-myr peak at 177 Ma. Other maxima in lamp shell species occurred near 91.3-myr theoretical peaks at 450, 359, and 85 Ma.

Of all the 91.3-myr downturns, the one after 268 Ma exerted the greatest impact. A news release from the University of Southern California on October 7, 2008 discusses the proliferation of extinctions following the 268 Ma EUWS peak. Entitled *Extinction by Asteroid a Rarity*, the article relays the findings by a USC Earth scientist, David Bottjer:

> **"In geology as in cancer research, the silver bullet theory always gets the headlines and nearly always turns out to be wrong. For geologists who study mass extinctions, the silver bullet is a giant asteroid plunging to earth. But an asteroid is the prime suspect only in the most recent of five mass extinctions, said USC earth scientist David Bottjer. The cataclysm 65 million years ago wiped out the dinosaurs. 'The other four have not been resolvable to a rock falling out of the sky,' Bottjer said. For example, Bottjer and many others have published studies suggesting that the end-Permian extinction 250 million years ago happened in essence because 'the Earth got sick.'"** [Marziali, 2008]

Perhaps a "sick Earth" best describes the effect that declining EUWS cycles deliver to our planet. The 91.3-myr cycle also appears in the evolution of genes.[Ding *et al.*, 2006] It becomes easily visible after transforming the data from Ding *et al.* with a comparison of 50 and 110 million year moving averages. See **Chart 15E**.

Ding's data series reveals 20 complete cycles spanning 2 billion years. The highest rates of genetic change closely corresponded to theoretical peaks of the 91.3-myr cycle. Additionally, from 1.00 Ga to .54 Ga the TPD Principle came into play. During that time a shift occurred in which each of the evolutionary peaks arrived approximately one 30.4-myr sub-cycle early. After .54 Ga, the evolutionary maxima returned to the time predicted by 91.3-myr theoretical peaks.

Chart 15F shows a 91.3-myr cycle in meteorite strikes. (Note that this cycle contradicts the assumptions previously stated by USC geologist David Bottjer.) This cycle distinguishes itself by the absence of meteorite strikes near its theoretical peaks, while troughs coincide with a large number of meteorite strikes.

Shortly after the Earth formed at 4.54 Ga, a 91.3-myr cycle appeared in the formation of detrital zircons in Western Australia.[Cavosie *et al.*, 2004] See **Chart 15G**. This represents the oldest known geological evidence showing the impact of EUWS cycles. The chart shows maxima in zircon formations closely corresponded to theoretical peaks of the 91.3-myr cycle.

Like its parent cycles, the 91.3-myr cycle left its mark in many ways. These markings include fluctuations in global climate, oscillations in atmospheric carbon, the emergence and extinction of species, the evolution of genetic traits, cycles in meteorite impacts, and changes in the rate of mineral formation.

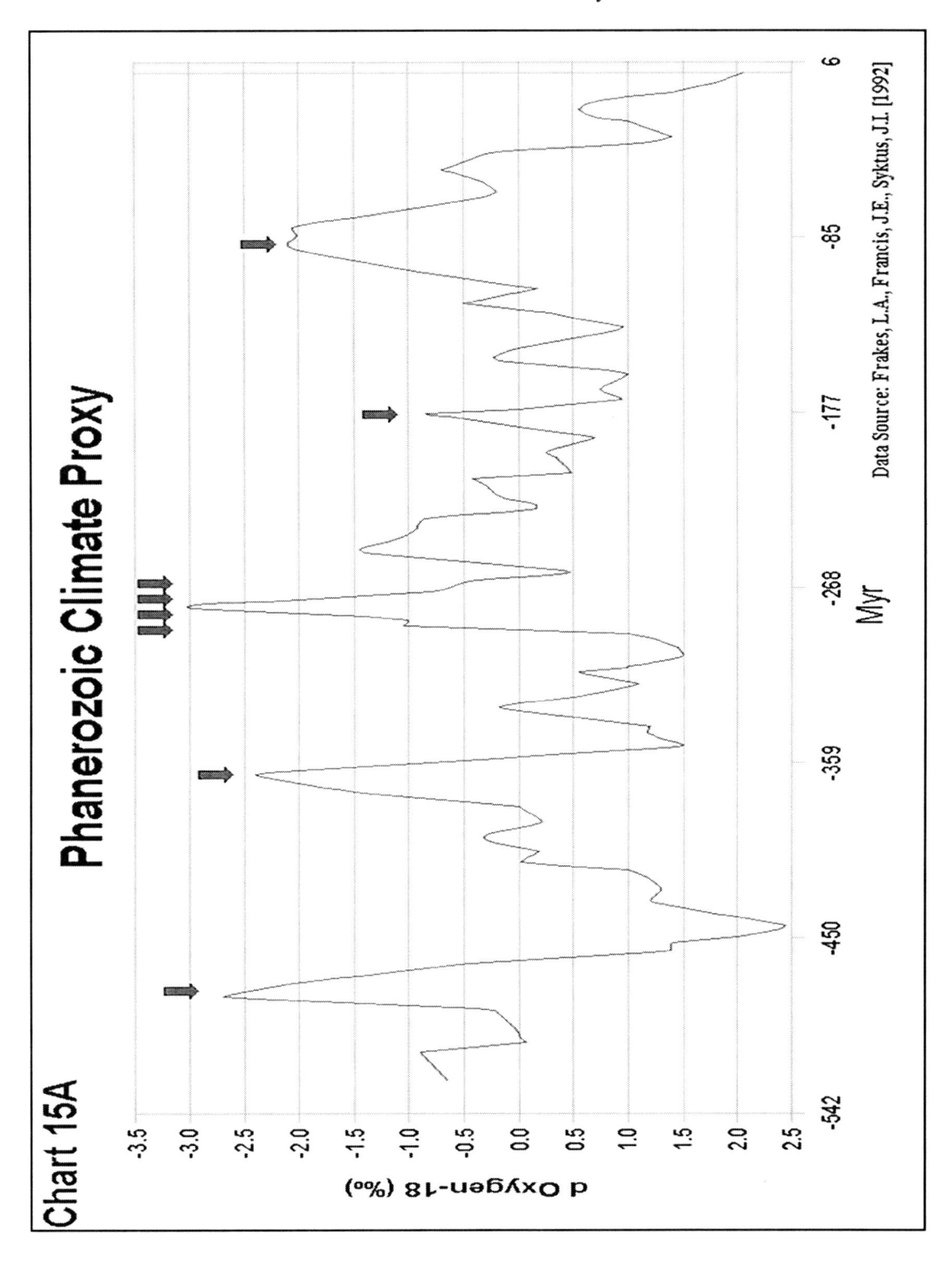

91.3 Million Year EUWS Cycle

Chart 15A — Phanerozoic Climate Proxy

Data Source: Frakes, L.A., Francis, J.E., Syktus, J.I. [1992]

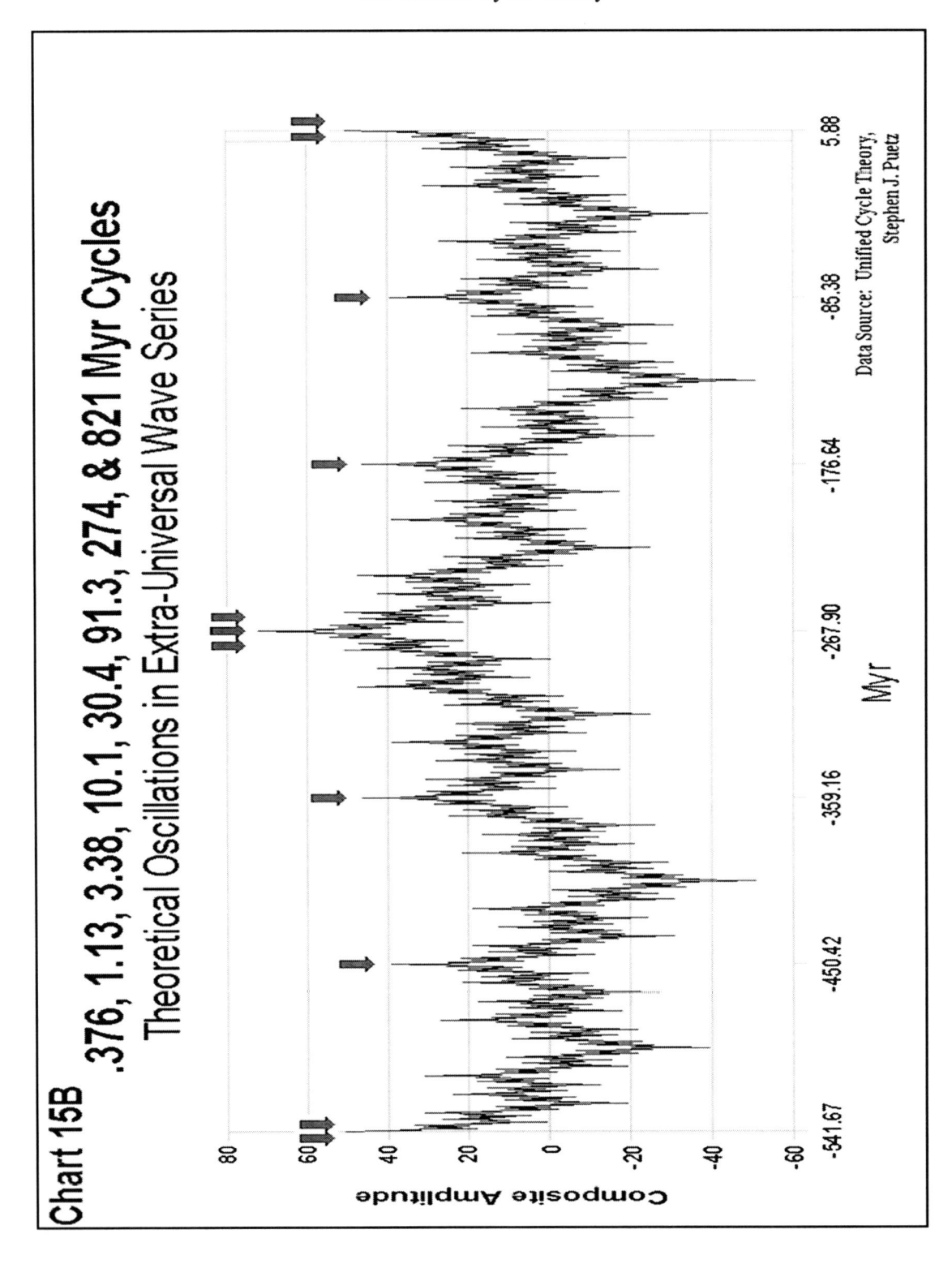

Chart 15B .376, 1.13, 3.38, 10.1, 30.4, 91.3, 274, & 821 Myr Cycles
Theoretical Oscillations in Extra-Universal Wave Series

Data Source: Unified Cycle Theory, Stephen J. Puetz

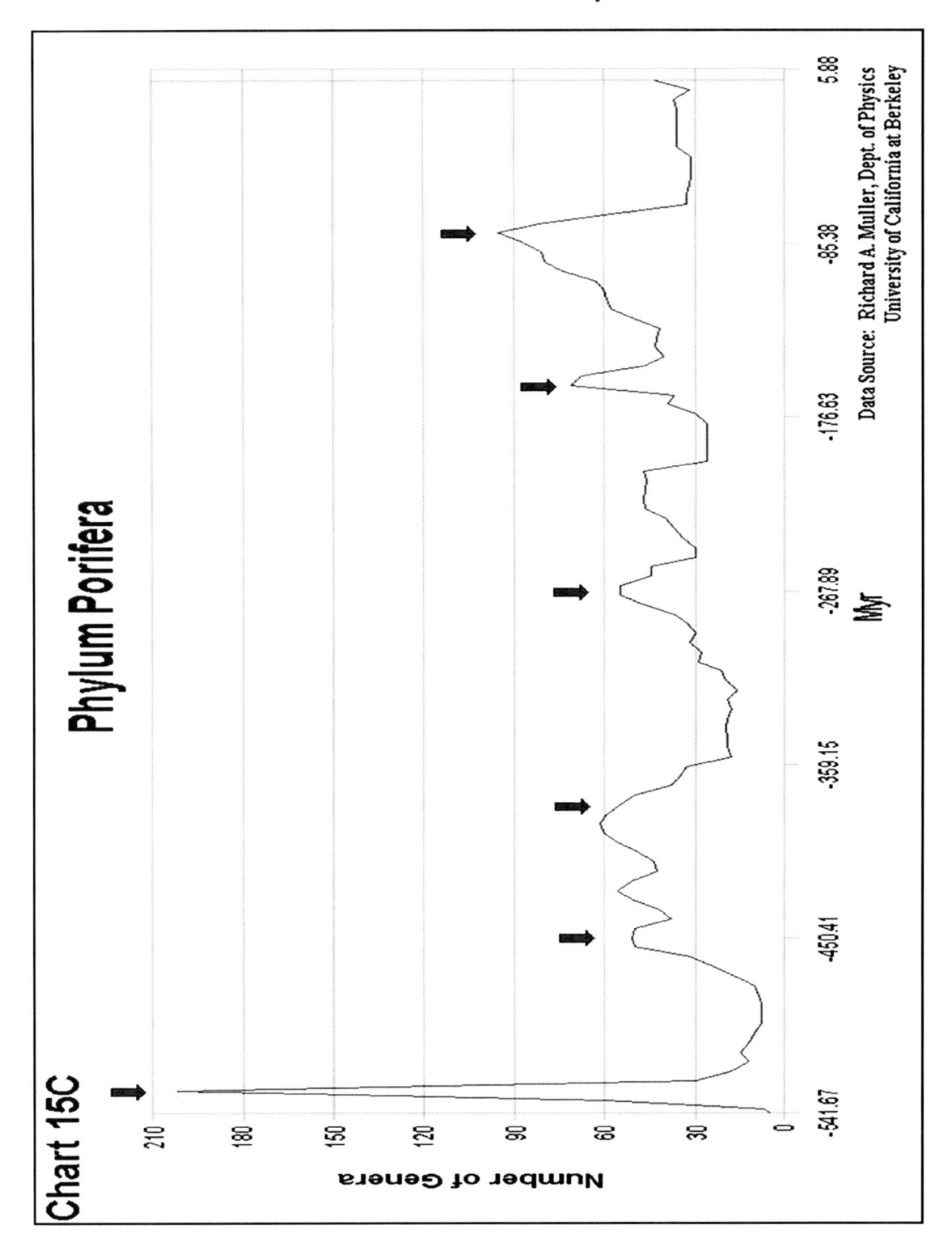

91.3 Million Year EUWS Cycle

Chart 15C — Phylum Porifera

Number of Genera vs. Myr

Data Source: Richard A. Muller, Dept. of Physics, University of California at Berkeley

Chart 15D
Phylum Brachiopoda

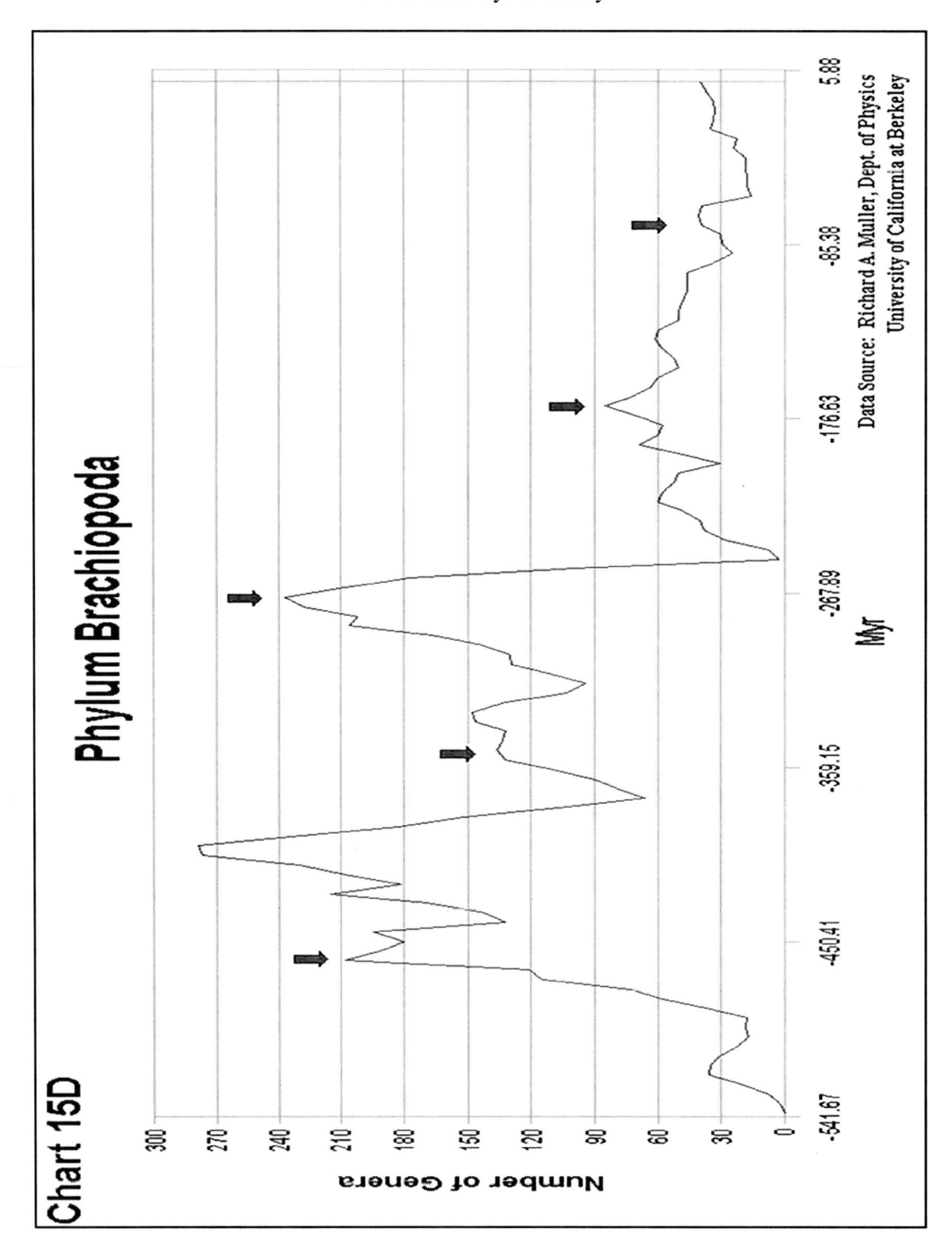

Data Source: Richard A. Muller, Dept. of Physics
University of California at Berkeley

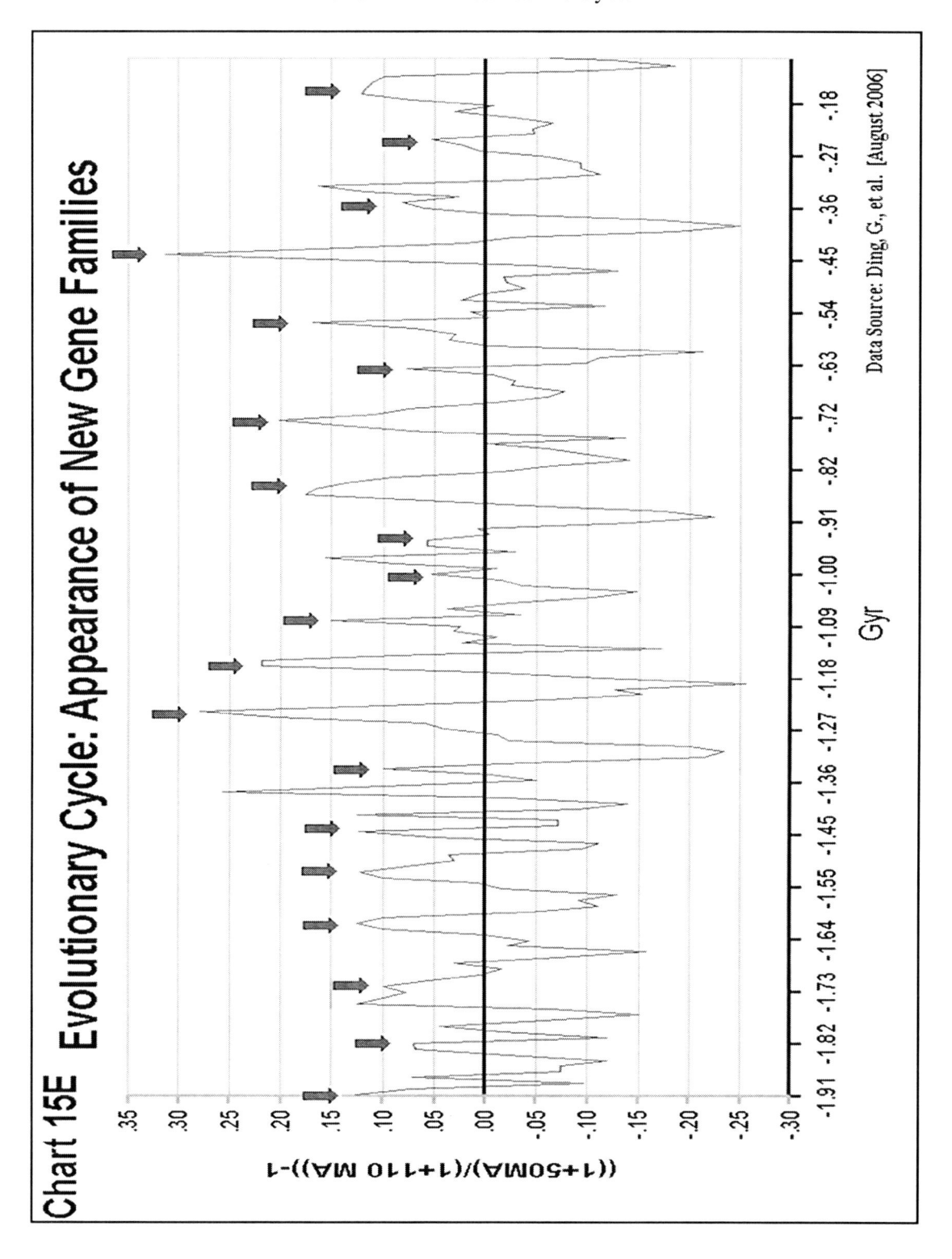

Chart 15E Evolutionary Cycle: Appearance of New Gene Families

91.3 Million Year EUWS Cycle

Data Source: Ding, G., et al. [August 2006]

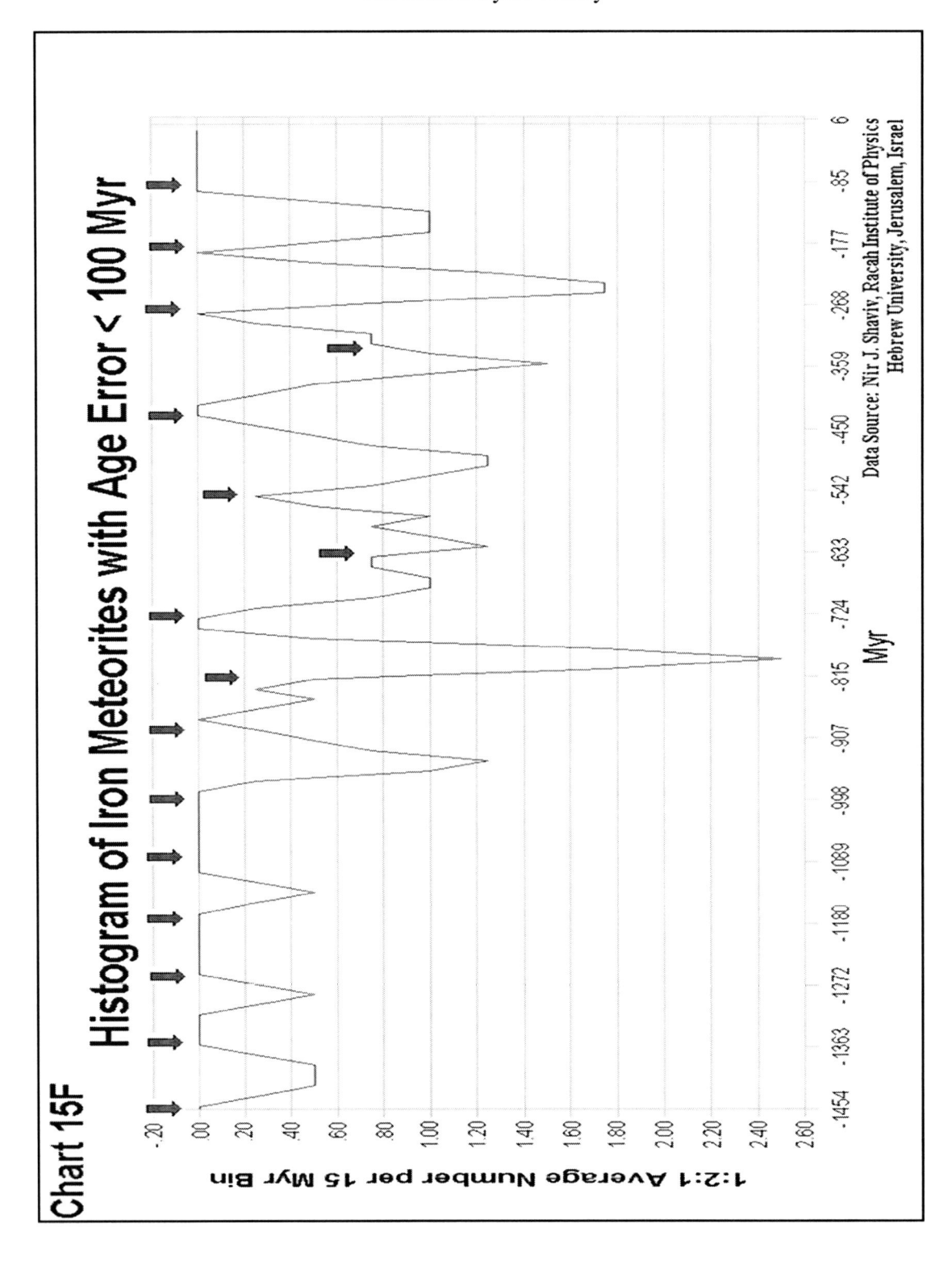

Chart 15F: Histogram of Iron Meteorites with Age Error < 100 Myr

Data Source: Nir J. Shaviv, Racah Institute of Physics, Hebrew University, Jerusalem, Israel

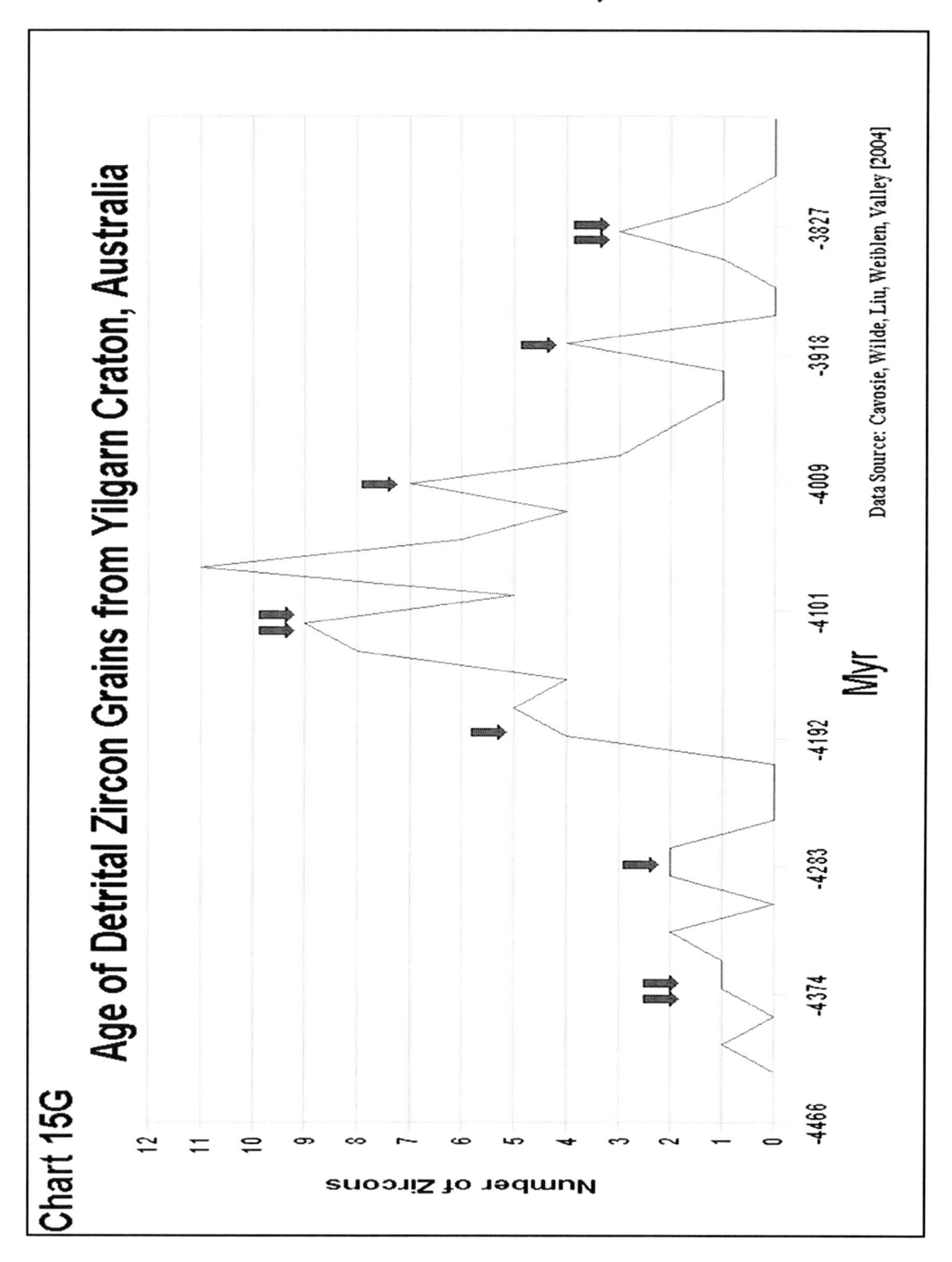

91.3 Million Year EUWS Cycle

Chart 15G: Age of Detrital Zircon Grains from Yilgarn Craton, Australia

Data Source: Cavosie, Wilde, Liu, Weiblen, Valley [2004]

Chapter 16
30.4 Million Year EUWS Cycle

Dividing the 91.3-myr cycle by three, it produces the 30.4-myr EUWS cycle. At the beginning of the prior chapter, **Table 15.1** showed how theoretical peaks of the 30.4-myr cycles closely matched geological periods. The 30.4-myr cycle also appears in the climate-proxy from Frakes *et al*. See **Chart 16A**. Amazingly, the chart shows 17 distinct climate peaks occurring through 17 successive occurrences of the 30.4-myr cycle!

Chart 16B combines all of the EUWS cycles between 376 thousand years and 821 million years into a composite model projecting how global temperatures should have behaved during the last 542 million years. The model follows Phanerozoic temperatures closely – depicted in **Chart 16A**. That's especially true after 420 Ma.

The Phanerozoic climate history from Frakes *et al*. provides value with its duration. However, to achieve better accuracy, a shorter but more precise 70 million year history comes from a research team led by University of California Santa Cruz paleoclimatologist James Zachos. See **Chart 16C**. During that 70 million year period, global temperatures showed two distinctive crests – both times almost perfectly matching theoretical peaks of the 30.4-myr EUWS cycles at 54.96 Ga and 24.54 Ga!

Both cycles replicate the pattern often seen at EUWS peaks – a sudden, dramatic rise to the top followed by a quick reversal downward. The Zachos team leaves out intricate details regarding the Late Oligocene Warming around 24.54 Ma. However, for the Late Paleocene Thermal Maximum, they beautifully describe the magnitude of the environmental shock:

"The most prominent of the climatic aberrations is the Late Paleocene Thermal Maximum, which occurred at 55 Ma near the Paleocene/Eocene boundary. This event is characterized by a 5° to 6°C rise in deep-sea temperature ... in less than 10 kyr. Sea surface temperatures as constrained by planktonic isotope records also increased, by as much as 8°C at high latitudes."[Zachos *et al.*, 2001]

Later in the abstract, Zachos' team zooms in further to more precisely identify the timing of the Late Paleocene Thermal Maximum:

"The time scale is based on the cycle stratigraphy of Site 690 ... with the base of the excursion placed at 54.95 Ma."[Zachos *et al.*, 2001]

Quite remarkably, this estimate for the Late Paleocene Thermal Maximum matches the corresponding 30.4-myr theoretical peak to four digits – with the theoretical EUWS peak calculated at 54.959 Ma. As already seen in other chapters, the number of direct hits between theory and actual data comes with unexpectedly high regularity.

The 30.4-myr cycle exhibits the same diversity as its ancestral cycles. However, each cycle holds its own distinguishing characteristic. The 22.2-gyr cycle makes an impact on big bangs. The 7.39-gyr cycle coincides with catastrophic events on Earth. The 2.46-gyr cycle regulates mantle plume events. The 821-myr cycle modulates juvenile crust formation. The 91.3-myr cycle coincides with global temperature cycles. And the 30.4-myr cycle governs variations in evolution and extinctions.

After discovering the original core set of six cycles, the 30.4-myr cycle caught my attention next. Even though it fit into the harmonic sequence established by the core set, a large gap of missing cycles sat in between. Furthermore, researchers labeled this repetition as an extinction cycle (as opposed to an evolutionary cycle). And the reasons given for extinctions varied widely. The list included:

- Solar system movements through the Milky Way's galactic plane.
- Solar system periodic movement through the Milky Way's spiral arms.
- Periodic episodes of volcanic eruptions.
- Continental drift mechanics.
- Gamma ray bursts from Supernova explosions.
- Hydrogen sulfide emissions from the oceans.
- Methane gas emissions.
- Global warming and cooling cycles.
- Sea level changes.
- Asteroid impacts.

30.4 Million Year EUWS Cycle

With so many seemingly independent contributing factors, at first glance, the extinction cycle appeared to result from mere chance. As a result, ideas about the EUWS harmonic sequence extending beyond a few thousand years were discarded. However, after discovering most of the missing cycles in between, the 30.4-myr cycle came to the forefront again. But the second time around, with a better understanding of how these oscillations function, the analysis changed to investigating it as an evolution-extinction cycle – instead of a cycle involving only extinctions.

The original research on this subject concentrated on the extinction side of the equation. Thirty years ago, Arthur Fischer and Michael Arthur started the inquiry with their publication *Secular Variations in the Pelagic Realm*. They suggested a 32-myr cycle in mass extinctions based on a 250 million year study of fossil records.[Fischer and Arthur, 1977]

About a decade later, two geologists from the University of Chicago, David Raup and John Sepkoski, Jr., came along to extend the investigation. In their abstract entitled *Periodicity of Extinctions in the Geologic Past*, the duo notes the important contributions from Fischer and Arthur, as well as enhancements to the original study:

> "Virtually all species of animals and plants that have ever lived are now extinct, and the known fossil record documents some 200,000 such extinctions. It has been generally assumed that extinction is a continuous process in the sense that species are always at risk and that mass extinctions simply reflect relatively short-term increases in that risk. Following this view, the extinction process is often described mathematically as a time homogeneous process using standard birth-death models.... There is increasing evidence, however, that many extinctions are actually short-lived events of special stress, separated by periods of much lower, or even negligible, risk. Fischer and Arthur departed from convention by arguing that the major extinction events of the past 250 million years occurred periodically at nearly constant intervals of 32 Ma.... Their study used a limited data base, and no statistical testing was done. The purpose of this paper, therefore, is to test the proposition of periodicity in the record of marine extinctions over the past 250 Ma ... by using as rigorous a methodology as present data permit."[Raup and Sepkoski, 1984]

Raup and Sepkoski isolated fossils of marine vertebrates, invertebrates, and protozoans for their research. They came to the conclusion that a 26 million year frequency regulated extinction events. However, they also show the fickle nature of statistical analysis. While deciding that a frequency of 26-myr held the greatest significance, their tests repeatedly revealed a 30-Myr cycle. However, they concluded that more data was needed to validate the 30-Myr cycle:

> "This spectrum shows a peak in power near the first harmonic (reflecting the high extinction percentages near the two ends of the time series) and a pronounced peak at the eighth harmonic, suggesting a periodicity in the neighborhood of 30 million years.... The Fourier results were corroborated by a standard autocorrelation analysis. The correlogram showed statistically

significant autocorrelation (P < 0.01) for cycles between 27 and 35 ma. However, we do not consider this conclusive because autocorrelation gives undue weight to the long intervals of background extinction between peaks."[Raup and Sepkoski, 1984]

Raup and Sepkoski conclude:

"The significant cycle at ... 30 should be interpreted only as suggestions for future exploration when stronger data bases are available. The cycle at 30 Ma may be real but cannot be confirmed with the present time series."[Raup and Sepkoski, 1984]

A year later, a pair of physicists from the University of Louisiana at Lafayette, John Matese and Daniel Whitmire, observed the 30 million year period in extinctions, and they theorized that an undiscovered "Planet X" periodically disturbed the Kuiper Belt – consequently causing meteorites to bombard Earth.[Whitmire and Matese, 1985] However, this theory has been mostly discredited.

More recently, in their abstract entitled *Cycles in Fossil Diversity*, Robert Rohde and Richard Muller determined a frequency of 62 million years by studying fossils since the start of the Cambrian Explosion. Although, they properly noted the 30-myr cycle:

"Several authors have reported cycles of 26–32-myr in diversity or extinctions. Although the 62-myr cycle is the dominant cycle in Sepkoski's diversity data, the existence of secondary features in the middle of some cycles (Silurian, Upper Carboniferous, Lower Jurassic and Eocene) might have influenced previous reports of this ≈ 30-myr cyclicity."[Rohde and Muller, 2005]

Almost certainly, Rohde and Muller observed conflicting amplitudes related to the interplay between the 91.3-myr and 30.4-myr EUWS cycles. The midpoint between these two cycles acts as a naturally occurring secondary harmonic. At 60.84-myr, this secondary EUWS harmonic closely matches the 62-myr estimate of Rohde and Muller.

While relatively rare events such as gamma ray bursts and asteroid impacts provide possible explanations for a few mass extinctions, they don't fit well as the ultimate causes of these cycles. For example, in the abstract *Extraterrestrial Cause for the Cretaceous-Tertiary Extinction*, a team led by Luis Alvarez showed how a tremendous asteroid strike hit the Yucatan Peninsula 65 Ma. This catastrophic impact created a sudden climate change that lead to the extinction of dinosaurs, along with many other species.[Alvarez et al., 1980]

Yet, by collecting a large set of data on the existence of various sub-species, Muller showed that evolutionary patterns persist in spite of catastrophic external events. While fluctuations in meteorite occurrences do follow EUWS patterns, meteorites don't appear to act as the primary cause of extinction cycles. Muller's data shows that the evolution-extinction cycle peaked at 85

Ma. And it was already in a long-term downtrend when the Yucatan asteroid strike hit at 65 Ma. Hence, the Yucatan strike amplified the number of extinctions already taking place. But the asteroid strike certainly wasn't something that started the Late-Cretaceous extinctions.

In an abstract entitled *Evolution as a Self-Organized Critical Phenomenon*, an international research team led by Princeton University physicist Kim Sneppen gives a plausible explanation of how the evolution-extinction cycle operates. Sneppen's team more accurately describes this oscillation as a macro-evolutionary cycle:

> **"The fundamental cause of evolutionary change is explained by Darwin's theory which locates it to natural selection operating by struggle among individual organisms for reproductive success. Darwin's theory may thus be thought of as the 'atomic theory' for evolution. However, there is no theory deriving the consequences of Darwin's principles for macroevolution. This is the challenge to which we are responding.... We present a simple mathematical model of biological macroevolution. The model describes an ecology of adapting, interacting species. The environment of any given species is affected by other evolving species; hence, it is not constant in time. The ecology as a whole evolves to a self-organized critical state where periods of stasis alternate with avalanches of causally connected evolutionary changes. This characteristic behavior of natural history, known as punctuated equilibrium, thus finds a theoretical explanation as a self-organized critical phenomenon. The evolutionary behavior of single species is intermittent. Also, large bursts of apparently simultaneous evolutionary activity require no external cause. Extinctions of all sizes, including mass extinctions, may be a simple consequence of ecosystem dynamics."**[Sneppen *et al.*, 1995]

While disagreeing with the Sneppen-team assessment that macro-evolution requires no external cause, their concept of a fluctuating ecosystem seems on target. Pinpointing the timing of these ecosystem dynamics, EUWS cycles consistently trigger the fluctuations. Organisms continually respond to their changing environment. Natural selection permits some species to mutate positively, allowing descendants to survive modified ecosystems. Other species are already adapted for a wide range of environments, and they survive as well. Others fail to adapt, and become extinct. In this way, EUWS cycles modulate the macro-evolutionary environment, as suggested by Sneppen *et al*.

As an example of how EUWS cycles affect macro-evolution, study **Chart 16D** depicting Muller's data for genera existing less than 45 Myr. Notice how the Phanerozoic climate in **Chart 16A** moderately correlates with the fluctuations in **Chart 16D**. The inflection points and amplitudes match in both charts. Double arrows mark 91.3-myr peaks, while single arrows denote 30.4-myr peaks.

While Muller's 62-myr oscillation serves as a good estimate, its true frequency equals 60.8-myr. As stated previously, the 60.8-myr extinction cycle functions as a secondary harmonic – making it less important when viewed from the perspective of the entire EUWS spectrum.

Within the biological classification system, the Cephalopod class includes mollusks such as octopuses, squid, cuttlefish, and nautilus. The 30.4-myr EUWS cycle makes an especially strong imprint on fluctuations of evolution-extinction for Cephalopods. See **Chart 16E**.

In conclusion, the 30.4-myr cycle shares the same traits as its EUWS ancestors in previous chapters. However, its evolutionary-extinction influence stands out as its distinguishing feature.

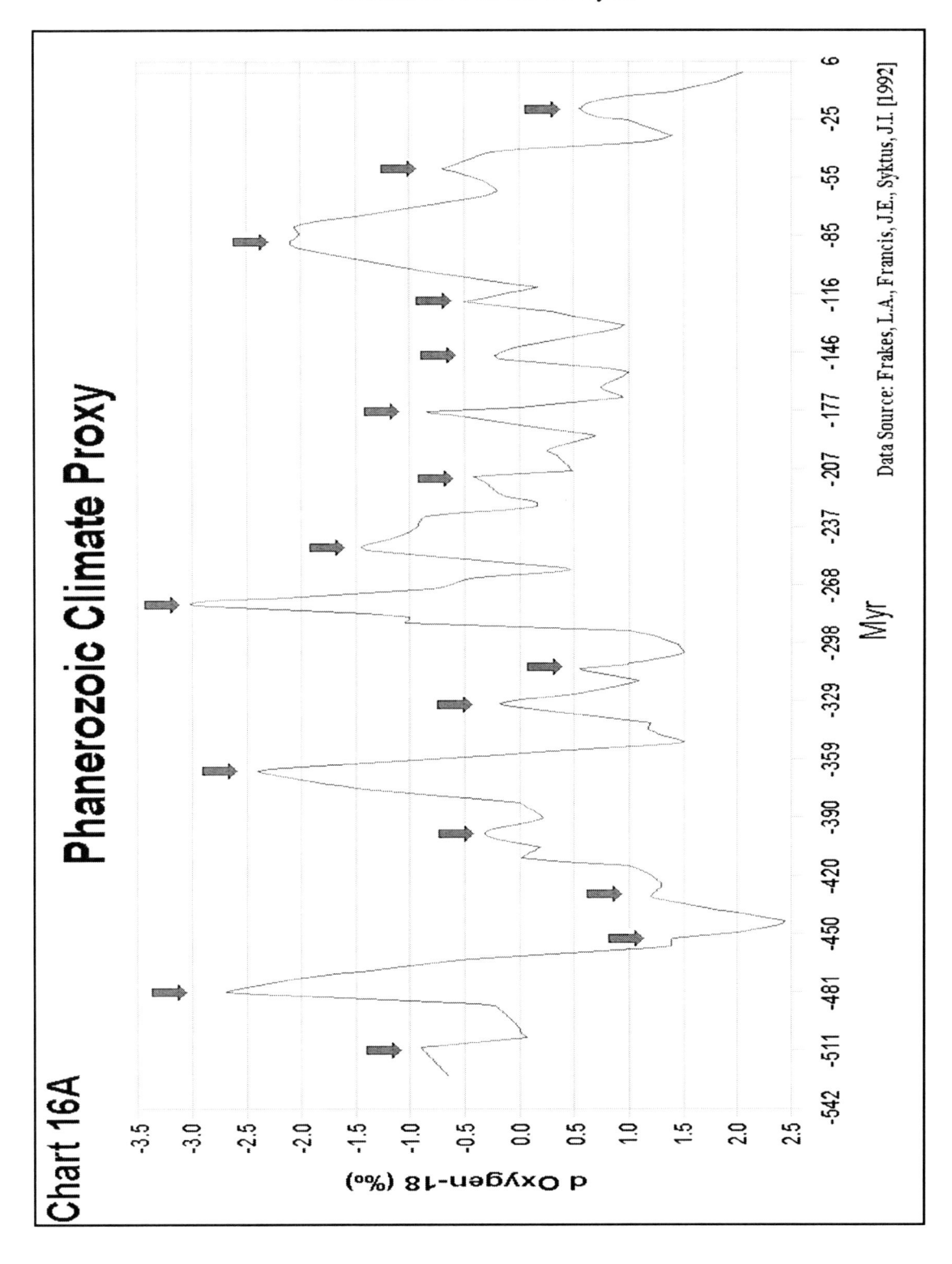

30.4 Million Year EUWS Cycle

Chart 16A — Phanerozoic Climate Proxy

Data Source: Frakes, L.A, Francis, J.E, Syktus, J.I. [1992]

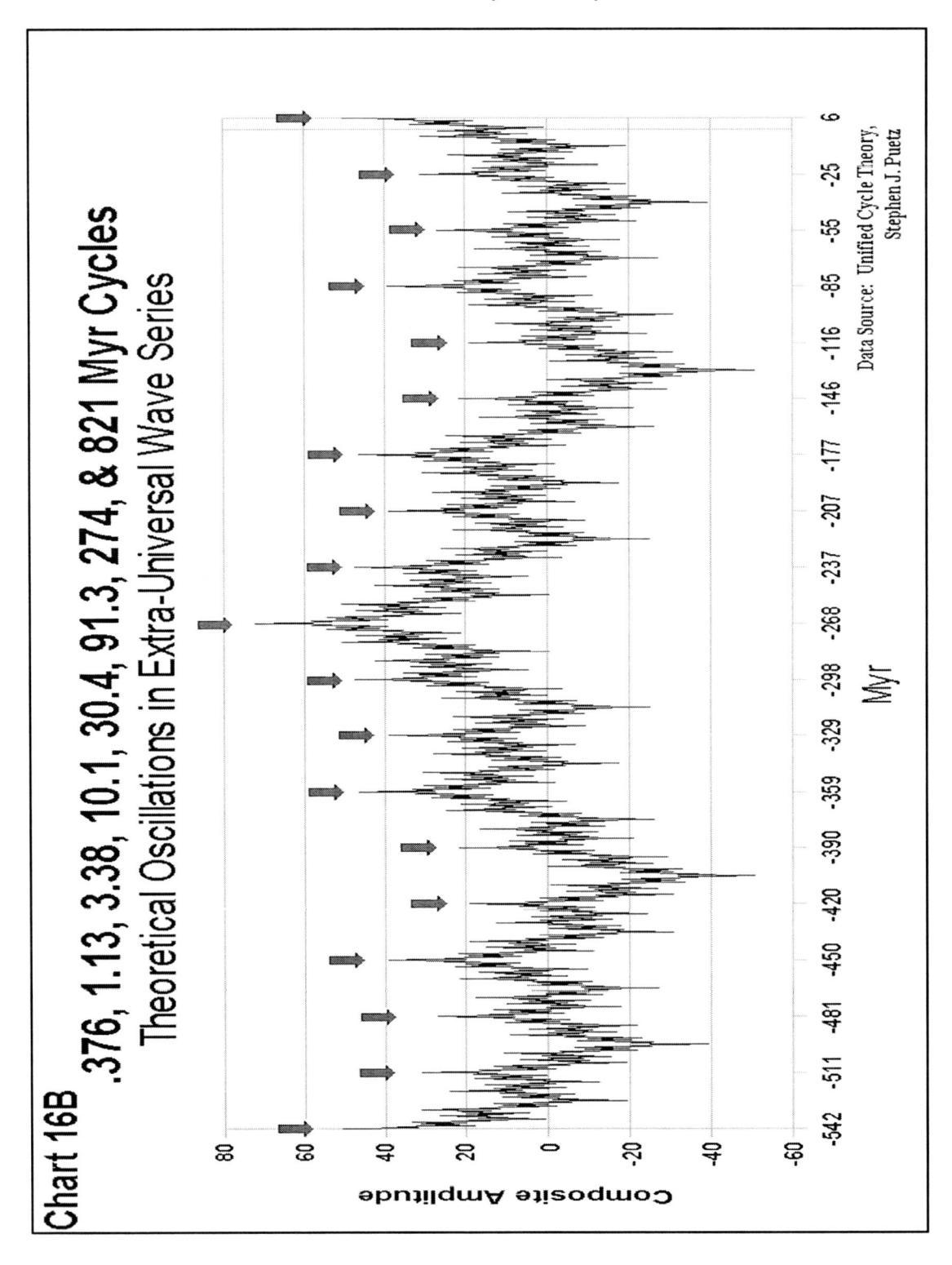

Chart 16B
.376, 1.13, 3.38, 10.1, 30.4, 91.3, 274, & 821 Myr Cycles
Theoretical Oscillations in Extra-Universal Wave Series

Data Source: Unified Cycle Theory, Stephen J. Puetz

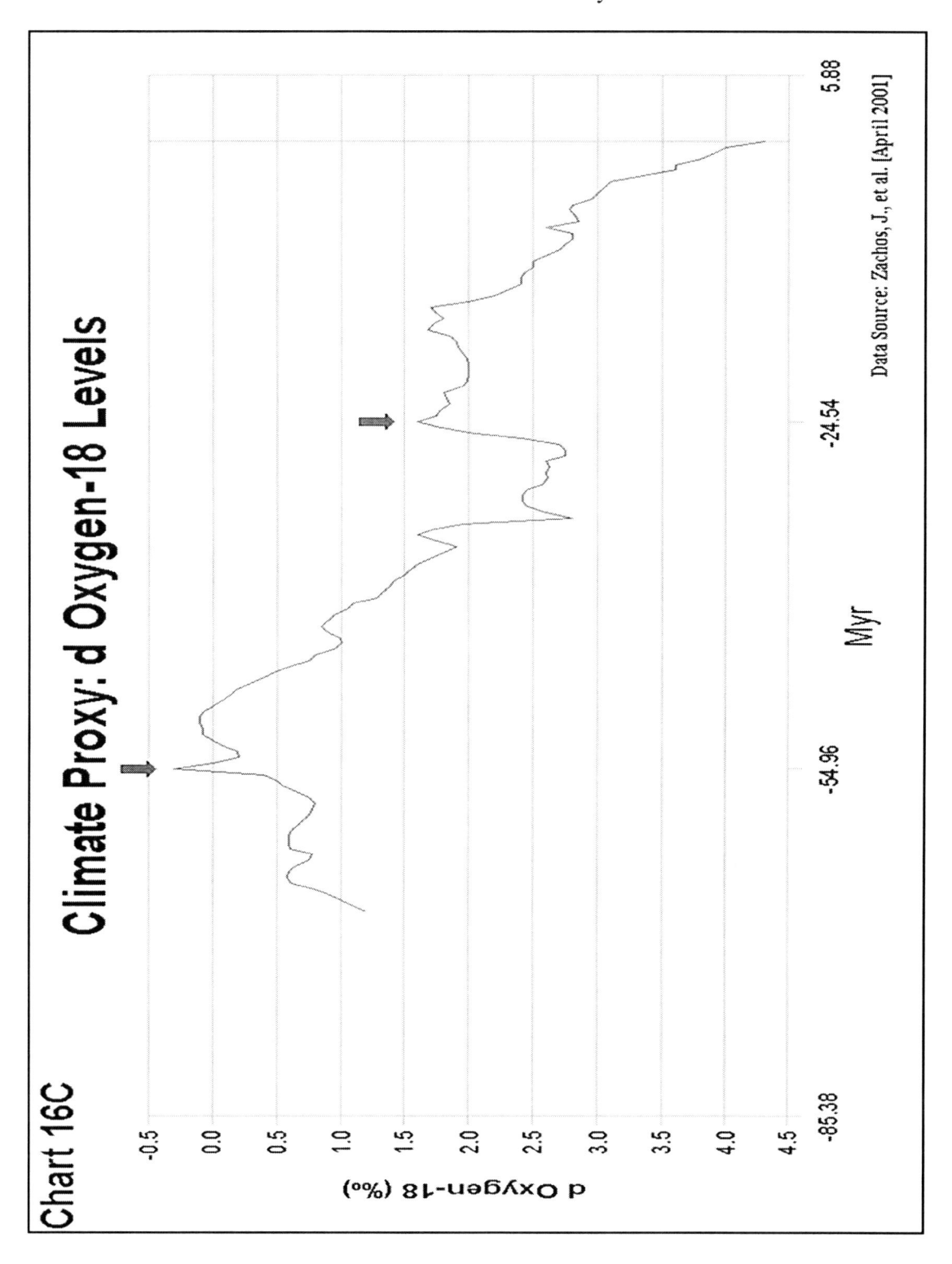

30.4 Million Year EUWS Cycle

Chart 16C — Climate Proxy: d Oxygen-18 Levels

X-axis: Myr (−85.38, −54.96, −24.54, 5.88)
Y-axis: d Oxygen-18 (‰), range −0.5 to 4.5

Data Source: Zachos, J., et al. [April 2001]

Chart 16D

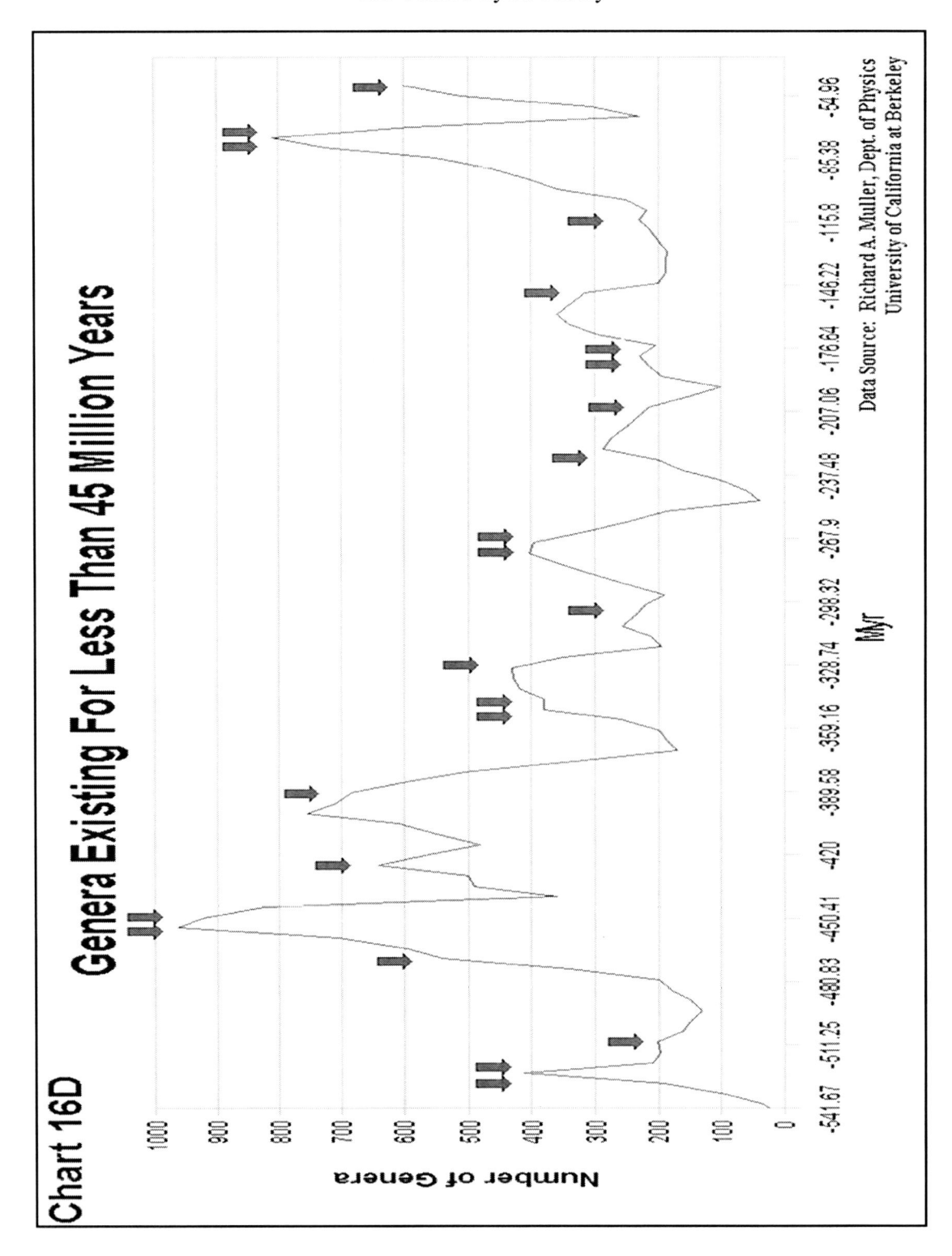

Genera Existing For Less Than 45 Million Years

Data Source: Richard A. Muller, Dept. of Physics, University of California at Berkeley

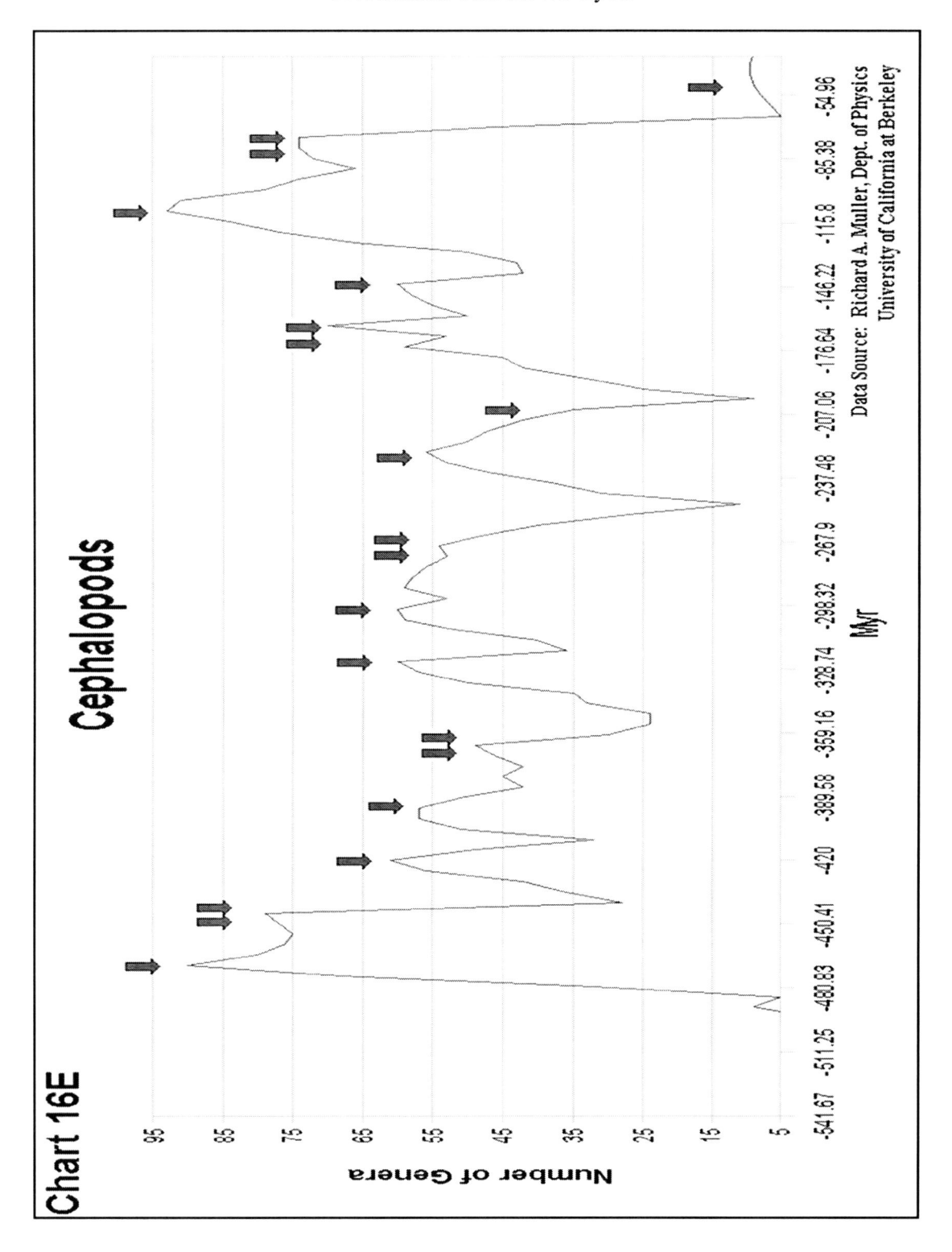

30.4 Million Year EUWS Cycle

Chart 16E — Cephalopods

Data Source: Richard A. Muller, Dept. of Physics
University of California at Berkeley

Chapter 17
10.1 Million Year EUWS Cycle

Dividing by three, the 30.4-myr cycle gives birth to its 10.1-myr offspring. And similar to its ancestors, the 10.1-myr cycle corresponds to the geological timescale. In this particular case, the 10.1-myr peaks correspond to geological epochs. See **Table 17.1**.

Throughout the Phanerozoic eon, all but three epoch-breakpoints closely correspond to 10.1-myr peaks, with an unusually large number of those also equaling 30.4-myr peaks. (The table marks 30.4-myr peaks in bold, followed by a double asterisk.) Looking at the 32 epochs in more detail, 11 of the 32 corresponded to 30.4-myr EUWS theoretical peaks, 18 of the 32 correlated with 10.1-myr EUWS theoretical peaks, while 3 epochs failed to match either of these EUWS maxima.

Similar to other EUWS cycles, the geological timescale provide interesting matchups that certainly do not happen by chance. However, climate cycles provide undeniable evidence of the link. By zooming in on the climate proxy from Zachos *et al.* (also used in Chapter 16), six of the seven 10.1-myr EUWS theoretical peaks correspond almost perfectly with global climate changes during the last 70 million years. See **Chart 17A**. The only significant miss came at 4.26 Ma.

Table 17.1 – Geological Epochs vs. 30.4-Myr and 10.1-Myr Theoretical Peaks.

Geological Epochs	Timescale (Ma)	EUWS Peaks	Comment
Pliocene	5.332 to 2.588	4.26 Ma	
Miocene	23.03 to 5.332	**24.54 ****	
Oligocene	33.9 to 23.03	34.68 Ma	Almost a direct hit.
Eocene	55.8 to 33.9	**54.96 ****	Almost a direct hit.
Paleocene	65.5 to 55.8	65.10 Ma	Almost a direct hit.
Late Cretaceous	99.6 to 65.5	95.52 Ma	
Early Cretaceous	145.5 to 99.6	**146.22 ****	Almost a direct hit.
Late Jurassic	161.2 to 145.5	166.50 Ma	
Middle Jurassic	175.6 to 161.2	**176.64 ****	Almost a direct hit.
Early Jurassic	199.6 to 175.6	196.92 Ma	
Late Triassic	228.0 to 199.6	227.34 Ma	Almost a direct hit.
Middle Triassic	245.0 to 228.0	247.62 Ma	
Early Triassic	251.0 to 245.0		No corresponding EUWS peak.
Lopingian	260.4 to 251.0	257.76 Ma	
Guadalupian	270.6 to 260.4	**267.90 ****	
Cisuralian	299.0 to 270.6	**298.32 ****	Almost a direct hit.
Late Mississippian	306.5 to 299.0	308.46 Ma	
Middle Mississippian	311.7 to 306.5	318.60 Ma	
Early Mississippian	359.2 to 311.7	**359.16 ****	Almost a direct hit.
Late Devonian	385.3 to 359.2	**389.58 ****	
Middle Devonian	397.5 to 385.3	399.72 Ma	
Early Devonian	416.0 to 397.5		No corresponding EUWS peak.
Pridoli	418.7 to 416.0	**420.00 ****	
Ludlow	422.9 to 418.7		No corresponding EUWS peak.
Wenlock	428.2 to 422.9	430.14 Ma	
Alexandrian	443.7 to 428.2	440.28 Ma	
Late Ordovician	460.9 to 443.7	460.55 Ma	Almost a direct hit.
Middle Ordovician	471.8 to 460.9	470.69 Ma	
Early Ordovician	488.3 to 471.8	490.97 Ma	
Furongian Cambrian	501.0 to 488.3	501.11 Ma	Almost a direct hit.
Middle Cambrian	513.0 to 501.0	**511.25 ****	
Early Cambrian	542.0 to 513.0	**541.67 ****	Almost a direct hit.

10.1 Million Year EUWS Cycle

The climate data perfectly matched the six theoretical peaks from 65.10 Ma to 14.40 Ma. And even the slight miss at 4.26 Ma, may have been a direct hit – depending on the data series used. A climate proxy based on Eastern Equatorial Pacific water temperatures show a distinct peak precisely at 4.26 Ma.[K.T. Lawrence, 2006]

In addition to near-perfect timing at theoretical peaks, temperature <u>amplitudes</u> during the last 70 million years coincided very well with the relative strength of EUWS cycles. That is, the two 30.4-myr peaks recorded the greatest spikes prior to their tops, while the less powerful 10.1-myr peaks displayed smaller rises.

A different climate-proxy, constructed by University of Ottawa geologist Jan Veizer, shows similar temperature patterns over the last 200 million years. See **Chart 17B**. Starting at 187 Ma, every third 30.4-myr cycle exhibits stronger amplitude. In between, the 10.1-myr cycles fill the gaps with less intensity. The only major deviation from the pattern came at the 116 Ma peak of the 30.4-myr cycle.

Without a doubt, the most interesting example of the 10.1-myr EUWS cycle came from the geological record 2.7 billion years ago! In his PowerPoint presentation entitled *What on Earth Happened 2.7 Billion Years Ago*, Kent Condie provides the data for **Chart 17C** – showing the estimated distribution of granitoid zircon ages at the time of the Catastrophic Mantle Plume Event.[Condie, 2003] While **Chart 17C** may seem rather normal at first glance, it contains several unusual qualities. Because of these special qualities, the messages revealed from this chart require further explanation. In particular, note these traits:

- The huge peak at 2.7015 Ga closely coincides with the theoretical peak of the 22.2-gyr cycle.
- The 2.7015 Ga top represents a 30.4-myr sub-cycle deviation from the theoretical peak of the 22.2-gyr cycle at 2.7319 Ga. This provides a rare example <u>of a near direct hit (deviating less than 0.3%)</u> in which the deviation equaled a frequency six generations deep in the subdivision.
- The chart shows indications of 30.4-myr peaks at 2.7319 Ga and 2.7015 Ga.
- The ten arrows represent sub-cycle tops corresponding to theoretical 10.1-myr peaks.
- As descendants of the 22.2-gyr cycle, the 10.1-myr oscillations remained visible seven generations deep!
- At 2.7015 Ga, the 22.2-gyr peak did not form as a smooth normally distributed curve. Instead, it stair stepped its way up to the summit in 7th-generation sub-cycle increments. Then, after the climax, it stair stepped down in the same 10.1-myr increments.

These characteristics highlight the intricate dependency of the EUWS cycles. They link themselves together at a high level. Then the EUWS cycles display their inter-dependency by subdividing by a factor of 3 one sub-cycle after another – before finally reducing their frequency to an infinitesimally tiny fraction of the original cycle.

Similar to the other EUWS cycles, the 10.1-myr frequency exhibits its own distinguishing features. In this particular case, the near-perfect correlation with climate cycles makes it stand out.

However, it also exhibits evolutionary tendencies. By the time the series shrinks to 10.1-myr, evidence of evolution-extinction cycles begin to fade, but not completely. In their abstract *Insights into the Coupling of Duplication Events and Macroevolution from an Age Profile of Animal Transmembrane Gene Families*, Ding *et al.* provide mild evidence for both the 10.14 and 30.42-myr cycles:

> **"We identified 3 potential peaks, which are 60.92-myr, 27.29-myr, and 10.32-myr cycles."** [Ding *et al.*, 2006]

Because of risk associated with a false reading, Ding's team threw out the 10.32-myr cycle because its frequency roughly equaled twice the size of the 5-myr bins used in their histogram. In this particular case, caution resulted in discarding a valid cycle. By putting the 10.32-myr cycle back into the mix, the three genetic cycles identified by Ding *et al.* closely match equivalent EUWS cycles. See the following comparison.

Ding *et al*.	**EUWS Frequencies**
60.92-myr	60.84-myr (2 x 30.42, a secondary EUWS harmonic)
27.29-myr	30.42-myr
10.32-myr	10.14-myr

In conclusion, the 10.1-myr EUWS cycle left its mark on Earth in a variety of ways. These imprints include geological records, climate histories, and evolution – the same as its ancestor cycles.

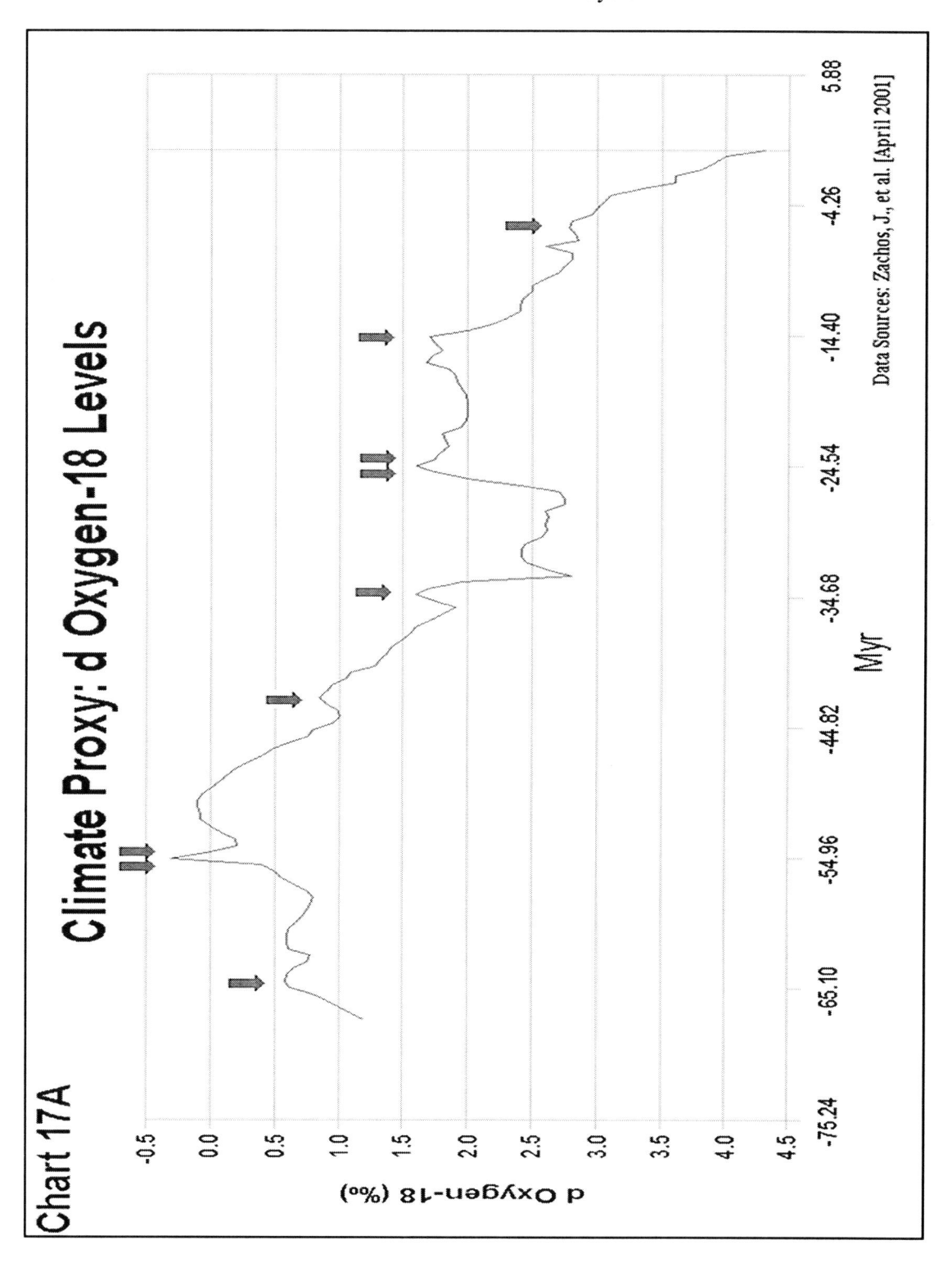

10.1 Million Year EUWS Cycle

Chart 17A

Climate Proxy: d Oxygen-18 Levels

Data Sources: Zachos, J., et al. [April 2001]

Chart 17B

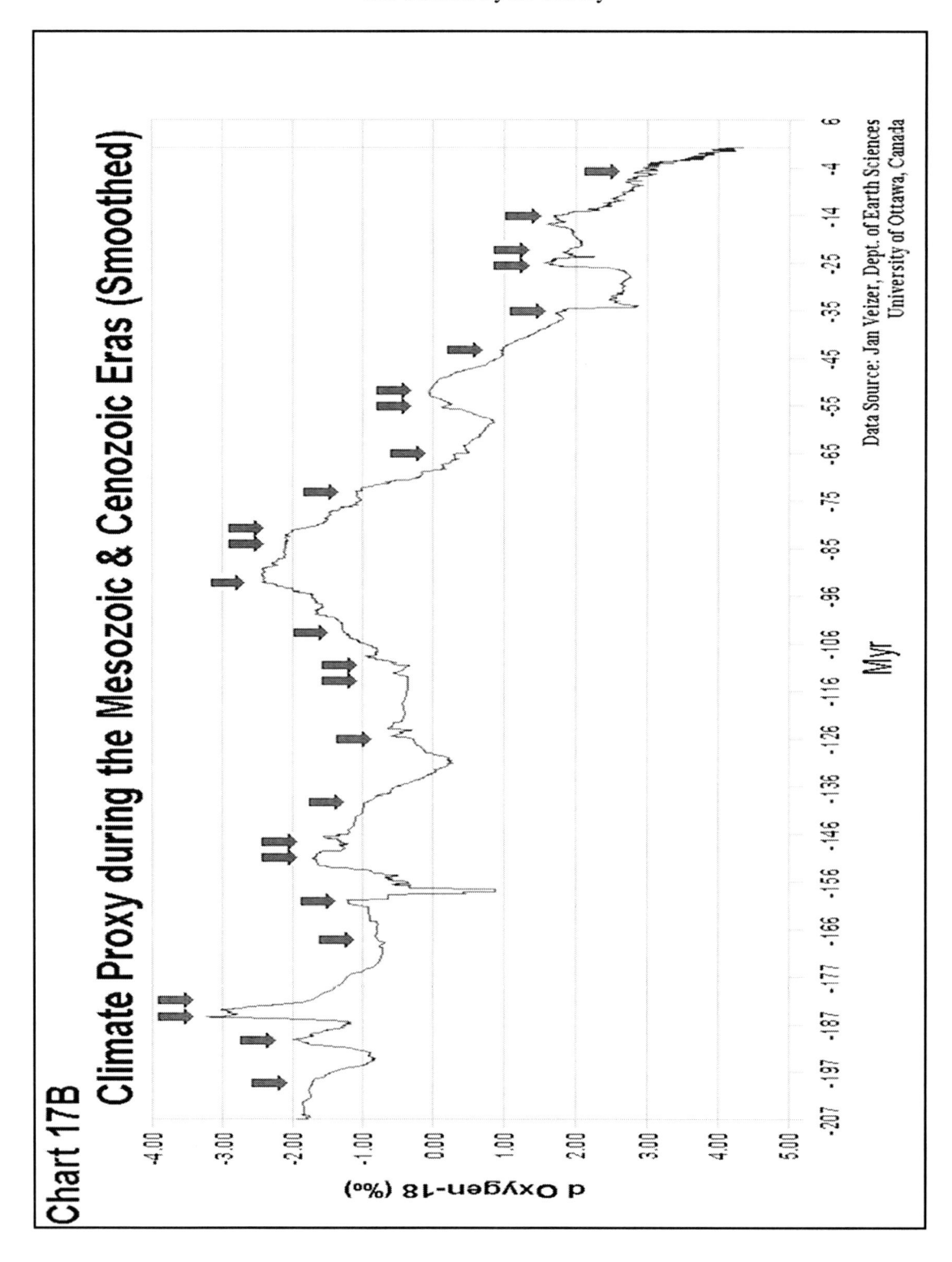

Climate Proxy during the Mesozoic & Cenozoic Eras (Smoothed)

Data Source: Jan Veizer, Dept. of Earth Sciences, University of Ottawa, Canada

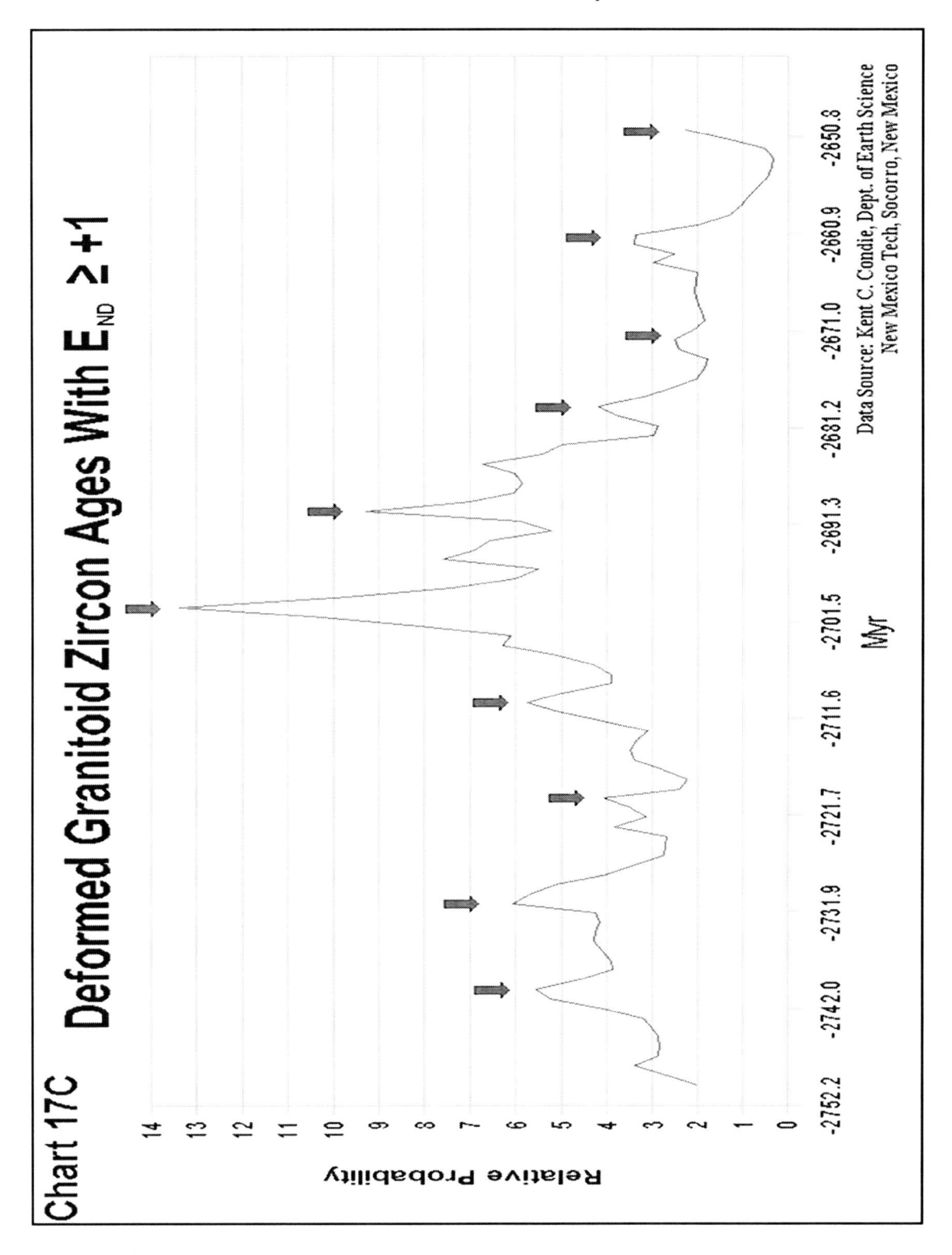

10.1 Million Year EUWS Cycle

Chart 17C — Deformed Granitoid Zircon Ages With $E_{ND} \geq +1$

Data Source: Kent C. Condie, Dept. of Earth Science, New Mexico Tech, Socorro, New Mexico

Chapter 18
3.38 Million Year EUWS Cycle

Zooming in further on the EUWS sub-cycles, the 3.38-myr frequency comes next. And like its ancestor cycles, the 3.38-myr frequency corresponds to the geological timescale – in this case, matching with geological ages. However, the correlation appears in a different manner. (Geological ages were only assigned to the Phanerozoic eon.) First, the correlation between geological ages and the 3.38-myr cycle results primarily from their spacing. Breakpoints for geological ages tend to fall between <u>the time of a EUWS theoretical peak</u> and <u>one million years afterwards</u>. See **Table 18.1**, which contains a small subset of the entire spectrum of geological ages.

Second, the frequency of geological ages fluctuates with time. Surrounding windows of 91.3-myr troughs (at 40, 131, 222, 314, 405, and 496 Ma), the spacing between ages increases substantially. For example, in **Table 18.1**, notice how three EUWS cycles had no corresponding geological age. Essentially, these three geological ages lengthened to roughly 7-myr around the time of the 91.3-myr trough-area from 29 Ma to 52 Ma. Geological ages become even more infrequent around the 274-myr troughs of 131 Ma and 405 Ma. In short, EUWS peaks tend to mark geological divisions in a more consistent manner than EUWS troughs.

Third, excluding trough areas, the correlation between geological ages and 3.38-myr peaks becomes a 1-to-1 relationship. **Table 18.1** shows 18 EUWS theoretical peaks, and 15 geological ages. After eliminating the 91.3-myr trough area from 29 to 52 Ma, the matchup between geological ages and 3.38-myr cycles turns into a perfect 11-for-11 relationship.

Table 18.1 – Geological Ages versus the 3.38-Myr EUWS Cycle.

Geological Ages	Timescale (Ma)	EUWS Peaks	Comment
Messinian	7.246 to 5.332	7.64 Ma	
Tortonian	11.608 to 7.246	11.02 Ma	
Burdigalian	13.65 to 11.608	14.40 Ma	
Serravallian	15.97 to 13.65	17.78 Ma	
Langhian	20.43 to 15.97	21.16 Ma	
Aquitanian	23.03 to 20.43	24.54 Ma	
Chattian	28.4 to 23.03	27.92 Ma	
		31.30 Ma	No corresponding geological age.
Rupelian	33.9 to 28.4	34.68 Ma	
Priabonian	37.2 to 33.9	38.06 Ma	
Bartonian	40.4 to 37.2	41.44 Ma	
		44.82 Ma	No corresponding geological age.
Lutetian	48.6 to 40.4	48.20 Ma	
		51.58 Ma	No corresponding geological age.
Ypresian	55.8 to 48.6	54.96 Ma	
Thanetian	58.7 to 55.8	58.34 Ma	
Selandian	61.7 to 65.5	61.72 Ma	
Danian	65.5 to 61.7	65.10 Ma	

For the remaining geological ages not listed in **Table 18.1**, the same rules tend to hold, albeit less perfectly. This has not been verified; however, the *Unified Cycle Theory* predicts the existence of unconformities in larger numbers at the troughs of major EUWS cycles. The greatest number of unconformities should appear around the time when geological ages skipped a beat – for example, around 31.3 Ma, 44.82 Ma, and 51.58 Ma. Major EUWS troughs appear to produce both geological unconformities and skipped beats in geological ages. Hence, the two are probably linked.

Other than the geological timescale, additional evidence supporting the 3.38-myr cycle appears sparse. As part of a study involving Pacific Ocean sediments, geologists from Cardiff University, led by Caroline Lear, observed a temperature cycle approaching 3 million year. Their abstract *Late Eocene to Early Miocene Ice Sheet Dynamics and the Global Carbon Cycle* details the activity between 37 and 28 Ma:

> **"The paired Mg/Ca and oxygen isotope records are used to calculate seawater d18O (dw). Calculated dw suggests that a large Antarctic ice sheet formed during Oi-1 and subsequently fluctuated throughout the Oligocene on both short (<0.5 Myr) and long (2–3 Myr) timescales."** [Lear *et al.*, 2004]

Chart 18A shows the huge plunge in global temperatures at 34.68 Ma, which coincided with the start of an Antarctic ice sheet formation. Marked with double-arrows in the chart, 10.1-myr peaks distinctly stand out. In between the large cycle, the 3.38-myr peaks display smaller amplitudes – exactly as they should. Hence, the climate history through the Cenozoic provides strong support for the 3.38-myr cycle.

Depicted in **Chart 18B**, notice how the EUWS model correlates moderately well with the global temperatures shown in **Chart 18A**. The major difference between the two charts comes from the <u>downward slope</u> exhibited by global temperatures versus the <u>fluctuating (but level) trend</u> predicted by the model.

A third set of temperature data comes from the National Oceanic and Atmospheric Administration's NCDC Paleoclimatology Program, via Lafayette College geologist Kira Lawrence.[Lawrence, 2006] The data reflects equatorial temperatures in the Eastern Pacific Ocean, covering the last 5 million years. Unfortunately, the data series spans such a short period that it disallows adequate testing for the 3.38-myr cycle. See **Chart 18C**. Nonetheless, for the two theoretical peaks it does cover, significant temperature maxima occurred close to the theoretical peaks at 4260 Ka and 880 Ka.

This scant, yet supportive evidence now comes to a conclusion. Additional data must be located to solidify the validity of the 3.38-myr EUWS cycle.

The Unified Cycle Theory

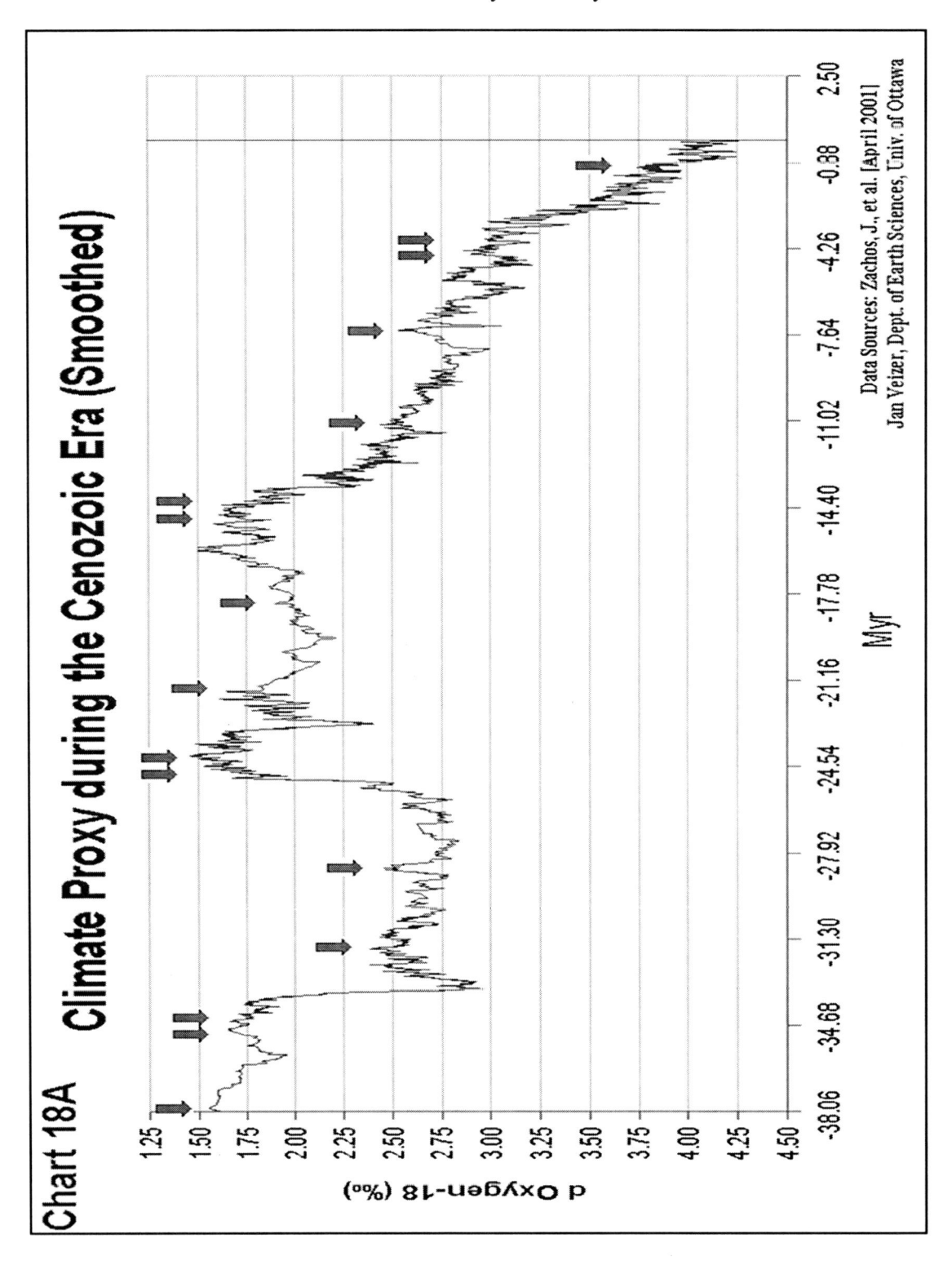

Chart 18A Climate Proxy during the Cenozoic Era (Smoothed)

Data Sources: Zachos, J., et al. [April 2001]
Jan Veizer, Dept. of Earth Sciences, Univ. of Ottawa

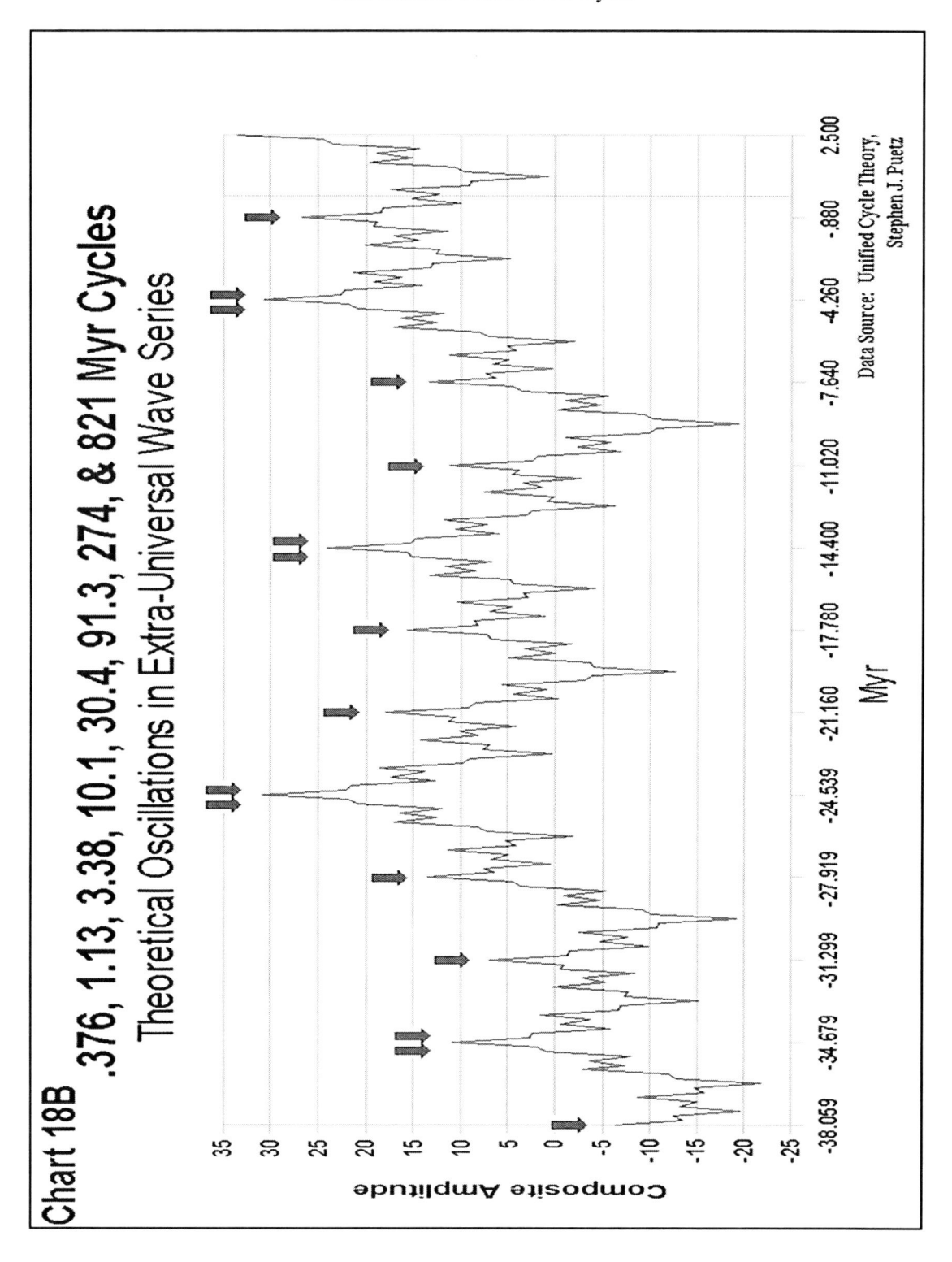

3.38 Million Year EUWS Cycle

Chart 18B .376, 1.13, 3.38, 10.1, 30.4, 91.3, 274, & 821 Myr Cycles
Theoretical Oscillations in Extra-Universal Wave Series

Data Source: Unified Cycle Theory, Stephen J. Puetz

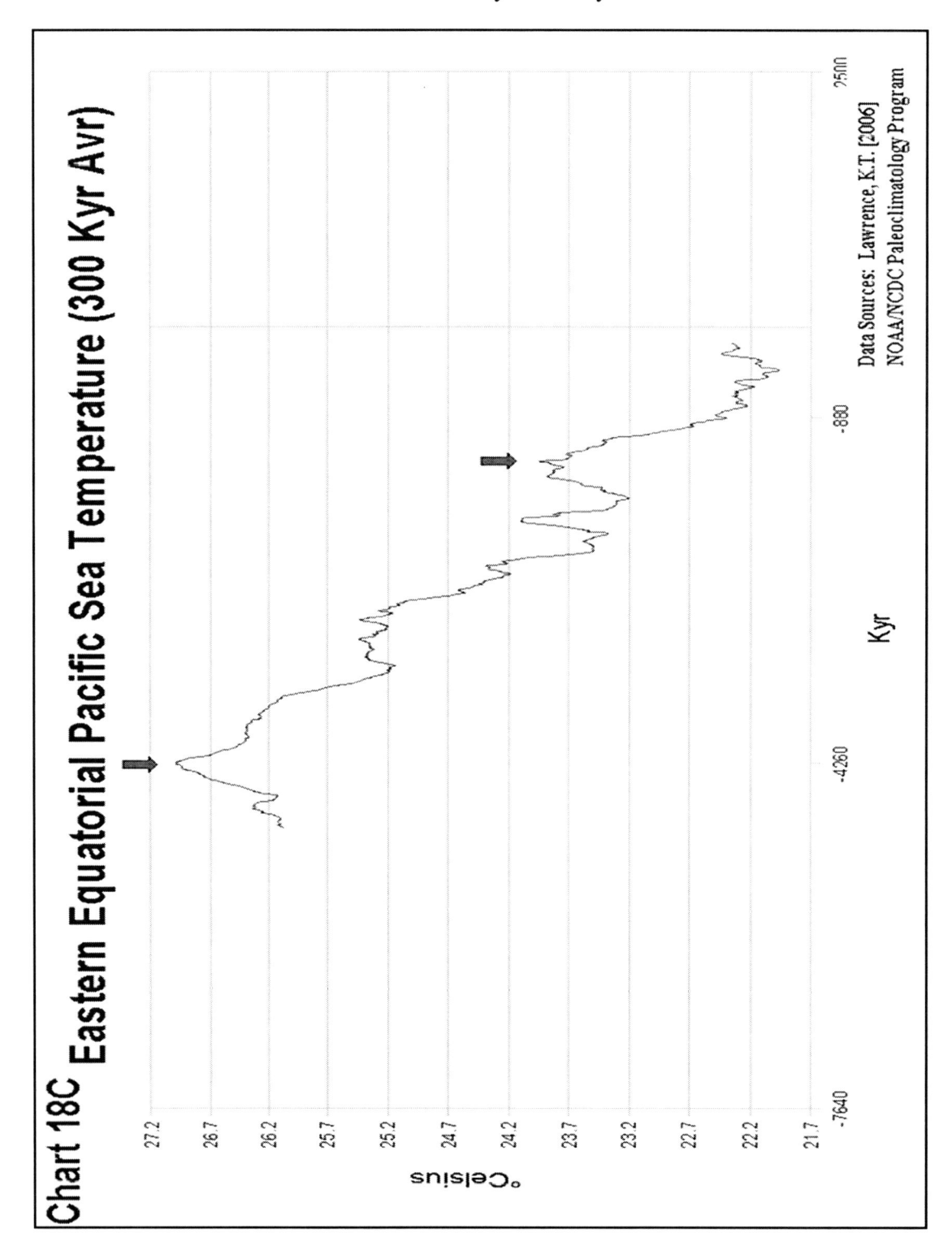

Chart 18C Eastern Equatorial Pacific Sea Temperature (300 Kyr Avr)

*Data Sources: Lawrence, K.T. [2006]
NOAA/NCDC Paleoclimatology Program*

Chapter 19
1.13 Million Year EUWS Cycle

Zooming in still further, the 1.13-myr cycle comes next in this never-ending series. With the reduction to 1.13-myr, the EUWS approaches a frequency comparable to Milankovitch oscillations. Previously mentioned in Chapter 11, a European research team identified a 1.2-myr climate cycle which they attributed to Milankovitch dynamics.[Wade & Palike, 2004]

However, instead of observing Milankovitch-related cycles, Wade and Palike actually saw the 1.13-myr EUWS cycle hard at work. Similar misinterpretations occur for all the Milankovitch cycles. Global temperature data in **Chart 19A** shows that maxima consistently coincide with 1.13-myr EUWS theoretical peaks.

The chart labels 3.38-myr peaks with double-arrows. Prior to temperature highs, 3.38-myr cycles should exhibit stronger upsurges than 1.13-myr cycles. Even though they fail to do that on the upside, 3.38-myr tops display sharp declines afterward. In this way, the 3.38-myr cycles stand out from their 1.13-myr sub-cycles.

For the 16 cycles in the Neogene period, the 1.13-myr cycle leaves a strikingly regular trail. Just as the 3.38-myr cycle behaved, the 1.13-myr cycle shows its power not so much by its ascent prior to peaks, but by the relatively sharp declines following theoretical peaks. Actually, this type of behavior should be expected during a time when global temperatures trended lower.

In addition, the climate proxy from **Chart 19A** matches the model in **Chart 19B** fairly well. Similar to the 3.38-myr cycle, the major difference between the two charts involves their slopes. However, the relative direction and amplitude within each subdivision matches quite well.

Over the past 5 million years, equatorial temperatures in the Eastern Pacific Ocean coincided perfectly with 1.13-myr theoretical peaks at 4.260, 3.133, and 2.006 Ma. For the 880 Ka cycle, the TPD Principle came into play. For that cycle, the sea temperature peaked one 376-kyr cycle early, at 1.255 Ma. See **Chart 18C**.

Evidence for the 1.13-myr cycle appears fairly supportive; however, a lengthier history would definitely be more desirable.

1.13 Million Year EUWS Cycle

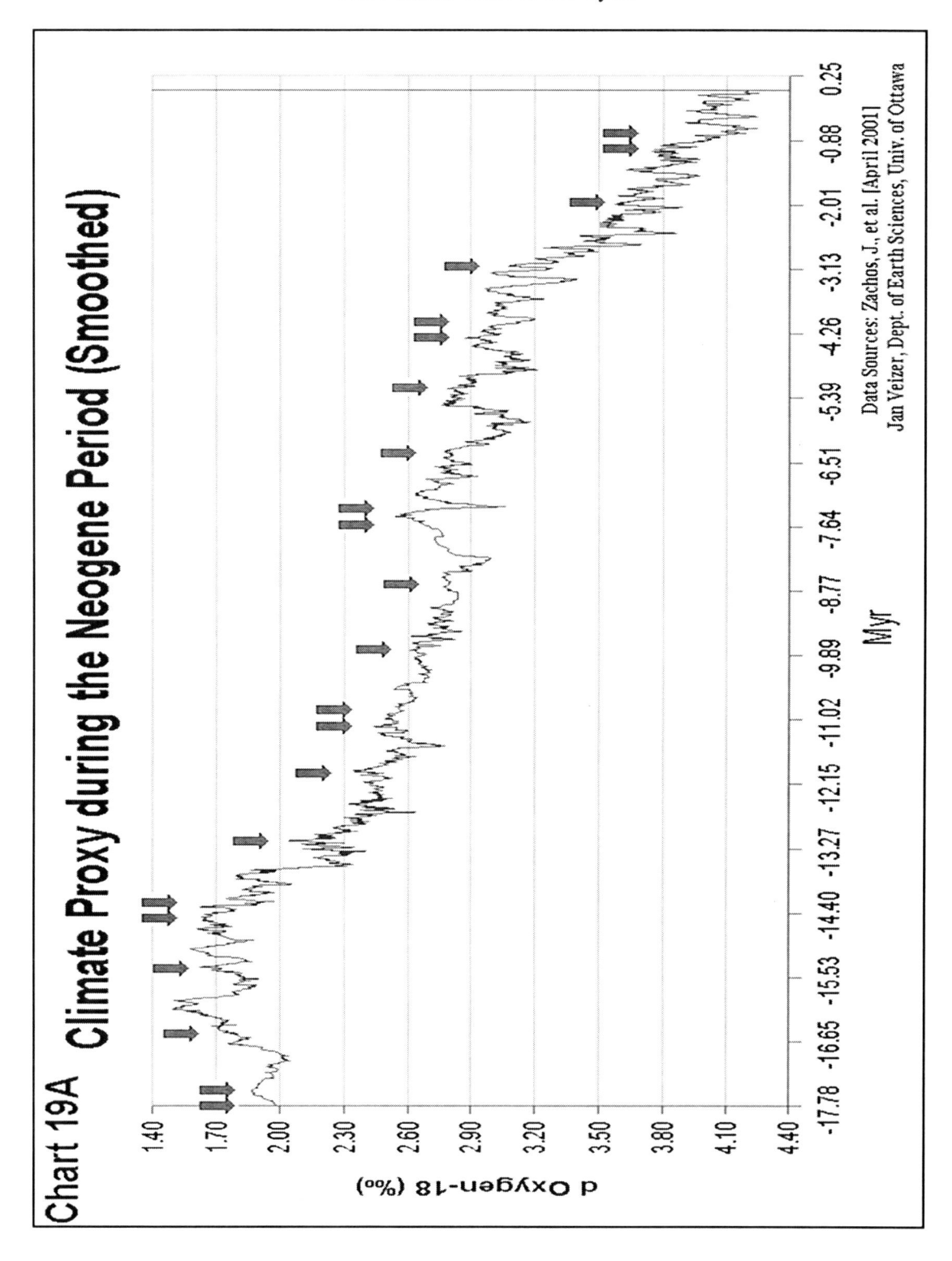

Chart 19A Climate Proxy during the Neogene Period (Smoothed)

Data Sources: Zachos, J., et al. [April 2001]
Jan Veizer, Dept. of Earth Sciences, Univ. of Ottawa

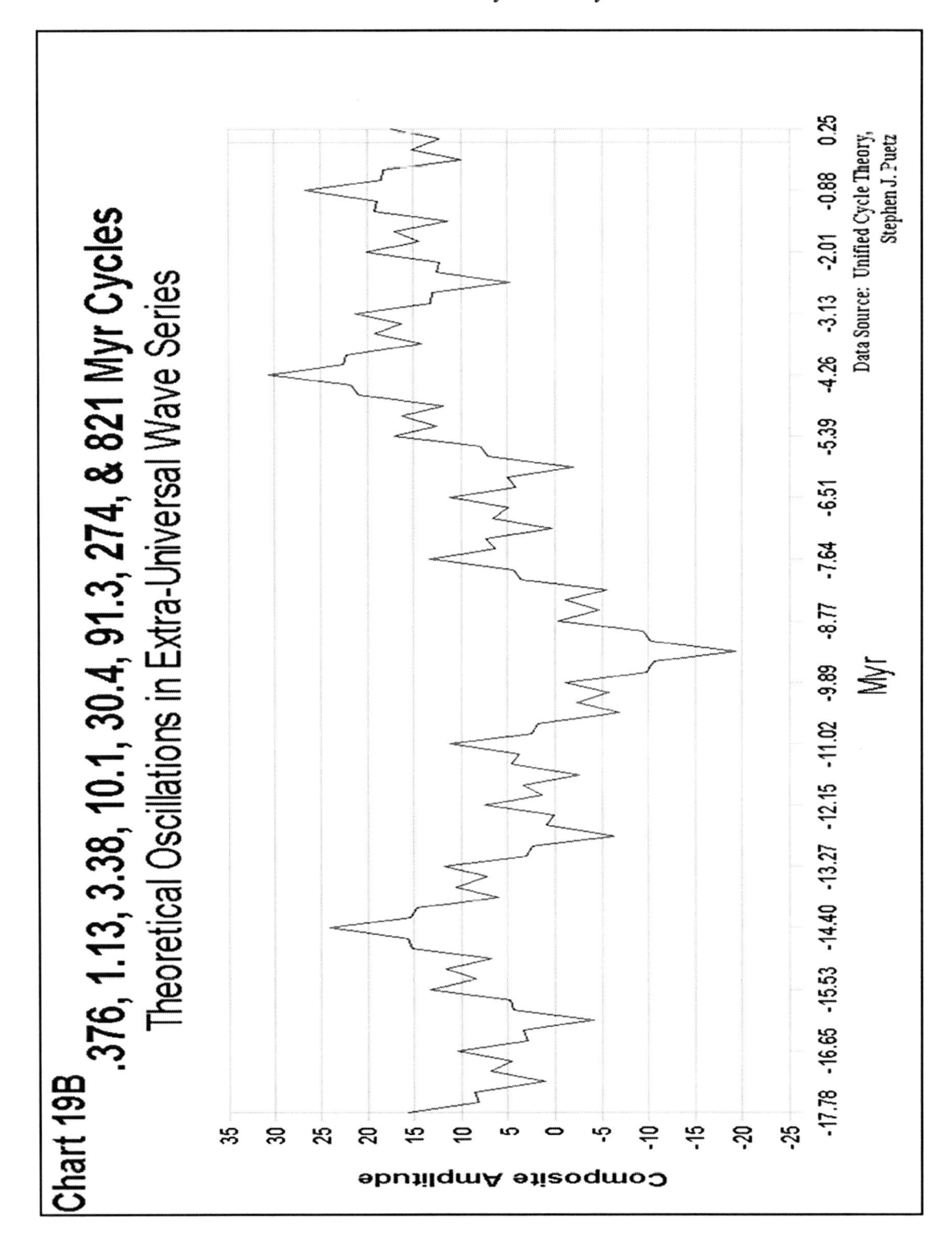

Chart 19B .376, 1.13, 3.38, 10.1, 30.4, 91.3, 274, & 821 Myr Cycles
Theoretical Oscillations in Extra-Universal Wave Series

Data Source: Unified Cycle Theory, Stephen J. Puetz

1.13 Million Year EUWS Cycle

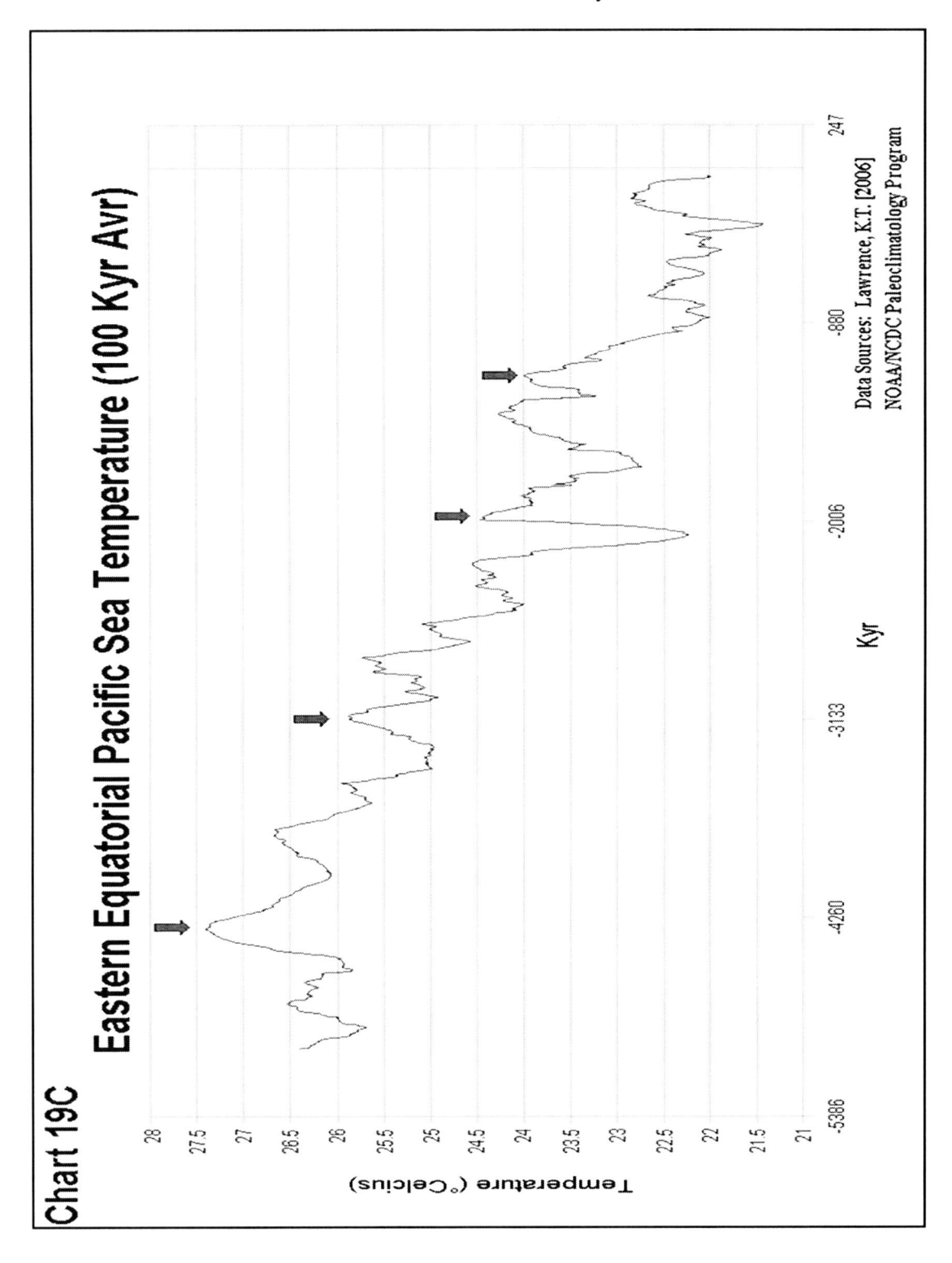

Chart 19C — Eastern Equatorial Pacific Sea Temperature (100 Kyr Avr)

Data Sources: Lawrence, K.T. [2006]
NOAA/NCDC Paleoclimatology Program

Chapter 20
376 Thousand Year EUWS Cycle

Dividing the 1.13-myr cycle by three, its 376-kyr child appears next in the Extra-Universal Wave Series. As depicted in **Chart 20A**, the 376-kyr frequency left prominent imprints on Earth's climate proxy during the last 8 million years. Double arrows mark 1.13-myr maxima, while single arrows denote 376-kyr tops. **Chart 20B** shows the 376-kyr cycle in Eastern Pacific sea temperatures, where it appears even more clearly.

In earlier chapters, numerous examples showed how geomagnetic cycles interfere with EUWS frequencies less than 20 years. For timescales from 10 thousand to 1 million years, similar competition develops between EUWS and Milankovitch cycles.

The *Unified Cycle Theory* isolates every major naturally occurring cycle. By doing so, the task of estimating each individual contributor becomes possible, albeit, still difficult. The theory identifies the EUWS as the dominant set of cycles in our universe, but not the sole source of cyclic behavior.

Milankovitch cycles undoubtedly influence global climate. However, infatuation with the Milankovitch Theory has detracted from understanding by providing researchers with a convenient mechanism to attribute all global temperature fluctuations. Ultimately, for oscillations in the 20-kyr to 400-kyr range, Milankovitch cycles may account for 10% to 40% of global climate changes. The contribution almost certainly amounts to something less than 50%, but determining the precise percentage may prove difficult.

Chart 20C shows a model consisting of the 405-kyr Milankovitch cycle working in combination with four EUWS cycles.

Notice how **Chart 20B** correlates with **Chart 20C**. While not perfect, both the magnitude of the moves and the timing of their reversals synchronize quite well. Under its umbrella, the *Unified Cycle Theory* accepts any and all cycles that exhibit a true physical origin. Milankovitch and EUWS cycles both satisfy those criteria. The only point of contention centers on their degree of climatic influence.

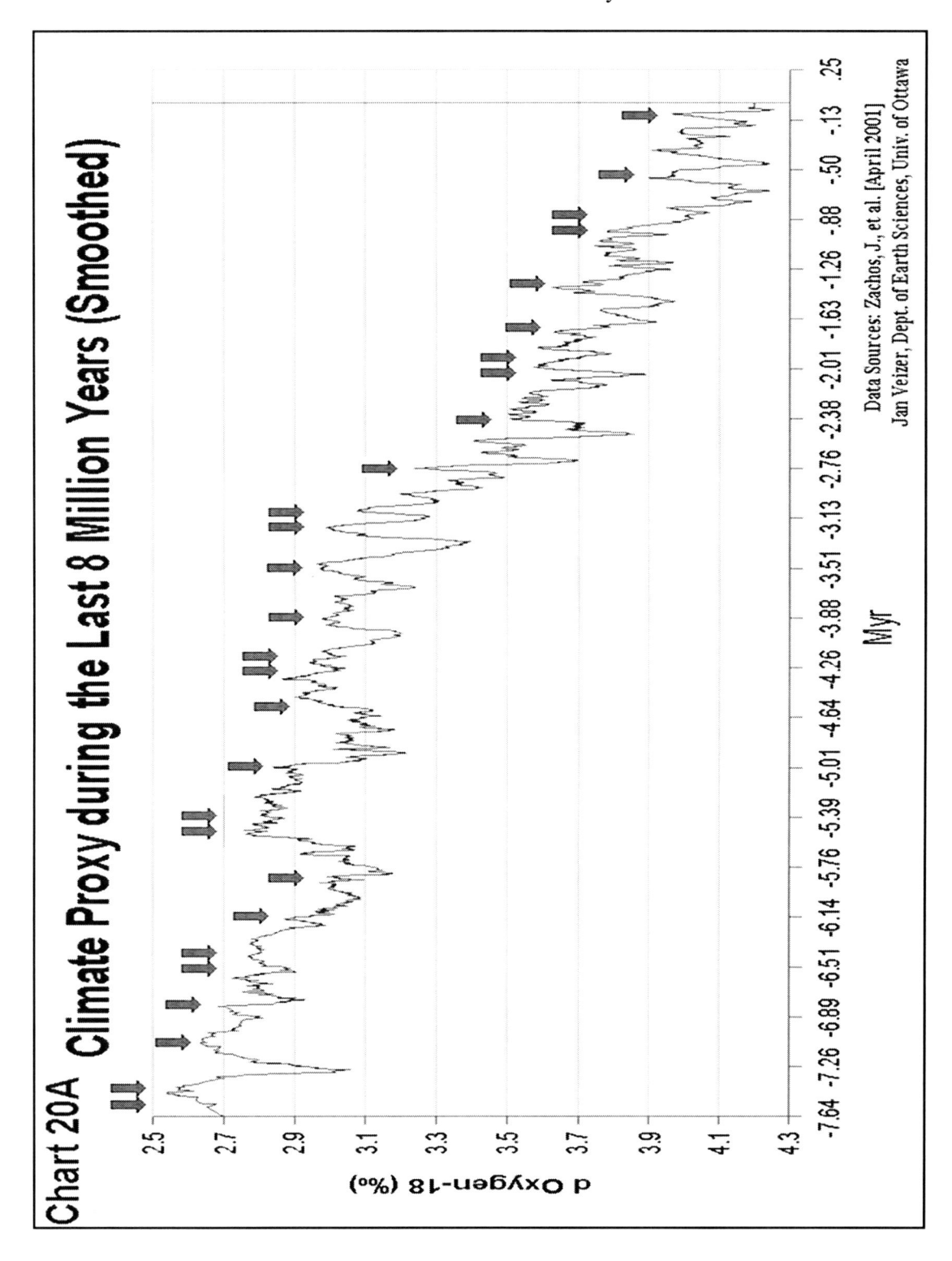

376 Thousand Year EUWS Cycle

Chart 20A Climate Proxy during the Last 8 Million Years (Smoothed)

Data Sources: Zachos, J., et al. [April 2001]
Jan Veizer, Dept. of Earth Sciences, Univ. of Ottawa

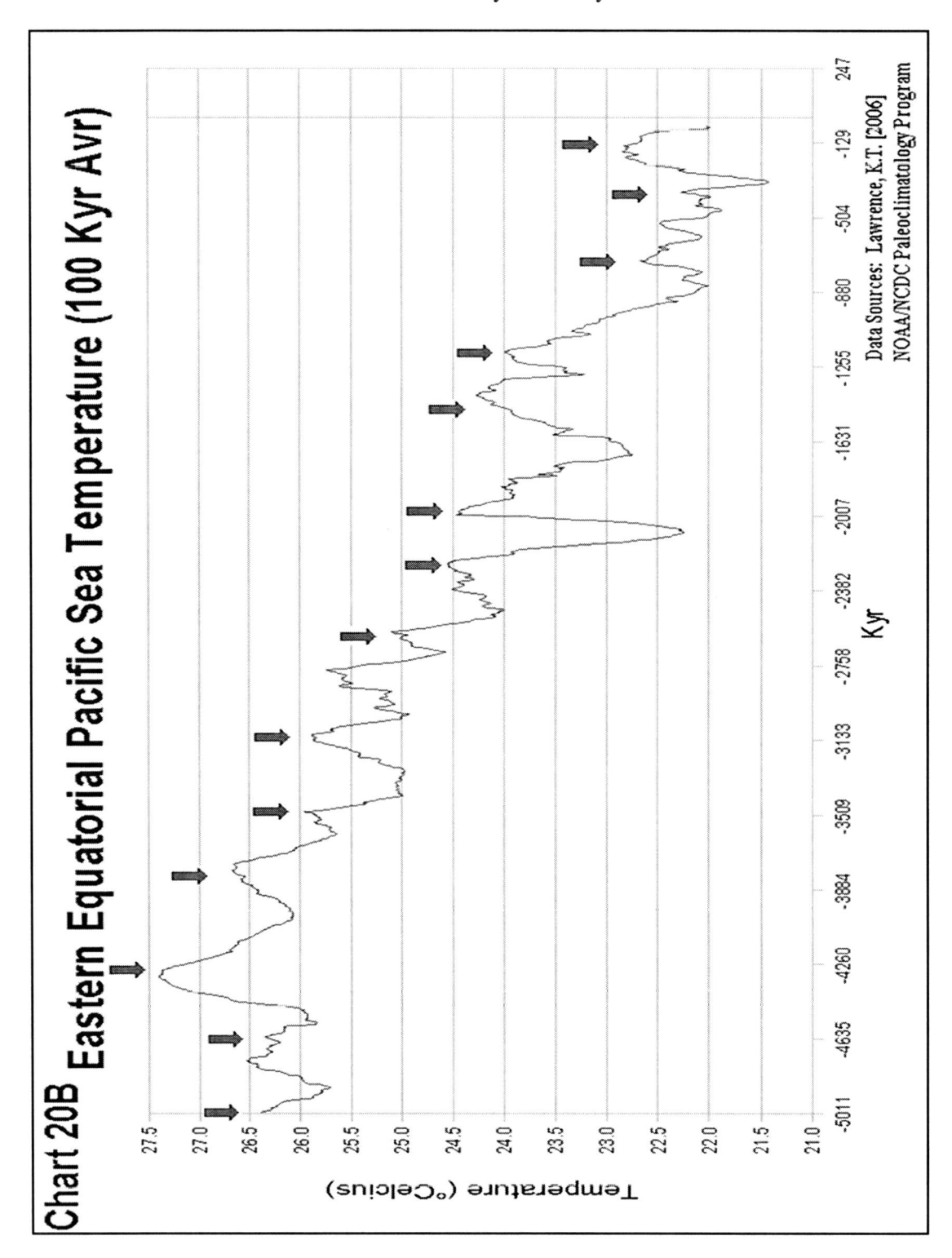

Chart 20B Eastern Equatorial Pacific Sea Temperature (100 Kyr Avr)

Data Sources: Lawrence, K.T. [2006]
NOAA/NCDC Paleoclimatology Program

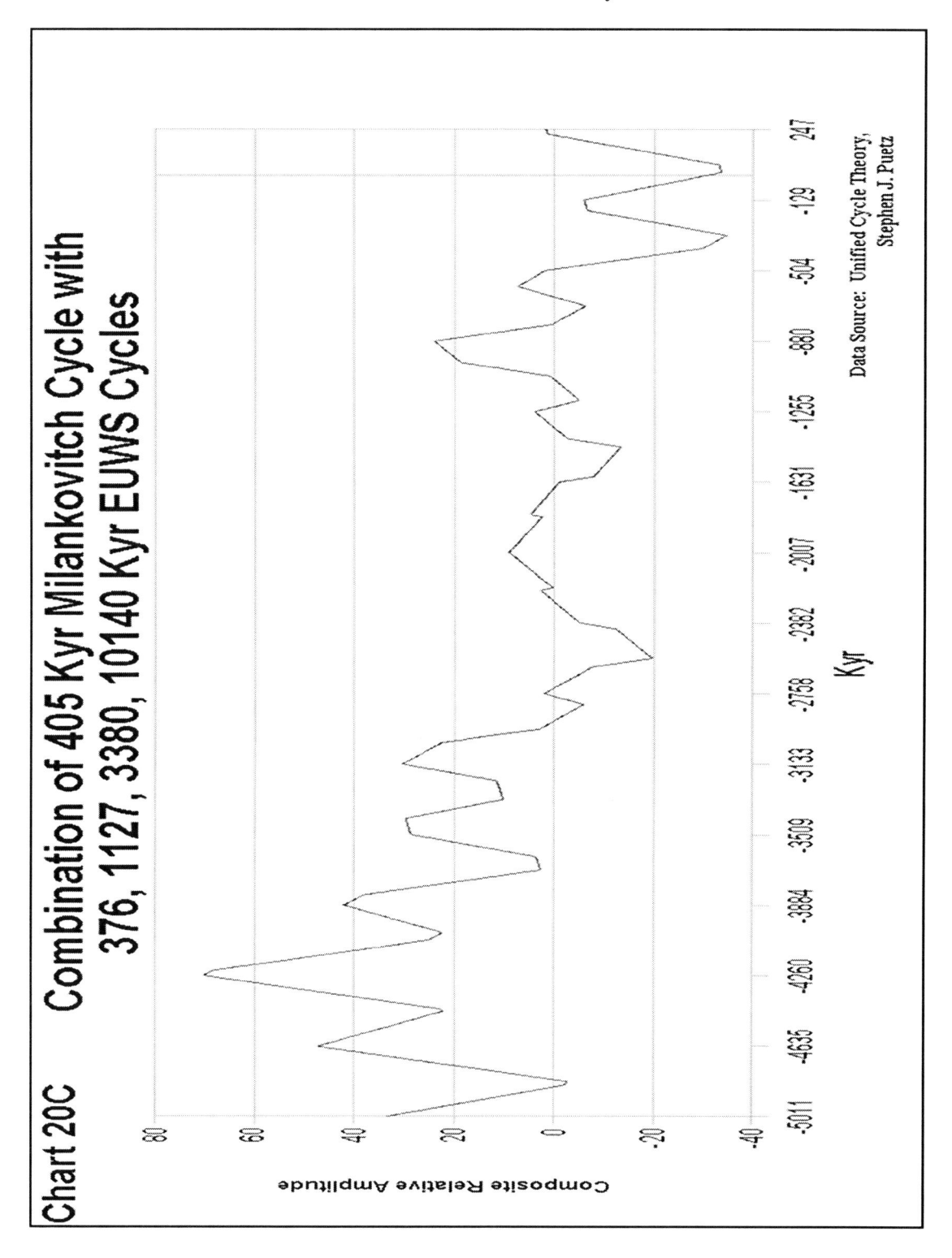

376 Thousand Year EUWS Cycle

Chart 20C Combination of 405 Kyr Milankovitch Cycle with 376, 1127, 3380, 10140 Kyr EUWS Cycles

Data Source: Unified Cycle Theory, Stephen J. Puetz

Chapter 21
125 Thousand Year EUWS Cycle

As the EUWS cycles shrink to the 100 thousand year range, the Milankovitch-EUWS battle reaches the height of its competition. In Chapter 11, physicists from the University of California identified the 125-kyr cycle as a strong peak in the eccentricity spectrum.[Muller and MacDonald, 2002] The 125-kyr cycle plays a key role in modulating ice-age cycles. Acting as a near-perfect match, the frequency of the 125.184-kyr EUWS cycle differs from the possible eccentricity cycle by only 0.1%. **Charts 21A** (global climate) and **21B** (Eastern Equatorial Pacific temperatures) show the 125-kyr EUWS cycle.

The scientific community has heavily documented the 125-kyr cycle from the paleoclimate record. And because the Milankovitch-EUWS frequencies essentially match, the task of determining which one contributes the most to ice-age cycles becomes more complex.

As an example of the lengths gone to construct Milankovitch-based climate models, a pair of geologists, working at the Wave Propagation Laboratory at the University of North Carolina, identified sideband frequencies at 75, 85, 107, 110, 123, 125, and 143-kyr. In their abstract entitled *Understanding Nonlinear Responses of the Climate System to Orbital Forcing*, Jose Rial and Cheri Anaclerio conclude these sidebands result from modulation coming from long-term Milankovitch cycles:

"This is interpreted as frequency modulation of the main carrier of period 95 Kyr

by the 413 Kyr modulating signal."[Rial and Anaclerio, 2000]

In the abstract *The Timing of Major Climate Terminations,* Columbia University's Maureen Raymo describes the 100-kyr ice-age cycle as an alternating beat between cycles of 80-kyr and 120-kyr.[Raymo, 1997] Ed Mercurio, a climatologist at Hartnell College, observed the same beats in the ice-age cycle:

> **"Inclination with relation to the invariable plane plus inclination with relation to the solar equator establishes the basic beat frequency seen in paleotemperatures of alternating ~80,000 and ~120,000 year periods."**[Mercurio, 2001]

However, Mercurio proposes that these alternating beats originate from a different source – galactic cosmic rays. Whatever the explanation, Milankovitch frequencies cannot fully explain global temperature fluctuations. **Chart 21C** depicts a model based on three Milankovitch frequencies, plus seven EUWS cycles.

Chart 21C (the model) correlates with **Chart 21B** (Eastern Pacific Ocean temperatures) fairly well, with the exception of activity around 880 Ka. At that time, the expected 1.13-kyr EUWS peak at 880 Ka shifted one 376-kyr sub-cycle away – to 1255 Ka. These shifts, as defined by the TPD Principle, create havoc with models. However, until the source of these shifts can be found, we must live with the imperfections they create.

The comparison expands to the last 5.1 million years in **Chart 21D** (Eastern Pacific Ocean temperatures) and **Chart 21E** (the model). In the model, notice how the various cycles tend to cancel each other out between 4400 Ka and 3600 Ka. Within this 800 thousand year period, the smoother long-term movements in the model closely matched the less erratic behavior of Pacific Sea temperatures. On either side of this 4400-to-3600 Ka timeframe, Pacific temperatures oscillated in a wilder fashion. These larger fluctuations also matched the model in **Chart 21E**.

Up to this point, the model focuses only on Milankovitch and EUWS cycles. As mentioned earlier in this chapter, Mercurio suggested a GCR origin. Furthermore, in an abstract entitled *Variations in Solar Magnetic Activity during the Last 200,000 years: Is there a Sun-Climate Connection*, Mukul Sharma, a geologist from Dartmouth, proposes these cycles may be related to effects from both GCRs and our Sun:

> **"The production of Be-10 in the Earth's atmosphere depends on the galactic cosmic ray influx that, in turn, is affected by the solar surface magnetic activity and the geomagnetic dipole strength. Using the estimated changes in Be-10 production rate and the geomagnetic field intensity, variations in solar activity are calculated for the last 200 Ka. Large variations in solar activity are evident with the Sun experiencing periods of normal, enhanced, and suppressed activity. The marine δO-18 record and solar modulation are strongly correlated at the 100 Kyr timescale. It is proposed that variations in solar activity control**

> the 100 Kyr glacial–interglacial cycles. However, the Be-10 production rate variations may have been under-estimated during the interval between 115 Ka and 125 Ka and may have biased the results."[Sharma, 2002]

This book already demonstrated how EUWS cycles influence many facets of our universe. Now, the EUWS reveals another offshoot to its extensive power – a correlation to solar strength! As the EUWS frequencies shorten, their correlations with solar activity increases. Upcoming chapters present evidence showing that EUWS cycles as short as 172 years significantly influence sunspot activity.

With causation claims coming from EUWS, Milankovitch, GCR, and solar sources, determining cause-and-effect becomes all the more difficult. Suffice it to say that a global temperature model based on multiple factors will be required to fully explain the wide climate fluctuations that persist throughout all timescales.

In previous chapters, the longer-term EUWS cycles demonstrated dramatic effects on life and evolution. In a similar manner, the smaller EUWS cycles also affect life on Earth. Looking at the trough of the last 125-kyr cycle, spanning from 50 to 80 Ka, a great number of extinctions took place. Mercurio describes the evolutionary environment, and the implications to mankind. In particular, he notes the extinction of several species of large animals (mega fauna) that humans depended on for meat:

> "The start of civilization coincides with the decline of the heterogeneous mosaic savanna and the extinction of the mega fauna that was probably a consistently dependable resource. This 'fall from Eden' was probably a major factor in the start of agriculture and a sedentary existence.... The peak on the combined inclinations and obliquities curve at ~50,000 years ago during the last glacial was somewhat lower than the peak that correlates to deglaciation and was short of the interglacial threshold.... The extinction of the Pleistocene mega fauna of Australia occurred around this time (Roberts *et al.* 2001).... Both paleontology and human genetics point to a rapid emergence and expansion of modern humans out of Africa and into most of the Old World at this time."[Mercurio, 2001]

The larger EUWS cycles demonstrated increasing rates of extinctions during their down-phases – with extinctions reaching their maxima at cyclic troughs. The down-phase of the last 125-kyr cycle covered the time from its peak at 128.6 Ka to its trough at 66 Ka. Consistent with the larger EUWS cycles, the most recent 125-kyr down-phase coincided with numerous extinctions of large animals. According to some scientists, it also involved the near extinction of Homo-Sapiens. Stanley Ambrose, an anthropologist from the University of Illinois, proposes that a volcanic eruption caused the near-extinction of humans around 70 ka:

> "Climatic and geological evidence suggest an alternative hypothesis for Late Pleistocene population bottlenecks and releases. The last glacial period was

preceded by one thousand years of the coldest temperatures of the Later Pleistocene (~71–70 ka), apparently caused by the eruption of Toba, Sumatra. Toba was the largest known explosive eruption of the Quaternary. Toba's volcanic winter could have decimated most modern human populations, especially outside of isolated tropical refugia. Release from the bottleneck could have occurred either at the end of this hyper cold phase, or 10,000 years later, at the transition from cold oxygen isotope stage 4 to warmer stage 3.... Volcanic winter may have reduced populations to levels low enough for founder effects, genetic drift, and local adaptations to produce rapid population differentiation. If Toba caused the bottlenecks, then modern human races may have differentiated abruptly, only 70 thousand years ago."[Ambrose, 1998]

Some researchers believe the entire human population shrank as low as 2,000 people during this tragic period. But mankind survived. An international team, led by Lev A. Zhivotovsky of the Russian Academy of Sciences, isolated the period of the small band of survivors to somewhere between 71 and 142 ka:

"We study data on variation in 52 worldwide populations ... to infer a demographic history of human populations.... When a stepwise mutation model is used, a population tree based on T estimates of divergence time suggests that the branches leading to the present sub-Saharan African populations of hunter-gatherers were the first to diverge from a common ancestral population (~71–142 thousand years ago)."[Zhivotovsky *et al.*, 2003]

This wide range of years covers the entire down-phase of the of the 125-kyr cycle from 129 to 66 Ka. Yet, the bulk of the evidence indicates the near-extinction of humans came closer to the 66 Ka trough. Up to this point, evidence centers on the extinction side of the equation. However, once past a trough, EUWS cycles enter an expansionary phase. Geneticists from the University of La Laguna in Spain studied the distribution of human mitochondrial DNA variations to determine how mankind populated Earth. The team concluded:

"The first detectable expansion occurred around 59,000-69,000 years ago from Africa, independently colonizing western Asia and India and, following this southern route, swiftly reaching East Asia.... Around 39,000-52,000 years ago, the western Asian branch spread radically, bringing Caucasians to North Africa and Europe, also reaching India, and expanding to North and East Asia."[Maca-Meyer *et al.*, 2003]

Using the center of their estimate as a best guess, the human population explosion started from Africa about 64,000 years ago. Once again, this estimate closely coincides with the 66 Ka trough of the 125-kyr cycle. And according to the *Unified Cycle Theory*, troughs represent the optimal time for extinctions to end and evolutionary periods to begin. And so it happened with humans around the time of the 66 Ka trough.

Before concluding, the cause of extinctions requires further discussion. Could bitter cold temperatures around 70 Ka have caused the extinction of numerous species of animals and the near-extinction of mankind? Or did the Toba eruption create climate changes that threatened life on Earth? Going back to the extinction of dinosaurs at 65 Ma, climate change brought on by a massive asteroid strike provides a plausible explanation for the sudden disappearance of life at that time.

On the surface, explaining extinctions with climate change seems quite believable. However, a more basic force seems to be at work. In studying extinctions, they overwhelmingly occur during down-phases of the EUWS cycles – during times when temperatures change from warm-to-cold. Yet, by assuming that most organisms are well adapted to their current environment, then dramatic increases in temperature should also cause significant extinctions. These temperature increases could develop from either cold-to-cool, or cool-to-warm, or warm-to-hot, or hot-to-sweltering. But significant extinctions fail to happen under any of these warming scenarios. They only happen in large numbers when temperatures decrease.

Drastic environmental changes occurred regularly throughout Earth's history. These sudden shifts appear equally distributed between rising and falling temperatures. During up-phases when temperatures rise, all species exhibit a great ability to adapt, evolve, and survive – no matter how extreme the climatic jump becomes. As EUWS cycles move toward their peaks, species easily overcome all environmental challenges. However, immediately after EUWS cycles turn lower, for some mysterious reason, this ability to adapt disappears. Because of this lost ability to mutate, extinctions rise dramatically during EUWS down-phases. Hence, rapid climate changes alone cannot consistently explain extinctions. This inconsistency shows that something more basic causes extinctions. Looking for a consistent explanation, it appears that EUWS cycles turn on evolutionary switches during up-phases and then turn off these switches during down-phases. These evolutionary switches seem to operate the same way in all living organisms.

The Unified Cycle Theory

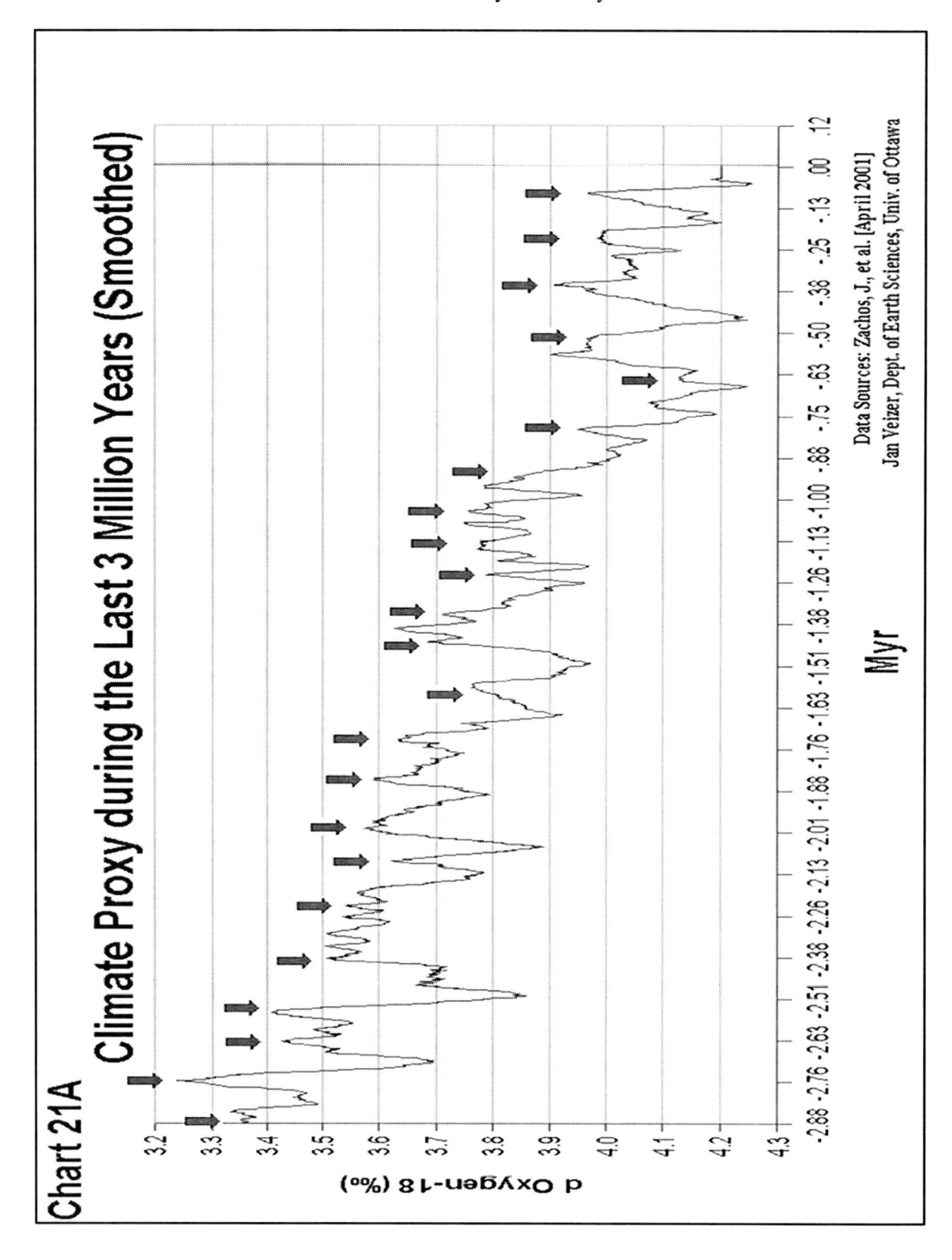

Chart 21A

Climate Proxy during the Last 3 Million Years (Smoothed)

Data Sources: Zachos, J., et al. [April 2001]
Jan Veizer, Dept. of Earth Sciences, Univ. of Ottawa

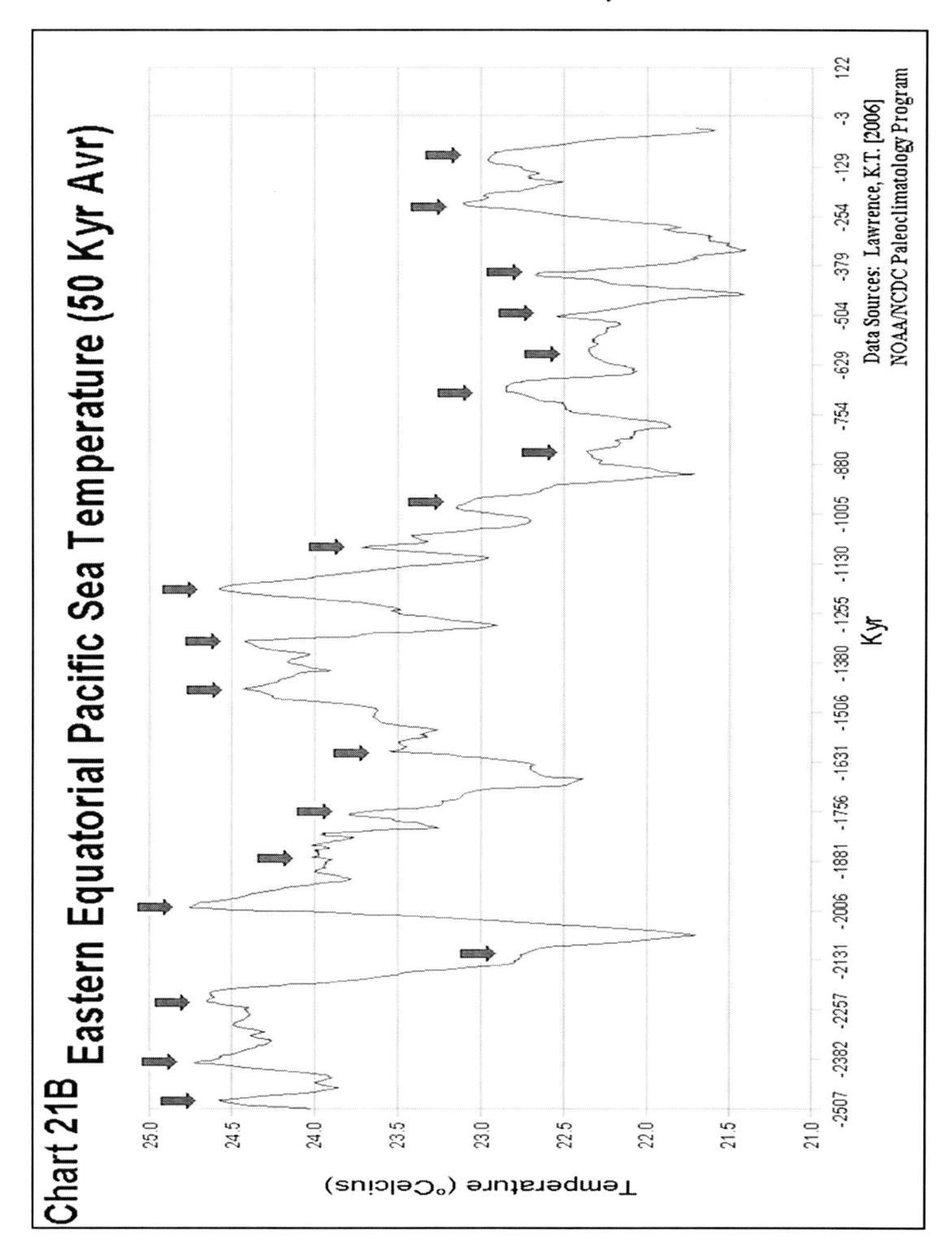

Chart 21B Eastern Equatorial Pacific Sea Temperature (50 Kyr Avr)

125 Thousand Year EUWS Cycle

Data Sources: Lawrence, K.T. [2006]
NOAA/NCDC Paleoclimatology Program

243

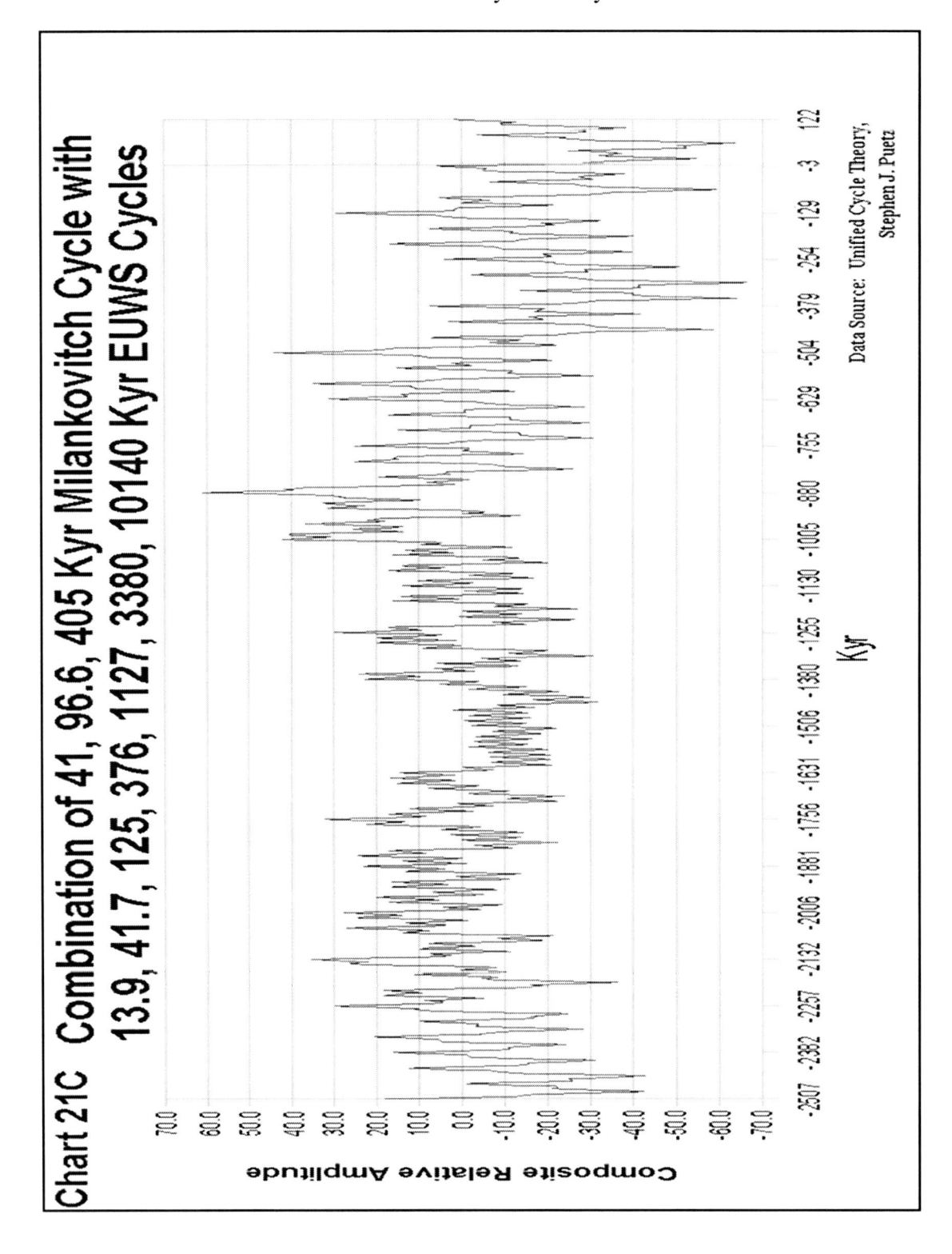

Chart 21C Combination of 41, 96.6, 405 Kyr Milankovitch Cycle with 13.9, 41.7, 125, 376, 1127, 3380, 10140 Kyr EUWS Cycles

Data Source: Unified Cycle Theory, Stephen J. Puetz

Chart 21D

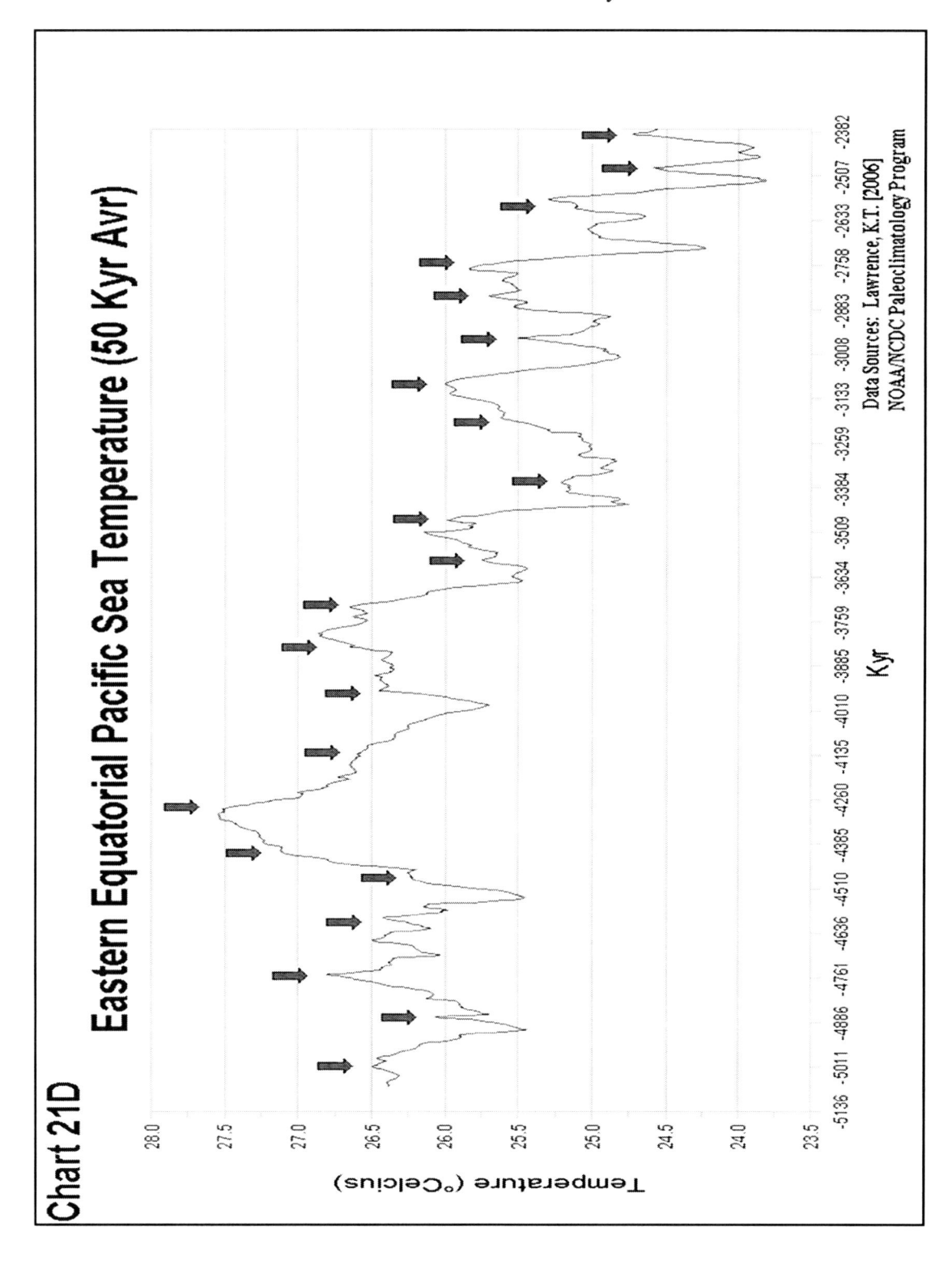

125 Thousand Year EUWS Cycle

Eastern Equatorial Pacific Sea Temperature (50 Kyr Avr)

Data Sources: Lawrence, K.T. [2006]
NOAA/NCDC Paleoclimatology Program

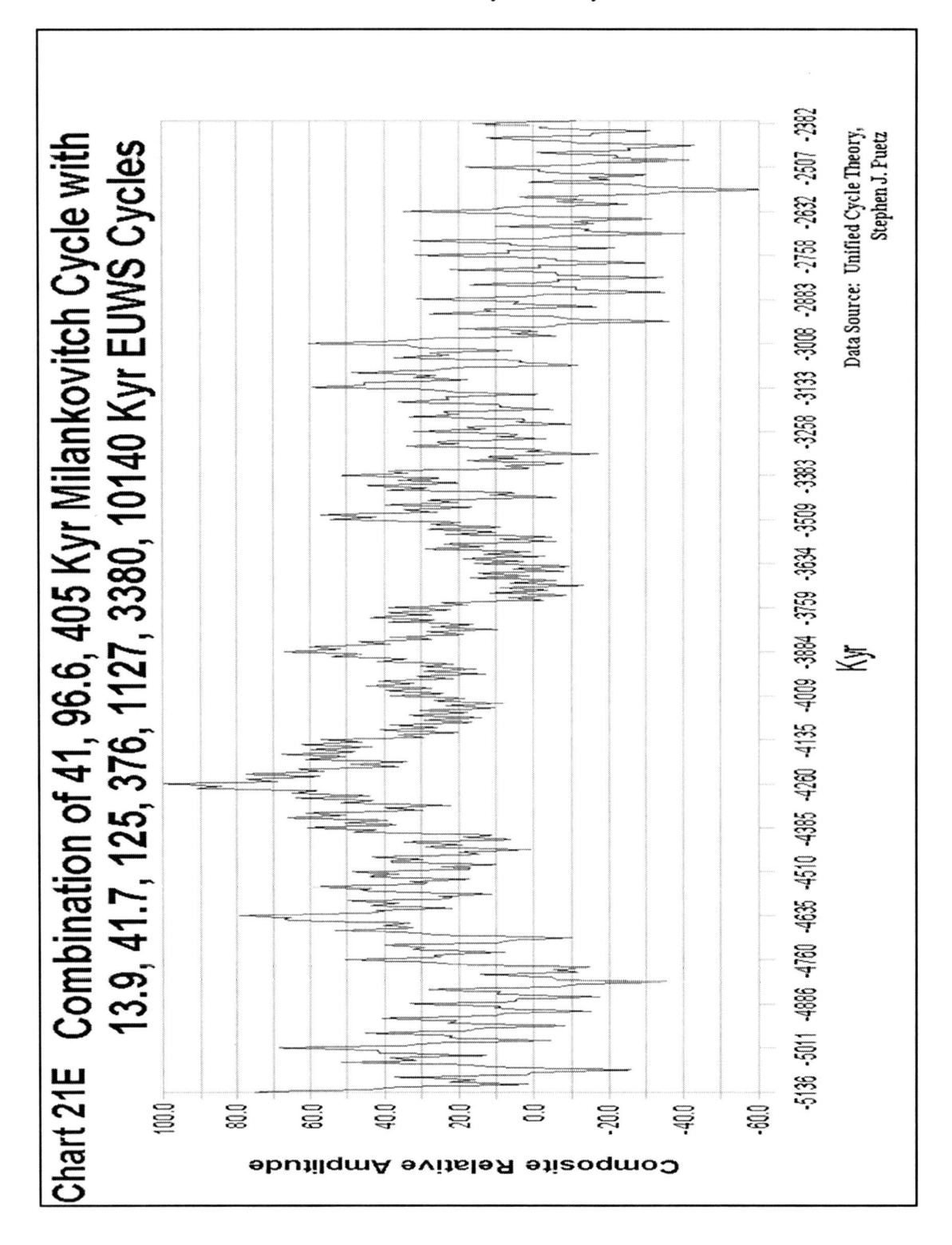

Chart 21E Combination of 41, 96.6, 405 Kyr Milankovitch Cycle with 13.9, 41.7, 125, 376, 1127, 3380, 10140 Kyr EUWS Cycles

Data Source: Unified Cycle Theory, Stephen J. Puetz

Chapter 22
41,728 Year EUWS Cycle

Dividing by three, the 125.184-kyr EUWS cycle gives birth to its 41,728 year child. The Milankovitch-EUWS battle reaches its climax with the 41.7-kyr cycle. Chapter 11 presented a list of problems associated with the Milankovitch Theory. In this chapter, the issues related to the 41.7-kyr EUWS cycle will be addressed.

Milankovitch Problem 2 – The 41,800 Year Frequency and Obliquity. As already noted in chapter 11, the V-1 borehole from the Pannonian Basin shows the presence of a 41,800-year cycle from 2.58 to .99 Ma – covering 39 cycles over 1.6 million years.[Nador et al., 2002] These oscillations provide a near-perfect match with the EUWS period of 41,728 years. This suggests that the 41.7-kyr EUWS cycle plays a more dominant role than obliquity in modulating Earth's temperature. That's because physicists generally estimate the obliquity frequency in a range from 40,000 to 41,000 years.

Milankovitch Problem 4 – Disappearing Cycles. Physicist John Wilkins noted that the five million year record from Lake Baikal showed evidence of the 41,000 year cycle that was particularly strongly 1.8 to 0.8 million years BP. The 100,000 year cycle only showed strength during the last 800,000 years. The record also showed major cooling episodes around 1.7 and 2.7 million years BP (corresponding to troughs of the 1.13-myr EUWS cycle). Finally, a 65 million year record showed effects that were quite varied. The 100,000 year cycle being very pronounced sometimes, and not large at others.[Wilkins, 2008]

The Unified Cycle Theory

The *Unified Cycle Theory* solves the conundrum of disappearing cycles. Two cycles with slightly different frequencies and roughly equivalent amplitudes produce alternating periods of visibility. When the two frequencies oscillate in sync, the combined amplitudes produce a highly visible cycle. However, when the cycles move out of sync, their amplitudes cancel, causing the cycles to gradually disappear. **Chart 22A** shows the cancellation/amplification effect from two cycles with frequencies of 41,000 and 41,728 years.

However, a good model replicating the cancellation/amplification effect cannot be built until the frequencies for both cycles become more accurately defined. Physicists generally place the frequency of the obliquity cycle between 40,000 and 41,000 years. As an example of the dramatic effect from a slight change in frequency, a 41,728 EUWS cycle oscillating with an obliquity frequency of 40,300 years will produce a cancellation/amplification cycle spanning 1.2 million years. In contrast, an obliquity frequency of 41,000 years produces a cancellation/amplification cycle of 2.4 million years. In other words, an increase of 700 years in the obliquity frequency produces a cancellation/amplification cycle that doubles in size! Hence, a good climate model cannot be built until these frequencies achieve better accuracy.

Milankovitch Problem 10 – Variable Frequencies, Dotted with Rapid Changes. If the gravity and motion periods associated with Milankovitch Theory served as the sole source of climate cycles, then ice-age cycles should produce temperature curves similar to sine-waves. For example, they should produce smooth, consistent curves similar to the daily and annual temperature cycles. But they fail to do that.

Instead, ice-age cycles exhibit erratic behavior – resembling the magnetic waves produced by the Schwabe sunspot cycle. Even though ice-age cycles show periodicity, individual periods vary quite significantly from one cycle to the next. In extreme cases, one cycle can cover a period twice as long as its predecessor. Furthermore, ice-age turning points tend to reverse with spikes. If gravity and motion acted as the sole source, they should exhibit smooth, sine-wave reversals.

For all frequencies, the erratic behavior displayed by long-term climate cycles indicates they originate from some type of electromagnetic source. Ed Mercurio, a professor with the Natural Sciences Department at Hartnell College, concentrated his efforts on galactic cosmic rays. At a minimum, a cosmic ray origin would at least satisfy the constraint that ice-age cycles possess an electromagnetic association.

First, Mercurio identified <u>solar magnetic cycles</u> with lengths of approximately 11, 22, 80, and 2400 years. Known effects on global weather result from these cycles. Next, he identified a separate set of cycles with approximate lengths of 13,000, 100,000, and 412,000 years. He noted the 100,000 year cycle operated in beats of 80,000 and 120,000 years. He classified these oscillations as <u>geomagnetic cycles</u>. Because it originates from gravity and motion, Mercurio separated the 41,000 year Obliquity Cycle from the others. Finally, Mercurio postulates that cycles in solar magnetism and geomagnetism modulate Earth-bound GCRs, thus creating climate cycles.[Mercurio, 2001]

41,728 Year EUWS Cycle

It's interesting that the lengths of Mercurio's cycles closely match the theoretical lengths of equivalent EUWS cycles. By throwing in the 1.2-myr cycle Wade and Palike attributed to eccentricity and obliquity, a suspicious looking comparison appears in **Table 22.1**.

Table 22.1 – EUWS Cycles Compared to Mercurio's Geomagnetic Cycles.

EUWS Freq. (years)	Mercurio's Ice-Age / Geomagnetic Period	Dev.	Comments
13,909	≈ 13,000 years	6.5%	B Angle Timing Cycle – span not given.
41,728	≈ 41,000 years	1.8%	Obliquity Cycle over 3 million years.
125,184	beat between ≈ 80,000 and ≈ 120,000 years	4.3% 4.1%	80,000 ≈ two 41,728 year EUWS cycles. 120,000 ≈ one 125,184 year EUWS cycle.
375,552	≈ 412,000 years	9.7%	Eccentricity Cycle. Only noted 2 cycles.
1,126,655	≈ 1.2 million years	6.1%	Wade & Palike – eccentricity/obliquity.

Every cycle mentioned by Mercurio (or Wade and Palike) comes within 10% of the estimated frequency of a corresponding EUWS cycle. By throwing out the 412-kyr cycle (because Mercurio only observed it for two cycles), the largest deviation amounts to only 6.5%! Also note that two 41.7-kyr cycles ≈ one 80-kyr alternating beat (in the 100-kyr ice-age cycle).

Once again, we're thrown into a case of multiple, moderate correlations adding complexity to cause-and-effect determinations. Before commenting on these correlations, another observation from Mercurio provides important clues for the investigation:

> **"At least nine ice-age [epochs] can be seen in the earth's history, with a variable spacing. They have an average spacing of approximately 300 million years and an average duration of approximately 40 million years. Ice age [epochs] seem to occur at times when the earth has high mountain ranges in certain configurations.... This produces a continental interior that can get both cold enough and wet enough to build and sustain continental glaciers of great size and area. This continent may then become a 'thermal pacemaker' for the entire world as North America has been for the ice-age we are currently in. There appears to be a cycle of worldwide tectonic relief that corresponds to the ice-age cycle with a periodicity that also averages ~300 million years."**[Mercurio, 2001]

Mercurio makes great observations; however, his conclusions appear questionable. Already presented in Chapter 14, the 274-myr EUWS cycle showed 13 oscillations between .268 and 3.827 Ga, with periods of high tectonic activity clustered around the theoretical peaks of the 274-myr cycle.

Earth expands near EUWS peaks. During the 22.2-gyr peak at 2.7 Ga, great volcanic explosions took place. High mountains form at major EUWS peaks as Earth attempts to expand. However, the formation of mountain ranges seems unrelated to the development of subsequent ice-age epochs. Their formation simply serves as an identifiable symptom of major EUWS peaks. During EUWS down-phases, mountains erode away as Earth cools and contracts.

Summarizing the evidence, the EUWS cycles tend to associate themselves with two basic forces:

A) **Gravitational tendencies:**

- All of the Milankovitch cycles correlate either directly to EUWS cycles, or as secondary harmonics of the EUWS cycles. This fact opens the possibility that EUWS cycles somehow regulate gravitational cycles.
- Directly corresponding to EUWS frequencies, Earth attempts to explode at peaks, and implode at troughs. This opens the possibility that EUWS cycles alter the state of gravity. Instead of being a constant, perhaps the gravitational force varies with phases of the EUWS cycles.
- Similar implosion-explosion cycles appear for every other 2.46-gyr EUWS oscillation corresponding to the star formation rate. Also, a potential Big Crunch - Big Bang reversal occurred shortly after the most recent 22.2-gyr trough. These occurrences suggest that EUWS cycles modulate gravitational cycles in some manner.

B) **Magnetic tendencies:**

- Through all timescales, global temperatures fluctuation in sync with EUWS cycles. And when they reverse direction, they do so in a spiked manner. This behavior indicates the EUWS cycles come from an electromagnetic origin.
- Through all timescales, global temperatures fluctuate with a periodicity that can vary substantially at times. This tendency indicates that EUWS cycles come from an electromagnetic origin.
- The EUWS cycles correlate closely to GCR cycles. This correlation doesn't directly imply that EUWS cycles are magnetic. However, it reinforces their association with electromagnetic cycles.
- The EUWS cycles correlate with oscillations in extinctions, evolution, and life through all timescales. Electromagnetic forces affect all forms of life, unlike the other basic physical forces. This correlation to life strongly implies that EUWS cycles contain an electromagnetic component.

Ultimately, EUWS cycles may originate from neither gravitational nor electromagnetic sources. They could come from something more basic – perhaps something that modulates gravity and magnetism. Alternatively, the global explosions that occur near EUWS peaks could result from tremendous geomagnetic storms rather than some type of gravitational cycle. That being the case, the evidence suddenly tilts heavily in favor of the EUWS cycles originating from electromagnetism.

41,728 Year EUWS Cycle

Milankovitch Problem 11 – A Catch All for Every Climate Cycle. At this point, repeating the harsh criticism given in the abstract *A Causality Problem for Milankovitch* seems worthwhile:

> **"If we allow every discordant measurement to have its own explanation, we do not have a theory that can make predictions, and that means it really isn't a theory at all."**[Karner and Muller, 2000]

Without a doubt, Milankovitch mechanics receive too much credit for climate variability. And by approximating Milankovitch frequencies so closely, the EUWS cycles surely act as the invisible forces that create most global climate changes – but with Milankovitch cycles receiving the credit! Eventually, a better climate model will consist of a combination of both Milankovitch cycles and EUWS cycles. As an example, **Chart 22B** combines three Milankovitch cycles with seven EUWS cycles.

Notice the large drop in global temperatures predicted for the next 50,000 years. Following that, the model projects a substantial warm-up until around 250,000 AP – bringing global temperatures well above current levels. After that, the coldest ice-age in the last 600 million years should develop around 400,000 AP!

Getting back to the 41.7-kyr cycle, **Chart 22C** shows fluctuations in Eastern Pacific Sea Temperatures that closely match corresponding EUWS peaks and troughs. Even though the 41.7-kyr cycle holds the distinction of being one of the most significant and documented cycles, it doesn't appear with enhanced visibility. Similar to the other EUWS cycles, occasional shifts – mixed in with a few cycles displaying weak amplitudes – make it hard to identify for some cycles. Yet, after these occasional deviations, the fact that it always returns to its theoretical turning points identifies it as EUWS. And that summarizes how the 41.7-kyr cycle behaved in **Chart 22C**.

The next chart strengthens the case that EUWS cycles posses a magnetic origin. Unfortunately, reliable geomagnetic records only cover a relatively short span of Earth's history – only 0.01% of Earth's existence, to be exact. Working for the Institut de Physique de Globe, Yohan Guyodo and Jean-Pierre Valet authored an abstract entitled *Global Changes in Intensity of the Earth's Magnetic Field during the Past 800 Kyr*.

In spite of its brevity, their data showed a distinct 41,728 year cycle in geomagnetism.[Guyodo and Valet, 1999] See **Chart 22D1**. Furthermore, these geomagnetic cycles lead changes in Pacific Sea Temperatures (**Chart 22C**) by about 20 to 40 thousand years. In comparing the charts, notice how even small jig-jags in paleointensity correlate to tiny climate fluctuations – with paleointensity leading temperature changes by 20 to 40 thousand years. This amazing lead-time correlation elevates the prospects of Earth's magnetic field acting as either a direct or indirect cause of global climate change.

Chart 22E shows the 41,728 frequency in Pacific Sea Temperatures between 1 and 2 million years ago. During this period, the oscillations match theoretical EUWS peaks with even more precision than the prior period. If its frequency equaled 41,000 years, rather than 41,728, then the entire sequence should have shifted by a complete cycle during the past 2 million years. But it didn't. This indicates that the widely reported 41-kyr cycle actually equals the 41.7-kyr EUWS cycle. This fact further damages the Milankovitch interpretation.

Chart 22F shows the model consisting of 3 Milankovitch cycles and 7 EUWS cycles. Notice how the model tends to lead changes in Pacific Sea temperatures by 20,000 to 40,000 years.

Recently, a German research team from the University of Heidelberg, led by M. Christl, dedicated an abstract to the growing conviction that magnetic field fluctuations greatly contribute to global temperature changes. Unfortunately, reliable geomagnetic records only cover the last million years of Earth's history. In their abstract entitled *Evidence for a Link between the Flux of Galactic Cosmic Rays and Earth's Climate during the Past 200,000 Years*, the researchers review the evidence:

> **"Numerous reconstructions of GPI from magnetic remanence measurements in sediment cores are available.... The question whether large climatic changes during the last hundreds of thousands of years may be associated with changes of the Earth's magnetic field has been subject to many investigations ... but led to ambiguous results: Orbital frequencies that indicate a link between the Earth's magnetic field and climate have been found in several paleointensity (PI) records.... For the Ontong–Java Plateau, PI was found to correlate with temperature ... and a correlation between PI and climate may be suspected. However, Worm (1997) rejected the suggested link because the discrepancy in results between the different records.... In contrast, a recent high-resolution study by St-Onge *et al*. (2003) suggests that a link between GPI, GCR and climate existed throughout the Holocene on centennial to millennial timescales.... The different results highlight the complications associated with the development of a reliable PI record based on magnetic remanence measurements."**[Christl *et al.*, 2003]

In spite of these complications, by linking known geomagnetic data with the 41,728 year EUWS cycle, a clear connection emerges. But this correlation fails to prove anything. It merely serves as another item in a long list of correlations to EUWS cycles. More than likely, all correlations act as symptoms derived from a more basic origin at the core of the EUWS. And their nature (especially their exceedingly long frequencies) indicates their core lies outside of our known universe – originating from inside an infinitely large multiverse.

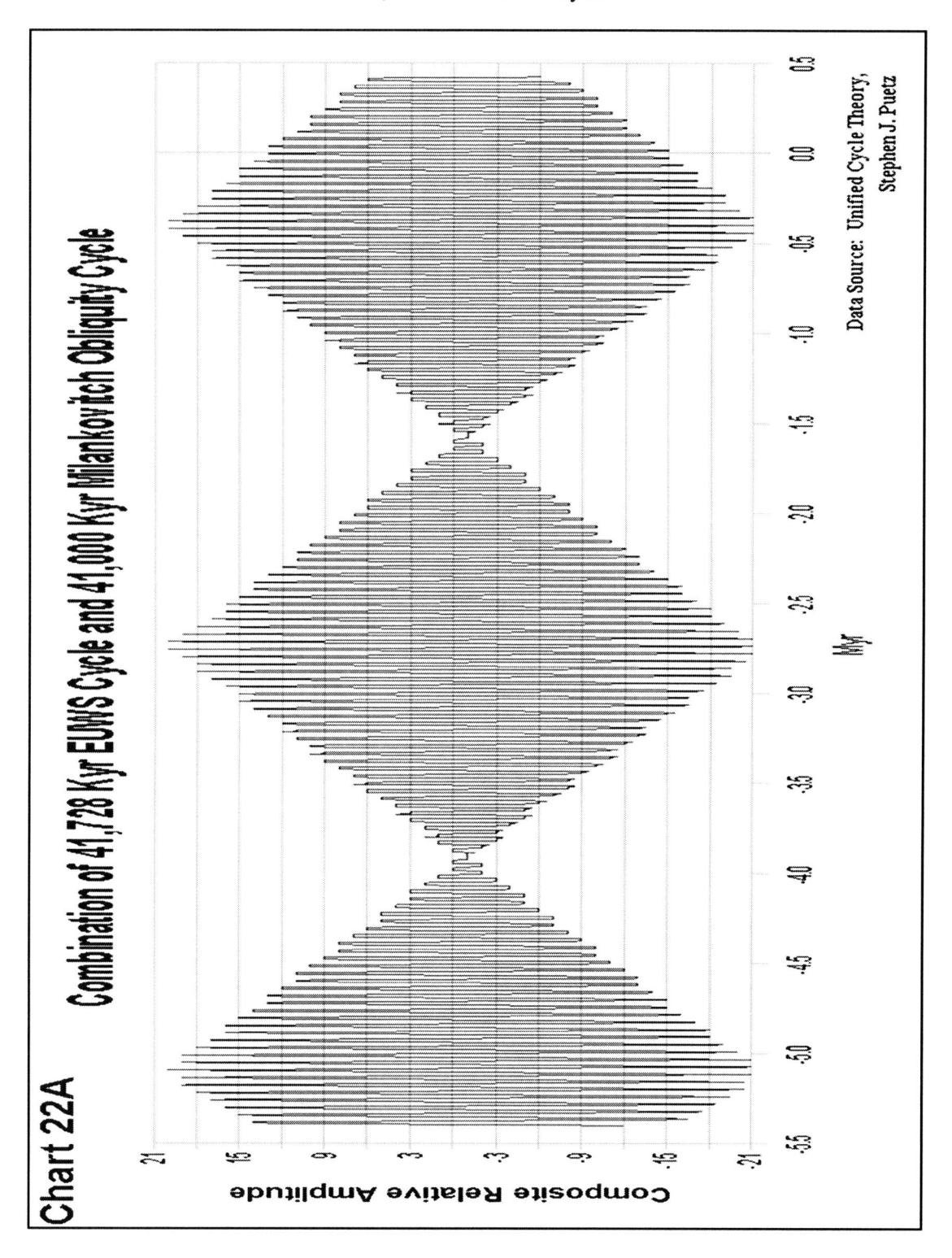

41,728 Year EUWS Cycle

Chart 22A Combination of 41,728 Kyr EUWS Cycle and 41,000 Kyr Milankovitch Obliquity Cycle

Data Source: Unified Cycle Theory, Stephen J. Puetz

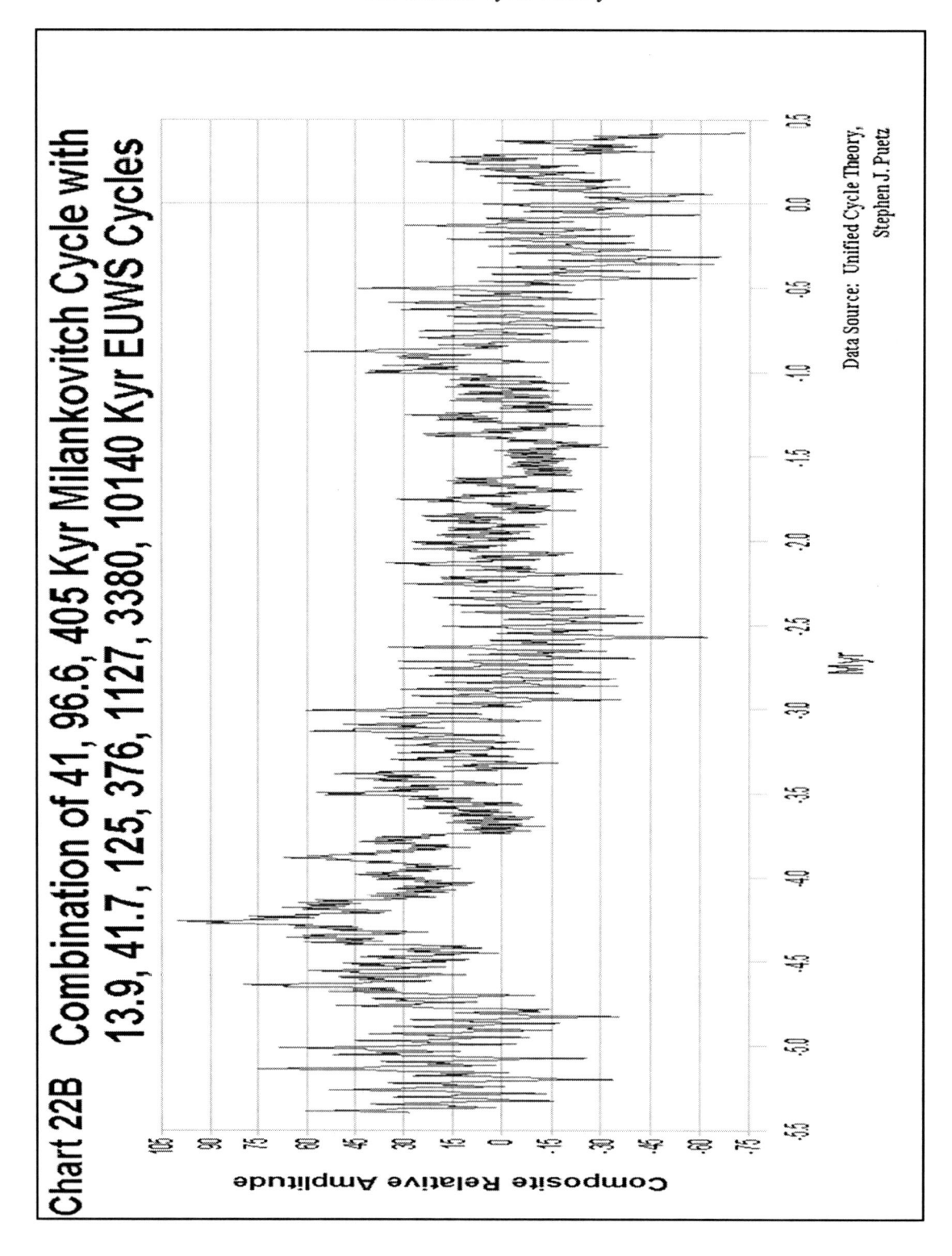

Chart 22B Combination of 41, 96.6, 405 Kyr Milankovitch Cycle with 13.9, 41.7, 125, 376, 1127, 3380, 10140 Kyr EUWS Cycles

Data Source: Unified Cycle Theory, Stephen J. Puetz

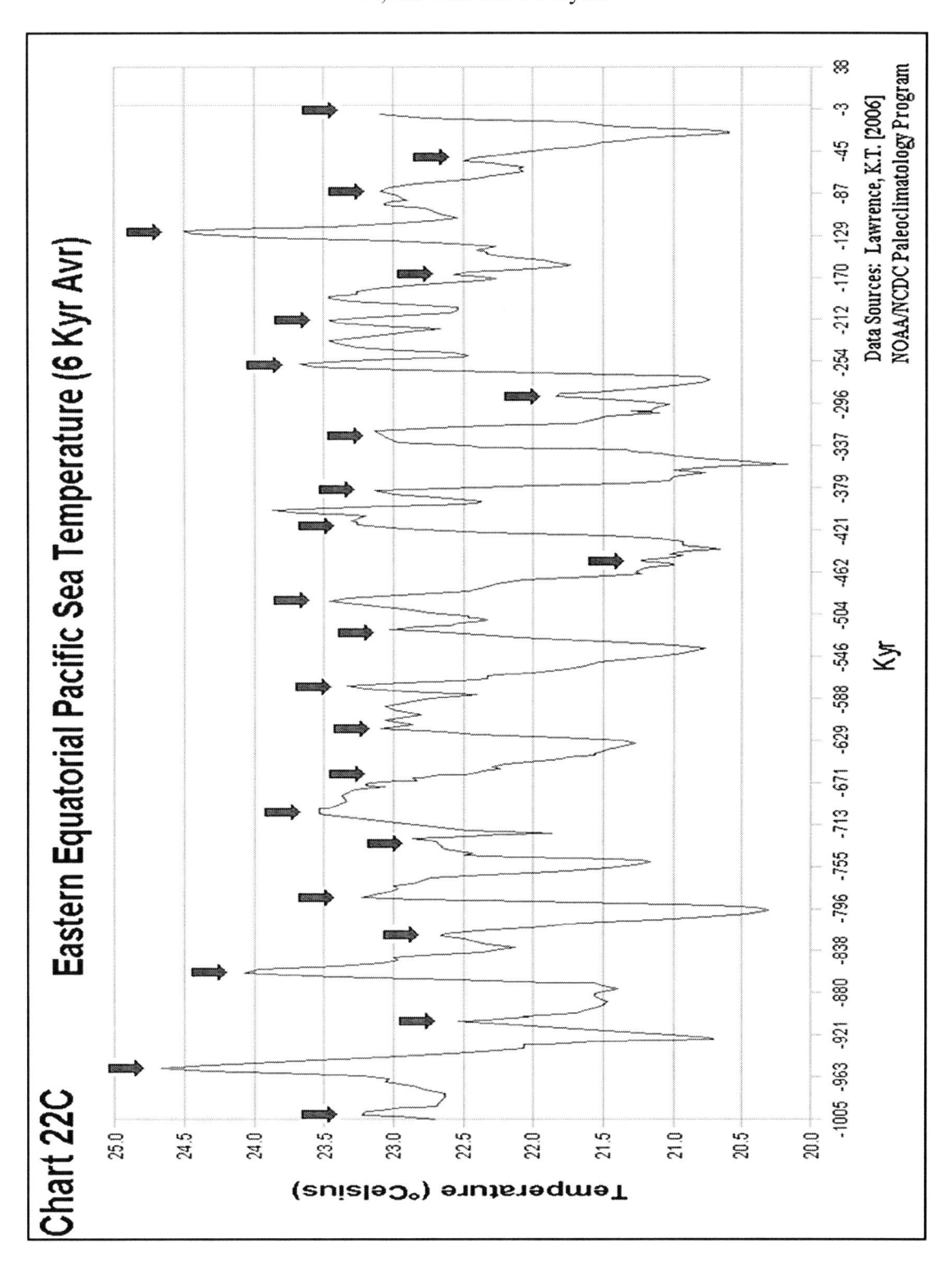

41,728 Year EUWS Cycle

Chart 22C — Eastern Equatorial Pacific Sea Temperature (6 Kyr Avr)

Data Sources: Lawrence, K.T. [2006]
NOAA/NCDC Paleoclimatology Program

The Unified Cycle Theory

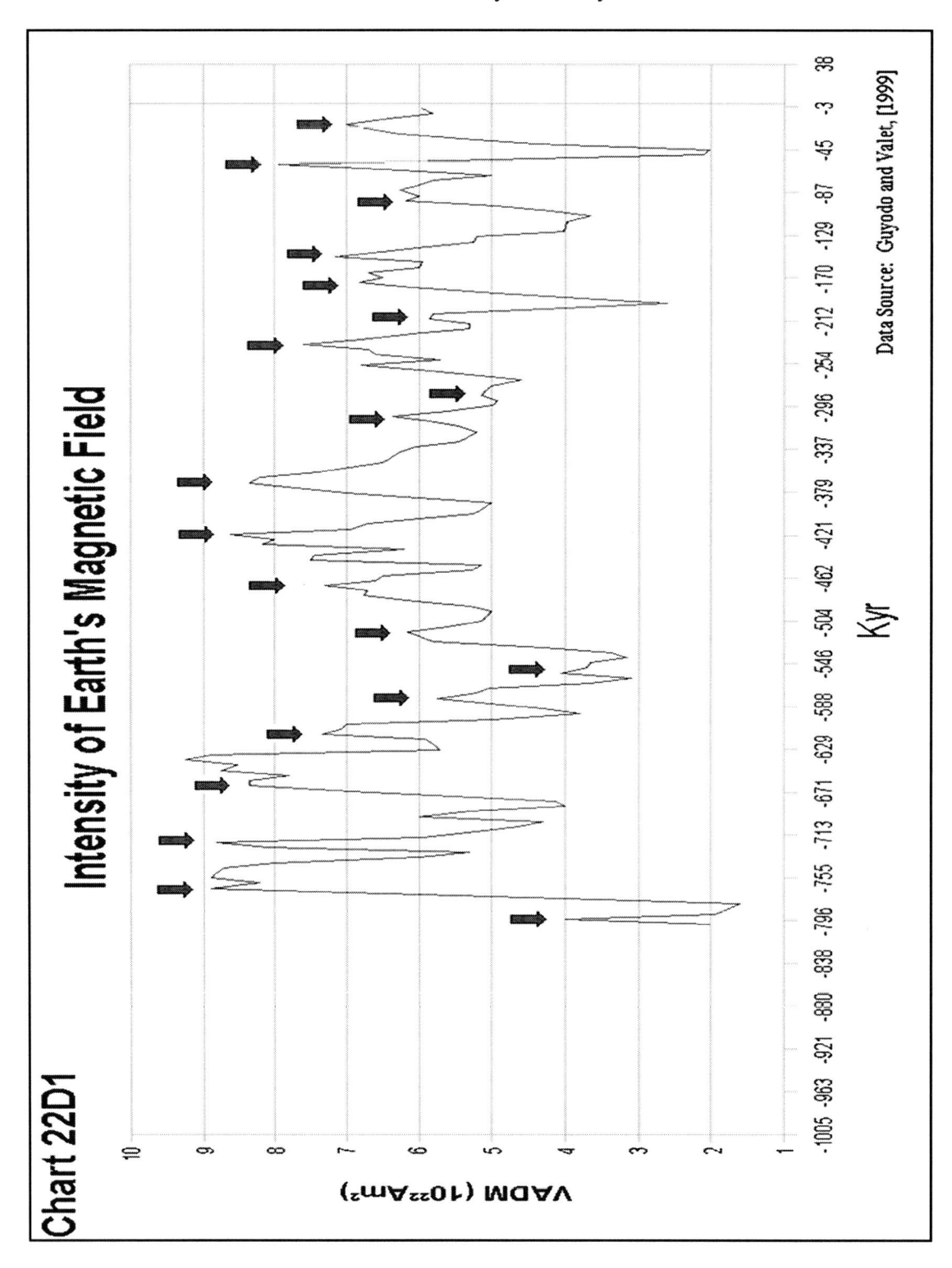

Chart 22D1: Intensity of Earth's Magnetic Field

Data Source: Guyodo and Valet, [1999]

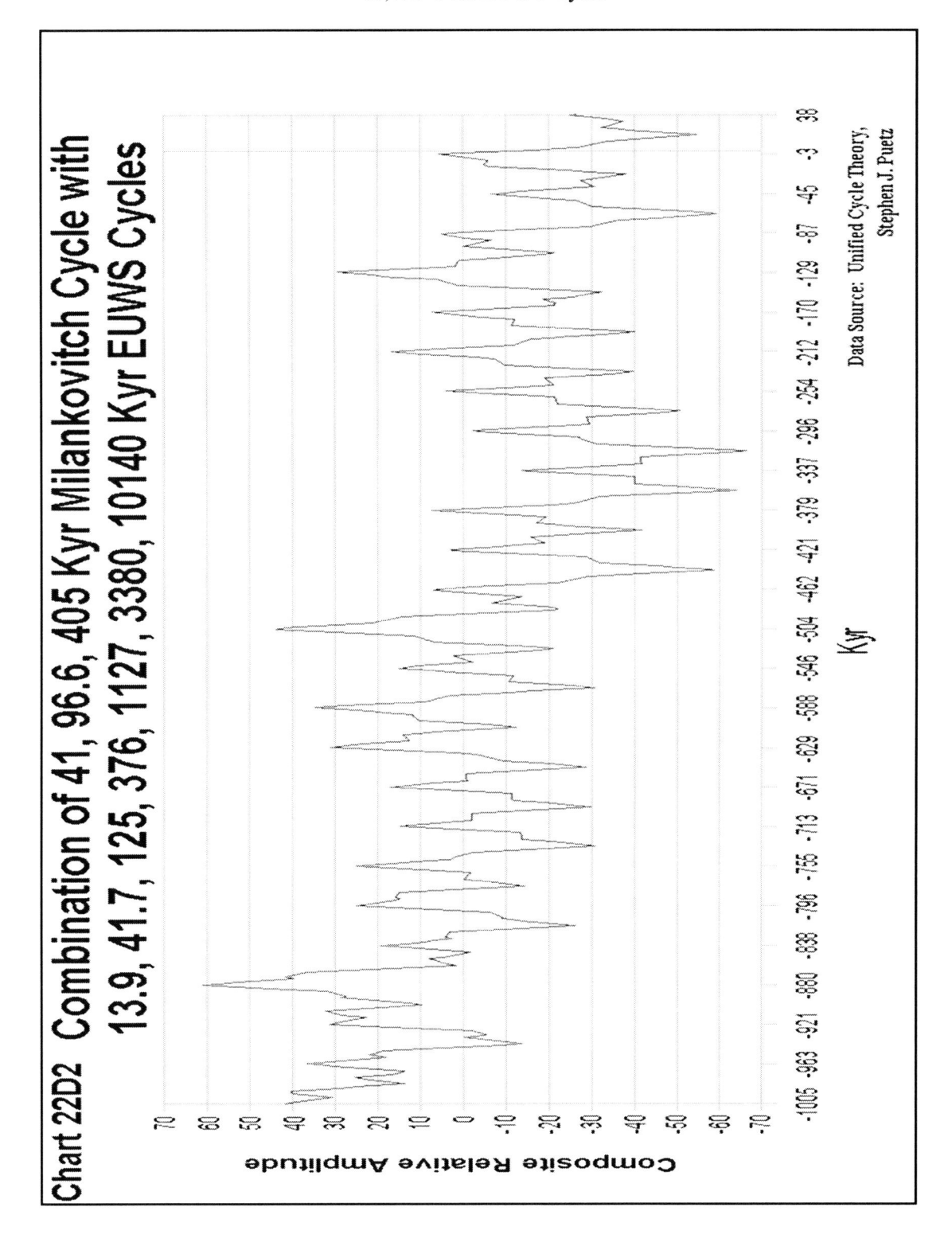

41,728 Year EUWS Cycle

Chart 22D2 Combination of 41, 96.6, 405 Kyr Milankovitch Cycle with 13.9, 41.7, 125, 376, 1127, 3380, 10140 Kyr EUWS Cycles

Data Source: Unified Cycle Theory, Stephen J. Puetz

The Unified Cycle Theory

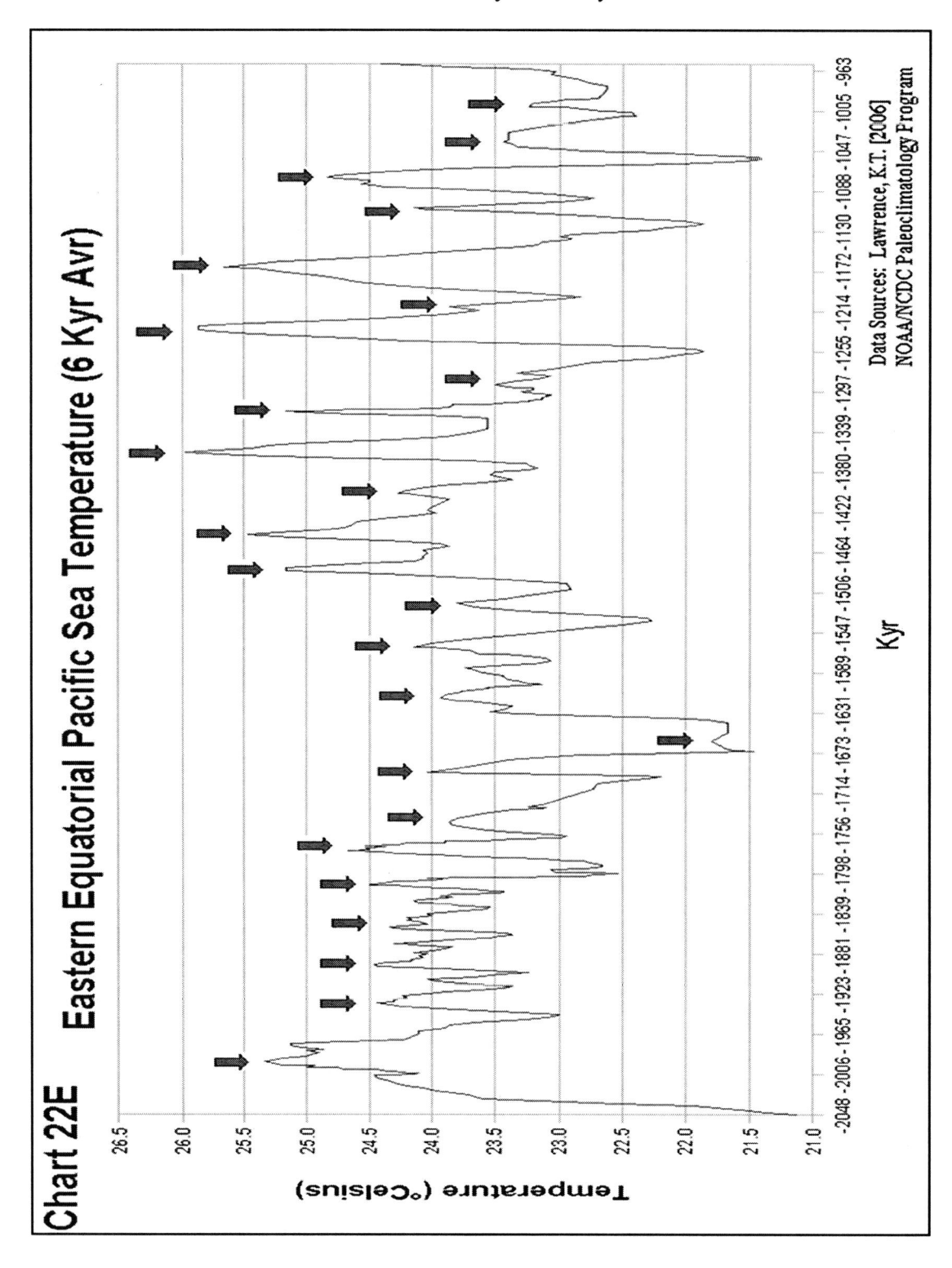

Chart 22E — Eastern Equatorial Pacific Sea Temperature (6 Kyr Avr)

Data Sources: Lawrence, K.T. [2006]
NOAA/NCDC Paleoclimatology Program

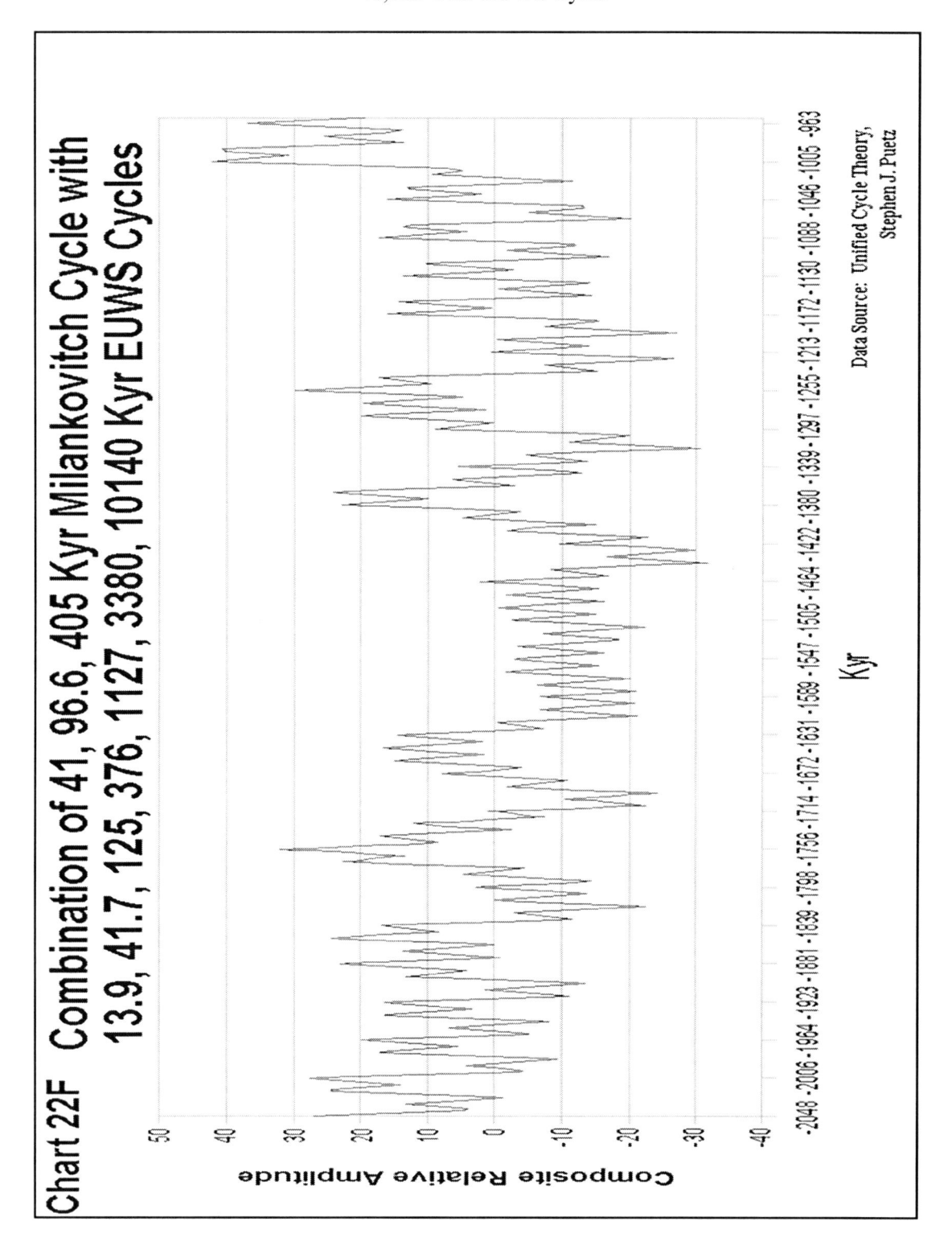

41,728 Year EUWS Cycle

Chart 22F Combination of 41, 96.6, 405 Kyr Milankovitch Cycle with 13.9, 41.7, 125, 376, 1127, 3380, 10140 Kyr EUWS Cycles

Data Source: Unified Cycle Theory, Stephen J. Puetz

Chapter 23
13,909 Year EUWS Cycle

The great attention given to the 41,728 year cycle quickly fades away when focus shifts to its child – the 13,909 year EUWS cycle. References to the 13,909-year cycle rarely appear. This lack of attention may be attributable to the obsession with the Milankovitch cycles that operate directly above it.

However, as indicated in Chapter 22, Ed Mercurio, a professor with the Natural Science Department at Hartnell College, emphasizes the geomagnetic nature of a 13,000 year cycle.

> **"Changes in geomagnetism are used to explain glacial-interglacial and ~13,000 year cycles.... The longest climatic cycles that appear to be modulated by GCRs are glacial-interglacial cycles and a ~13,000 year cycle. In these cycles, GCRs are modulated by geomagnetism instead of solar magnetism."**[Mercurio, 2001]

Chart 23A shows the 13.9-kyr cycle in the Greenland Ice Core (GISP2). The charts mark 41.7-kyr peaks with a double arrow. Single arrows represent 13.9-kyr peaks. Actual tops closely match theoretical peaks, except for two shifts at 31.2 Ka and 17.3 Ka.

Also notice the temperature plateau that formed during the Holocene. This is highly unusual. Normally, EUWS tops appear as spikes – not plateaus. When climatologists look at the stability in global temperatures during the past 11 thousand years, they observe an anomaly.

Perhaps this deviant behavior results from the 125-kyr theoretical peak that occurred 3427 years ago. In terms of a 125-kyr cycle, 3 thousand years amounts to virtually nothing. It will take another 60-kyr before the 125-kyr cycle hits bottom. **Chart 23B** shows the previous 125-kyr peak at 129 Ka. The chart marks the 129 Ka top with four arrows because it also represents a 376-kyr maximum.

Summarizing the projected trend for EUWS oscillations in the coming 60 thousand years, the 376-kyr cycle peaked at 129 Ka, and it will continue heading lower for another 60 thousand years. The 125-kyr cycle peaked 3427 years ago, and it also reaches its nadir in 60 thousand years. And the 41.7-kyr and 13.9-kyr cycles also peaked 3427 years BP (1427 BC). The 13.9-kyr cycle projects substantially cooler temperatures for the next 3500 years. Working in combination, the model indicates that EUWS cycles will push global temperatures significantly lower over the next 60,000 years – as always, doing so by ratcheting down in a stair-step fashion.

The 13,909 year EUWS cycle appears more prominently in the Eastern Equatorial Pacific Ocean record than it does in the Greenland ice core. **Charts 23B and 23C** show the Pacific Ocean history over the past 615 thousand years. Three arrows mark 125-kyr peaks; double arrows mark 41.7-kyr peaks; while single arrows indicate 13.9-kyr peaks. In the charts, notice how some of the major cycles deviate from their theoretical peaks by one 13.9-kyr sub-cycle. These deviations happen because ocean temperatures often fluctuate with a 20 to 40 thousand year delay from theoretical turning points – as previously noted.

Even though it lacks the extensive documentation enjoyed by its 41,728 year parent, the 13,909 year EUWS cycle leaves an impressive trail – closely matching theoretical peaks through 44 cycles in Pacific Ocean temperatures.

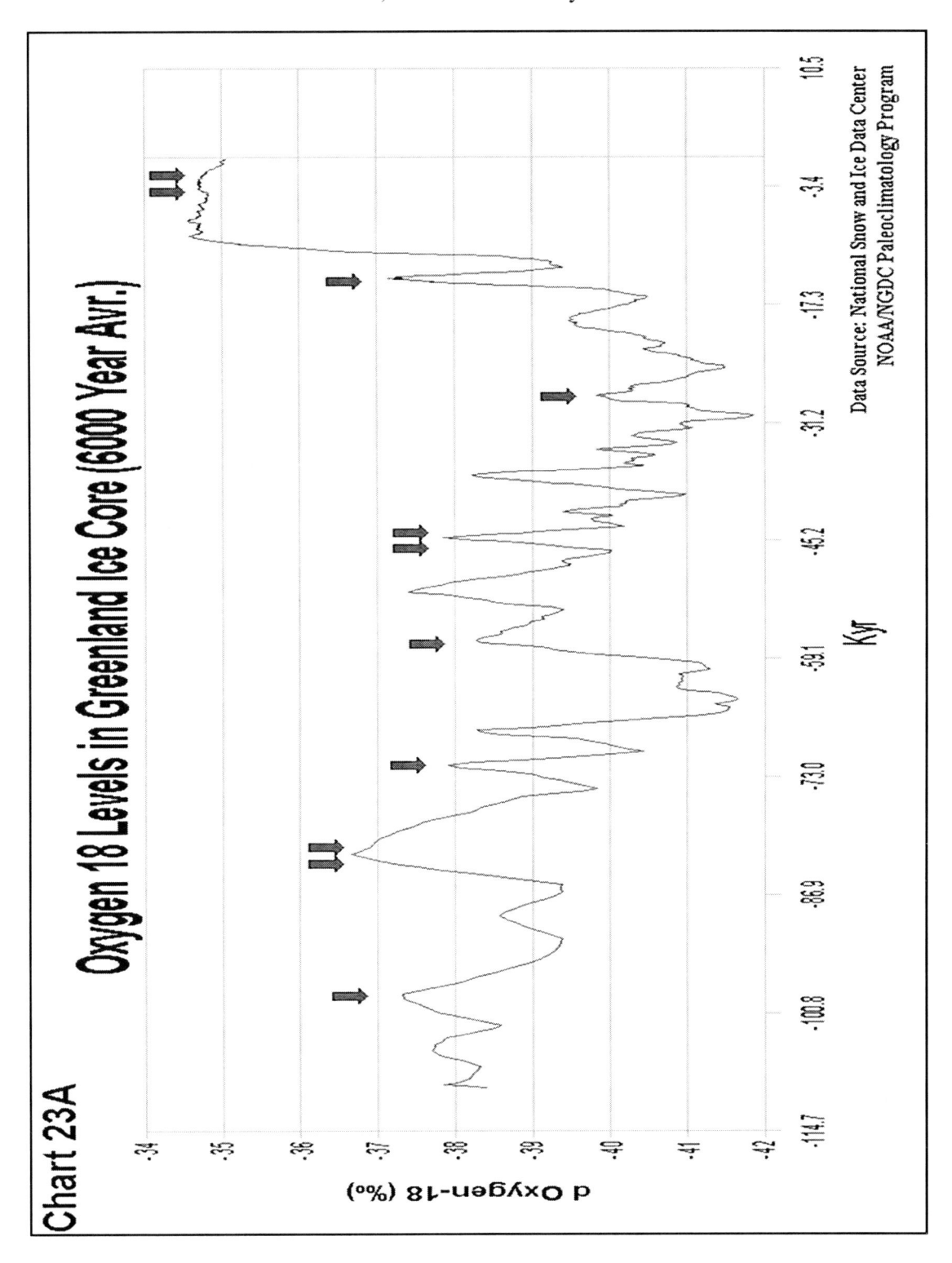

13,909 Year EUWS Cycle

Chart 23A — Oxygen 18 Levels in Greenland Ice Core (6000 Year Avr.)

Data Source: National Snow and Ice Data Center
NOAA/NGDC Paleoclimatology Program

Chart 23B

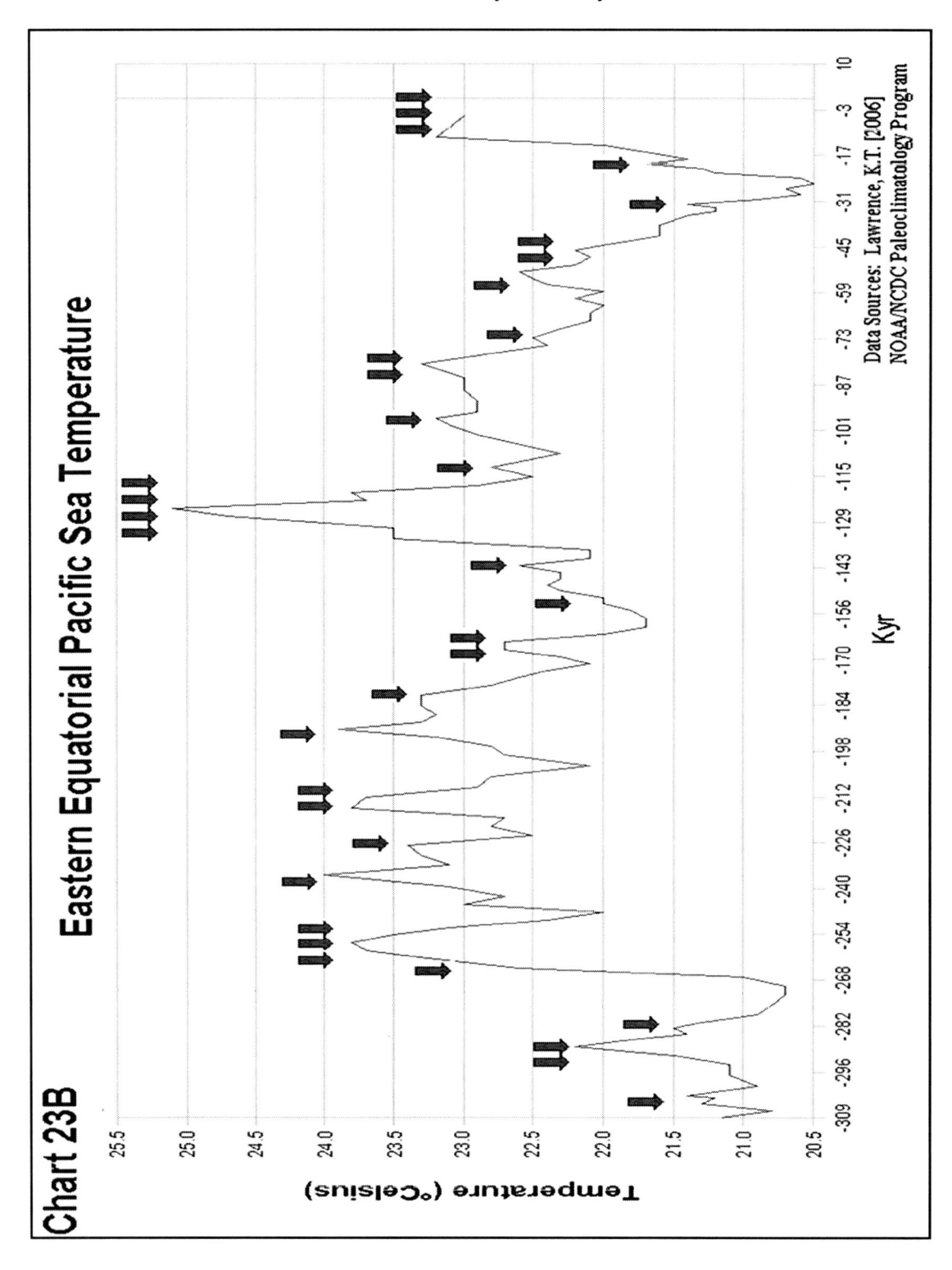

Eastern Equatorial Pacific Sea Temperature

Data Sources: Lawrence, K.T. [2006]
NOAA/NCDC Paleoclimatology Program

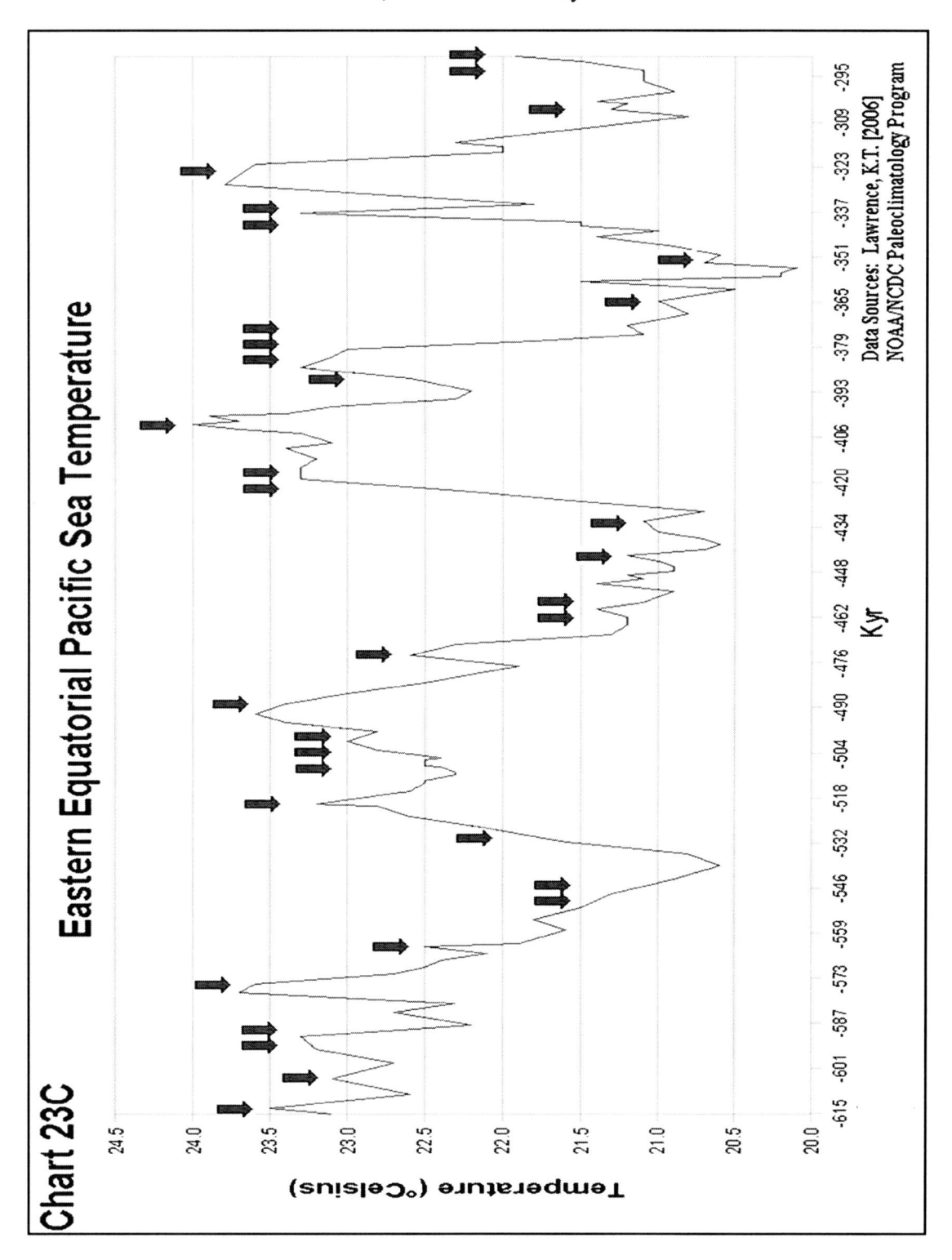

13,909 Year EUWS Cycle

Chapter 24
4,636 Year EUWS Cycle

Next in the sequence comes the 4636-year EUWS cycle. This cycle enjoys moderate attention.

A research team led by Gerard Bond of the Lamont-Doherty Earth Observatory of Columbia University, studied North Atlantic Ocean sediments. In their abstract, entitled *A Pervasive Millennial-Scale Cycle in North Atlantic Holocene and Glacial Climates*, the team found the following cycles:

> "**For the glacial interval, the mean pacing is 1536 ±563 years, essentially the same as a 1450-year cycle identified in the glacial portion of the Greenland Ice Sheet Project 2 (GISP2) geochemical series.... Thus, the spacings of the Holocene and glacial events are the same statistically, and together the two series constitute a cyclic signal centered on ~1470 ±532 years.... In addition, F variance ratio tests reveal lines with .95% probability at 4670 ... years.**"[Bond *et al.*, 1997]

Cycles in the range of 1500 years often go by the name of Dansgaard-Oeschger cycles. They will be reviewed in the next chapter. For this chapter, focus centers on the cycle Bond *et al.* identified at 4670-years – which comes within 0.7% of the 4636-year EUWS cycle. Cycles in the range of 4500 to 6000 years, named after their discoverer, are often called Bond cycles.

Led by climatologist Andre Berger from the Catholic University de Louvain in Belgium, another research team studied climate conditions near the equator. In their abstract entitled *Equatorial Insolation: From Precession Harmonics to Eccentricity Frequencies*, they summarized the results:

"The spectrum ... shows the 400, 100, 41, ~10 and ~5-kyr quasi-periods, all being significant, especially those of ~100 and ~5-kyr." [Berger *et al.*, 2006]

Similar to other spectral sequences, the equatorial climate cycles identified by Berger's team roughly fall into the EUWS pattern of 376, 125, 41.7, 13.9, and 4.6-kyr cycles. While the 5000-year cycle isolated by Berger's group deviates more than desired, the errors associated with these estimates puts it within tolerance of the 4636-year EUWS frequency.

In addition to the large number of other physical influences, the EUWS cycles appear to modulate sunspot cycles. To observe the 4636-year cycle in solar activity, NOAA's Paleoclimatology Program provides reconstructed sunspot numbers covering the last 11,400 years.[Solanki *et al.*, 2005] Constructing the series involved Carbon-14 dating of tree rings. While it only represents 1½ cycles, a smoothed version of the data in **Chart 24A** shows that sunspot peaks and troughs almost perfectly match recent theoretical turning points of the 4636-year EUWS cycle.

Chart 24B shows the 4636-year cycle in temperatures constructed from Greenland ice cores.[Alley, 2004] In this particular sequence, temperatures topped precisely at the theoretical peaks at 45.15, 8.06, and 3.43 Ka. However, in between those times, all of the 4636-year tops shifted forward by approximately one 1545-year sub-cycle – providing an example of the TPD Principle behaving as a coordinated shift. This extensive shift could explain why some analysts perceive this as a 5-kyr to 6-kyr cycle, rather than its true frequency at 4636 years.

While evidence supporting the 4636-year EUWS frequency lacks the desired longevity, the data that can be found matches the expected patterns rather well. However, from this point forward, EUWS evidence mushrooms to a point where it cannot be presented in its entirety. Especially interesting in all the remaining chapters, the EUWS cycles exert a tremendous influence on human behavioral patterns.

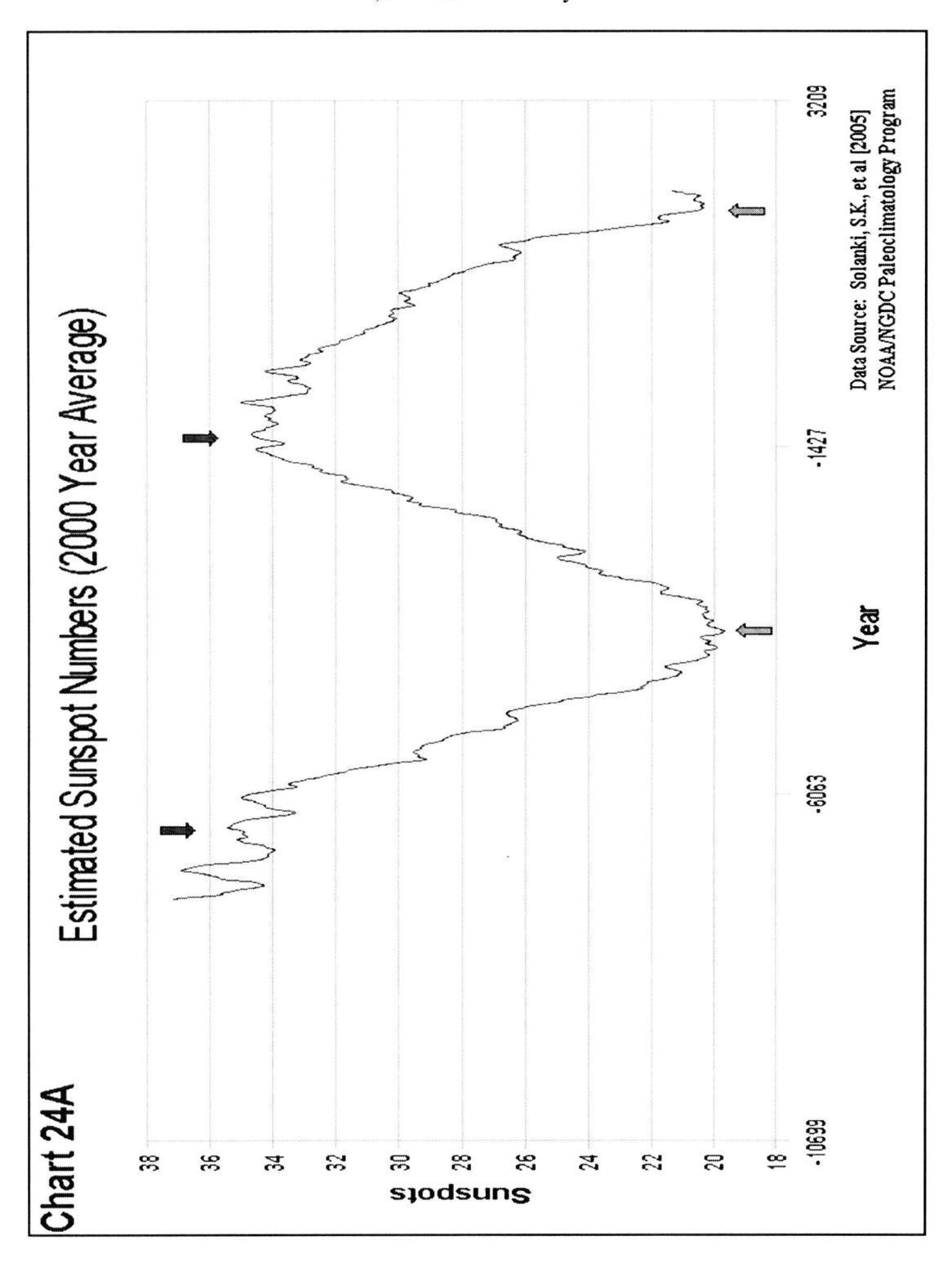

4,636 Year EUWS Cycle

Chart 24A — Estimated Sunspot Numbers (2000 Year Average)

Data Source: Solanki, S.K., et al [2005]
NOAA/NGDC Paleoclimatology Program

The Unified Cycle Theory

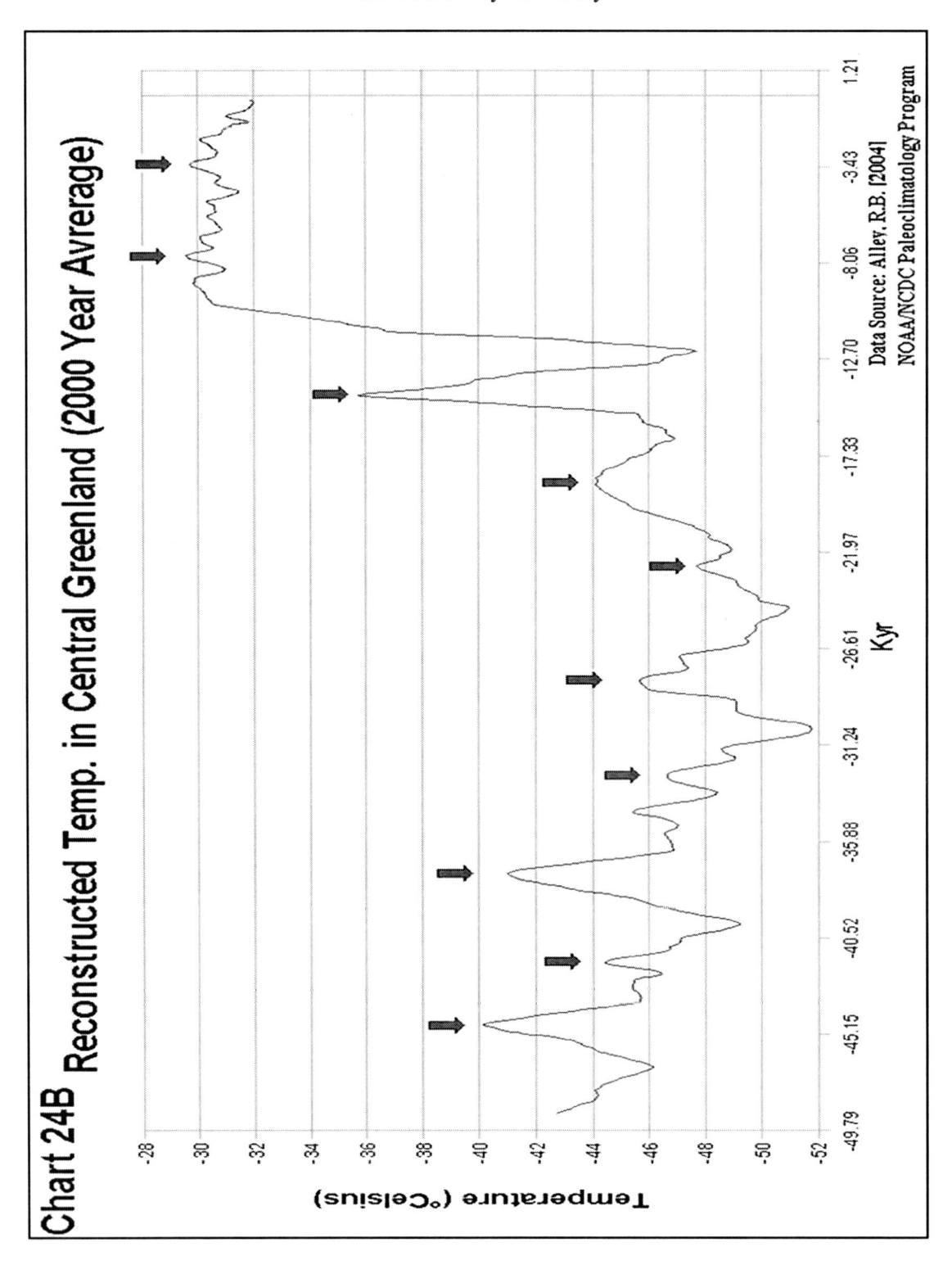

Chart 24B Reconstructed Temp. in Central Greenland (2000 Year Average)

Data Source: Alley, R.B. [2004]
NOAA/NCDC Paleoclimatology Program

270

Chapter 25
The 1545-Year EUWS Cycle

Next in the EUWS comes the 1545-year cycle. Climatologists often refer to this as the Dansgaard-Oeschger cycle – named in honor of Danish paleoclimatologist Willi Dansgaard, a Professor Emeritus of Geophysics at the University of Copenhagen, and German paleoclimatologist Hans Oeschger, Founder of the Division of Climate and Environmental Physics at the University of Bern. The pair consistently came up with innovative ways of determining ancient climate conditions, and both were heavily involved in the Greenland ice core projects. They became the first scientists to notice continual fluctuations in Greenland temperatures.

In addition to Greenland ice cores, the 1545-year cycle appears in numerous other records. A team of geologists, led by Yemane Asmerom from the University of New Mexico, presented the first high resolution climate proxy for the Southwest United States in an abstract entitled *Solar Forcing of Holocene Climate: New insights from a Speleothem Record, Southwestern United States*. They conducted the study by analyzing delta Oxygen-18 variations in a speleothem spanning the entire Holocene, and then matching it with existing Carbon-14 records. After analyzing both sets of data, the team concluded:

> **"The two records have matching periodicities at 1533, 444, 170, 146, and 88 yr above the 95% confidence interval."**[Asmerom *et al.*, 2007]

The 1533 and 170 year cycles almost perfectly match the EUWS cycles of 1545 and 172 years. However, the 444 year frequency falls a little short of the 515-year EUWS cycle (which comes in the next chapter). And the 88-year frequency matches the well-known Gleissberg solar cycle.

Ed Mercurio also observed this cycle in his abstract *The Effects of Galactic Cosmic Rays on Weather and Climate on Multiple Time Scales*. In the abstract, Mercurio noted the variability associated with the cycle:

> **"The periodicities of and variations in intensities of maxima and minima often give the appearance of 1400 to 1600 year cycles in paleorecords."**[Mercurio, 2001]

A John Hopkins University website, entitled *Chronos – Cyclostratigraphy Online Database & Research Center*, describes the mystery surrounding this cycle, which also applies to the other EUWS cycles:

> **"Paleoclimatologists have become increasingly focused on the fact that the high amplitude 'Dansgaard-Oeschger' climate oscillations of the Last Glacial Cycle in the Greenland ice core records appear to have an unusually precise timing... Some researchers doubt that such a stable cycle derives from internal oscillatory behavior in the climate system, thought instead to be comparatively irregular, but an external source of this 1470-year 'clock' has yet to be identified. Once an origin has been identified, this millennial cycle may also qualify as a high-precision calibrator of geological time."**[John Hopkins Univ., 2008]

Taking these Chronos observations, and applying then to all EUWS cycles, the preceding statements can be modified ever so slightly to read as follows:

(1) Yes, all EUWS cycles exhibit unusually precise timing.
(2) Yes, the EUWS source is external to the climate system.
(3) The EUWS source, while still not fully identified, does appear to originate from outside our universe.
(4) Yes, the long-term EUWS cycles do correlate well with geological timescales, and they probably can be used for better geological calibration.

Up to this point, the preceding estimates equaled 1470, 1500 (1400 to 1600), and 1533 years – all slightly below the 1545-year frequency of the EUWS cycle now under review. However, on the other side of the world, researches place the Dansgaard-Oeschger period at 1550 years – within five years of the 1545-year EUWS cycle! A group led by Anil Gupta studied the Indian monsoon in an abstract entitled *Solar Influence on the Indian Summer Monsoon during the Holocene*. The team detailed their findings:

> **"It is remarkable to observe the 1550 year cycle in the summer monsoon record, which has been noted in numerous climate records during the last glacial as**

well as the present interglacial... The presence of this periodicity, a part of Dansgaard/Oeschger cycles, in the North Atlantic ..., as well as the summer monsoon record strengthens the sun-monsoon-North Atlantic link."[Gupta et al., 2005]

Gupta *et al.* attribute the monsoon cycle to the Sun. And, in fact, by taking a 750-year average of NOAA's Paleoclimatology sunspot estimates (discussed in the previous chapter), **Chart 25A** confirms a 1545-year periodicity in solar activity.

However, it's necessary to distinguish between internal solar cycles (such as the Schwabe and Gleissberg cycles) and those being modulated by an external source – more than likely modulated by EUWS cycles. A growing minority of scientists believe that GCRs serve as the modulating external source. But once again, GCR cycles and solar cycles may correlate without one causing the other. Conceivably, CO_2 cycles, long-term solar frequencies, and GCRs oscillations correlate because they share the same common cause – with the common source being various EUWS cycles. However, for now, simply note the correlations, and wait for the complete set of evidence.

In **Chart 25A**, the actual sunspot turning points represent near-perfect hits of the theoretical peaks and troughs, with only three exceptions. After the 2972 BC theoretical peak, the actual top procrastinated by 475 years. The following low came quickly afterwards, 185 years ahead of the theoretical low at 2200 BC. After that, oscillations returned to their normal pattern, until the post-118 peak. Around 118, the sunspot cycle began a long-term plunge that lasted until 1735. (That coincided with the final down-phase of the 4636-year cycle.) Around 1135, a brief recovery materialized around the time of a theoretical low. However, the long-term trend obliterated this potential cyclical low. If the cycle returns to its normal pattern, the next low should develop around the year 2436. Following that, the next peak comes in the year 3209.

In their abstract *Persistent Solar Influence on North Atlantic Climate during the Holocene*, Columbia University's Lamont-Doherty Earth Observatory team also attributes this cycle to the Sun:

> **"Those drift-ice cycles compose part of an enigmatic, at best quasi-periodic, 1500-year cycle that appeared to persist across the glacial termination and well into the last glaciation, suggesting that the cycle is a pervasive feature of the climate system.... At least the Holocene segment of the North Atlantic's 1500-year cycle appears to have been linked to variations in solar irradiance."**[Bond et al., 2001]

In their book *Unstoppable Global Warming: Every 1500 Years*, authors Fred Singer and Dennis Avery take this cycle a step further. They use it to challenge the Global Warming Hypothesis.[Singer and Avery, 2007]

In Chapter 28, this book enters the 20th Century Global Warming controversy with a close examination of short-term EUWS cycles. For now, note the 1545-year cycle from the Greenland ice-core record. See **Charts 25B, 25C, and 25D**. They depict Greenland temperatures during the past 50,000 years. Even though fluctuations often deviate from theoretical peaks, only one cycle appears for each period. These patterns reinforce both the 1545-year periodicity and the frequent variability in these cycles as indicated by the TPD principle. Note: The five down arrows in **Chart 25B** represent the 125-kyr peak at 1427 BC.

Perhaps the most interesting aspect of the 1545-year cycle comes from its association with human activity. In particular, up-phases correspond with great technological advancements and civilization-building, while its down-phases coincide with dark-ages. In his book *The Recurring Dark Ages: Ecological Stress, Climate Changes, and System Transformation*, sociologist Sing Chew theorizes that mankind creates these cycles through ecological mismanagement associated with deforestation, which leads to ecological disaster, which causes great contractions in human population:

> **"Excessive ecological degradation leads to environmental collapse, and along these lines, there are certain phases of environmental collapses that occur mutatis mutandis with civilization demises. This relationship between environmental collapses and civilization demises suggests that when societal relations with the natural environment become exploitative over time, a social system crisis is triggered. As the natural environment plays a part in social system reproduction, we need therefore to widen our gaze for other factors that engender a social system crisis/transition beyond those that are social, political, and economic in nature."**[Chew, 2006]

The preceding section dealt with peaks. However, troughs provide the most notable characteristic of the 1545-year dark-age cycle. But dark-age troughs present a great challenge. By their very nature, dark-ages represent the flip side of golden ages. Historians record volumes during golden ages. Conversely, when a dark-age arrives, strong central governments, written language, and knowledge vanish. <u>Lack of information</u> generally reflects the arrival of a dark-age.

Symptoms of a dark-age often include, but are not limited to, the following:

1. Rising mortality rates and falling fertility (population losses)
2. Economic decline
3. Political unrest and civil war
4. Collapse of centralized forms of government
5. Abandonment of rule by written law
6. Disappearance of monumental architecture
7. Reduced literacy, and loss of knowledge
8. Smaller settlement sizes (de-urbanization)
9. Abandonment of earlier religious forms

10. Increase in intra-group violence
11. Reduced regional trade

The demise of the Roman Empire provides a perfect example of economic decline, collapse of centralized government, disappearance of great architecture, and de-urbanization associated with a dark-age. At its peak, during the 2nd Century golden age, Rome's population reached about one million people. By 410, the population shrank to 500,000. Around the year 500, the population declined further to between 40,000 and 60,000. By 547, only 30,000 inhabitants remained in Rome.[Williams, 2008]

As another example of dark-age symptoms, after the wealthy Mycenaean civilization vanished around 1400 BC, Greek writing disappeared from archaeological records for nearly 500 years. Historians call this period the Greek Dark Ages. However, because of the absence of written accounts, good documentation about the causes of the Mycenaean decline vanished with their civilization. And that's the general problem with all dark-age periods. Lack of evidence prevents good knowledge of these eras. Presumably, difficulties associated with survival consume all human energies, and very little extra wealth remains to pay historians to record events.

Dark ages tend to arrive about 400 years after theoretical peaks of the 1545-year cycle, and they tend to last another 375 years after that. The following section reviews the last four dark-age cycles, covering 6500 years of human history. And it includes the fifth cycle, which started over 300 years ago.

Theoretical Peak 1, at 4518 BC – Around 5000 BC, mankind invented proto-writing systems (the use of symbols in writing), and by 4500 BC, proto-writing spread throughout Europe and Asia. Also around 5000 BC, farming techniques emerged. Farmers cultivated rice in Southeast Asia, maize in Mexico, and a variety of crops in Europe. By 4500 BC, ploughs were used in Europe. Also around 4500 BC, the invention of the wheel in Mesopotamia acted as another great technological breakthrough. These huge breakthroughs (in communications, agriculture, and transportation) allowed people to document a variety of important activities, grow an abundance of food, and more easily engage in longer-distance trade.

Theoretical Dark Age 1, at 3746 BC – Around 4000 BC in Egypt, the Tasian and Badarian cultures ended.

Theoretical Peak 2, at 2973 BC – Around 3500 BC, the Mesopotamians developed the first functional writing system. By 3300 BC, the Bronze Age began in the Near East. Between 3300 and 3000 BC, more advanced cultures spread throughout Europe. Between 3100 and 3000 BC, construction began on the first phase of Stonehenge. Around 3200 BC, the Cycladic civilization formed in Greece. The first great civilization in recorded history, the Early Dynasty of ancient Egypt, appeared around 3200 BC. Then, it reached its zenith under the pharaohs Narmer, Djer, Djet, and Den, who ruled from approximately 3100 BC to 2937 BC.

Theoretical Dark Age 2, at 2200 BC – The span from 2200 to 1700 BC coincided with dark-ages in Mesopotamia, India, and West Asia.[Chew, 2006] The years 2183 to 2055 BC marked the First Intermediate Period in Egypt. Other than its ultimate collapse one dark-age cycle later, this period coincided with ancient Egypt's worst economic depression.

Theoretical Peak 3, at 1427 BC – The golden age of ancient Egypt coincided with this time, under the rule of two of Egypt's greatest pharaohs, Thutmose III and Amenhotep III. The good times lasted another 170 years, until the reign of Ramesses the Great. In another land not far away, sometime between the years 1400 BC and 1300 BC, a period of great wealth among the Mycenaean Greek civilization reached its climax.

Theoretical Dark Age 3, at 655 BC – The time from 1200 BC to 700 BC coincided with dark-ages in Asia and the Mediterranean.[Chew, 2006] The Greek Dark Ages lasted from 1100 BC to 750 BC. And from 1069 to 732 BC, Egypt entered the Third Intermediate Period – a period that started ancient Egypt's collapse to oblivion.

Theoretical Peak 4, at 118 AD – The golden age of ancient Rome came to an abrupt end on August 9, 117, with the death of one of its most respected leaders, Trajan. Amazingly, Trajan's death came ½ year prior to the theoretical peak of the 1545-year cycle, which occurred during February 118. Trajan's death coincided with the territorial peak of Rome. In eastern Asia, one of the greatest periods in ancient China climaxed – almost in perfect harmony with the peak at Rome. The Han Dynasty started expanding rapidly after 206 BC. At its peak, during the rule of Emperor Han Wudi from 140 BC to 87 BC, the population reached 55 million. The political and cultural influence extended to Korea, Mongolia, Vietnam, Japan, and central Asia. After a downturn, the Han Dynasty strengthened again – until the end of the reign of Emperor Zhang from 75 to 88 AD. After 88 AD, the Han Dynasty weakened substantially. This decline came thirty years ahead of the theoretical EUWS peak at 118 AD. By 220 AD, the Han Dynasty deteriorated to the point where it split in three separate kingdoms.

Theoretical Dark Age 4, at 891 AD – In Europe, a dark-age period spanned the years from 476 to 900. On the other side of the globe, the Meso American Dark Ages lasted from 700 to 1200.

Theoretical Peak 5, at 1663 AD – The European Renaissance reached its climax ahead of this peak. This also coincided with the last great period for Imperial China. During the 300 years following this peak, the great majority of all ancient cultures and civilizations vanished. World domination by the Southern European powers immediately shifted northward and westward around 1663 – toward England and America.

Theoretical Dark Age 5, at 2436 AD – In place of ancient civilizations, Western-style civilizations emerged. The roots of Western Civilization came from ideas involving individual liberties and rights, centered in England. These English ideas were then copied in America, thus spawning the political philosophies for modern Western civilization. During this down-phase, power shifted away from central governments and toward individualism – symptomatic of dark-ages. During the past 50 years that long-term trend temporarily reversed in both Europe and America. However, a trend back toward individualism should resume – as the final 400 years of

The 1545-Year EUWS Cycle

the current cycle takes us to the bottom of the next dark-age trough in 2436.

Similar to its parent cycles, the 1545-year cycle oscillates in sync with global temperatures, the carbon cycle, and solar cycles. In addition, it significantly impacts human behavior. And its impact on mankind gets even more interesting as the EUWS cycles shrink further.

The Unified Cycle Theory

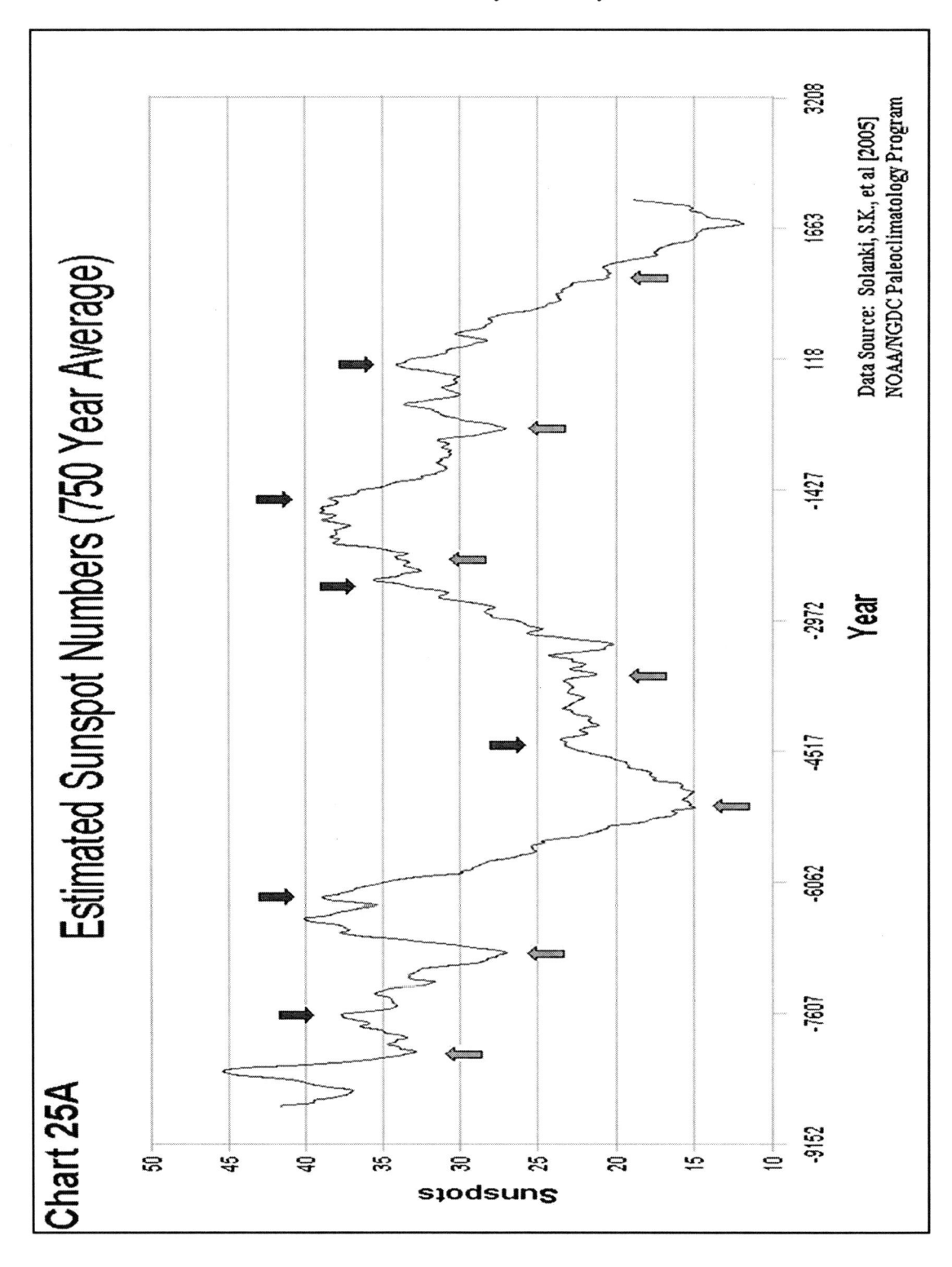

Chart 25A — Estimated Sunspot Numbers (750 Year Average)

Data Source: Solanki, S.K., et al [2005]
NOAA/NGDC Paleoclimatology Program

278

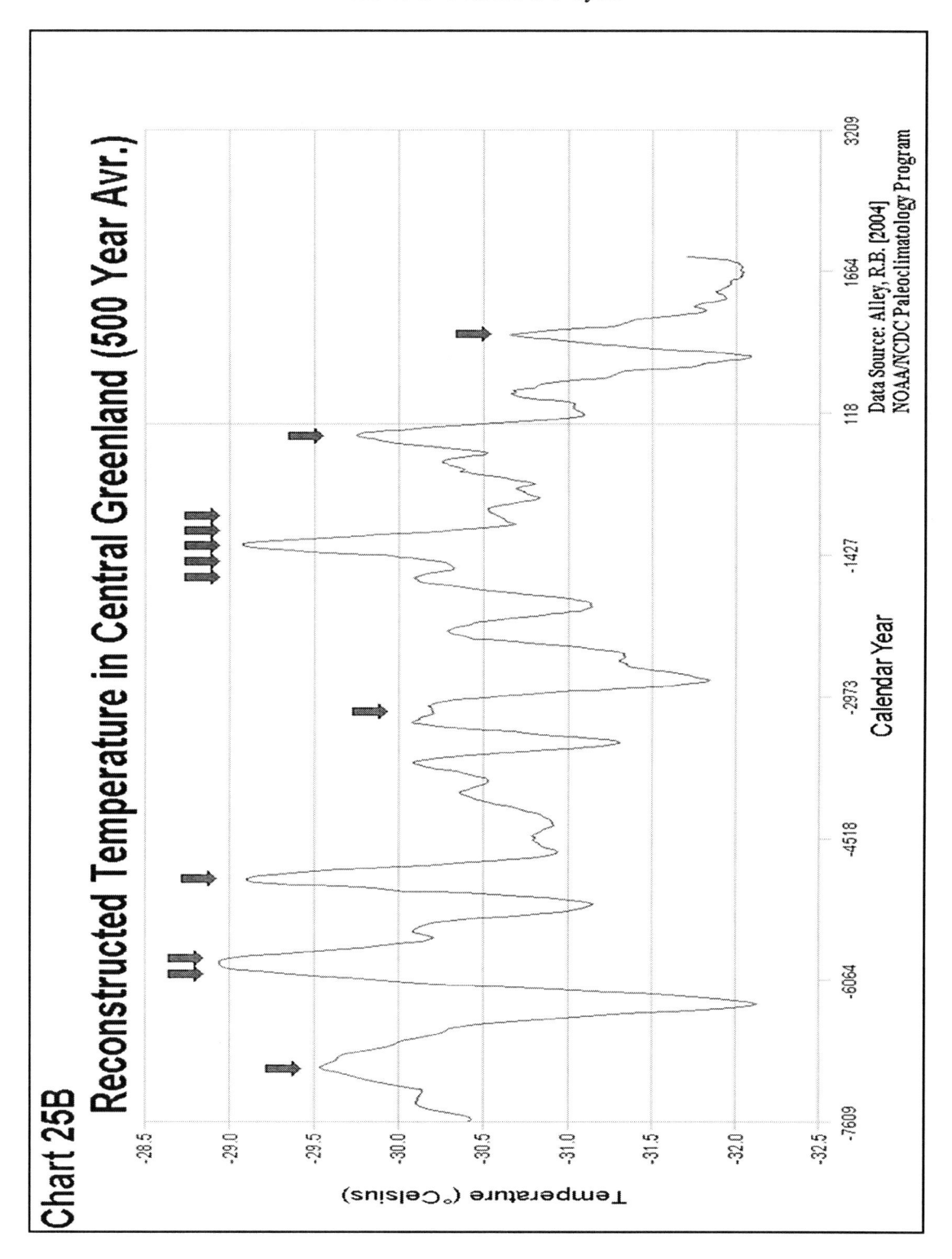

The 1545-Year EUWS Cycle

Chart 25B

Reconstructed Temperature in Central Greenland (500 Year Avr.)

Data Source: Alley, R.B. [2004]
NOAA/NCDC Paleoclimatology Program

The Unified Cycle Theory

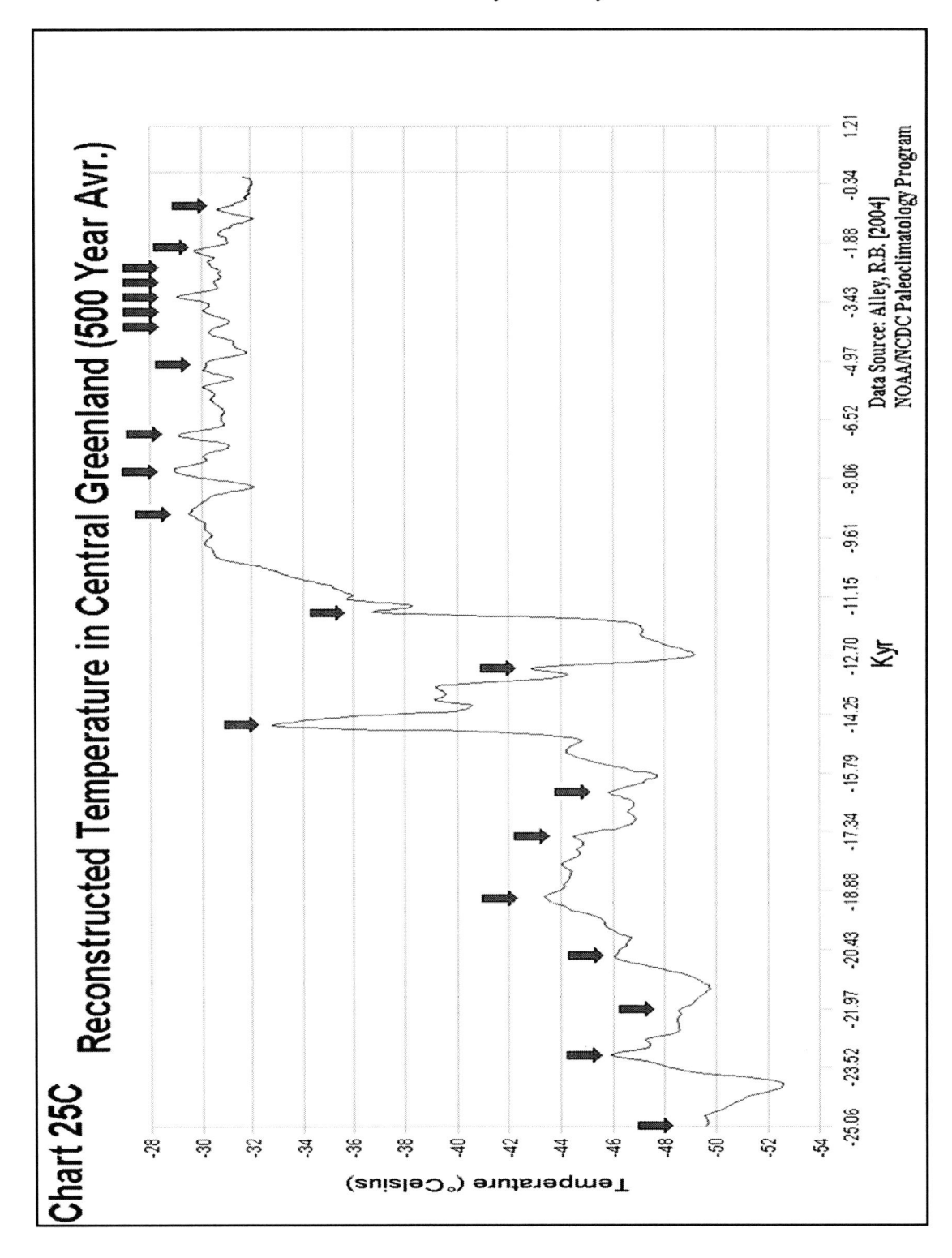

Chart 25C Reconstructed Temperature in Central Greenland (500 Year Avr.)

Data Source: Alley, R.B. [2004]
NOAA/NCDC Paleoclimatology Program

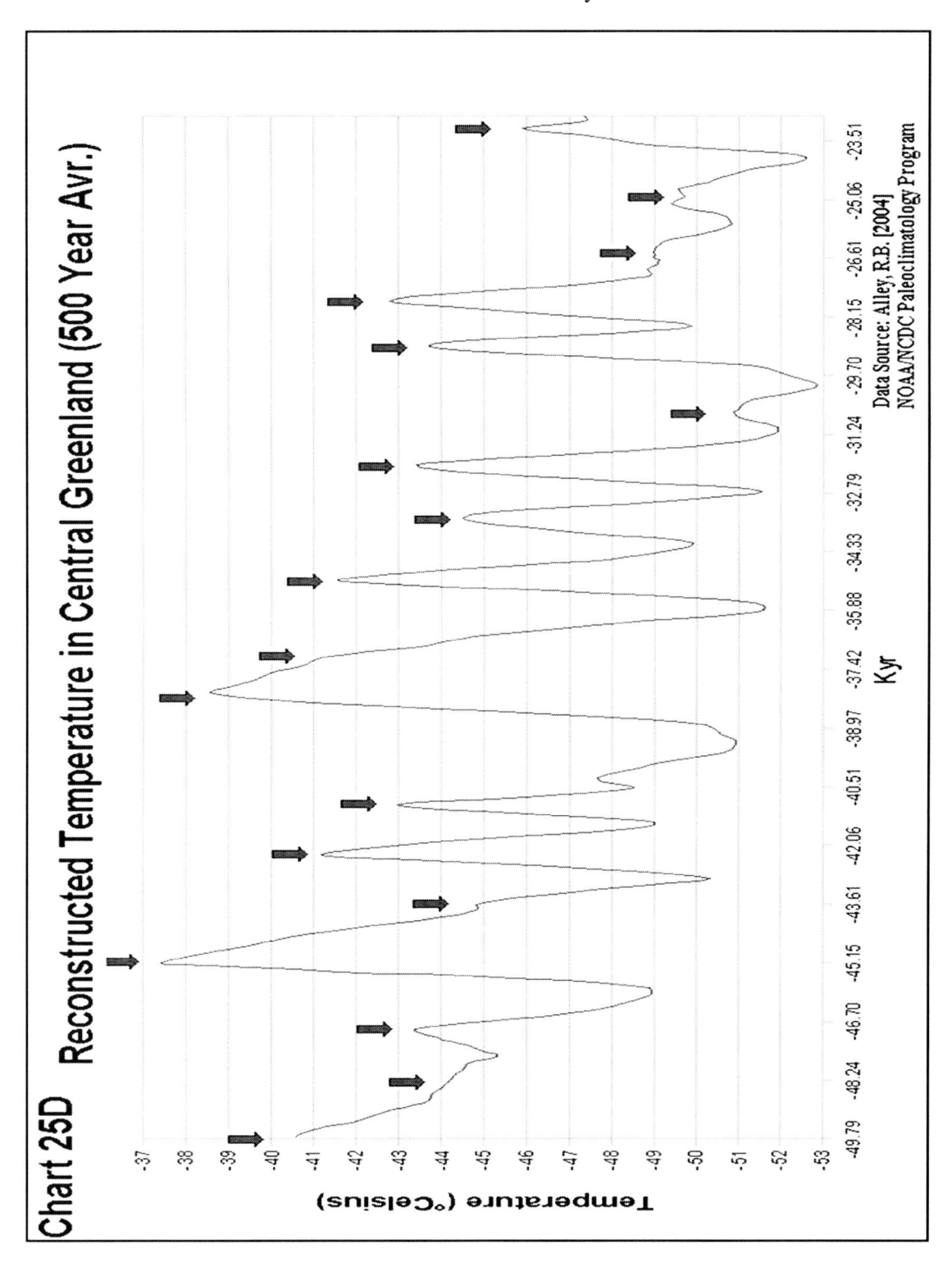

The 1545-Year EUWS Cycle

Chart 25D Reconstructed Temperature in Central Greenland (500 Year Avr.)

Data Source: Alley, R.B. [2004]
NOAA/NCDC Paleoclimatology Program

Chapter 26
The 515-Year EUWS Cycle

Dividing the 1545.48-year dark-age cycle by three, the 515.16-year EUWS cycle appears. This cycle greatly impacts social organization. New civilizations generally form 200 to 300 years before its theoretical peaks. Conversely, dominant civilizations tend to decline, then collapse, in the 100 to 300 years following its peaks.

The list below contains theoretical peaks of the 515-year cycle for the last 6500 years. Note that two asterisks mark years that double as 1545-year dark-age peaks.

4518 BC **	2973 BC **	1427 BC **	118 **	1664 **
4003 BC	2458 BC	912 BC	633	2179
3488 BC	1942 BC	397 BC	1148	

This chapter covers all aspects of the 515-year cycle, however, the main focus centers on its civilization impact. Supporting evidence spans more than 5000 years, covering many empires. The selection includes empires from ancient China, Meso-America, Egypt, Greece, Rome, Medieval Europe, and modern Anglo-Saxon nations. Validation methods included identifying golden ages, substantial economic declines, civil unrest, civil war, and weakened military strength.

Especially in ancient times, histories seldom provide the desired perfection to pinpoint key events to a specific year or decade. Nonetheless, even rough estimates for these early times provide good clues about cycles. In the end, the critical piece of information becomes the approximate number of years spanned by each civilization.

When analyzing the 515-year civilization cycle, important fluctuations and deviations often occur around its 172-year sub-cycle. Hence, this section also serves as a good example of the TPD Principle. Because of its scope, this book condenses exceedingly interesting histories into small paragraphs – as they relate to the 515-year and 172-year EUWS cycles. The references cited provide more detailed overviews.

Important note: Unless specified otherwise, the following sections use starting and ending dates for civilization cycles from the *Encyclopedia Britannica*.

Chinese Civilization

Validation of the 515-year civilization cycles starts with China. An excellent record of Chinese economic and political history goes back more than 4000 years. This ancient culture provides a perfect starting point for the analysis. This investigation begins with the 515-year theoretical peak that occurred during 1942 BC.

1942 BC Theoretical Peak.

2100-1600 BC – **Xia Dynasty**. Written accounts of the Xia Dynasty do not exist. Yet, a rich oral tradition has been handed down for generations. Historian Yun Kuen Lee notes the challenge of separating fact from fiction.[Lee, 2002] Yet, archaeological discoveries confirm that the Xia Dynasty achieved economic success significantly above pre-Xia cultures. In essence, the Xia Dynasty served as an evolutionary stage to the Shang Dynasty.[Poon, 2008] Most importantly, the Xia Dynasty spanned roughly 500 years – close to the length of one civilization cycle. However, evidence cannot be found indicating when the Xia Dynasty reached its golden age.

1427 BC Theoretical Peak.

1600-1046 BC – **Shang Dynasty**. The Shang Dynasty formed when a rebel leader overthrew the last Xia ruler – the tyrant Jie. During this period, written language appeared.[Poon, 2008] By 1400, the Shang Dynasty flourished.[Evansville, 2008] This marked its golden era, which occurred during the reign of Pan Geng, who ruled from approximately 1398 BC to 1371 BC. Xiao Xin succeeded Pan Geng, and the Shang Dynasty started its decline. The last 70 years of the dynasty coincided with a monumental collapse under Di Yi and the despot Di Xin. The end of the Shang Dynasty came with defeat at the hands of Zhou Wuwang (Ji Fa) in 1027 BC.[China Guide, Shang, 2008] In summary, the Shang Dynasty lasted about 554 years, and its golden age came within 50 years of a theoretical peak – thus, closely coinciding with a typical 515-year civilization cycle.

The 515-Year EUWS Cycle

912 BC Theoretical Peak.

1046-771 BC – **Western Zhou**. The years from 1122 BC to 1046 BC involved transition. During that time, the Shang Dynasty weakened while the Western Zhou Dynasty strengthened. The Zhou dynasty actually lasted about 900 years, but it's generally broken into two parts, with the first part being the Western Zhou.[Poon, 2008] After the defeat of Di Xin, the Ji family architected the Western Zhou Dynasty and gained firm control over China. Prosperity reached its height under Zhou Kangwang, whose reign ended around 996 BC. This golden age came approximately 85 years ahead of the theoretical peak of the corresponding civilization cycle. Then the Western Zhou Dynasty declined under Zhou Zhaowang, who reigned from 995 BC to 977 BC. The situation stabilized under the next two emperors. However, the decline intensified during the time of Zhou Yiwang, who ruled China from 899 BC to 892 BC. After that, the Western Zhou deterioration continued until its demise in 771 BC.[China Guide, 2008] While this cycle fell well short of the typical 515-year civilization span, its final collapse (from 899 BC to 771 BC) started just thirteen years after the 912 BC theoretical peak.

397 BC Theoretical Peak.

771-221 BC – **Eastern Zhou**. Following the 771 BC overthrow, the Eastern Zhou remained somewhat turbulent until a period of exhaustive warfare ended with a disarmament pact in 579 BC – involving Qi, Qin, Jin, and Chu. After the peace conference in 579 BC, a period of stability emerged. Confucius wrote and taught during this tranquil time. Historians refer to the first half of the Eastern Zhou Dynasty, from 771 BC to 476 BC, as the Spring and Autumn Period.[Britannica: Zhou, 2008] Near the time of Confucius' death, a transition period evolved with increasing conflicts within the Jin. As the theoretical peak of the 515-year cycle arrived in 397 BC, the relative tranquility ended. The Warring States Period followed. Historians disagree on the start of this period of fighting. Some put it at 476 BC, others at 403 BC (6 years before the theoretical peak). With the tripartition of the Jin in 403 BC, the three remaining elite families in Jin – Zhao, Wei and Han – splintered the Eastern Zhou. From that point forward, the Eastern Zhou Dynasty slowly descended into oblivion until its complete collapse in 256 BC.[Hunan China, 2008] In this particular case, from start to finish, the Eastern Zhou Dynasty endured 550 years. And the decline started in 403 BC – six years before the theoretical peak of a 515-year cycle!

118 AD Theoretical Peak.

This particular civilization cycle consisted of three dynasties – Qin, Han, and the Three Kingdoms.

221-206 BC – **Qin Dynasty**. Starting in 256 BC, the Qin state began conquering and annexing its rival states, thus bringing the Warring States Period to an end. By 221 BC, the last conquest reunified China under Shi Huangdi. This marked the beginning of Imperial China.[China Guide, 2008] However, after Shi Huangdi died, rebellion brought the dynasty to a quick end.[Britannica: Qin, 2008] In a sense, the Qin Dynasty acted as a unifying phase paving the way for transition from Eastern Zhou feudalism to Han imperialism. Near the end of the Qin Dynasty, an economic boom unfolded, highlighted by the start of construction of the Great Wall of China.[Evansville, 2008]

206 BC-220 AD – **Han Dynasty**. Historians consider the Han Dynasty as one of the great periods in China. Its population expanded to 55 million people. Its political and cultural influence extended to Korea, Mongolia, Vietnam, Japan, and central Asia. Authoritarian Han rulers often resorted to book burnings as a means of stamping out dissent.[Britannica: Han, 2008] The first peak of the Han period spanned the years from 140 BC to 49 BC – during the reigns of Han Wudi, Han Zhaodi, and Han Xuandi.[China Guide, 2008] A secondary peak took place during the rule of Emperor Zhang from 75 to 88. From there, the dynasty descended rapidly for the next 130 years. Hence, the collapse of the Han Dynasty began 30 years before the theoretical peak of the 515-year cycle.

220-280 – **Three Kingdoms**. Following the death of Emperor Ling in 189, a period of internal strife ensued. Around 220, the Han Dynasty split into three parts – Wei, Wu, and Shu. After the split, countless battles raged among the three kingdoms.[China Guide, 2008]

This civilization cycle lasted 501 years. It started with the unification under Qin. It reached its peak during the Han Dynasty. And it descended to a nadir during the Three Kingdoms period.

633 AD Theoretical Peak.

This civilization cycle consisted of four dynasties – Jin, Southern & Northern, Sui, and Tang.

265-420 – **Jin Dynasty**. Before the Three Kingdoms period ended, the Jin Dynasty already formed. A difficult road to recovery followed the collapse of the Han Dynasty. A brief period of moderate unity spanned the Jin Dynasty years. However, fiction greatly romanticized this period. The Jin could not contain invasions from nomadic peoples. In 317, the Jin court fled from Luoyang and reestablished itself in the south.[Poon, 2008]

420-589 – **Southern & Northern Dynasties**. The period of the Southern and Northern Dynasties coincided with an age of unrest, civil war, and political disunity. The transfer of the capital from Luoyang in 317 coincided with China's political fragmentation into a succession of dynasties that lasted until 589.[Poon, 2008]

589-618 – **Sui Dynasty**. This period marked the re-unification of Southern and Northern China. It ended four centuries of division between rival regimes. In many ways, the Sui Dynasty mimicked the Qin Dynasty eight hundred years earlier – mostly because of its second ruler, the tyrannical Emperor Yang.[China Guide, 2008] Like the Qin Dynasty, the Sui Dynasty only lasted about twenty years, re-unified China, used ruthless tactics, and acted as a transition phase to a more admirable government.

618-907 – **Tang Dynasty**. The Tang Dynasty reached its golden age during its first one hundred years.[Britannica: Tang, 2008] From 627 to 649, the greatly respected Emperor Taizong ruled China. As Taizong's rule progressed, the Tang Dynasty reached unparalleled prosperity. The economy flourished, the social order stabilized, corruption never existed, and the national boundaries opened to foreign countries. Between Taizong's death and the year 712, the economy marked time. Then, during the early part of Emperor Xuanzong's reign, from 712 to 735, Tang China re-ascended to a level comparable to Taizong's prosperity. Its capital, Chang'an, became the largest and most prosperous metropolis in the world.[China Guide, 2008] However, toward the end Xuanzong's reign, Tang power ebbed. Domestic economic instability and military defeat in 751 (from Arabs in Central Asia) marked the beginning of the end. Misrule, court intrigues, economic turmoil, and popular rebellions weakened the empire, allowing northern invaders to destroy the dynasty in 907.[Poon, 2008]

Similar to several other Chinese civilization cycles, the one centered around 633 AD consisted of multiple dynasties. At the start of the cycle, the Jin, Southern, Northern, and Sui Dynasties served as transitions from chaos following the Han Dynasty collapse. The great Tang Dynasty surrounded the 633 AD theoretical peak. The civilization cycle ended with the Tang Dynasty disintegrating into 10 kingdoms. The complete cycle covered 642 years.

1148 AD Theoretical Peak.

This civilization cycle consisted of six dynasties – Five Dynasties, Liao, Song, Jin, Western Xia, and Yuan.

907-960 – **Five Dynasties & Ten Kingdoms.** This era coincided with political unrest, when five would-be dynasties followed one another in quick succession. Ten kingdoms materialized from the chaos of the Tang Dynasty collapse.[Britannica: Five & Ten, 2008]

907-1125 – **Liao Dynasty**. One of the ten kingdoms eventually developed into the Liao Dynasty. Unlike the others, it survived, expanded, and existed along with the Song Dynasty. Around 1125, the Liao Dynasty split into two, forming the Jin Dynasty and the Western Xia Dynasty. This happened when the Jin army captured the last emperor of the Liao Dynasty, Emperor Tianzuo.[China Guide, 2008]

960-1279 – **Song Dynasty**. From the splinters of the Five Dynasties and Ten Kingdoms, the Song Dynasty arose. Although it never achieved the glory of either the Han or Tang dynasties, some historians consider the Song Dynasty as another 'golden age' period in China.[China Guide, 2008] At the height of its glory in 1127, as the theoretical peak of the civilization cycle approached, the Song Dynasty split after losing control of northern China. At that point, one remnant became the Southern Song Dynasty. The Jin Dynasty merged the other remnant with its expanding territory.

1038-1234 – **Jin Dynasty & Western Xia Dynasty**. The Southern Song, Western Xia, and Jin Dynasties all peaked at the same time – around the year 1128. After 970, availability of rice price records provides an additional benchmark for gauging historical accounts of prosperity. See **Chart 26A**. Rising rice prices from 1000 to 1128 reflect the general prosperity at the start of the second millennium – which coincided with the early portions of the Jin, Song, and Western Xia dynasties. The massive deflation after 1128 came twenty years ahead of the theoretical peak of the civilization cycle. A secondary, lower peak came around 1200. See **Chart 26A** again. This secondary peak applied to the Jin Dynasty. During the reigns of Emperor Shizong (1161-1189) and Emperor Zhangzong (1190-1208), the national strength of the Jin Dynasty reached its zenith.[China Guide, 2008]

1271-1368 – **Yuan Dynasty**. The Yuan Dynasty formed as the Western Xia, Jin, and Southern Song dynasties collapsed, and the Mongols took control. These dynasties fell in order of their proximity to Mongolia. Western Xia came first; Jin territory followed; and the takeovers ended with Southern Song lands falling to the Mongols in 1279.[China Guide, 2008] Starting around the time of the Mongolian invasion, rice prices records ceased for nearly 100 years – from 1260 to 1360. Notice the large gap in **Chart 26A**. Rice records resumed around 1360, which coincided with the collapse of the Yuan Dynasty, and the ouster of the Mongols. At that same time, rice prices touched their lowest level of the millennium! These low prices reflected a huge depression in the years around 1368 – coinciding with the bottom of this civilization cycle.

This civilization cycle began in 907 with the chaos surrounding the 5 Dynasties and 10 Kingdoms. It reached its summit between 1128 and 1220. It collapsed with the Mongolian occupation from 1271 to 1368. This particular civilization cycle spanned 461 years – a little shy of the ideal 515-year period. Rice prices confirm this civilization's economic ascent from 1000 to 1128, and its subsequent collapse.[IISH: Prices and Wages, 2008]

1663 AD Theoretical Peak.

This civilization cycle consisted of two dynasties – Ming and Qing.

1368-1644 – **Ming Dynasty**. After ousting the Mongolians, power in China returned to the Hans in the form of the Ming Dynasty. The economy continually improved until it approached the theoretical peak in 1663. Review **Chart 26A** again. Notice the steadily rising rice prices from approximately 1400 to 1640. The turning point came during the reign of Emperor Shenzong. During his early rule, under the assistance of a skillful chancellor, Zhang Juzheng, the national economy, agriculture, water conservancy, and military affairs improved. However, after the death of Juzheng in 1582, the emperor began to neglect state affairs.[China Guide, 2008] Long wars with the Mongols, incursions by the Japanese into Korea, and harassment of Chinese coastal cities by the Japanese weakened Ming rule. Similar to earlier periods of sovereign rule, the Ming Dynasty became ripe for an alien takeover by the Manchurians in 1644. [Poon, 2008]

1644-1911 – **Qing Dynasty**. During its first hundred years, the Qing Dynasty prospered somewhat. The feudal economy of the Qing Dynasty reached its zenith during the reigns of Kangxi, Yongzheng, and Qianlong – often called "the golden age of three emperors." But by the middle 1700s, the Chinese system faced increasing pressure both internally and from the outside world. Under the later rulers, various rebellions and uprisings broke out. In 1840 when the Opium War broke out, the Qing court faced troubles at home and aggression from abroad. Reforms included the Westernization Movement, the Reform Movement of 1898, and the Taiping Rebellion, but none of them succeeded in saving the Qing Dynasty. Finally, the Revolution of 1911 led by Sun Yat-sen resulted in the overthrow of the Qing Dynasty. With it, two thousand years of Chinese feudal monarchy came to an end.[China Guide, 2008]

Centered on the 1663 theoretical peak, this Chinese civilization cycle lasted 543 years. It advanced during the Ming Dynasty, and then declined throughout most of the Qing years.

Meso-American Civilization

Within the past 4000 years, lands of present-day Mexico and Central America housed many great civilizations. Their importance to the *Unified Cycle Theory* lies with the fact that these Meso-American civilizations flourished in isolation – well removed from the influence of ancient empires in Europe, the Middle East, and Asia. Hence, Meso-American cultures provide a completely independent test of civilization cycles. Unfortunately, historical records concerning Meso-America lack the quality of Eurasian accounts. Nonetheless, they contain enough accuracy to allow adequate judgments about their civilization oscillations.

Note: In the vast majority of cases, Britannica dates serve as the civilization delimiters. However, for the Olmec civilization, three significantly different ranges appear. Wikipedia gives 1400 BC to 400 BC; Encarta denotes 1500 BC to 600 BC; while Britannica indicates 1200 BC to 400 BC. These wide divergences probably imply that Olmec civilization appeared and disappeared gradually, rather than abruptly. These three sources generally agree quite closely on all of their dates. The Olmec disagreement represents an exception. In this case, because of the large differences, median starting and ending points were selected for the Olmec analysis.

912 BC Theoretical Peak.

1400-900 BC – **Olmec Civilization, San Lorenzo Phase**. The Olmec civilization first emerged between 1500 BC and 1200 BC. Slightly after the 912 BC theoretical peak of the civilization cycle, for an unknown reason, the Olmec capital at San Lorenzo was destroyed.[Encarta: Olmec, 2008] Estimates place the San Lorenzo destruction at about 900 BC. Providing additional evidence of the decline, around 950 BC a wholesale destruction of many San Lorenzo monuments occurred, which may point to an internal uprising or, less likely, an invasion.[Wikipedia: Olmec, 2008] The San Lorenzo phase of the Olmec civilization lasted approximately 500 years, with its collapse coming almost precisely at the theoretical peak of the civilization cycle. However, the Olmec civilization survived. Whatever caused the San Lorenzo destruction, and in spite of the hardships associated with its demise, the Olmec people picked up their lives and moved to La Venta.

397 BC Theoretical Peak.

900-400 BC – **Olmec Civilization, La Venta Phase**. After the turmoil from 950 BC to 900 BC, the magnitude of the downturn remains unclear. Whatever its severity, the disturbances failed to destroy Olmec civilization. With their capital moved to La Venta, prosperity reappeared. It lasted until at least 600 BC. Then, a gradual decline may have taken place between 600 BC and 400 BC. After the 397 BC theoretical peak of the civilization cycle, the Olmec civilization faded rapidly. Olmec stylistic influence disappeared after about 400 BC. The Olmec people didn't abandon all their sites, but their culture changed. And the area ceased to be the cultural leader of Mesoamerica.[Britannica: Olmec, 2008]

In spite of the uncertainties surrounding the Olmec civilization, two items stand out. First, the Olmec civilization split into two distinct phases, with each phase lasting roughly 500 years. Second, each civilization phase suffered from the demise of their capitals around the time of a theoretical peak of the 515-year cycle.

118 AD Theoretical Peak & 633 AD Theoretical Peak.

After the fall of the Olmec, two new civilizations filled the vacuum – the Teotihuacan and Maya civilizations. The height of these two civilizations marked the Meso-American golden age.

200 BC-800 AD – **Teotihuacan Civilization**. Around 200 BC, the Teotihuacan civilization rose to power in a timetable nearly identical to the ascent of the Roman Empire in the Mediterranean and the rise of the Han Dynasty in China. Builders completed construction of its largest pyramid, the Pyramid of the Sun, around the year 100, marking the start of the golden age of Teotihuacan. However, unlike Rome and China, the Teotihuacan civilization held up fairly well once the 118 peak passed. At its height, toward the close of the 6th Century, it was probably the largest city in pre-Columbian America. It covered about eight square miles and may have housed 150,000 to 200,000 inhabitants.[Britannica: Teotihuacan, 2008] In essence, its size rivaled Rome at its height. However, once the civilization cycle peaked in the year 633, the Teotihuacan civilization completely fell apart. Its decline lasted until the year 800. *All Empires Online History* perfectly captures the sentiment shift during the downturn:

> "The theme of Teotihuacan art is ... incredibly upbeat, though just before the end it took on a dour and sinister turn.... Why Teotihuacan fell is still a hotly debated mystery, but most of its 200,000 inhabitants abandoned the city leaving it to the jungle.... Over population beyond the resources is a popular solution, loss of faith is another. Teotihuacan art also took on a military edge in its final death throws, so military historians look to either barbarians or civil war for answers. Much of the city shows signs of destruction, statues of deities defaced, and structures burned to the ground. It is known the exodus from the city was gradual... Also, the city lost domination of its outer provinces quite a while before its fall, whether ... caused by an outside invader preying on the dying city or done by the priests symbolizing the gods abandoning them is, perhaps,

something we will never know."[All Empires: Teotihuacan, 2008]

250-900 – **Maya Civilization, Classic Period**. The Maya civilization first emerged as early as 1500 BC.[Britannica: Maya, 2008] However, the stages of advancement remain vague until the year 250 – when the Mayan culture ascended into a golden age. And the golden age lasted at least 400 years. Sometime after the year 633, possibly as late as 700, the Maya civilization started to spiral downward. For unknown reasons, the Mayan cities of the southern lowlands entered a long-term decline between the years 700 and 900. And after 900, the Maya completely abandoned these cities.[Coe, 2002] Hence, the Maya classic period spanned 650 years. And its collapse started sometime within 67 years after the 633 theoretical peak of the 515-year cycle.

1148 Theoretical Peak.

900-1519 – **Maya Civilization, Post-Classic Period**. Even though the Mayan people abandoned the southern lowlands, their civilization continued in the Yucatan highlands. Primary Mayan cities in the post-classic period included Chichen Itza, Uxmal, and Mayapan.[Britannica: Maya, 2008] While this civilization cycle covered 619 years, the timing of its descent remains unclear.

1663 AD Theoretical Peak.

1200-1520 – **Aztec Civilization**. About two hundred years after the Mayan and Teotihuacan golden ages ended, the Aztecs emerged. Their civilization continually expanded from 1200 until 1519. By then, the Aztec empire consisted of 400 to 500 small states, comprising up to 6 million people, spread over 80 thousand square miles.[Britannica: Aztec, 2008] But after 1519, conditions changed abruptly. This case serves as an example of a civilization collapsing well ahead of its theoretical peak. Spanish conquistadors, led by Cortez, brought an abrupt end to the Aztec expansion. However, rather than military might, diseases carried by the Spanish invaders may have caused the Aztec collapse. A Spanish soldier infected with the smallpox virus transmitted the disease to an Aztec warrior. Within two weeks the disease infected the Aztec Empire and one fourth of the population died.[Cohn, 1989] However, this premature ending to a civilization cycle – coming 150 years ahead of a theoretical peak – must be classified as an exception to the 515-year cycle, rather than the rule.

Ancient Egyptian Civilization

This section dates Egyptian civilization using the *University College London* chronology for the first three cycles, and *Encyclopedia Britannica* chronology for the remainder. University College London provides better consistency in their dating, and they assigned years to early phases of Egyptian civilization – which *Encyclopedia Britannica* fails to do.

4003 BC Theoretical Peak.

4000-3500 BC – **Naqada I, Amratrian Culture**. In Egypt, the Naqada culture controlled the nation from 4000 BC until the age of the pharaohs. Over hundreds of years, the Naqada people slowly expanded from a small tribe into a powerful culture controlling the Nile Valley. Around the theoretical civilization cycle peak at 4003 BC, Egyptian technology advanced enough to indicate increased social organization. Painted pottery first appeared, and settlements became more densely populated.[Britannica: Amratrian, 2008] The Amratrian Culture distinguished itself by engaging in regional trade. They imported copper from the Sinai, gold from Nubia, and exchanged pottery with other parts of the region.[Grimal, 1992] After significant advancements around 4000 BC, the Naqada I culture stabilized for about 500 years – until the next civilization cycle arrived.

3488 BC Theoretical Peak.

3500-3200 BC – **Naqada II, Gerzean Culture**. Around the next civilization cycle peak at 3488 BC, technology in Egypt made another step forward. These new technologies included abrasive tubular drills for stone cutting; pear-shaped mace-heads, ripple-flaked flint knives, and advanced metallurgy.[Britannica: Gerzean, 2008] The building of mud-brick homes became widespread. And a number of small cities appeared with populations around 5,000.[Redford, 1992] Lasting nearly 300 years, Naqada II culture ended near the trough of the civilization cycle. Its demise paved the way for Naqada III and the first great dynasty of ancient Egypt.

2972 BC Theoretical Peak.

This civilization cycle consists of the Naqada III culture and the Early Dynasty.

3200-3100 BC – **Naqada III**. Archaeologists discovered extremely wealthy individuals buried in some Naqada III cemeteries. This reveals both increasing wealth and different social classes at that time. Also, writing first appeared in Egypt at this time, acting as another indicator of the sudden emergence of a more advanced civilization.[UCL: Egyptian Chronology, 2008] Egypt became more unified during this period, with indications that kings ruled large states inside Egypt.[Britannica: Gerzean, 2008]

3100-2686 BC – **Early Dynasty**. The first pharaohs gained control over Egypt during the Early Dynasty – which actually consisted of the first 2 dynasties. This civilization cycle reached its zenith during the 1st Dynasty, under the pharaohs Narmer, Djer, Djet, and Den. Their rule approximately spanned the years 3100-2937 BC. During the 1st Dynasty, papyrus was invented, record-keeping and writing expanded, and prosperity surged. Thousands of tombs of all levels of wealth have been found throughout the country.[Britannica: Egypt, 2008] And this period of prosperity coincided with the theoretical peaks of the civilization cycle at 2972 BC. Immediately afterward, trouble appeared. Late in the 1st Dynasty, Anedjib put down several uprisings in Lower Egypt. And the second half of the 2nd Dynasty was a time of conflict with rival lines of kings.[Britannica: Egypt, 2008] Record-keeping faltered as the 2nd Dynasty progressed.

The 515-Year EUWS Cycle

The civilization cycle associated with Naqada III and the Early Dynasty lasted 514 years, and it peaked very close to its 2972 BC theoretical peak.

2457 BC Theoretical Peak.

2686-2130 BC – **Old Kingdom**. As the Old Kingdom period began, economic prosperity returned. Around 2686 BC, Djoser's construction of the Step Pyramid at Saqqara provided the first evidence of the resurgence. Prosperity continued throughout the 3rd and 4th Dynasties. The 4th Dynasty lasted until 2498 BC, ending 40 years before the theoretical civilization cycle peak at 2457 BC. During this time, Sneferu built the Bent Pyramid and the Red Pyramid – both in Dahshur. Khufu built the Great Pyramid of Giza. Khafra and Menkaura built the second and third largest pyramids at Giza. And Djedefra built the Sphinx of Giza as a monument to his father. Clearly, with all of this construction taking place, Egypt enjoyed tremendous prosperity. But with the passing of the civilization cycle peak, conditions deteriorated during the 5th Dynasty. The pyramids became smaller and less solidly constructed than those of the 4th Dynasty.[Britannica: Egypt, 2008] The economy declined even more during the 6th Dynasty, after Pepi II's 90-year reign ended in 2152 BC. Within the 17 year period of the 7th and 8th Dynasties, fragmentary records indicate that at least 18 kings ascended the throne with nominal control over the country. The crisis not only reformed the monarchy but also instilled a spirit of social justice and laid the foundation for compassion in future regimes.[Hassan, 2008]

The civilization cycle associated with the Old Kingdom lasted 556 years, and it peaked about 40 years ahead of its 2457 BC theoretical peak.

1942 BC Theoretical Peak.

This civilization cycle consists of the 1st Intermediate Period and the Middle Kingdom.

2130-2025 BC – **1st Intermediate Period**. Following the collapse of the Old Kingdom, conditions worsened. A one hundred year period of chaos, disunity, and economic hardship followed. After the Old Kingdom's disintegration, incidents of famine hit Egypt, and local violence sprung up. The country became impoverished and decentralized from this episode.[Britannica: Egypt, 2008] But, similar to other times, difficulties during the 9th and 10th dynasties ran their course. Chaos and hardship gave birth to confidence and prosperity.

2025-1630 BC – **Middle Kingdom**. At the start of the 11th Dynasty in 2025 BC, Egypt reunited as the Middle Kingdom. The apex of the Middle Kingdom came during the middle portion of the 12th Dynasty. Under the rule of Sesostris III and his successor, Amenemhet III, prosperity returned around 1850 BC. The prosperity created by peace, conquests, and agricultural development appears in monuments belonging to both royalty and the minor elite.[Britannica: Egypt, 2008] But after Amenemhet III's reign, government weakened, and the subsequent decline continued into the 13th and 14th dynasties. During the 13th Dynasty, about 70 kings occupied the throne. Then toward the end of the 14th Dynasty, Egypt lost control of Lower Nubia.[Britannica: Egypt, 2008]

The 1st Intermediate-Middle Kingdom civilization cycle covered 500 years, and it peaked about 110 years after its 1942 BC theoretical peak.

1427 BC Theoretical Peak.

The 1427 BC civilization cycle consists of the 2nd Intermediate Period and the New Kingdom.

1630-1540 BC – **2nd Intermediate Period**. Following the collapse of the Middle Kingdom, the down-phase of the civilization cycle hit bottom during a ninety year span from 1630 BC to 1540 BC. The disarray during the 2nd Intermediate Period can be seen by the three separate dynasties that ran more-or-less concurrently. The Hyksos rulers from Asia led the 15th Dynasty, controlling most of the Nile valley. The 16th Dynasty maintained the area around Thebes. And the 17th Dynasty operated from upper-Egypt.

1540-1075 BC – **New Kingdom**. The 18th Dynasty of the New Kingdom, covering the years from 1540 BC to 1292 BC, is perhaps the best known Egyptian dynasty. This period began with Ahmose I expelling the Hyksos from Egypt. It included the legendary King Tut. And one of Egypt's greatest pharaohs ruled during the 18th Dynasty – Thutmose III. His reign spanned 1479 BC to 1426 BC (coinciding with a theoretical peak of the civilization cycle). Thutmose III created the core of Egypt's largest empire. He conquered lands as far away as Gaza, Yemen, Palestine, Babylonia, Assyria, and lower-Nubia. However, fifty years later, during the reign of Thutmose I, conquests in the Middle East and Africa allowed Egypt to reach its territorial maximum. But these territories were not firmly held.[Britannica: Egypt, 2008] For the most part, Egypt held its territorial gains into the 19th Dynasty and through the long reign of Ramses II (1279-1212 BC) – often called Ramses the Great. Under Ramses II, peace prevailed.[Britannica: Egypt, 2008] However, after his death, the New Kingdom faded quickly. Around 1200 BC, the rebel king Amenmesse led a short-lived secession of the southern part of the country. During 1153 BC, assassins murdered Ramesses III. In the far south, civil war broke out around 1080 BC. Economic crises, foreign bandit raids, and massive tomb-robbing accompanied these events.[Dodson, 2008] The unrest intensified and eventually led to the collapse of the New Kingdom around 1075 BC.

The 2nd Intermediate-New Kingdom civilization cycle covered 555 years, and it peaked about 50 years after its 1427 BC theoretical peak.

912 BC Theoretical Peak.

1075-664 BC – **3rd Intermediate Period**. Most of the 3rd Intermediate Period revolved around a fractured Egypt. From 1075-950 BC, the north came under control of the Tanite 21st Dynasty. And Theban priests ruled much of the southern Nile Valley. However, at the start of the 22nd Dynasty, Egypt enjoyed a brief revival. Influence in the Middle East increased after Sheshonk I plundered Jerusalem. Prestige from this exploit may have lasted through the reign of Osorkon II (929–914 BC).[Britannica: Egypt, 2008] This revival coincided with the theoretical peak of the civilization cycle at 912 BC. After that theoretical peak, Egypt soon split again. The latter part of the 22nd Dynasty ended with new fragmentation. A separate 23rd Dynasty ruled Thebes, but this dynasty endured a long-running civil war. The north faced numerous semi-independent areas ruled by Libyan chiefs. Upper-Egypt came under Tanite control. And Egypt faced threats from Nubian in the south. In fact, around 732 BC, Nubia took control of Egypt.[Dodson, 2008] The 3rd Intermediate Period ended with the 25th Dynasty – when the Assyrians conquered Thebes, and forced Taharqo to flee south to Nubia.

The 3rd Intermediate civilization cycle covered 411 years, and it reached its summit within two years of the 912 BC theoretical peak.

397 BC Theoretical Peak.

The 397 BC civilization cycle consists of the Late, Macedonian, and Ptolemaic periods.

664-332 BC – **Late Period**. In 656 BC, the second ruler of the 26th Dynasty, Psamtek I, reunited Egypt. But the unification only lasted thirty years. By 525 BC, Persian rulers controlled Egypt – starting the 27th Dynasty which lasted 120 years. Then, from 404 BC to 343 BC, the last series of relatively strong native Egyptian kings enjoyed reigns. This short span surrounded the theoretical peak of the 397 BC civilization cycle. After the fall of Nectanebo II in 343 BC, Egyptian power vanished permanently. The Persians regained control, thus ending the Late Period.

332-30 BC – **Macedonian and Ptolemaic Period**. Alexander the Great seized Egypt in 332 BC, but it regained its independence after the break-up of his empire in 310 BC. However, the new ruler, Ptolemy I, was a Macedonian Greek, and the ruling class of the state was now foreign, running the country as part of the Ptolemaic kings' wider Mediterranean agendas. The ancient religion and culture were supported and new temples built, but the dominant culture was now increasingly European, with Greek becoming the language of state. The increasingly bloody internal struggles of the ruling house brought Egypt within the orbit of the still-growing Roman Empire, culminating in the defeat of the last of Ptolemy's ruling descendants, Cleopatra VII, and her Roman lover Mark Antony in 30 BC, resulting in the country's absorption into the empire that same year. Egypt now became a mere province, with its primary goal to provide grain for the rest of the Roman Empire.[Dodson, 2008]

This final Egyptian civilization cycle covered 634 years, and it peaked during the years surrounding the 397 BC theoretical peak of the 515-year cycle!

Ancient Greek Civilization

As Egyptian civilization sank into obscurity, the merchant nation of Greece replaced it as the Mediterranean power. Ancient Greek civilization flourished in the area of the Aegean Sea since at least 7000 BC. The main Greek areas consisted of Crete, the Cyclades, other islands in the area, and the Greek mainland.

During the Bronze Age, various Greek civilizations were associated with specific locations. Crete civilization is called Minoan, after the legendary King Minos of Knossos. For the Cyclades, historians call the civilization Cycladic. For the Greek mainland, Helladic civilization prevailed. Mycenaean civilization refers to the people who inhabited the mainland under Cretan influence in the 16th Century – named after its main city.

Important: Civilization dates in the following section are approximate, but conventional. Correlations with Egypt reduce the uncertainty from around 3000 BC onward. The correlations make sense because Aegean civilizations traded extensively with Egypt. Datable Bronze Age pottery from the Aegean has been found in Egypt, and vice versa. The chronology can be established with a leeway of a few centuries. After 2000 BC, the uncertainty can be fixed within reasonably narrower limits.[Britannica: Aegean, 2008]

1942 BC Theoretical Peak.

2200-1700 BC – **Minoan Civilization**. The period between 2200-2000 BC marked the end of the Early Bronze Age on the Greek mainland, and the beginning of a new era. During this time, new movements of people into the Cyclades and the southern part of the mainland shattered the unity in the Aegean area. Toward the end of the 3rd millennium, fire destroyed many of the mainland settlements. The houses built afterward were of a different type and more primitive – long and narrow, only one story high, and gable-roofed. Evidently, foreign invaders built the new houses to replace the ones they destroyed.[Britannica: Aegean, 2008] While unaffected by invasion, significant changes occurred in Crete toward the end of the 3rd millennium as well. The Minoans built great palaces around large rectangular open courts. This happened within a comparatively short time at the leading centers of Knossos, Phaistos, and Mallia. At the beginning of this Palatial Period, definitive writing patterns first appeared in Crete.[Britannica: Aegean, 2008] The time leading up to the 1942 BC theoretical peak coincided with great prosperity in ancient Greece, especially for the Minoans. However, conditions quickly changed. Various disasters occurred in Crete during the 17th Century BC. Possible invaders inflicted severe damage on the palaces at Knossos and Mallia. At Monastiraki, fire destroyed a building that may have been the residence of a local ruler. Fire burned the palace at Phaistos so violently that an enormous layer of virtually impenetrable vitrified mud brick formed. The mud served as the base for the new palace built on top of it – testimony to the massive destruction.[Britannica: Aegean, 2008] A great deal of uncertainty exists about the cause of these destructions. Historians speculate on accidents, earthquakes, internal warfare, and foreign invasions as possible agents. Evidence does not support an invasion of Crete. However, two things appear certain. This phase of Minoan Civilization lasted about 500 years, and it came to an end around 1700 BC.

The 515-Year EUWS Cycle

1427 BC Theoretical Peak.

1700-1200 BC – **Mycenaean Civilization**. Recovery quickly followed. The next two to three centuries coincided with the most flourishing period of the Aegean Bronze Age. On Crete, the Minoan civilization reached its zenith. The palaces at Knossos, Phaistos, and Mallia were restored with greater splendor than before.[Britannica: Aegean, 2008] On the mainland, the Mycenaean civilization arose to even greater heights. Evidence of great wealth comes from Shaft Graves at Mycenae dating 1600-1450 BC. Caretakers packed these graves with gold, silver, and bronze, showing a preference for portable gold and weapons. At the same time, wealthier families (possibly royalty) built beehive tombs. Two of these tombs, the Treasury of Atreus and the Tomb of Clytemnestra displayed splendid facades, with engaged half columns. Builders constructed these facades in two tiers with exotic, colored stones.[Britannica: Aegean, 2008] This era of great Mycenaean wealth climaxed around the time of the 515-year theoretical peak during 1427 BC. After that, Mycenaean fortunes reversed abruptly. By 1200 BC, Mycenaean society almost totally disappeared.[Encarta: Europe, 2008] Similar to most others, this civilization cycle lasted approximately 500 years.

912 BC Theoretical Peak.

1200 BC to 750 BC – **Greek Dark Ages.** As the Mycenaean civilization came to an end, the Greek Dark Ages began. In the northern part of the mainland, the palace at Thebes fell first. Then slightly later, around 1200 BC, all other mainland palaces burned within a short time. Mycenae and Tiryns continued to be inhabited, but Mycenaeans deserted Pylos. Residents inhabited Athens, but without much wealth. New centers of refuge and of independence appeared on the inner shore of Euboea. Small fortified settlements dotted the coast.[Britannica: Aegean, 2008] Historians failed to find a satisfactory explanation for the collapse of the palace systems and the movements of populations. Climatic change and drought, harvest failure, starvation, epidemic, civic unrest, economic decline, and invasion have all been proposed. The Mycenaeans completely abandoned their civilization between 1200 BC and 1100 BC. The populations of their once-mighty cities dwindled rapidly. Entire cities were destroyed. All of the great Mycenaean craftsmen faded without populations to support them. Most critically, they abandoned *writing*. Without writing, they left no history. We have, instead, only five centuries of mystery.[Hooker, 2008] At any rate, the stable states of the wealthy Mycenaean civilization collapsed into near chaos. The decline lasted until 1100 BC, but as late as 1000 BC in some places and 900 BC in others. After that, new people settled the area.[Britannica: Aegean, 2008] This sparked a mild recovery ahead of the 912 BC theoretical peak, but it failed to achieve strong follow-through. In spite of limited social organization during this time, great improvements in the smelting of iron paved the way for the Iron Age. In spite of the brief recovery in the 10th Century BC, the key element of writing remained absent from Greece archeology until sometime between 750-700 BC. As evidence, a cup, bearing the Greek inscription "I am the cup of Nestor" dates before 700 BC.[Britannica: Greek, 2008] The reemergence of writing signified the end of the Greek Dark Age, and the beginning of the Archaic Period. This particular civilization cycle, containing a weak recovery in the century ahead of the 912 BC theoretical peak, covered about 450 years.

The Unified Cycle Theory

397 BC Theoretical Peak.

This civilization cycle centered around 397 BC contained Greece's Archaic, Classical, and Hellenistic periods.

750-480 BC – **Archaic Greek Period**. During this period, Greece produced startling innovations. On the political front, the self-reliant population wanted to avoid strong central authority, although sometimes tyrants temporarily seized sole power of city-states. The Greeks tried to share rule, sometimes as an oligarchy, and other times as a form of democracy. In a few areas, they also devised a league – a loose alliance of small groups who agreed to share laws and defense. During the Archaic Period, the economy improved. The population grew rapidly. By 500 BC, Greeks founded numerous colonies in southern France, Spain, Italy, North Africa, and along the Black Sea.[Encarta: Archaic, 2008] Their rapidly improving economy paved the way for even greater glory.

480-323 BC – **Classical Greek Period**. At the beginning of the Classical Period, the city-state of Sparta, with its fearsome army, held the most power inside the Greek community. During 480 BC, at the narrow pass at Thermopylae, a small band of Spartans led by king Leonidas I achieved legendary status by holding off an enormous Persian army led by Xerxes I. This tactic gave residents time to evacuate before the Persians reached Athens and burned it. In addition to saving lives, the delay gave the Greeks time to prepare for a naval fight. The Athenian general Themistocles defeated the Persian navy by luring Persian ships into a narrow channel, where heavier Greek ships proceeded to ram and sink them. The following year, the Greeks used superior tactics to defeat the Persians at Plataea. This string of Persian War victories preserved Greek independence.[Encarta: Archaic, 2008] Furthermore, Greeks gained confidence as a world-power.[Encarta: Classical, 2008] After the Persian sacking, Athens quickly recovered with the formation of the Delian League, which gave it funding to build a powerful navy. As the theoretical peak of the civilization cycle approached, the rivalry between Athens (with its superior navy) and Sparta (with its dominant army) came to a head. Tensions erupted in 431 BC when Athens pressured Corinth and Megara, crucial Spartan allies, for seagoing trade. Sparta came to the defense of its allies, and the fighting escalated into civil conflict, resulting in the Peloponnesian War. An Athenian politician, Pericles, devised a strategy of periodic surprise naval raids on Spartan positions. Whenever Sparta counterattacked, the Athenians hid behind the walls of their city.[Encarta: Classical, 2008] Instead of victory, a stalemate ensued. Both sides paid a heavy price. Finally, in 404, incompetent Athenian admirals lost their naval fleet and the war. The Athenian loss ended the Delian League. The victorious Spartans then installed a brutal puppet government in Athens, but that regime failed within a year, and rebels restored democracy in Athens by 403 BC. Athens soon rebuilt its strength, competing with Sparta, Corinth, and Thebes for leadership. But none of the states were strong enough to dominate the others. Once the civilization cycle passed its 397 BC theoretical peak, the constant internal warfare drove the Greek civilization to exhaustion during the first half of the 4th Century BC. The interstate rivalries created dangerous instability.[Encarta: Classical, 2008] Two Macedonian kings, Philip II and Alexander the Great, quickly filled the power vacuum created by the internal bickering. The people of Macedonia, in a mountainous region north of the Greek heartland, never embraced the city-state form of Greek government. With the Greek states weakened from a century of fighting, Philip II's lethal army forced the Greek city-states to acknowledge him as their leader in 338 BC. This ended the era of independent Greek city-states. After a noble murdered Philip in 336 BC, his son Alexander succeeded him. Alexander led the most astonishing military campaign in ancient history. While still in his twenties, he conquered all the lands between present-day Turkey, Egypt, and Afghanistan. His greatness consisted of his ability to motivate his men to follow him into hostile regions. However, after a fatal illness struck Alexander in 323 BC, his generals tore apart the Greek Empire – as each general tried to secure his own power. [Encarta: Classical, 2008]

323-146 BC – **Hellenistic Greek Period**. The Hellenistic Period coincided with the downfall of ancient Greece. After the death of Alexander the Great, Greek domination rapidly eroded. One hundred years later, the emergence of Roman power became troublesome. Like many declining civilizations, it becomes difficult putting an exact date to the end of Greek power. The end came through a long series of events. To minimize the threat from Rome, Macedonia allied itself with Carthage in 215 BC. However, Rome defeated Carthage thirteen years later. Then, in 191 BC, Rome annihilated Antiochus at Thermopylae. In 171 BC, Rome declared war on Macedonia, and defeated it three years later. During 146 BC, the Greek peninsula became a Roman protectorate. For all practical purposes, Greek civilization ended at that point. In 88 BC, the Greeks rebelled against Rome, but Pompey the Great finally defeated the rebellion twenty years later. Finally, during 27 BC, Greece became a Roman province.

This civilization cycle, consisting of Greece's Archaic, Classical, and Hellenistic periods, spanned 604 years. It reached its pinnacle around the time of the 397 BC theoretical peak of the 515-year cycle.

Ancient Roman Civilization

After ancient Greek civilization collapsed, Rome emerged as the next major civilization center. Roman historical documents survived in large numbers, providing great detail. In addition to supporting the 515-year cycle, these records show its 172-year sub-cycle. In spite of tremendous power at its zenith, the Roman Empire only lasted two cycles. For these reasons, Roman civilization will be examined from the perspective of both the 515-year cycle and its 172-year sub-cycle.

397 BC Theoretical Peak of the 515-Year Cycle.

509-340 BC – **Early Roman Republic.** The Roman Republic formed in 509 BC. The republic built a strong army and expanded until the theoretical peak arrived in 397 BC. But in 390 BC, Rome suffered a rare military defeat against the Gauls in the battle of Allia River. After the defeat, the Gauls sacked Rome, and then demanded a huge ransom to prevent its complete destruction.[Britannica: Rome, 2008] As a consequence, the republic barely survived. However, Romans endured these hardships. After the sacking of Rome, the Early Republic struggled until

340 BC. As evidence of the economic difficulties encountered between 390-340 BC, the Licinio-Sextian Rogations of 367 BC provided for the alleviation of indebtedness, and the Genucian Law of 342 BC temporarily suspended interest charges on loans.[Britannica: Rome, 2008]

This sub-cycle lasted 169 years, with strong evidence of a significant economic downturn for the sixty years following the 397 BC peak.

225 BC Theoretical Peak of the 172-Year sub-cycle.

This sub-cycle includes the Italian conquest and the Middle Republic periods.

340-264 BC – **Italian Conquest.** Rome's expansion inside Italy began around 340 BC. By 326 BC, Rome still only controlled 10,000 square kilometers of land – an area roughly 62 miles long and 62 miles wide. But Rome's military might allowed it to grow rapidly. As a consequence of victory in the Latin War (340-338 BC), Rome annexed the states of the Latin League into its territory.[Britannica: Rome, 2008] Between 343-290 BC, the Romans fought three bitter campaigns against the Samnites in the mountainous regions of central Italy. Despite serious losses, Rome ultimately prevailed.[Encarta: Rome, 2008] From 290-280 BC, Rome put down unrest in northern Italy. In the south, a quarrel involving the Greek city Tarentum brought the dangerous army of King Pyrrhus of Epirus into Italy to fight against Rome. Pyrrhus prevailed, and then defeated the Romans a second time in 279 BC at Asculum. Then Pyrrhus left Italy, only to return again in 275 BC. This time, Rome claimed victory at Beneventum. Then Rome took Tarentum itself by siege in 272 BC.[Britannica: Rome, 2008] These victories made Rome the unquestioned master of Italy.

264-201 BC – **First two Punic Wars.** By 264 BC, Rome began to extend its territory beyond natural Italian borders. At the invitation of mercenary soldiers in the Sicilian city of Messina, Rome became involved in a conflict involving Carthage and Syracuse. Fighting related to this Roman-Messina alliance initiated the 1st Punic War, which lasted 23 years. In 242 BC, a Roman naval commander defeated a Punic fleet in stormy seas, capturing or sinking 120 Carthaginian ships. By 241 BC, Carthage surrendered Sicily to Rome along with a payment of 3,200 talents. That victory made Rome the leading power in the western Mediterranean.[Encarta: Rome, 2008] The 2nd Punic War broke out in 218 BC, when a Carthaginian force led by Hannibal boldly marched across Spain, through Gaul, over the Alps, and into northern Italy. During 217 BC, Hannibal moved around inside Italy, eventually ending in the south. The Carthaginian army avoided Rome. Instead, Hannibal remained in southern Italy for more than a decade. He was able to do so chiefly due to the exhaustion of Rome. Agriculture in Italy had collapsed, and the Romans had to import their food supply from Sardinia and Sicily.[Britannica: Rome, 2008] With Hannibal creating havoc, Rome experienced difficult times during the two decades after the 225 BC theoretical peak. However, in 203 BC Hannibal left Italy, in order to engage in an African expedition. Hannibal's departure helped the Roman situation, and Rome finally won the 2nd Punic War in 201 BC. In spite of its victory, hard times surrounded the 2nd Punic War –

The 515-Year EUWS Cycle

which marked the low point for this cycle.

This sub-cycle lasted 139 years, with moderate weakness developing for twenty years after the 225 BC theoretical peak of the 172-year cycle.

53 BC Theoretical Peak of the 172-Year Cycle.

201-31 BC – **Late Roman Republic.** With its defeat in the 2nd Punic War, Carthage became an insignificant force in the Mediterranean, while Rome's prosperity accelerated. Yet, the influential Roman senator Marcus Porcius Cato still feared Carthage, and for decades he ended every speech with the statement: **"And Carthage must be destroyed."** Rome finally seized on a minor offense to initiate the 3rd Punic War. After a difficult 3-year siege, Carthage fell to the Roman army in 146 BC.[Encarta: Rome, 2008] With the threat from Carthage effectively eliminated after the 2nd Punic War, Rome began directing attention eastward, toward Greek territories. In the winter of 200 BC, Roman legions marched into the Balkans to start the 2nd Macedonian War. During 197 BC, Rome went to war against Philip V of Macedon, and they defeated the Greeks at Cynoscephalae. As part of the settlement, Rome granted freedom to all Greek cities and placed them under Roman protection. This involvement with Greek affairs eventually drew Rome into conflict with the region's most powerful king, Antiochus III. Rome defeated Antiochus at Thermopylae in 191 BC. After these victories, Roman commanders became increasingly arrogant and ruthless. By 168 BC, Rome defeated Macedonia and its Greek allies at Pydna, and they enslaved 150,000 men, women, and children in the process.[Britannica: Rome, 2008] With a strong foothold in the Greek territories and the destruction of Carthage in the 3rd Punic War, the Roman Republic dominated southern Europe, northern Africa, and the middle East by 146 BC. With all external foes defeated, the Roman Republic fell victim to its greatest threat – internal discontent. A period of persistent civil war started with the Social War in 91 BC. At its conclusion, Rome granted citizenship to all free inhabitants of Italy. During 88 BC, Sulla marched on Rome against Gaius Marius to retain his command of legions headed for Athens. After securing victory in 83 BC, Sulla once again battled in Rome to squelch his opposition. These civil wars resulted in the death of 200,000 free Romans and Italians. As a consequence, the frightened Senate appointed Sulla dictator.[Encarta: Rome, 2008] After Sulla's retirement, various power struggles emerged between Crassus, Pompey, Cato, and Julius Caesar. By 53 BC, the Senate favored Pompey over Caesar. During January 49 BC, humiliated by the Senate's actions, Caesar proclaimed: **"The die is cast."** At that moment, civil war began. Caesar and his legions aggressively pursued Pompey's. During the next decade, Caesar defeated Pompey, and upon his return to Rome in 44 BC, the Senate appointed him dictator for life. But Caesar only enjoyed his triumph for a few days. Led by Brutus, a group of senators stabbed Caesar to death. Following the assassination, another civil war broke out. This time it pitted Caesar's deputy, Mark Antony, against Caesar's adopted grandnephew, Octavian. After an extensive conflict, Octavian defeated Antony in 31 BC.[Encarta: Rome, 2008] This victory paved the way for the demise of the Republic, and Octavian's cunning ascent to emperor of the new Roman Empire.

This sub-cycle lasted 170 years. The declining phase started 40 years before the 53 BC theoretical peak. Starting in 91 BC, sixty years of continual civil war resulted in the collapse of the Roman Republic in 31 BC.

118 AD Theoretical Peak of the 515-Year Cycle.

31 BC – 265 AD – **Golden Age of the Roman Empire.** After becoming emperor, Octavian eventually received the title Augustus Caesar, and he ruled the new empire until his death in 14 AD. Following the 60 years of civil war, Romans enjoyed an extended period of peace and prosperity under the new regime. After enduring several incompetent and/or tyrannical emperors, Rome ascended to its glory during the reign of the five good emperors – Nerva (96-98), Trajan (98-117), Hadrian (117-138), Antoninus Pius (138-161), and Marcus Aurelius (161-180).[Encarta: Rome, 2008] With Trajan's conquest of Dacia, the Roman Empire reached its territorial peak. At the time of Trajan's death, the empire covered 2.5 million square miles (6.5 million km^2).[Scarre, 1995] A few months later, during February 118, the theoretical peak of the 515-year civilization cycle arrived. And from that point forward, the Roman Empire started a descent into oblivion. For the first fifty years after the theoretical peak, the Roman economy stabilized at a high plateau. When Commodus became emperor in 180, the age of the good emperors came to an end. A century of turmoil ensued. The downturn caused a collapse of political institutions. It weakened the army. And it created economic disaster. After the murder of Commodus in 192, civil wars between rival claimants to the throne penetrated every corner of the empire.[Encarta: Rome, 2008] Monetary conditions also took a turn for the worse. During the reign of Augustus Caesar, a silver denarius weighed 5.7 grams, with 99% purity. By 193, the denarius had dropped to 4.3 grams, with only 70% purity.[Encarta: Rome, 2008] After 211, Caracalla debased the denarius an additional 25%.[Britannica: Rome, 2008] Attacks, rebellions, and civil wars followed. Rome fell into the position of defending its empire, rather than expanding it. The situation reached its worst during the 260s, but the entire period from 235 to 284 brought the empire close to collapse.[Britannica: Rome, 2008] As a reflection of the political instability during that span, 25 different emperors ruled. However, all cycles eventually come to an end, and the 3rd Century crisis ended around 265.

This sub-cycle lasted 296 years. It spanned the Golden Age of the Roman Empire – from 31 BC until 192 AD. However, the territorial peak of the Roman Empire occurred within a year of the 118 theoretical peak of the 515-year civilization cycle.

290 Theoretical Peak of the 172-Year Cycle.

265-410 – **Roman Empire Contraction.** The extraordinary recovery of the 4th Century reversed the seemingly hopeless situation of the previous decades. Leadership came from Diocletian, who ruled from 284 to 305. Diocletian instituted reforms that restored stable government and prosperity to the empire racked by fifty years of civil unrest.[Encarta: Rome, 2008] Diocletian took the dramatic step of naming a co-emperor, and two junior emperors to ensure a peaceful succession. This rule of four, called a tetrarchy, divided the administration of the empire. A key development took place in 330, when Constantine I dedicated the new city of Constantinople.[Britannica: Rome, 2008] This move paved the way for a split in the empire. Permanent division into the Eastern Roman Empire (later known as the Byzantine Empire) and the Western Roman Empire came in 395.[Encarta: Rome, 2008] Associated with the split, economic conditions worsened after 350 in the Western Empire. Archaeological data retains the degree of wealth in public buildings by their materials, size, and architecture. This archaeological record shows a noticeable decline throughout the European provinces. The decline tended to affect the cities earlier than the rural areas and it's detectable by 350.[Britannica: Rome, 2008] The economic decline persisted. By 410 BC the weakness became so pronounced that the Visigoths, led by Alaric I, descended on Rome and sacked and pillaged it for three days.[Britannica: Rome, 2008] With the sacking, the empire hit bottom again.

This sub-cycle lasted 145 years. Recovery took place from 265 to 350; however, the Roman economy declined sharply during the sixty years that followed.

462 AD Theoretical Peak of the 172-Year Cycle.

410-476 – **Fall of the Western Roman Empire.** After the sacking of Rome by the Visigoths, a recovery came during the rule of Valentinian III (423-455). Under the leadership of General Aetius, Rome took back Arles and Narbonne from the Visigoths in 436. Then, in 450, Aetius defeated Attila's Huns. However, after the death of Valentinian III in 455, the empire became vulnerable to the intrigues of German chieftain Odoacer. After the 462 theoretical peak passed, the remaining portions of the Western Empire fell quickly. Sixteen years later, in 476, Odoacer deposed the last Roman emperor, Romulus Augustus.[Britannica: Rome, 2008] Britannica concludes the Roman Empire history with this piece:

> "The causes of the fall of the empire have been sought in a great many directions and with a great deal of interest, even urgency, among historians of the West; for it has been natural for them to see … parallels between Rome's fate and that of their own times."[Britannica: Rome, 2008]

In searching for the timing of civilization collapses, the 515-year EUWS cycle provides good answers. In searching for the cause of civilization collapses, the *Unified Cycle Theory* proposes that the EUWS cycles trigger emotional swings in humans that create oscillations of <u>confidence and trust during up-phases</u> turning to <u>fear and suspicion during down-phases</u>.

Medieval European Civilizations

Even after the fall of the ancient Greek and Roman civilizations, important power centers remained in Southern Europe for more than a thousand years. Centered in present-day Turkey, the Byzantine Empire emerged as the first dominant post-Roman civilization.

633 Theoretical Peak.

395-867 – **First Civilization Cycle for the Byzantine Empire.** After the death of Theodosius I in 395, the Roman Empire officially split in two. The western part collapsed within one hundred years of the split, while the eastern portion survived. After the fall of the west, historians generally refer to the Eastern Roman Empire as the Byzantine Empire – named after its capital, Byzantium. Byzantium was renamed to Constantinople after Emperor Constantine's death. One of Constantine's early successes involved establishing the gold solidus. This gold coin went by the nickname of Byzant. Unlike the Romans, the Byzantine rulers restrained from debasing the Byzant. It remained the monetary standard throughout Europe for centuries.[Britannica: Byzantine, 2008] With a strong monetary base, the new empire thrived economically. Under Justinian I (527-565), Byzantine territory expanded tremendously to include North Africa, Italy, Sicily, Sardinia, and parts of Spain.[Encarta: Byzantine, 2008] As the 7th Century arrived, Byzantine fortunes improved further under Heraclius. In 622, he liberated Asia Minor. During 627, his armies destroyed the main Persian host at Nineveh. And in 628, he deposed his enemy, Khosrow, after occupying Dastagird.[Britannica: Byzantine, 2008] But then the theoretical peak of the civilization cycle arrived in 633. Immediately afterward, the empire's direction reversed! Perhaps exhaustion from its pre-633 conquests left Byzantine vulnerable. Less than one year passed before Byzantine began losing a large portion of its territory to the Arabs. Between 634 and 642, the Arabs took land in Palestine, Syria, Mesopotamia, and Egypt. Constantinople weathered major Arab sieges in the 670s and in 717-18. And Asia Minor territories endured constant Arab raids.[Encarta: Byzantine, 2008] The retrenchment lasted until the death of Michael III in 867.

The first cycle of the Byzantine Empire covered 472 years. It reached its summit in the same year as the 633 EUWS theoretical peak. During the following year, the empire shrank as Arab armies began capturing some of its territories.

1148 Theoretical Peak.

867-1453 – **Second Civilization Cycle for the Byzantine Empire.** Byzantine fortunes began to improve after the Macedonian, Basil I, murdered his way to the throne in 867 – killing Michael III. Under the Macedonians, at least until the death of Basil II in 1025, the empire enjoyed a golden age. Its armies regained territories lost to the Arabs in the East, and its missionaries extended Byzantine influence into Russia and the Balkans.[Britannica: Byzantine, 2008] After the death of Basil II, the empire enjoyed economic expansion and prosperity but suffered from a series of mediocre emperors. During 1054, the Byzantines lost their last foothold in Italy. And in 1071, it lost most of Asia Minor to the Seljuk Turks.[Encarta: Byzantine, 2008] After that, the empire rose to new heights. Initially, the First Crusade benefited Byzantine. In 1097, the crusaders captured Nicaea and handed it to the Emperor. In 1098, they captured Antioch. But the crusaders refused to turn over Antioch. The final turning point came during the reign of Manuel I (1143-1180). Early in his regime, Manuel enjoyed a few small victories. However, in 1176 the Turks annihilated Manuel's army at Myriocephalon.[Britannica: Byzantine, 2008] This loss served as a crippling blow to Byzantine. Crusaders allied with Venice took advantage of

The 515-Year EUWS Cycle

Byzantine weakness to seize and plunder Constantinople in 1204, establishing their own Latin Empire. However, Michael VIII recaptured Constantinople from the Latins in 1261.[Encarta: Byzantine, 2008] In 1302, Turkish warriors defeated the Byzantine army in northwestern Anatolia. Its leader, Osman I, organized the strengthening Ottoman movement. In 1348, the Serbian Empire swallowed Byzantine territory in northern Greece. About the same time, the Black Death decimated the population of Constantinople and other parts of the empire. During 1354, the Ottomans occupied Gallipoli. In 1387, Turks drove deep into Macedonia. By 1393, Bayezid completed his conquest of Bulgaria. When Murad II became sultan in 1421, the days of Constantinople were numbered. By 1430, the sultan captured Thessalonica. During 1449, Sultan Mehmed II prepared for the final assault on Constantinople. Mehmed II laid siege to the walls in April 1453. He finally captured Constantinople on May 29, 1453.[Britannica: Byzantine, 2008]

The second cycle of the Byzantine Empire covered 585 years. It reached the peak of its golden age between 1054 and 1176 – surrounding both sides of the 1148 theoretical peak of the 515-year cycle. During the following 300 years, the Byzantine Empire completely collapsed.

1663 Theoretical Peak.

1299 to 1923 – **Ottoman Empire.** The Ottoman Empire (Turkish Empire) formed in 1299, when Osman I declared himself sovereign ruler. By 1453, the armies of Mehmed II toppled the Byzantine Empire. During the century following the reign of Mehmed II, the Ottoman Empire reached its zenith. New conquests extended its domain well into central Europe and throughout the Arab world.[Britannica: Ottoman, 2008] Ottoman fortunes declined after the death of Suleyman in 1566. However, the changes were imperceptible at first. Ottoman weakness began to show itself in the 17th Century against both the Habsburgs and Iran. Under Sultan Murad IV (1623-1640), the empire experienced one final period of power. Murad strengthened the eastern Ottoman flank by capturing Baghdad.[Encarta: Ottoman, 2008] After his death the empire fell victim to a severe internal crises. This downturn came just 23 years before the 1663 theoretical peak of the civilization cycle. During 1656, Koprulu Muhammad Pasha restored political order after he assumed the office of grand vizier. However, by 1683, Polish armies routed Ottoman forces at Vienna. Military losses mounted. In 1697, an Austrian commander ambushed the Ottoman army in northern Serbia, inflicting great losses.[Encarta: Ottoman, 2008] The empire continued losing territory during the 18th Century. By 1808, the Ottoman situation appeared desperate. Within the empire, the central government exhibited minimal authority. Control of North Africa had long since faded.[Britannica: Ottoman, 2008] The empire continued its decline during the 19th Century. On November 1, 1922, the Grand National Assembly abolished the sultanate, formally ending the Ottoman Empire. In the following year, the Republic of Turkey replaced it.[Encarta: Ottoman, 2008]

The Ottoman Empire lasted 624 years. Its power climaxed between the years 1566 and 1683 – surrounding both sides of the 1663 theoretical peak. During the next 240 years, the Ottoman Empire completely collapsed.

1148 Theoretical Peak.

1000-1479 – **Spanish Empire, Phase 1.** Spain emerged as a power in two phases. At the turn of the 1st Millennium, Muslims dominated the Iberian Peninsula. At that time, Spain's Christian rulers took advantage of civil war among the Muslims. These civil wars enabled Ramon Borrell to sack Cordoba in 1010. Alfonso V of Leon (999–1028) exploited the situation to restore his kingdom in 1017.[Britannica: Spain, 2008] Additional gains came in 1085, when the Muslims surrendered Toledo to Alfonso VI – thus giving the Christians control over northern Spain. Further gains came when James I captured Valencia in 1238. Alfonso IX captured Merida and Badajoz in 1230. Ferdinand III conquered Cordoba in 1236, Murcia in 1243, Jaen in 1246, and Seville in 1248. After this string of victories, Spain dominated the peninsula, with Muslim control restricted to Granada.[Britannica: Spain, 2008] But the cycle turned downward after that. The Black Death plague devastated Spain during 1348-49. By 1385, John I invaded Aljubarrota and suffered defeat at the hands of the Portuguese. Royal prestige and authority suffered terribly during the long reign of John II (1406–54). During the reign of Henry IV (1454-74), the nobles continued to engage in an intense struggle for influence. However, Ferdinand's accession to the Aragonese throne in 1479 brought about a personal union of Aragon and Castile.[Britannica: Spain, 2008] Thus, another civilization cycle came to a conclusion.

Phase 1 of the Spanish Empire covered 479 years. It climaxed one hundred years after the 1148 theoretical peak. Then it withstood 200 years of consolidation.

1663 Theoretical Peak.

1479-1978 – **Spanish Empire, Phase 2.** Phase 2 of the Spanish Empire began with the political unification under Ferdinand. More importantly, the remaining Muslim presence on the Iberian Peninsula finally gave way. On January 2, 1492, Muhammad XI surrendered his authority to Spain.[Britannica: Spain, 2008] With the surrender, Spain refocused on bigger and better things. Sailing under the Spanish flag, Columbus discovered America in 1492, thus opening the gateway for global colonization. The first great flowering of literature and art in Spain coincided with the country's brief dominance of Europe (and other parts of the world) – a period that lasted approximately from 1550 to 1650. The country also produced major works in architecture and philosophy during its Golden Age.[Encarta: Spain, 2008] However, conditions in Spain changed drastically slightly ahead of the civilization cycle peak of 1663. Toward the end of the reign of Philip III, an extended period of peace ended with the outbreak of the Thirty Years' War (1618-1648). Spanish civilization descended rapidly after the war. First, Spain lost Portugal in 1640. In 1648, it lost the Netherlands. During the 1640s, declining shipments of American silver also hurt Spain. The reduced shipments made it difficult to finance the war effort, and Spain declared bankruptcy in both 1647 and again in 1652.[Encarta: Spain, 2008] The War of the Spanish Succession (1701-1714) stripped Spain of its last European possessions. Under the settlement reached in the Peace of Utrecht, Spain lost Gibraltar, Milan, Naples, Sardinia, Sicily, Minorca, and the last of its territories in the Spanish Netherlands. By the 1800s, revolutionary movements took hold in many of Spain's American colonies. By 1826 only Cuba, Puerto Rico, the Philippine Islands, Guam, and several settlements in northern

Africa remained under Spanish rule. Then, in 1898, Spain lost most of its remaining overseas territories in the devastating Spanish-American War. Finally, after Franco's death, the government drafted a new democratic constitution for Spain. In 1978, parliament approved the new constitution. It established a constitutional monarchy, with the king serving as head of state and a symbol of national unity. It created an independent judiciary and placed significant restrictions on two of Spain's most historically important institutions: the military and the Catholic Church.[Encarta: Spain, 2008]

Phase 2 of the Spanish Empire covered 499 years. This civilization climaxed about forty years before the 1663 theoretical peak.

Modern Western Civilization

After the last civilization cycle turned down in 1663, all of the imperial powers in Asia and Europe started to crumble. New political ideas favoring individual freedoms emerged in England. These ideas paved the way for the formation of democracy and modern Western civilizations. Ideas about free markets, individual rights, and *laissez faire* government proved so successful that nations around the globe were forced to either accept these ideas or lag economically. After winning independence from England in 1783, the United States expanded on these ideas to include very limited government and government with checks and balances.

England and the United States emerged as the new global powers in the current civilization cycle. In an attempt to improve their own economies, one by one, ancient forms of government fell by the wayside. And the process continues to this day. Within the last twenty-five years, the former Soviet Union and Communist China (among others) turned toward freer markets as a means of stimulating their previously disastrous government-planned economies.

Ironically, the long-term global leader in free markets, the United States, now bucks the trend it initially set. During the past forty years, both the federal government and the Federal Reserve have taken numerous unprecedented steps to intervene, manage, and plan the USA economy. These moves away from free market operation flash strong danger signals. Moves toward socialism, large government, and massive budget deficits always act as key leading indicators for civilizations heading for collapse. And the United States recently embraced all three of these methods in a big way.

The theoretical peak of the next 515-year civilization cycle comes in 2178. If previous trends continue, the years surrounding that date should mark a peak in the current up-phase for Western Civilization. After the Ottoman and Spanish empires fell apart, Great Britain ascended to the position of the leading global power. The apex of British power came shortly after the theoretical peak of the 172-year cycle in 1835 – during the Victorian Age. However, a hundred years later (during the 1940s), Britain found itself under attack from Germany. Britain remained an international power after World War II. However, it relinquished its role as the Western leader to its former colony, the United States.

The United States maintained that role ever since. However, with the passing of another theoretical peak of the 172-year cycle, the leadership position held by the United States appears to be in jeopardy. The threat primarily comes from the poor state of USA finances. Basically, the federal government owes huge debts, with only its taxing power providing income. Compounding the problem, the household and corporate sectors of the economy are equally over-indebted. Taxing an over-indebted private sector will prove difficult.

Similar to previous great civilizations, military and technological superiority allowed the USA to consume an unusually high percentage of global resources. But overconsumption can only be maintained as long as the technological advantage persists. However, by 2008, most of the prior advantage had evaporated. A large portion of the world has caught up with the USA on the technology front. In fact, the USA imports a great deal of that technology from overseas. From this point forward, USA overconsumption will be extremely difficult to maintain.

The signs are many fold. Developing countries are now willing and able to pay for an increased share of limited natural resources. This makes the supply available for the United States increasingly expensive, and sometimes unavailable. Heavy debt-to-income ratios in the United Sates indicate Americans are living beyond their means. Heavy Federal Reserve intervention in the financial markets shows that the United States is turning away from free market policies and toward socialism – an ill-advised defensive tactic aimed at minimizing social unrest. The United States has allowed its currency to depreciate in foreign exchange markets. Dollar depreciation equates to coinage debasement – a sign of overspending that appeared prior to other civilization collapses.

The United States still holds a significant military advantage. However, the rapidly deteriorating financial condition almost guarantees that the USA will eventually lose that advantage, as well. And for the United States, it indicates rough economic times for decades to come. In order to replace current American dominance, the next global power must be a nation with a surplus of savings, with a strong technology base, and with an ability to implement existing military technologies now used by the USA. Leading candidates include Germany and Japan, with China and India not too far behind. Within twenty years, China and India could shrink the gap if they continue to advance at the rate they have been for the last thirty years.

Solar and Weather Cycles

So far, this chapter only shows how the 515-year cycle impacts civilizations. Similar to its parent cycles, the 515-year cycle appears in other phenomena as well. **Table 26.1** provides a list of researchers who have identified sunspot and/or weather-related cycles ranging between 500 years and 530 years.

The 515-Year EUWS Cycle

Table 26.1 – Sunspot and Climate Cycles Approximately 515 Years in Length.

Period	Type & Span	Research Team	Area	Research Method
520 yrs	Wet-dry 3,000 years	[Pederson, 2000]	Lake Canyon Utah, USA	Carbon-14 dating of sediment in two paleolakes in Lake Canyon
400 to 650 yrs	Solar activity, 14,000 years	[Sarnthein et al., 2003]	Barents Sea N. Europe	Carbon-14 dating of planktic foraminifera in the Barents Sea
500 yrs	Weather cycle	[Sonett & Suess, 1984]	S. California USA	Carbon-14 dating of bristle cone pine rings
510 yrs	Weather for 11,400 years	[Stager et al., 1997]	Lake Victoria Africa	Carbon-14 dating of sediment in Lake Victoria
530 yrs	Solar cycle, 16,500 years	[Stuiver et al., 1995]	Greenland	Oxygen-18 analysis of GISP2 ice core
512 yrs	Ocean for 12,000 years	[Stuiver & Brauzanias, 1993]	Global	Carbon-14 dating of marine samples
500 yrs	Monsoon for 30,000 years	[von Rad et al., 1999]	Arabian Sea by Pakistan	Hemipelagic sediments of the Oxygen Minimum Zone

By categorizing the cycle in Barents Sea planktic foraminifera as a range between 400 to 650 years, Sarnthein *et al.* capture the essence of the 515-year cycle.[Sarnthein *et al.*, 2003] The average of 400 and 650 equals 525 years – coming close to the 515-year frequency. And the wide range reflects the variable nature of its periods. Even though their turning points move around, their tendency to consistently come back to theoretical turning points indicates if a cycle belongs to the EUWS group, or not. And, in fact, the Barents Sea plankton cycles match 515-year EUWS oscillations quite well.

Chart 26B provides additional evidence of the 515-year solar cycle. To construct this estimated sunspot series, NOAA used radiocarbon dating techniques on various global datasets. In the chart, the oscillations appear as tremendously strong 515-year rhythms. The fluctuations occur over 22 successive cycles.

The down-arrows closely match theoretical peaks of the 515-year cycle. The only significant deviations came at 5547 BC, 1427 BC, and 1663. In all three cases, powerful 4636-year oscillations interfered with the performance of the 515-year cycle. Around 5547 BC and 1427 BC, down-phases of the 4636-year cycle squashed the expected tops at those times. In 1663, the up-phase of a 4636-year cycle virtually eliminated the expected downturn then. Other than these three exceptions, the sunspot cycle fluctuated in lock-step with expected 515-year EUWS turning points.

Even though sunspots fluctuate in a 515-year cycle, this particular cycle cannot be classified as a solar cycle. Its frequency varies from cycle to cycle, similar to the 11-year Schwabe cycle. However, both its frequency and its turning points match EUWS predictions. This fact implies that EUWS cycles influence solar activity. Hence, two sources exist for producing solar cycles –

one internal and the other external. Because of the interference problems this creates, and because of the jumpy nature of magnetic cycles, separating the various solar frequencies becomes a little more difficult.

Chart 26C shows how the 515-year cycle correlates with the Greenland ice core record. Similar to the other examples, Greenland temperatures closely match theoretical turning points of the 515-year EUWS cycle.

This chapter demonstrates the powerful influence the 515-year EUWS cycle exerts on human civilizations. Definitive evidence exists showing that the 515-year cycle correlates closely with human activity cycles, sunspot cycles, and weather cycles. As the book progresses, pay close attention these correlations (or lack thereof). They provide some of the most critical evidence concerning the nature of the EUWS cycles.

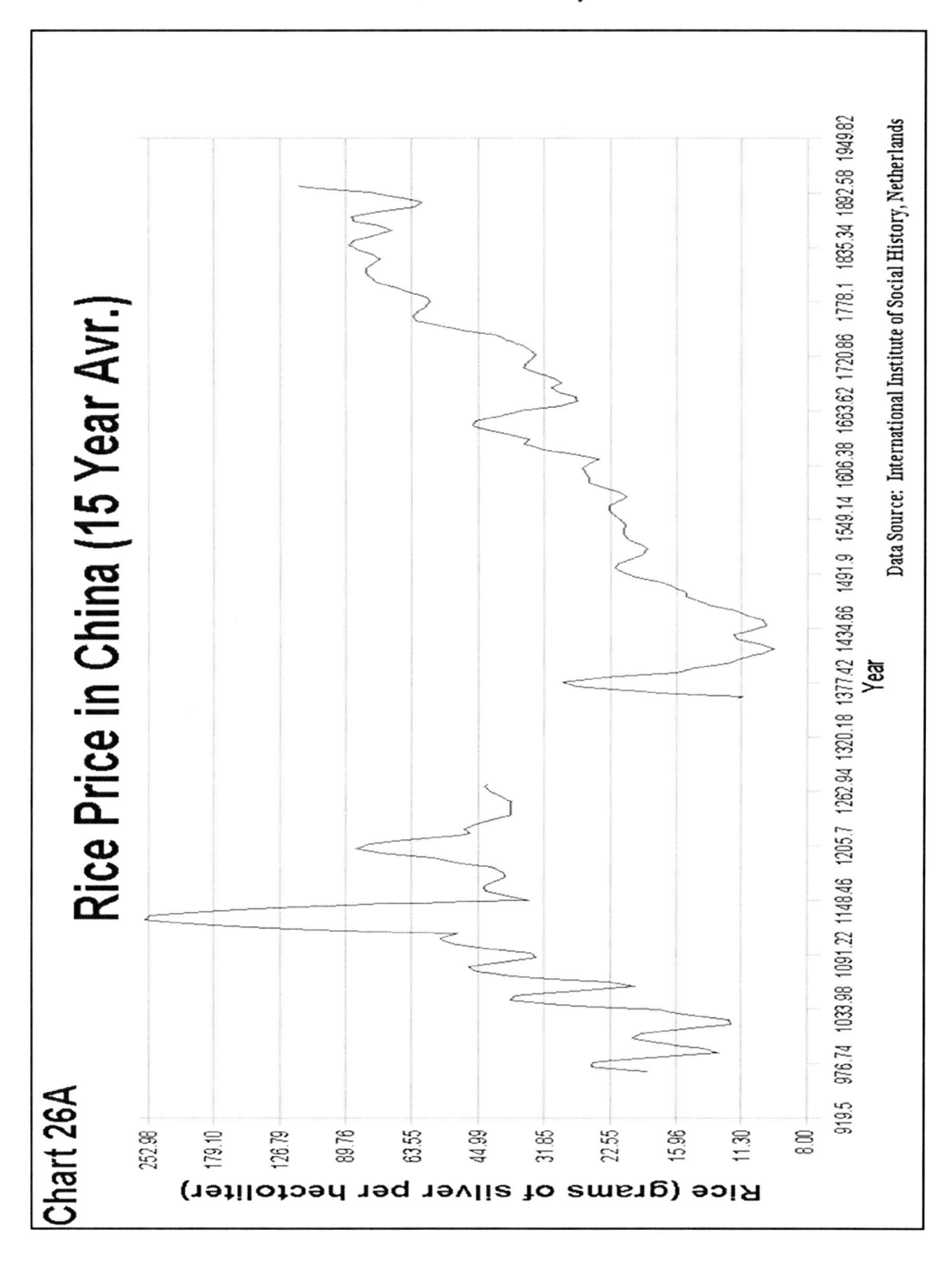

The 515-Year EUWS Cycle

Chart 26A: Rice Price in China (15 Year Avr.)

Data Source: International Institute of Social History, Netherlands

The Unified Cycle Theory

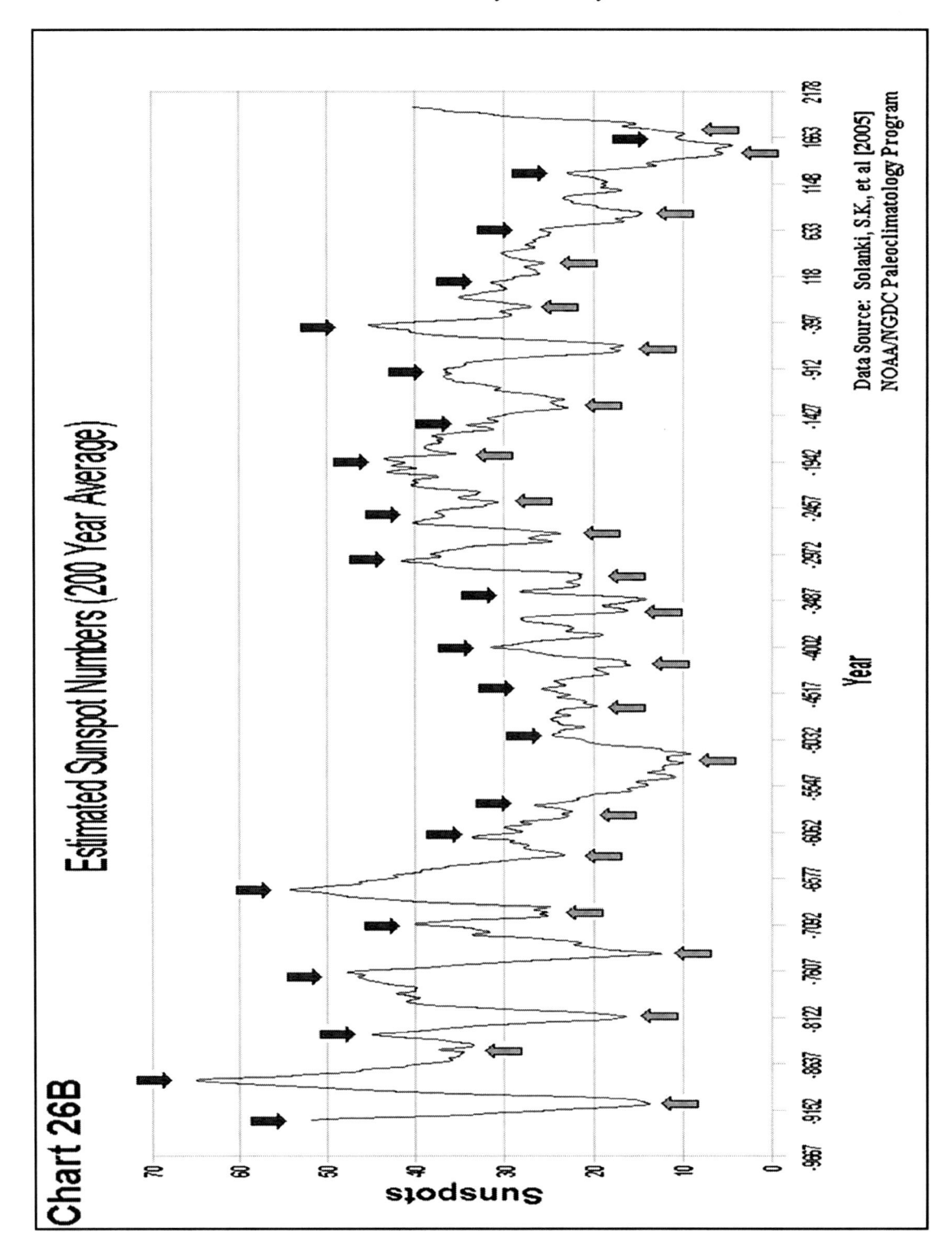

Chart 26B

312

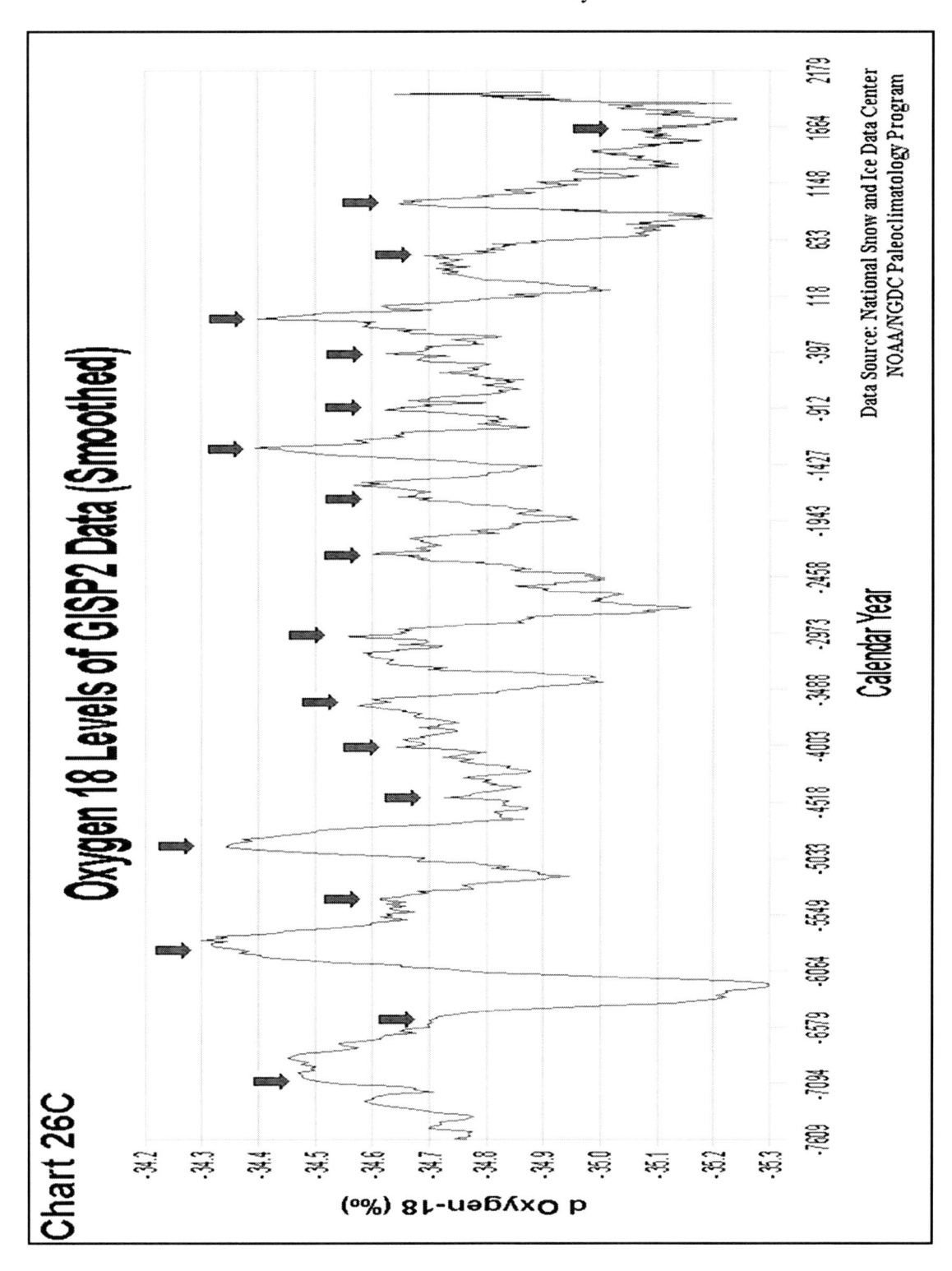

The 515-Year EUWS Cycle

Chapter 27
The 172-Year EUWS Cycle

Divided by three, the 515.16-year cycle produces the 171.72-year EUWS cycle. This will be referred to as the 172-year cycle. Similar to its parent, this cycle impacts civilizations and human activity. Chapter 26 showed how the 172-year cycle affected fluctuations in ancient Roman civilization.

The list below shows theoretical peaks of the 172-year EUWS cycle covering the last 3000 years. The years marked with a double asterisk (**) indicate theoretical peaks of the 515-year cycle -- 912 BC **, 740 BC, 568 BC, 397 BC **, 225 BC, 53 BC, 118 **, 289, 461, 633 **, 805, 976, 1148 **, 1320, 1491, 1663 **, 1835, 2007, and 2178 **.

The 172-year EUWS cycle also affects global climate. The correlation between climate and human behavior has already been noted. In Chapter 25, sociologist Sing Chew attributed the 1545-year dark-age cycle to climate changes caused by deforestation.[Chew, 2006] In his book *Climate: The Key to Understanding Business Cycles by Raymond Wheeler*, Michael Zahorchak interprets the data in the opposite way – attributing human activity cycles to climate changes.[Zahorchak, 1983] He noted that warm periods coincided with dictatorial rule and international wars, while cold periods produced civil unrest and democracy. He also noted a 170 year cycle in droughts and civil war. In addition, he observed that every third cycle produced more pronounced droughts and wars – producing a cycle that spanned approximately 510 years. Using Wheeler's data, Edward Dewey also identified a cycle in battles – occurring at 57 year

315

intervals.[Dewey, 1951]

In fact, reexamining the data from the last chapter, civilizations ascend along with rising global temperatures and increasing government control. At the peak of a civilization cycle, temperatures cool, war breaks out, and government control falls apart. The data from Chapter 26 confirms Wheeler's and Zahorchak's observations. However, instead of mankind causing the climate cycles (as Chew suggests) or temperature oscillations causing social cycles (as Wheeler and Zahorchak propose), more than likely, a third factor (EUWS cycles) produces the fluctuations in both climate and social behavior.

The cycles identified by Wheeler and Dewey almost perfectly match the 57.24, 171.72, and 515.16 year EUWS cycles. Of particular interest, do the EUWS cycles really produce cycles in war activity? The answer holds significant implications for mankind's future.

In an attempt to independently analyze war cycles, numerous difficulties arose. The study involved all EUWS cycle between 1545 and 57 years. The results mildly confirm the presence of war cycles. However, the following discussion concentrates more on the difficulties, uncertainties, and inconsistencies resulting from the study.

Analysis of the 1545-Year Cycle in Wars.

A 1545-year war cycle appears; however, it behaves exactly the opposite of what the EUWS cycle predicts. During the down-phases of the 1545-year dark-age cycle, war activity almost stopped. In reality, fighting probably intensified during these down-phases, but because written records often cease during dark-ages, historians weren't available to document the fighting.

In his book *War Before Civilization*, University of Illinois professor Lawrence H. Keeley describes the difference between documented wars conducted by established civilizations versus raids carried out by small uncivilized groups:

> "In civilized wars, because modern states have larger territories, redundant transportation networks, and a broad margin of productivity above the bare subsistence level, years of destruction and blockade may be necessary to reduce one to starvation. But ... pre-state societies had small territories and much slimmer margins of productivity. Primitive social units could be reduced to a famine by the consequences of a few days of raiding or even of a single surprise attack. Because the infrastructure and logistics of small-scale societies were more vulnerable to looting and destruction, the use of these methods was almost universal in primitive warfare.... In some regions of the American Southwest, the violent destruction of prehistoric settlements is well documented and during some periods was even common.... For example, the large pueblo at Sand Canyon in Colorado, although protected by a defensive wall, was almost entirely burned; artifacts in the rooms had been deliberately smashed; and bodies of some victims were left lying on the floors. After this catastrophe in the

late thirteenth century, the pueblo was never reoccupied."[Keeley, 1996]

Keeley infers that constant brutality existed before civilization emerged. He estimates 87% of tribal societies engaged in war annually. Going back further in time, he examines archaeological evidence related to circular ditches discovered during Europe's Neolithic period (from 5500-3000 BC). Rather than being religious or ceremonial sites (as some historians suggest), Keeley explains their defensive significance:

"A far different impression is conveyed by the reports of the archaeologists who have conducted extensive excavations of some of these enclosures. At several camps, the distribution of thousands of flint arrowheads, concentrated along the palisade and especially at the gates, provides clear evidence that they had quite obviously been defended against archery attack.... Moreover, the total destruction by fire of some of these camps seems to have been contemporaneous with the archery attacks. At one such site, intact skeletons of two young adult males were found at the bottom of the ditches, buried beneath the burned rubble of the collapsed palisade-rampart. In one poignant instance, the young man had been shot in the back by a flint-tipped arrow and was carrying an infant in his arms who had been crushed beneath him when he fell. Whatever ritual or symbolic functions the enclosures might have had, they were obviously fortifications, some of which were attacked and stormed."[Keeley, 1996]

Viewed from this perspective, any study of war cycles really boils down to a study of civilizations. When civilizations rise, documented war activity increases. When civilizations decline, documentation decreases, and evidence of fighting becomes obscure – even though fighting itself continues on.

Analysis of the 515-Year Cycle in Wars.

The rise and fall of civilizations in cycles of 515-years presents the same problem as the dark-age cycle. Documented conflicts decrease during the down-phases of the 515-year cycle. That runs counter to what the *Unified Cycle Theory* predicts. Furthermore, the number of wars inflates substantially with the approach of modern times, creating a bias toward recent data. Finally, the small number of documented wars fought more than 2000 years ago makes analysis of the 515-year cycle difficult, except for the last 4 cycles.

Analysis of the 172-Year Cycle in Wars.

For the 172-year cycle, no clear pattern appeared. However, my analysis included all types of wars. In contrast, Wheeler differentiated between international wars and civil wars. And that difference probably explains why Wheeler found the cycle. But numerous issues arise with this analysis.

Sometimes a fine line separates an international war from a civil war. That's especially true for expanding empires. An emperor may consider his provinces as parts of his empire, while rebels consider those same provinces as separate nations occupied by foreigners. In this scenario, should an outbreak of war be classified as civil or international? To varying degrees, a large number of wars fall into this gray area.

Defining a war by size becomes equally difficult. Should skirmishes, revolts, rebellions, coup d'états, insurgencies, conflicts, etc. be classified as wars? Should the length of fighting determine war conditions? For example, should a 2-day, intense battle be considered a war? And should a war require a minimum number of combatants? Unfortunately, these valid questions have no good answers.

Further complicating the analysis, adequate information about wars becomes fuzzy as their age increases. For example, historians often refer to the time from 481-221 BC as the Warring States Period in China. However, more than likely, some periods of peace split the fighting into several different phases. If so, then long-term wars, like the Warring States Period, need to be subdivided better for proper war-cycle analysis.

In a similar vein, ancient war information tends to be concentrated in the great civilizations. Hence, any war study could easily turn into a civilization study, rather than a war analysis. And the best documentation about ancient wars tends to come from the victors, thus creating another unwanted bias.

Analysis of the 57-Year Cycle in Wars.

My analysis confirmed Dewey's observation of a 57-year cycle in war activity. Furthermore, unlike the larger cycles, fighting during the 57-year cycle matched *Unified Cycle Theory* predictions – many wars occurred at the bottom of the cycles, with fewer wars at the top. The study included 42 cycles, giving each cycle equal weight. (This eliminated the bias resulting from more recorded wars during the last 500 years.) Each 57-year cycle consisted of six buckets. 21.9% of the wars started during the last bucket of the down-phases. Conversely, only 12.7% of the wars broke out during the last bucket in the up-phases. This pattern matches *Unified Cycle Theory* predictions.

That concludes the analysis of war cycles. Unfortunately, rather than solidly confirming the cycles reported by Wheeler and Dewey, the analysis only moderately confirmed the cycles. Additionally, the analysis identified a list of issues that need to be addressed before conducting conclusive research. However, a project such as this deserves considerable attention. Hopefully, this book will inspire historians to give war cycles due diligence.

The relative shortness of the 172-year cycle opens the door for moderate testing of another area of human behavior – on the stock market front. Equity markets first formed in Europe about 300 years ago. In the United States, the New York Stock Exchange formed humbly on May 17,

The 172-Year EUWS Cycle

1792 under the fabled Buttonwood Tree on Wall Street.

To remain consistent with other *Unified Cycle Theory* patterns, equity prices should move lower during down-phases of the 172-year cycle. Conversely, bull markets should materialize during its up-phases. Since the NYSE formed more than 200 years ago, only one complete 172-year cycle has transpired. Nonetheless, within that cycle, equities behaved closely to *Unified Cycle Theory* predictions.

May 5, 1835 Theoretical Peak of the 172-Year Cycle – Preceding this theoretical peak, stock market booms engulfed both England and the United States. For the USA, historians sometimes refer to this as the canal boom. Speculation in railroad stocks also contributed to the soaring stock market, as investors correctly foresaw the bright future for this new technology. However, like clockwork, once the cycle's theoretical peak arrived during May 1835, market conditions quickly changed. David Wheelock of the *St. Louis Federal Reserve Bank* gave this account of what happened next:

> "The boom was short-lived. Stock prices peaked in May 1835, and land sales peaked in the first six months of 1836.... In the absence of a well-functioning interregional reserves market, the ensuing outflow of reserves left the New York money market vulnerable to shocks and ... precipitated the Panic of 1837."[Wheelock, 2004]

In both countries, severe bear markets followed the 1835 theoretical peak. According to the *International Monetary Fund*, the bear market in England lasted from 1835 to 1839, and resulted in losses of 39.1% -- making it the second worst bear market for England in the 19th Century.[Helbling & Terrones, 2003] England's worst bear arrived a decade earlier, once again resulting from the after-effects of a great speculative fever. The 1825 English crash served as an early warning of the 172-year cycle's approaching peak:

> "The earliest and probably most infamous boom-bust in the modern era ended with the 1824–25 stock market crash in the United Kingdom.... After the Napoleonic wars and the successful resumption of the gold standard in 1821, the British economy enjoyed a period of rapid expansion, stimulated by both an export boom to the newly independent states of Latin America and investment in infrastructure projects (e.g., gas lighting, canals, and railroads). The sale of stocks to finance those ventures, in addition to gold and silver mines (some real, some fictitious) in Latin America, propelled a stock market boom fueled by the Bank of England's easy monetary policy.... The collapse triggered bank failures, which, once they reached important city banks, precipitated a full-fledged panic in early December. Only then did the Bank of England begin to act as a lender of last resort, and it was too late to prevent massive bank failures, contraction of loans, and a serious recession in early 1826."[Helbling & Terrones, 2003]

In the United States, a pair of bear markets followed that rivaled the Great Depression of the 1930s. The IMF reported a 46.6% loss in stocks during the years from 1835 to 1842, also making it the longest bear market in USA history. Eighteen years later, stocks underwent another severe bear market from 1853 to 1859. That decline resulted in losses of 53.4%.[Helbling & Terrones, 2003] This one-two punch started a string of financial panics that reoccurred every two decades – prompting Congress to created the Federal Reserve System in 1913 as a means of stopping the periodic panics. As it happens so frequently, Congress acted in a belated fashion with a poor solution. Just a few years later in 1921, the 172-year cycle reached its trough.

January 24, 2007 Theoretical Peak of the 172-Year Cycle – Only time will tell if the January 2007 theoretical peak of the 172-year cycle sets off another series of panics similar to the 1835 downturn. However, recent financial turmoil indicates that it has and that it will continue to do so. The January 2007 theoretical peak coincided exactly with the start of the subprime crisis. See **Chart 27A**. This chart shows the price of a contract reflecting the value of a package of subprime home equity loans. During January 2007, the contract traded close to par (100). By May 2008, its value sank to 5 cents on the dollar – a loss of 95%.

Initially, government officials downplayed the severity of the subprime crisis. On March 28, 2007, in testimony before the Joint Economic Committee of the U.S. Congress, Federal Reserve Board Chairman Ben Bernanke stated:

> **"Although the turmoil in the subprime mortgage market has created severe financial problems for many individuals and families, the implications of these developments for the housing market as a whole are less clear. The ongoing tightening of lending standards, although an appropriate market response, will reduce somewhat the effective demand for housing, and foreclosed properties will add to the inventories of unsold homes. At this juncture, however, the impact on the broader economy and financial markets of the problems in the subprime market seems likely to be contained."**[Bernanke, 2007]

It only took a little more than a year to prove Bernanke completely wrong. By September 2008, the top five investment banks in the United States had failed in one form or another. On May 30, 2008, to avoid a pending bankruptcy, Bear Stearns merged with JP Morgan/Chase. On September 15, 2008, Lehman Brothers announced their intent to file for bankruptcy. On that same day, Merrill Lynch engaged in a shotgun merger with Bank of America to avoid the same fate as Lehman Brothers. And on September 19, 2008, the two largest and strongest investment banks, Morgan Stanley and Goldman Sachs, applied to the Federal Reserve to be converted from investment banks to commercial banks. The switch gave Morgan Stanley and Goldman Sachs emergency funding sources not available to them as investment banks. Hence, in less than two years after the theoretical peak of the 172-year cycle, Wall Street's entire investment banking business model failed.

Certainly, the current financial crisis encompasses the worst credit conditions since the Great

The 172-Year EUWS Cycle

Depression. It will take several years before a definitive conclusion can be written. However, more than likely, it will end in another monumental depression.

Even though limited stock market histories prevent broad confirmation of the 172-year cycle, commodity markets provide a rich history of economic cycles. Prior to the year 900, records occasionally show the cost of various goods for various points in time. However, only one known source exists for a continuous series of commodity prices. The *International Institute of Social History* documents ancient Babylonian prices for five commodities on a fairly regular basis.[IISH: Prices and Wages, 2008] Price information on barley, dates, mustard, sesame, and wool allowed construction of a composite Babylonian Commodity Price Index.

The frequency of the data was less than desirable. Sometimes the Babylonians recorded prices twenty times within a year. Other times, price gaps lasted a decade. Nonetheless, filling the voids with interpolation permitted construction of a continuous series. Additionally, not all commodities experienced voids at the same time, thus enhancing the accuracy of the overall price trend. Hence, by using interpolation, and by giving equal weight to all five of the commodities, a fairly reliable commodity index emerged for ancient Babylonia. To further enhance its reliability, a 15-year average was employed. See **Chart 27B** for the end result.

The smoothed commodity index showed two prominent peaks – one at 339 BC and the other at 165 BC. That amounted to a 174 year separation between peaks. Hence, it yielded a near-perfect fit for the 172-year cycle. However, both of these commodity tops deviated from their ideal turning points by 57 years. First, it serves as an example of the TPD Principle, where the actual tops came one sub-cycle late. Second, it provides an example of a "shift" from the ideal turning point. Especially for the smaller EUWS cycles, shifts often span several cycles before their oscillations move back to the expected turning points. This can be seen on the chart, where the theoretical peaks came at 397 BC and 225 BC – approximately 57 years before the actual tops.

Next, the sunspot cycle exhibits a frequency that correlates with the 172-year EUWS cycles. And this sunspot cycle tends to lead commodity market prices by 20 to 60 years. Notice the rough correlation between **Chart 27 B** (Babylonian commodities) and **Chart 27C** (sunspots, shifted by one 57-year sub-cycle). Remember, the commodity index holds a degree of accuracy within about five years. The error in the sunspot estimates could be as high as thirty years. So the sunspot lead time might be less than what the charts show.

Similar sunspot cycle lead times appear in medieval and modern times, as well. The consistency of this behavior means that long-term cycles in solar activity act as leading indicators of major economic movements. However, that doesn't necessarily imply that solar cycles cause economic changes. Nonetheless, solar influence must be included in any list of potential causes of economic cycles.

After the Babylonian case, a thousand year gap follows before the next significant commodity history becomes available. Starting with the year 960, an extensive series of rice prices in China covers a millennium. See **Chart 27D**.[IISH: Prices and Wages, 2008]

The Unified Cycle Theory

Chart 27D shows that all significant tops in rice prices coincided with peaks of the 172-year cycle. The only exception came in 1320, when rice prices were unavailable. You can personally blame Kublai Khan for that absence, as his Mongolian-led invasion disrupted Chinese culture for nearly a century. Rice prices could have topped around 1320, but that piece of information will probably remain unknown.

Now, study **Chart 27E**, showing sunspot activity during that same interval. Notice how sunspot turning points tended to lead Chinese rice prices by anywhere from 20 to 60 years – exhibiting the same lead-time behavior as Babylonian commodities.

A more detailed summary of the 172-year cycle in Chinese rice prices follows....

Theoretical Peak 1, 976 – The actual peak in rice prices occurred around 973, about one 2.12-year cycle early. That was followed by a 45-year deflation ending around 1018.

Theoretical Peak 2, 1148 – The actual rice price peak took place around 1128, about one 19-year panic cycle early. A 100-year deflation followed, ending around 1248. The extreme nature of this deflation reflected the fact that it doubled as both a 515-year and a 172-year downturn.

Theoretical Peak 3, 1320 – Because of the 100-year absence of data during this cycle, the exact timing of the rice price decline remains unknown. Yet, prices did decline substantially sometime in this period. The data shows that rice prices stood at a level four times higher in 1260 than they were in 1360, when price information resumed. Presumably, the worst of the decline began sometime close to the theoretical peak of 1320, but we'll never know for sure.

Theoretical Peak 4, 1491 – A temporary top in rice prices developed in 1498. That was one 6.36-year market cycle late. The 1498 top interrupted a sustained fifty year rise in rice prices. A twenty year bear market followed, with prices hitting bottom in 1518.

Theoretical Peak 5, 1663 – During this cycle, prices topped around 1648. That happened about one 19-year panic cycle early. The deflation that followed lasted 25 years. While short in duration, price-wise, it represented the most severe deflation of the preceding 500 years. The unusually large deflation during this cycle matched the severity predicted by a downturn of the 515-year cycle at the same time.

Theoretical Peak 6, 1835 – Rice prices peaked around 1838, about one 2.12-year cycle late. A 45-year deflation followed, finally ending in 1883.

Theoretical Peak 7, 2007 – Following the death of Mao Tse-Tung in 1976, China remained under Communist control. However, the new leadership's adoption of a market economy prompted resurgence in Chinese power. Embracement of a market economy ended nearly 65 years of turmoil following the 1911 Revolution. Jumping forward to the present, in 2008, extensive price inflation has returned to China. The current price inflation falls into the same

The 172-Year EUWS Cycle

pattern displayed at previous 172-year peaks. If history repeats, the current inflation will soon reverse course, and end in another violent deflation.

At the other end of the Euro-Asian land mass, English historians actually recorded mankind's longest continual string of commodity prices. That's because, unlike the Chinese data, the English series consisted of gap-free data. **Chart 27F** shows English wheat prices from 1266 to 1935.[IISH: Prices and Wages, 2008] Over the long term, commodity prices behave differently in every country. Yet, in traveling to their long-term destination, prices fluctuate in distinctively clear 172-year cycles. The English wheat price history follows. (Turning points reflect a 15-year centered average, not actual turning points.)

Theoretical Peak 1, 1320 – Wheat prices peaked in 1316, two 2.12-year cycles early. The low came 23 years later, in 1339. However, this period also represented a down-phase for the civilization cycle. Reflecting the down-phase, after peaking in 1316, wheat prices bumped along the low for another 131 years, until 1447.

Theoretical Peak 2, 1491 – Wheat prices moved higher until 1484, but did so in a pathetic manner. After that, prices only declined for eight years, and then moved sideways for another eleven years. That was followed by one of the greatest commodity bull markets in history.

Theoretical Peak 3, 1663 – The great bull market in English wheat prices began in 1503 and lasted until 1655. It came to an end about one 6.36-year cycle early. (Without the moving average, the single-year top came in 1662, making the actual top virtually a direct hit!) An 89-year bear market followed, ending in 1744. Without a doubt, the civilization cycle up-phase from 1503 to 1663 heavily influenced the huge price upswing. And the subsequent civilization cycle down-phase contributed to the deflation that followed.

Theoretical Peak 4, 1835 – The 15-year average of wheat prices topped in 1807, more than one 19-year panic cycle early. The actual peak occurred in 1811 – slightly closer to the 19-year theoretical peak of 1816. A 92-year deflation followed, ending in 1899.

Theoretical Peak 5, 2007 – Once again, England finds itself engulfed in commodity price inflation. In addition to rising wheat prices, oil prices have risen ten-fold from levels prevailing during the 1990s. However, with the 172-year cycle now heading lower, the current commodity inflation should soon reverse course.

Chart 27G shows sunspot fluctuations from 1148 until 1935. Resembling other commodity markets, sunspot activity leads English wheat price trends anywhere from 20 to 60 years. This unusual behavior suggests that geomagnetic fluctuations influence humans like a battery – slowly charging while receiving power during up-phases, and then gradually dissipating as power fades during down-phases.

Ed Mercurio attributes these cycles to GCRs. Even though the GCR source is probably incorrect, his excellent abstract, entitled *The Effects of Galactic Cosmic Rays on Weather and*

Climate on Multiple Time Scales, contains many important details about these cycles. First, as the title implies, he recognizes that these cycles cover many timescales. Second, he understands that these cycles all originate from the same source. Third, he notices the wide variability in the cycles' frequencies. And fourth, he closely identifies the frequencies of the main cycles. The following passage from his abstract describes three of the EUWS cycles:

> "~2400 year solar cycles appear to be divided into overall lower GCR, warmer halves and overall higher GCR colder halves, but periodic maxima and minima, also correlated to GCR variations, occur (Eddy 1977). These maxima and minima are variable in strength and appear to occur in a variable periodicity alternating between average times of 220 years and 150 years apart (Damon and Sonnet 1992; Crowley and North 1991). The periodicities of and variations in intensities of maxima and minima often give the appearance of 1400 to 1600 year cycles in paleorecords."[Mercurio, 2001]

The 2400-year cycle roughly equals a secondary harmonic of the 4636-year EUWS cycle, at ½ of its frequency. The 1400-to-1600 year cycle translates into the 1545-year EUWS cycle. And the alternating beats of 150 and 220 years equate to the 172-year EUWS cycle.

Actually, difficulty in identification the 172-year solar cycle can be attributed to a confluence of cycles in its neighborhood. The 172-year EUWS cycle fights with the 208-year Suess cycle and the 87.8-year Gleissberg cycle to produce a mixture that makes all three oscillate somewhat erratically.

In an abstract entitled *Millennial to Century-Scale Variability in Gulf of Mexico Holocene Climate Records,* researchers from the U.S. Geological Survey in Reston, Virginia found evidence of both the solar Suess cycle and the 172-year EUWS cycle:

> "Abundance variations of G. saccuilfer show sub millennial-scale variability throughout most of the Holocene.... A 200 year cycle in RC 12-10 and a number of other climate records is characteristic of solar variability as monitored in 14C production and 10Be flux. Similarly, the 170 year cycle [is] found in the Gyre 97-6 PC 20 record."[Poore *et al.*, 2003]

All three cycles appear in **Charts 27H** through **27L**. These charts show sunspot fluctuations from 9000 BC to present.[Solanki *et al.*, 2005] The 172-year frequencies display the greatest amplitudes, normally making sharp spikes near theoretical peaks. One 87.8-year Gleissberg cycle generally appears as a smaller fluctuation in between each 172-year cycle. In addition, the 200+ year Suess cycle causes the 172-year EUWS cycle to move to-and-fro around its theoretical peaks. As you review these charts, notice the attributes of all three cycles. Down-arrows mark actual tops of the 172-year EUWS cycle.

One final note is in order. Multiplied by two, the 87.8-year Gleissberg cycle equals 175.6 years. That comes tantalizingly close to the 172-year EUWS cycle. It's possible that the Gleissberg cycle enhances the 172-year cycle. Equally possible, the Gleissberg cycle could act as a derivative of the 172-year EUWS cycle.

The final graphic, **Chart 27M**, shows the Greenland climate proxy from its ice core. The patterns in this chart exhibit distinctive differences from the solar cycles. The Gleissberg and Suess cycles failed to make much of an impact on Greenland temperatures. However, the 172-year EUWS cycle produced regular beats. Every third cycle shows greater amplitude – marking 515-year tops. But these 515-year tops come one sub-cycle earlier than they should. For example, the 54 BC top should be at 118, the 462 top should be at 633, the 977 top should be at 1148, the 1492 top should be at 1664, and the top forming at 2007 should be at 2178. Finally, every sixth cycle (a 1030-year secondary EUWS harmonic) dominates in this portion of the Greenland temperature proxy – with tops at 54 BC, 977, and 2007. If this pattern holds, Greenland temperatures should decline substantially for the next 500 years.

In summary, the 172-year EUWS cycle influences mankind in several ways. These include adjustments to the structure of civilizations, possible cycles in waging war, fluctuations in commodity prices, oscillations in the stock market, sunspot cycles, and global temperature cycles.

Chart 27A Home Equity Loans: ABX 2007 BBB- Rated

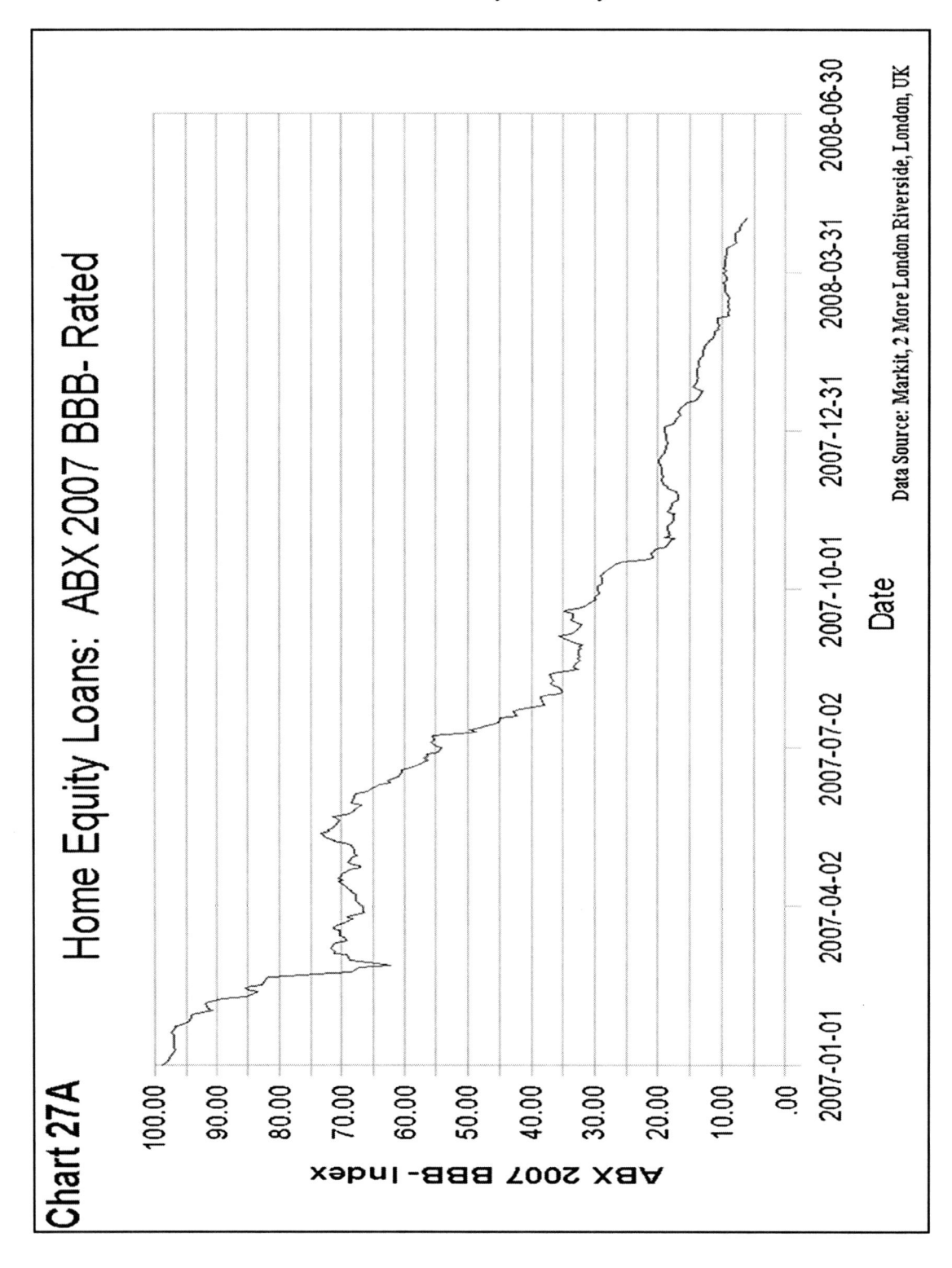

Data Source: Markit, 2 More London Riverside, London, UK

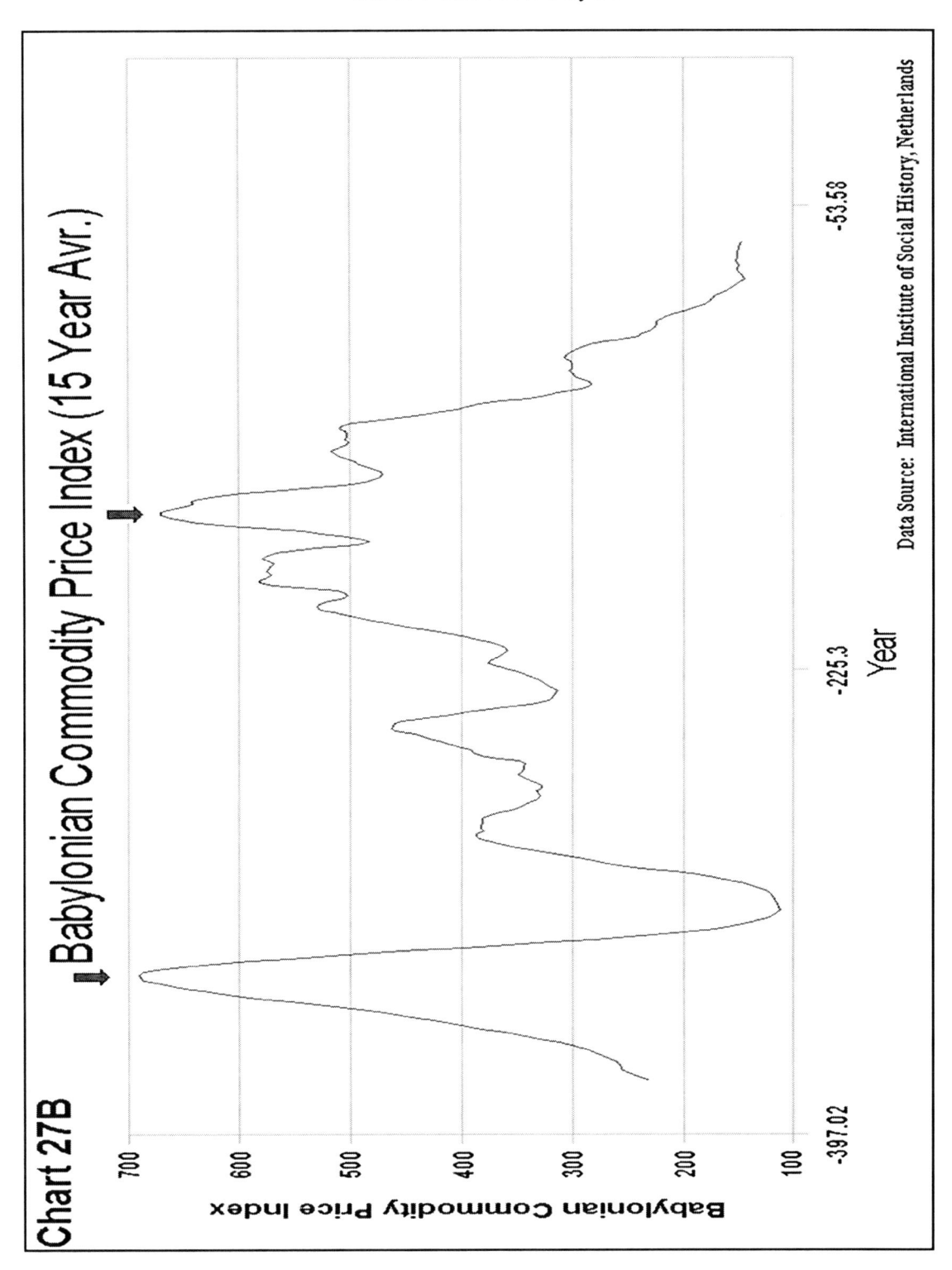

The 172-Year EUWS Cycle

Chart 27B — Babylonian Commodity Price Index (15 Year Avr.)

Data Source: International Institute of Social History, Netherlands

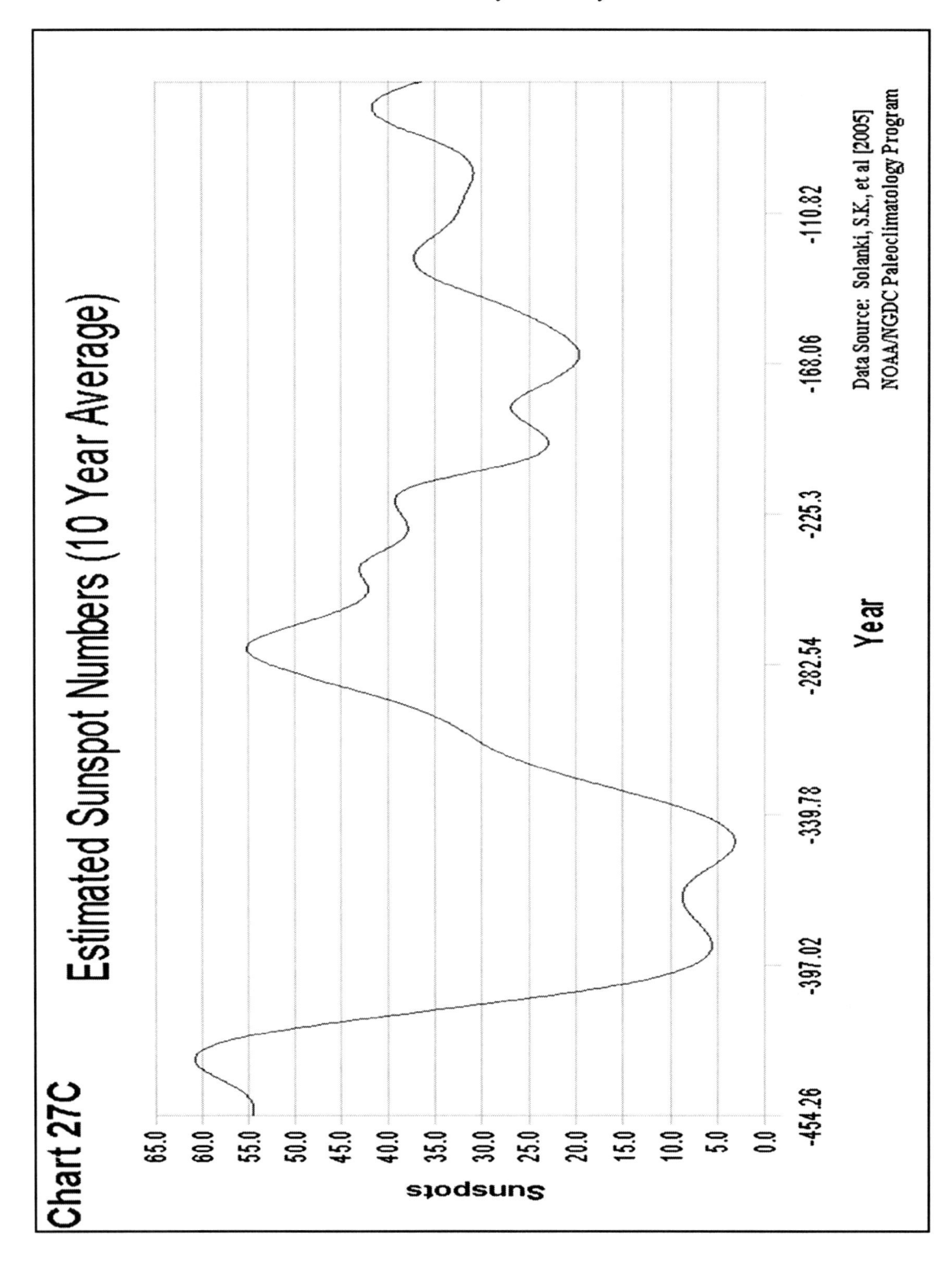

Chart 27C — Estimated Sunspot Numbers (10 Year Average)

Data Source: Solanki, S.K., et al [2005]
NOAA/NGDC Paleoclimatology Program

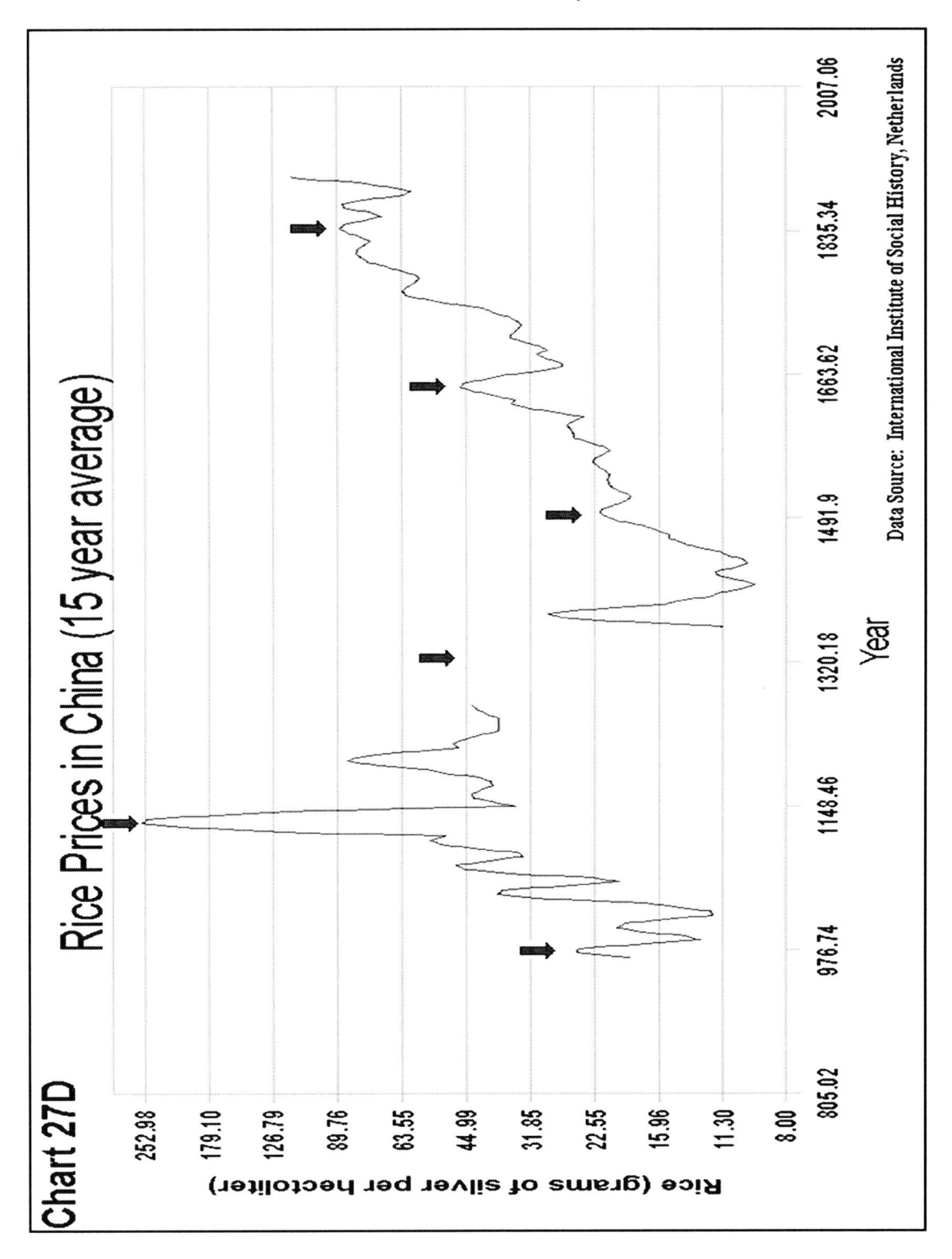

The 172-Year EUWS Cycle

Chart 27D Rice Prices in China (15 year average)

Data Source: International Institute of Social History, Netherlands

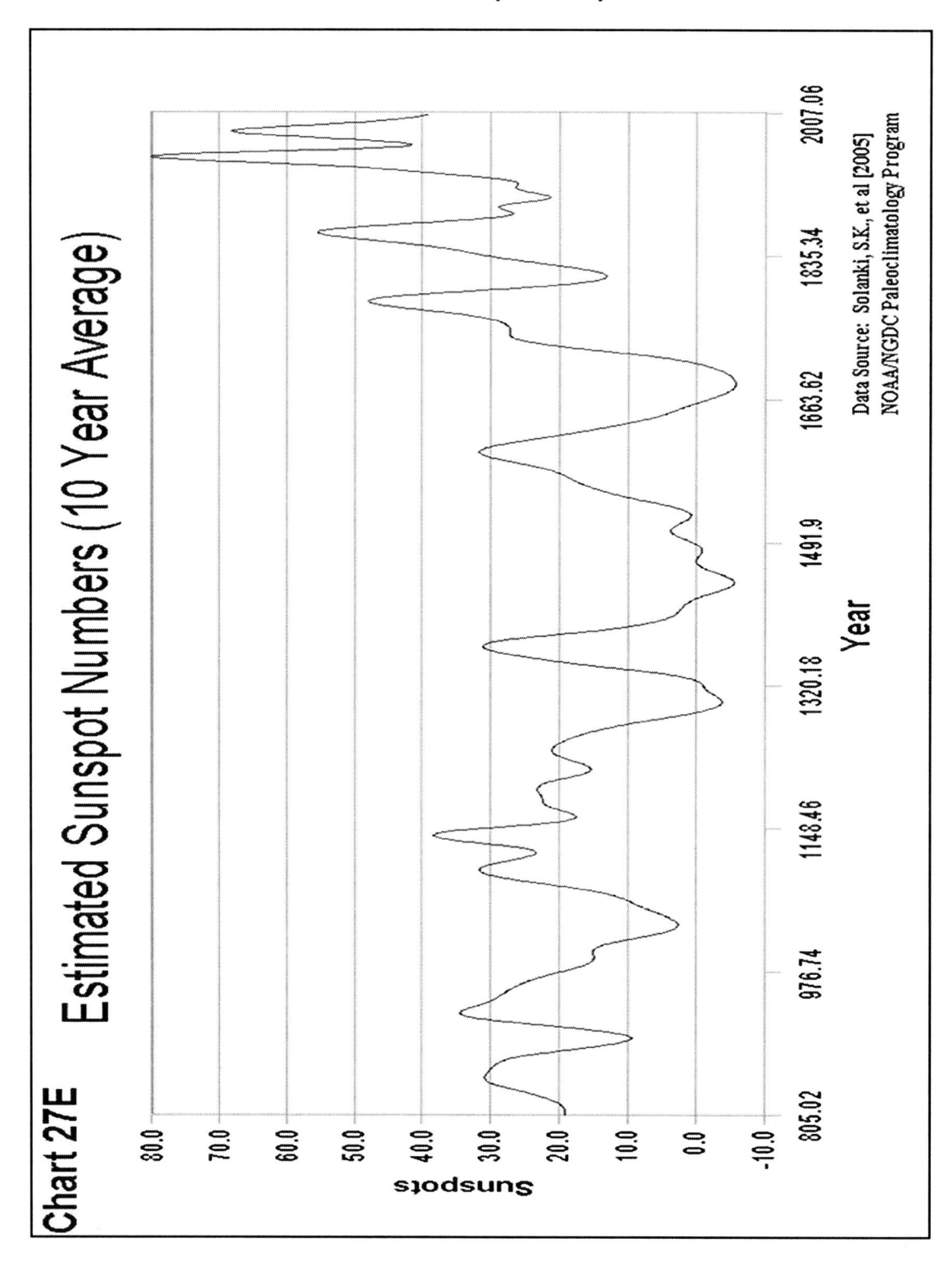

Chart 27E — Estimated Sunspot Numbers (10 Year Average)

Data Source: Solanki, S.K., et al [2005]
NOAA/NGDC Paleoclimatology Program

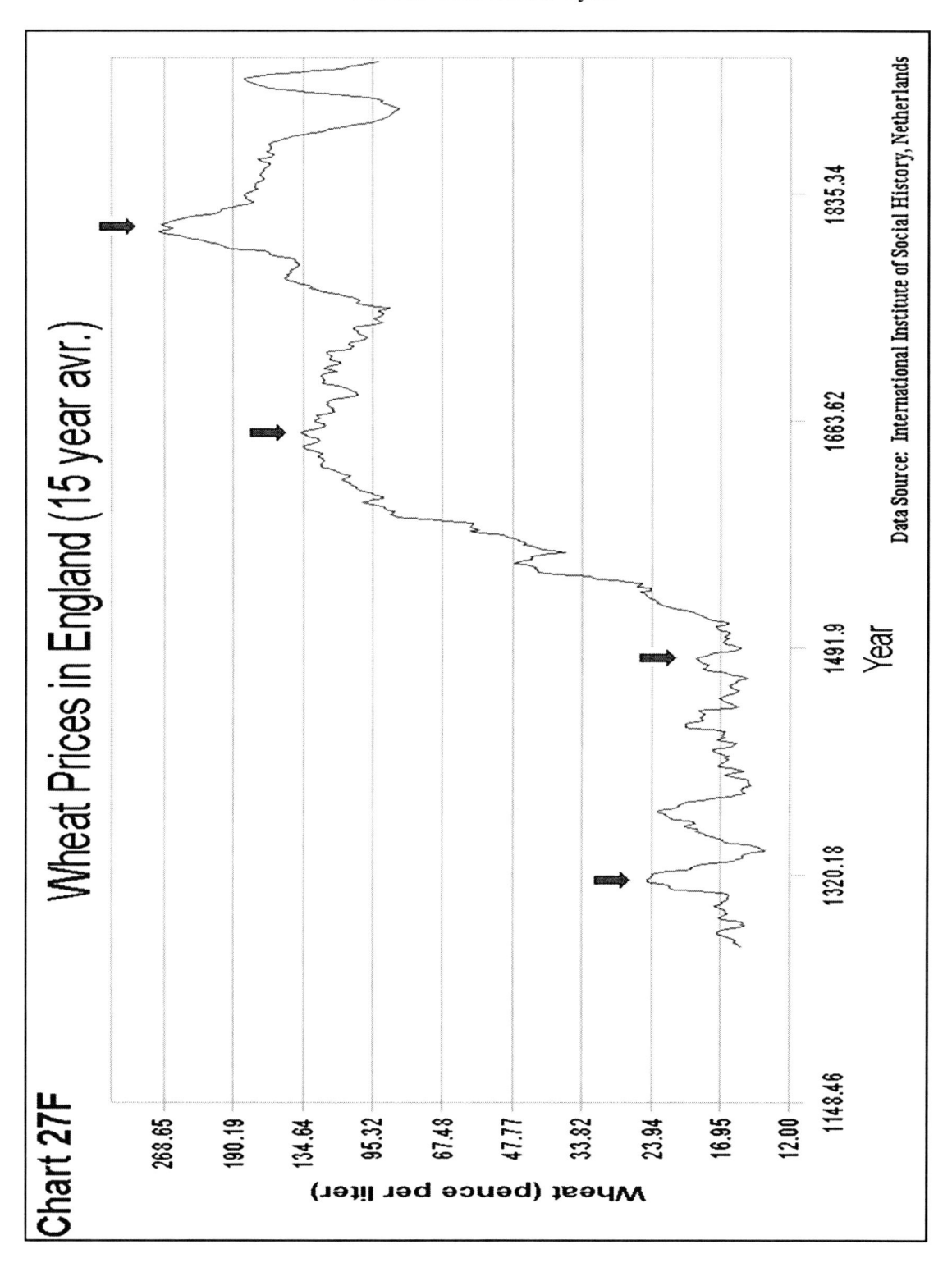

The 172-Year EUWS Cycle

Chart 27F

Wheat Prices in England (15 year avr.)

Data Source: International Institute of Social History, Netherlands

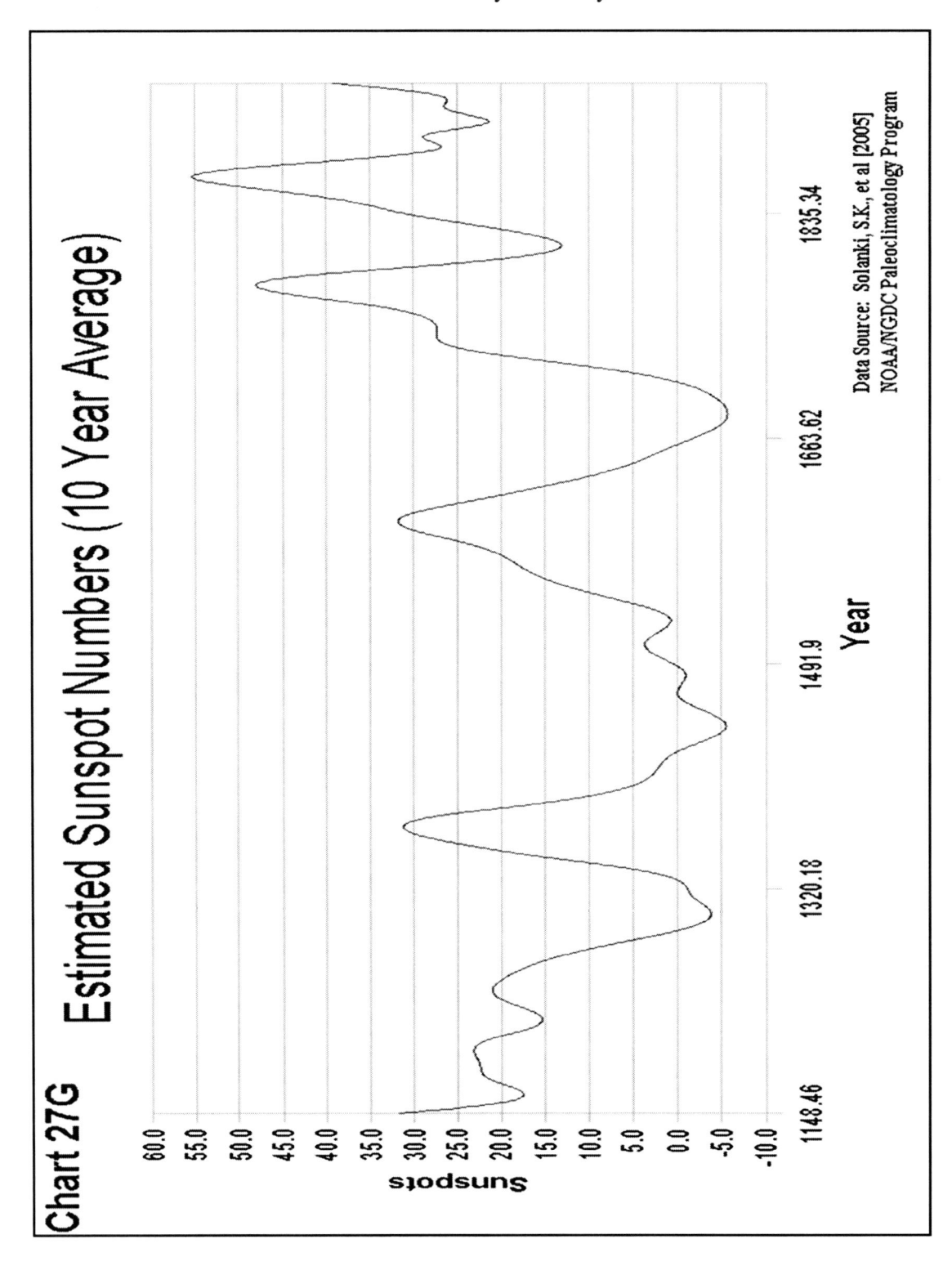

Chart 27G Estimated Sunspot Numbers (10 Year Average)

Data Source: Solanki, S.K., et al [2005]
NOAA/NGDC Paleoclimatology Program

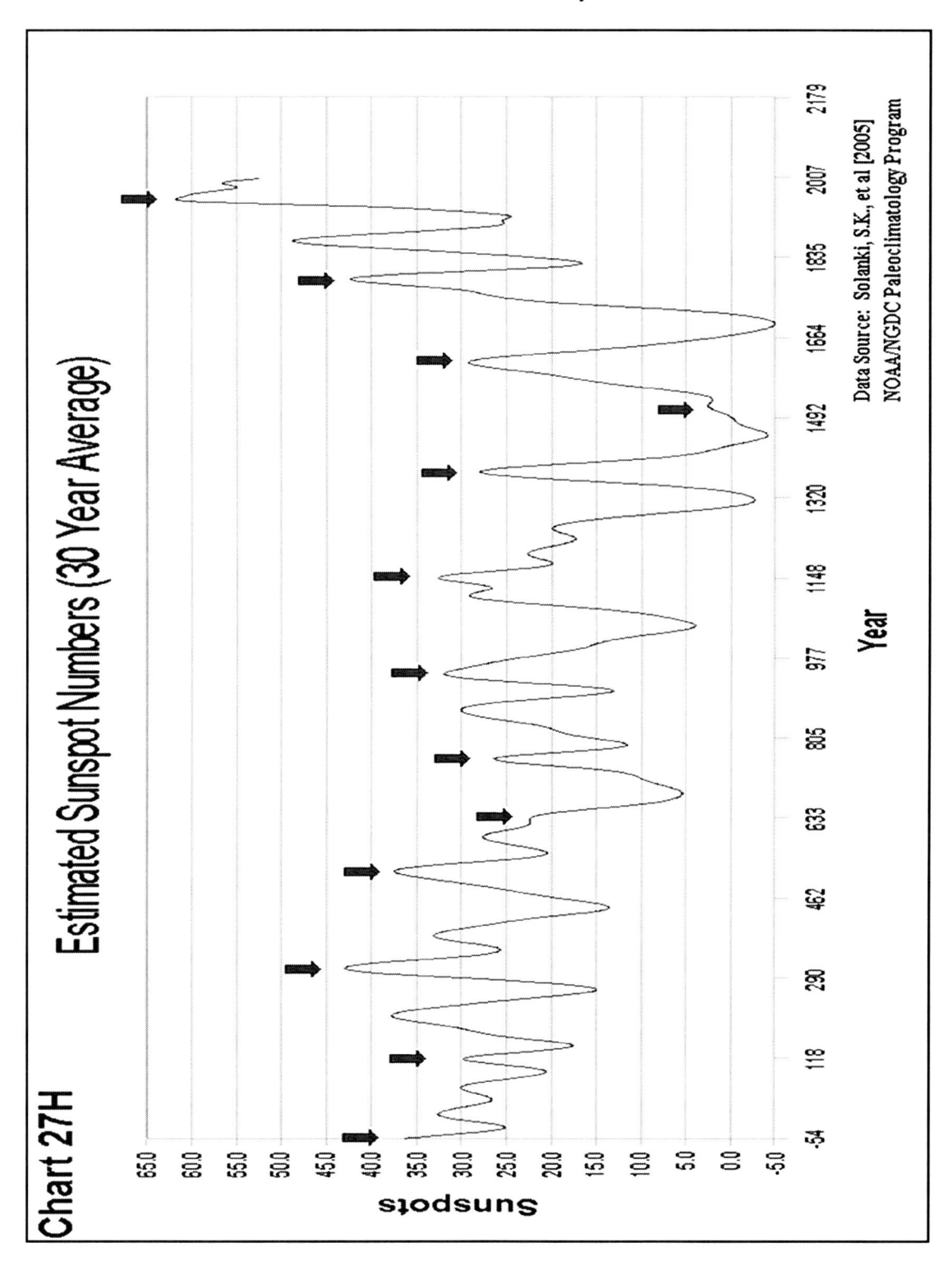

The 172-Year EUWS Cycle

Chart 27H — Estimated Sunspot Numbers (30 Year Average)

Data Source: Solanki, S.K., et al [2005]
NOAA/NGDC Paleoclimatology Program

The Unified Cycle Theory

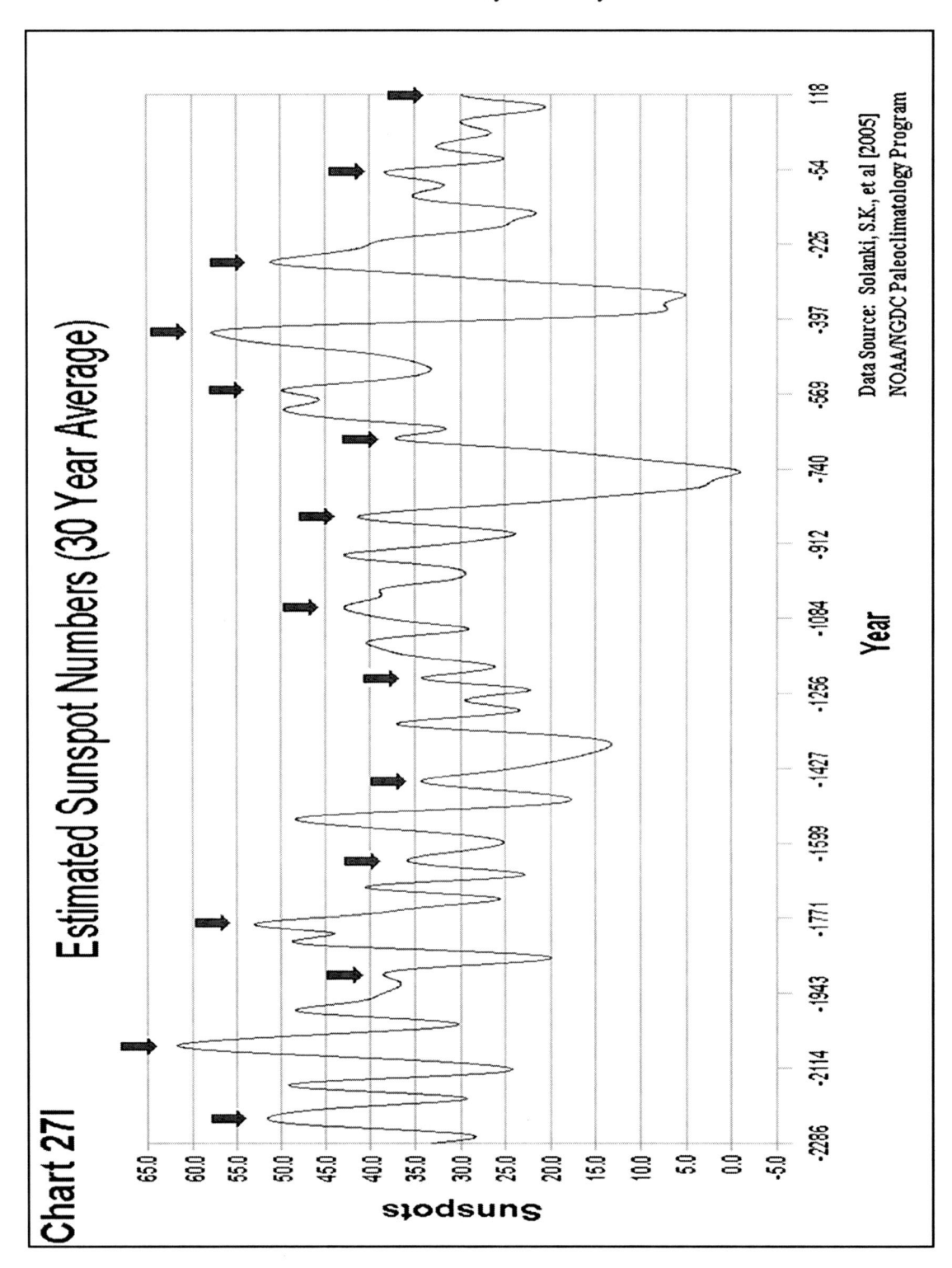

Chart 271 — Estimated Sunspot Numbers (30 Year Average)

Data Source: Solanki, S.K., et al [2005]
NOAA/NGDC Paleoclimatology Program

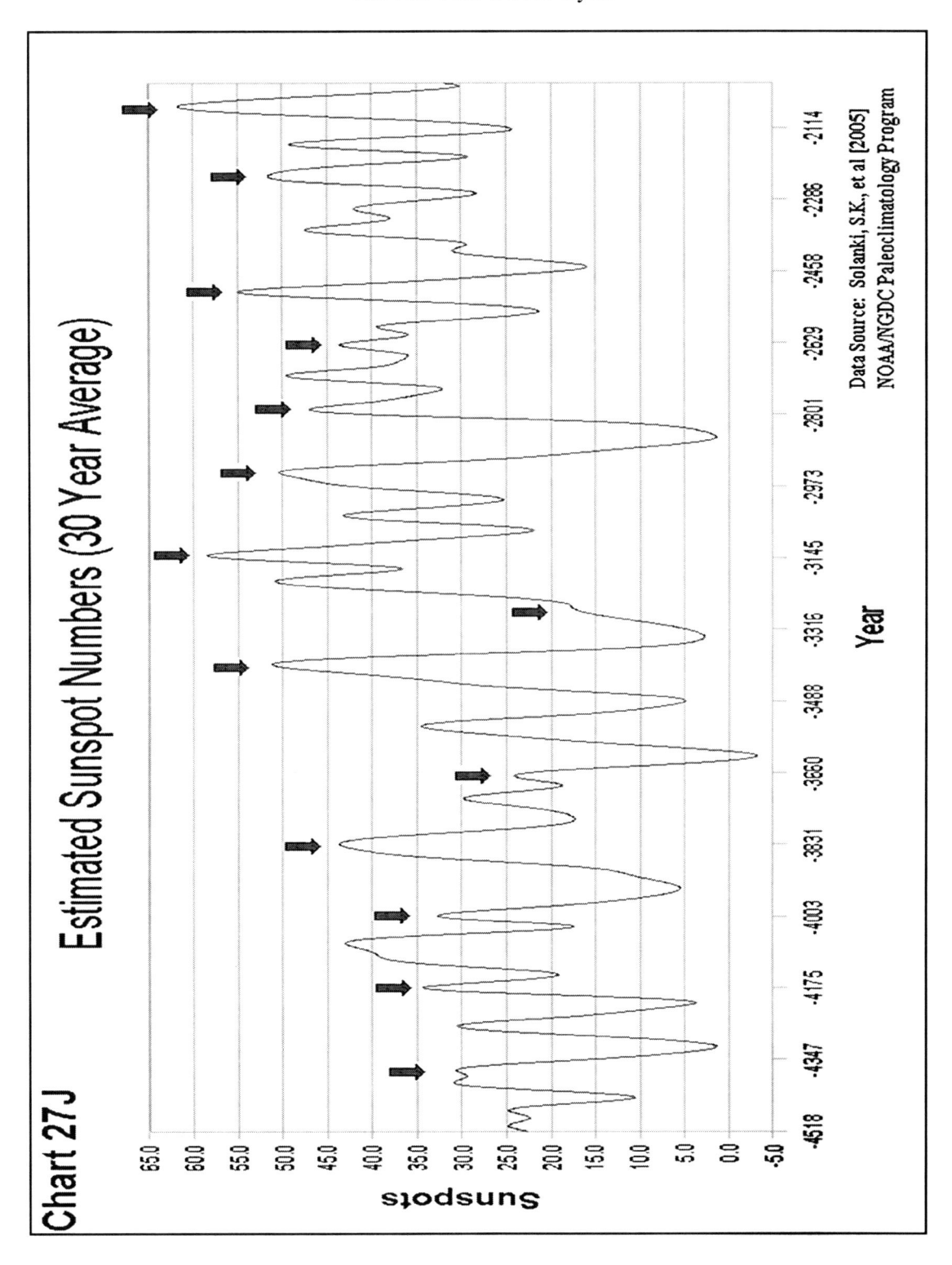

The 172-Year EUWS Cycle

Chart 27J — Estimated Sunspot Numbers (30 Year Average)

Data Source: Solanki, S.K., et al [2005]
NOAA/NGDC Paleoclimatology Program

The Unified Cycle Theory

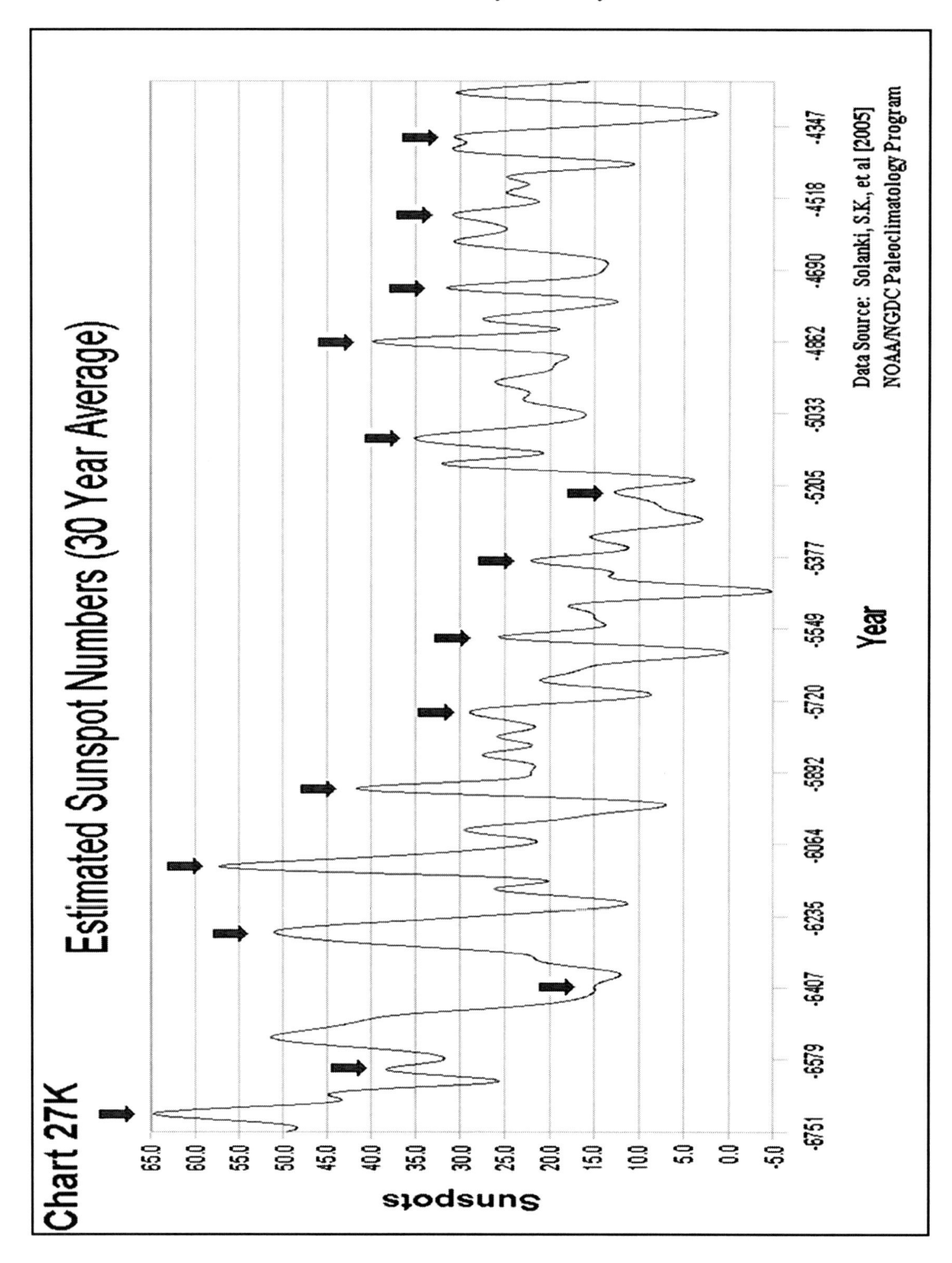

Chart 27K Estimated Sunspot Numbers (30 Year Average)

Data Source: Solanki, S.K., et al [2005]
NOAA/NGDC Paleoclimatology Program

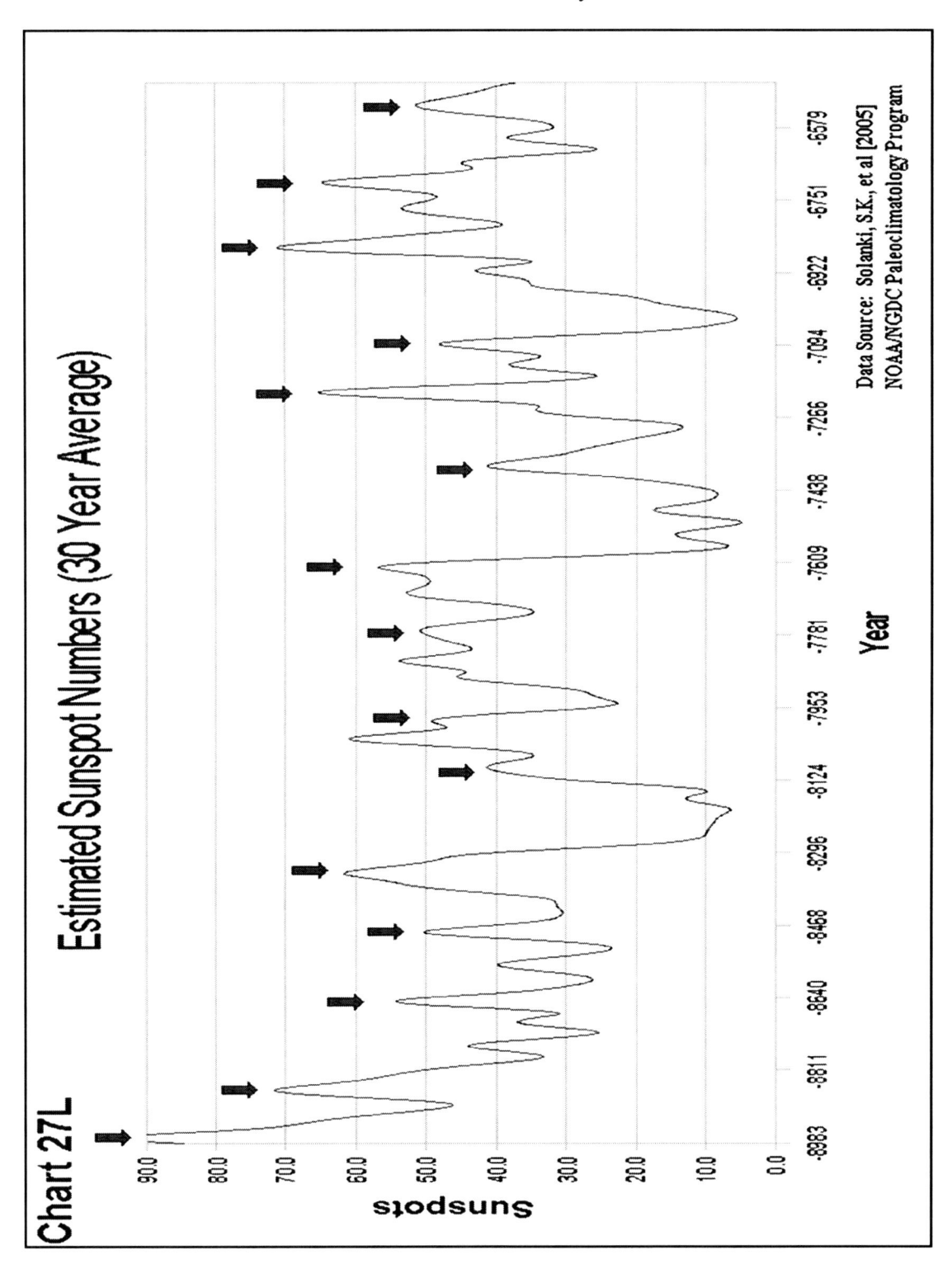

The 172-Year EUWS Cycle

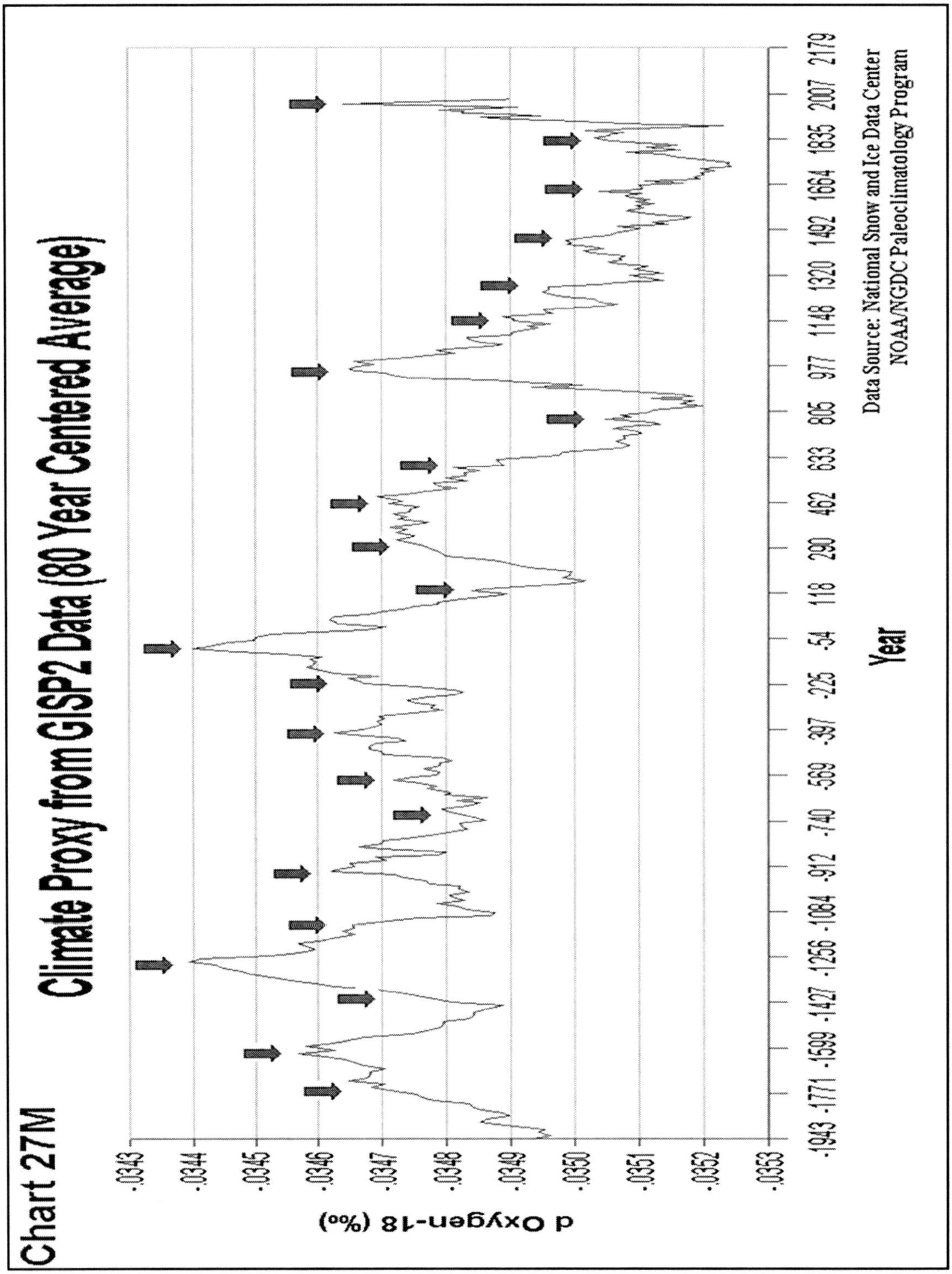
Chart 27M
Climate Proxy from GISP2 Data (80 Year Centered Average)
d Oxygen-18 (‰)
-.0343
-.0344
-.0345
-.0346
-.0347
-.0348
-.0349
-.0350
-.0351
-.0352
-.0353
-1943 -1771 -1599 -1427 -1256 -1084 -912 -740 -569 -397 -225 -54 118 290 462 633 805 977 1148 1320 1492 1664 1835 2007 2179
Year
Data Source: National Snow and Ice Data Center
NOAA/NGDC Paleoclimatology Program

Chapter 28
The 57-Year EUWS Cycle

Dividing the 171.72-year cycle by three, the 57.24-year cycle appears as the next frequency in the series. Perhaps this cycle is best known as the Kondratieff Wave – with its length now more precisely defined. It correlates with cycles in war, stock prices, credit markets, commodities, and economic depressions. The list below contains the last seven theoretical peaks for this cycle. Peaks marked with a double asterisk (**) indicate Economic Depression cycles that also serve as theoretical peaks of the 172-year cycle.

>August 14, 1663 **
>November 10, 1720
>February 5, 1778
>May 5, 1835 **
>August 1, 1892
>October 29, 1949
>January 24, 2007 **

Depressions aren't always easy to detect in historical records – especially after several centuries. In modern times, high unemployment typically indicates an economic depression. See **Chart 28A**. Price deflation and falling security prices also act as credible indicators of depressed business. However, the stock market decline during the 1890s wasn't especially severe. During the mid-1890s, the 20% unemployment rate reflected the depression more accurately than the bear market at that time. To detect a depression, a combination of commodity prices, security prices, and unemployment works well together.

Four recent economic depressions started precisely in the same year as theoretical peaks of the 57-year EUWS cycle. These four examples follow. (A precise starting point cannot be assigned to depressions that started around the time of the 1663 theoretical peak, so this analysis skips that turning point. However, all of the major ancient civilizations began collapsing at various times surrounding the year 1663.)

Theoretical Peak 1, 1720 – The modern form of security markets first developed in Europe around the year 1700. And the first two great security market crashes materialized shortly after that. During 1720, starting in the same year as a 57-year EUWS peak, the Mississippi Bubble crashed in France. The crash came in spite of John Law's (France's finance minister) efforts to support the price of Mississippi shares at 9000 livre during March and April. After a few weeks, supporting the bubble at 9000 became too expensive for the French government, and Law withdrew his support toward the end of May 1720. After that, Mississippi share prices resumed their collapse. By 1721, the company fell into bankruptcy.[Mackay, 1980] Also in 1720, across the Channel, the South Sea Bubble imploded in England. And bubble markets in Holland also burst that same year.[Mackay, 1980]

Theoretical Peak 2, 1778 – The next peak of the economic depression cycle came in February 1778. That coincided with the suffering and starvation endured by George Washington's Valley Forge troops that same month. The first depression in the USA lasted throughout the Revolutionary War. The first phase ended in 1781 with the collapse in the Continental Currency. This acted as our first harsh lesson in the dangers of using credit as a replacement for hard money (gold and silver). It also prompted our founding fathers to favor the stability of a gold dollar as the basis of our first monetary system. The second phase of the depression climaxed during 1786-87 with Shays' Rebellion.[Encarta: Shays, 2008]

Theoretical Peak 3, 1835 – The May 1835 theoretical peak of the economic depression cycle came next. As discussed in the previous chapter, that coincided, to the month, with the bursting of the bubble in railroad and canal stocks.[Wheelock, 2004] And the longest bear market in the history of the United States followed.

Theoretical Peak 4, 1892 – 57 years later, in 1892, another economic depression struck. A painful bear market followed. However, high unemployment best reflected the severity of the 1890s' depression. Once again, see **Chart 28A**. Until the Great Depression, the 1890s registered the high-water mark for unemployment.

Incredibly, these four depressions, arguably the four most severe depressions between the years 1700 and 1900, all started within the same year as theoretical peaks of the economic depression cycle! If economic depressions developed randomly, then their starting points should be scattered throughout this two hundred year period. Instead, 57-year intervals precisely separate them.

The 57-Year EUWS Cycle

If economic depressions and 57-year EUWS peaks were random events, then each depression would hold a 1-in-57 chance of starting in the same year as a theoretical peak in the economic depression cycle. The chance of this happening randomly four cycles in a row equals 1/57 x 1/57 x 1/57 x 1/57. That approximately equals 1 chance in 10 million of randomly occurring! Clearly, their regularity indicates a strong cyclical component to economic depressions.

As another example, look at the 57-year depression cycle between the troughs in 1807 and 1864. If depressions occurred randomly, then what are the odds of the May 1835 stock market crash starting in the same month and year as the theoretical peak on the 57-year depression cycle? With 12 months in a year, and 57 years in a depression cycle, the odds equal 1/57 x 1/12. In other words, there's 1 chance in 684 that these two events would occur randomly in the same month. During January 2007, the same type of coincidence happened when the subprime crisis erupted during the same year and month as a 57-year theoretical peak.

Prior to 1700, commodity markets offered the only available means of validating the 57-year cycle in economic activity. The International Institute of Social History has an extensive set of data allowing this analysis.[IISH: Prices and Wages, 2008] Before starting the review, it's important to remember that commodity prices often peak twenty years, or more, ahead of equities and the economy. For that reason, all commodity charts mark theoretical peaks one 19.08-year sub-cycle ahead of normal EUWS theoretical peaks.

One exception to this rule frequently appears. At theoretical peaks of the 172-year cycle, commodity markets often peak simultaneously with the stock market and the economy. Then, commodities collapse in-phase with everything else in a deflationary crash, as the down-phase of the 172-year cycle dominates. In the following charts, pay close attention to these deviations near 172-year peaks.

In the last chapter, ancient Babylonian commodity prices showed evidence of a 172-year cycle. This same price series reveals the 57-year cycle. In **Chart 28B**, the tick marks on the date-axis represent 57-year theoretical peaks moved ahead one 19-year sub-cycle.

The chart provides strong evidence of a depression cycle ruling ancient Babylon. Moderate deflations followed the commodity peaks at 301, 244, and 129 BC. The top of the first cycle developed at the normal theoretical peak, followed by a devastating deflation. The top of the fourth cycle formed as a double-top – with the first top coming at the theoretical commodity peak and the second top forming at the normal theoretical peak. Massive deflations followed each of the tops. These cases also provide additional examples of the TPD Principle at work. When a cycle fails to reverse at the expected point, it normally deviates by waiting until the next sub-cycle.

The next set of data comes from Chinese rice prices between the years 965 and 1455. See **Chart 28C**. The International Institute of Social History archived this series as decade averages. Timing inaccuracies associated with decade averages impact 515 and 172 year cycles minimally. However, when cycles shrink to 57 years, using decade averages significantly diminishes the reliability of identifying the timing. In spite of these potential inaccuracies, actual reversals matched theoretical turning points fairly well. During the first three cycles, rice prices reversed close to commodity theoretical peaks. During the final five cycles, prices shifted – topping closer to the normal EUWS peaks.

Chinese rice prices between the years 1415 and 1990 come next. See **Chart 28D**. For 8 of the 9 cycles, the actual tops arrived close to times of theoretical 57-year commodity peaks. For the second cycle (a 172-year peak at 1472), the top came near the normal peak in 1491. For the eighth cycle (a 172-year peak at 1816), the climax developed as a double-top – with the first top at the commodity peak and the second top at the normal EUWS peak.

Overall, for the entire 1000 year period, Chinese rice prices conformed very closely to expected behavior. Prices fluctuated with one distinctive top every 57 years. The only significant deviations occurred when several tops shifted away from theoretical commodity peaks and toward normal 57-year theoretical peaks.

Now, examination of the 57-year cycle continues with wheat prices in England. See **Chart 28E**, which depicts wheat prices from 1250 to 1930.[IISH: Prices and Wages, 2008] It provides good examples of 57-year oscillations moving in shifts. For cycles 1 through 4, the actual tops shifted toward the normal EUWS theoretical peaks. For cycles 5 through 9, the tops closely matched theoretical commodity peaks. And for cycles 10 through 12, the tops came about 10 years before theoretical commodity peaks.

As these cycles have shown time and again, once a EUWS shift begins, it often continues for several cycles before returning back to its theoretical turning points. This sequence of English wheat prices provides one of the best examples.

Next, an ancient index of Italian consumer prices also fluctuates in 57-year cycles. See **Chart 28F**, showing northern Italy prices from 1330 to 1900.[IISH: Prices and Wages, 2008] Unfortunately, a gap in prices occurred between 1620 and 1700. Many of these ancient price charts contain gaps. These gaps usually provide important clues about internal economic conditions.

In this particular case, missing data prior to 1330 coincided with the end of the dark-age in parts of Europe. The Italian Renaissance followed. The years from 1400 to 1600 marked the high point of the Renaissance. The price gap between 1620 and 1700 surrounds the civilization cycle peak at 1663. It coincided with a decline related to Spanish domination of Italy.

As an interesting side note, the two greatest periods in Italian history came in the centuries preceding the last two 1545-year EUWS peaks – coinciding with peaks in the dark-age cycle. From 200 BC to 118 AD, the Roman Empire ascended to its Golden Age. And from 1400 to 1600, Italy flourished during its Renaissance. But as the dark-age cycles turned lower, both of these glorious periods in Italy quickly faded.

The 57-Year EUWS Cycle

Getting back to the 57-year economic depression cycle, 8 of the 9 tops in **Chart 28F** matched theoretical commodity peaks. The only deviant turn came during the 1759 cycle, when muted tops came 9 years before and 9 years after a theoretical commodity peak.

Consumer prices in the Belgium/Netherlands area also exhibit a 57-year cycle in depressions. See **Chart 28G**, which shows Antwerp consumer prices from 1400 to 1900. 7 of the 9 consumer price cycles topped close to theoretical commodity peaks. During the 1472 cycle (a 172-year peak), the top came precisely at the normal EUWS peak.

The other deviation came in 1759. This deviation deserves special attention because of difficulty in observing the cycle. However, the muted cycle wasn't just limited to the Antwerp area. During this cycle, similar muted responses occurred in England, Italy, France, Germany, and Poland. (Visibility of the 1759 cycle appeared stronger in China and Istanbul.) Lack of visibility for the 1759 cycle suggests that the underlying force causing this cycle may have dissipated during this period.

Strasbourg, France comes next. Much like Antwerp, sitting on a border, Strasbourg consumer prices serve as a barometer for both France and Germany. See **Chart 28H**, which shows prices from 1380 to 1875. For half of the depression cycles, consumer prices topped near the theoretical commodity peaks – coming at 1530, 1587, 1644, and 1816. For the first two periods, Antwerp prices shifted – with their tops moving toward theoretical peaks of the normal EUWS cycles. For the cycles at 1701 and 1759 another shift occurred – with the tops coming about nine years ahead of schedule.

Following an economic crash in 1873, the German government began shifting away from free-trade policies toward protectionist measures that introduced tariffs on imports to protect German manufacturers.[Encarta: Germany, 2008] At that time, Bismark and the German government fought the normal deflationary response. Following the crash, Germany resorted to a series of monetary reforms that allowed inflation to run rampant. In 1873, Germany established the Goldmark, but issued currency in excess of its gold stock. At the start of World War I, Germany abandoned the gold standard, and created the Papiermark. Nine years later, ruinous inflation forced the Papiermark to be replaced with the Rentenmark in 1923. Just one year later, in 1924, unbelievable hyperinflation caused the Rentenmark to be abandoned in favor of the Reichsmark.

Before July 1922, the German currency had already dropped from about 4 to 493 to the dollar, but during the next 16 months it plummeted to 4.2 trillion to the dollar. The resulting inflation wiped out the savings, pensions, insurance, and other forms of fixed income of most middle-class and working-class Germans.[Encarta: Germany, 2008] The crisis reached its nadir during the mid-1930s when German citizens turned to Nazism as a solution to their impoverishment.

The German hyperinflation of the early 20th Century provides a prime example of the consequences of using currency depreciation to fight depression. In spite of this painful lesson,

The Unified Cycle Theory

on a global basis, currency depreciation became increasing popular in fighting economic weakness. That's especially true for the United States, where credit inflation ran rampant. However, as the German experience showed, the currency depreciation path not only fails to work, it delays the recovery process and intensifies poverty in the long-term.

Examination of the 57-year economic depression cycle now focuses on the Ottoman Empire at its height. Istanbul consumer prices between 1469 and 1783 provide the evidence.[IISH: Prices and Wages, 2008] During the Ottoman years, Istanbul went by the name of Constantinople. The Ottoman series represents an important set of data because of its relative distance from the other European countries. The diversity achieved by including both China and Istanbul adds to the strength of the evidence. See **Chart 28I**.

For Istanbul, 4 of the 5 cycles topped very close to theoretical commodity peaks. The lone exception came at a 172-year EUWS peak. At that time, the price-top shifted forward to the theoretical EUWS peak at 1663.

As a final note, the worst deflation in the Ottoman history started after the theoretical EUWS peak in 1720. In that same year, infamous stock market bubbles burst in France (Mississippi Bubble), England (the South Sea Bubble), and Holland. After 23 years of brutal deflation, prices finally hit bottom in 1749. See **Chart 28I** again. Forty years later, except for a small core around Istanbul, the Ottoman Empire disintegrated. And along with its demise, Istanbul price records vanished. The depression that struck in 1720 appears to have been global.

The final data series for analyzing the economic depression cycle comes from Krakow between the years 1400 and 1900. Until 1596, Poland established its capital at Krakow. Krakow's consumer price index adds to the diversity of this examination by representing the Eastern European area. See **Chart 28J**.

For 6 of the 9 depression cycles, Polish consumer prices topped close to the theoretical commodity peaks. For the other three cycles (1530, 1759, and 1873), mild responses appeared. Rather than triggering outright deflation, these three cycles only temporarily slowed the inflation rate.

Table 28.1 – The 57-Year Depression Cycle in the United States.

Theoretical Commodity Peaks	Actual Commod. Tops	Commodity Dev. from Theoretical	Major Wars after Stock Mkt Peaks	Theoretical Stock Mkt Peaks	Actual Stock Tops	Stock Mkt Dev. from Theoretical
1816	1816	direct hit	Civil War	1835	1835	direct hit
1873	1864	9 yrs early	WW I & II	1892	1892	direct hit
1930	1920	10 yrs early	Vietnam War	1949	1929	1 cycle early
1987	1980	7 yrs early	?	2007	2007	direct hit

The 57-Year EUWS Cycle

Now, the examination moves forward to the United States, covering the entire period since its formation over 200 years ago. **Table 28.1** summarizes key points for commodities, wars, and equities – as they pertain to the 57-year economic depression cycle.

Cycle 1, 1835 Theoretical Peak – The pinnacle for the commodity markets came in 1816, exactly one 19-year cycle ahead of the 1835 depression cycle peak. And the stock market scored a direct hit by climaxing during May 1835. Equities not only topped in the exact year of this 57-year cycle, they topped in the exact month of the theoretical peak. The longest bear market in USA history followed. The Civil War erupted in 1861, and it ended precisely at the bottom of the 57-year cycle. This matched the optimal time for war as described in the previous chapter. Specifically, 29% of all wars occurred during the first 19 years of a depression cycle downturn. 40% of all wars broke out during the 19 years surrounding depression cycle troughs. And 31% of all wars occurred during the final 19 years – corresponding with up-phases of the 57-year cycle.

Cycle 2, 1892 Theoretical Peak – Commodities topped in 1864, nine years ahead of the theoretical turning point. The stock market scored a direct hit, topping in 1892. However, the ensuing bear market did not mimic the crashes normally associated with economic depressions. Nonetheless, the unemployment rate rose to record levels. Since then, unemployment only exceeded the 1890s' maximum once – during the Great Depression. The United States entered World War I in 1917 – 56 years after the Civil War. And the war ended precisely at the trough of the 57-year cycle! Also during this cycle, the USA entered World War II in 1941, during the final up-phase of the 57-year cycle – a relatively unlikely time to become involved in war.

Cycle 3, 1949 Theoretical Peak – For this cycle, commodities reached their zenith in 1920. That occurred ten years ahead of the ideal peak in 1930. But the major deviation during this cycle occurred in the stock market. The 1929 crash and the ensuing Great Depression materialized one 19-year panic cycle early. Its theoretical peak came in 1949 – causing tremendous confusion and debate among Kondratieff Wave followers. Most K-Wave followers claim its frequency shortened. They believe the moderately high unemployment of the early 1980s marked the bottom of the K-Wave that followed the Great Depression. *Wikipedia* describes both the controversy and current thinking of K-Wave followers:

> "Long Wave Theory is not accepted by most academic economists, but it is one of the bases of innovation-based, development, and evolutionary economics.... Among economists who accept it, there has been no universal agreement about the start and the end years of particular waves. This points to another criticism of the theory: that it amounts to seeing patterns in a mass of statistics that aren't really there. Moreover, there is a lack of agreement over the cause of this phenomenon.... Most cycle theorists agree, however, with the 'Schumpeter-Freeman-Perez' paradigm of five waves so far since the industrial revolution, and the sixth one to come. These five cycles are...
> The Industrial Revolution--1771
> The Age of Steam and Railways--1829

The Age of Steel, Electricity and Heavy Engineering--1875
The Age of Oil, the Automobile and Mass Production--1908
The Age of Information and Telecommunications—1971."[Wikipedia: Kondratieff, 2008]

According to this model, the K-Wave frequency shortened to about 50 years, and the USA now resides in the early phase of a new up-swing. However, the EUWS cycles paint a completely different picture – as described in the next section.

Cycle 4, 2007 Theoretical Peak – For the current cycle, commodities registered a major top in 1980, seven years earlier than its theoretical commodity peak. But 2007 also coincided with a 172-year peak. Similar to many other 172-year theoretical peaks, a double-top formed in commodities for the current cycle. 1980 marked the initial top, while July 2008 acted as a secondary top. With the tops now in place, the *Unified Cycle Theory* projects massive deflation for decades to come. Market tops associated with the 57-year peak occurred in three phases.

Phase 1 – Credit markets broke first. The subprime crises erupted during January 2007 – matching the theoretical peak of the 57-year cycle to the year and month!

Phase 2 – Equity markets broke next. Almost every major global equity market recorded its all-time high during October 2007. This corresponded to the first 258-day sub-cycle peak after the January top!

Phase 3 – Commodity markets broke last. Global commodity markets topped, and then crashed during July 2008. This corresponded to the second 258-day sub-cycle peak after the January top! (A more detailed look at the 258-day EUWS cycle comes in Chapter 32.)

Given the magnitude of the 57-year and 172-year cycles, an extended stock market collapse should be expected. In fact, more than likely, an economic depression started during July 2008. And it will probably exceed the Great Depression in severity. The economic depression cycle paints a distressing picture for the next few decades.

The analysis now moves away from human activity and into the area of global climate. Not shy for controversy, the EUWS cycles projects a 100-year period of global cooling. See **Chart 28K**, depicting an index of global temperatures compiled by the *Climatic Research Unit* at the University of East Anglia in Norwich, UK.[CRU, 2008] Of course, a prediction of global cooling flies in the face of the predominant view today.

Nonetheless, compare the temperature oscillations in **Chart 28K** with theoretical turning points of the EUWS cycles. In the chart, single arrows mark 19-year peaks; double arrows pinpoint 57-year peaks; while triple arrows denote 172-year peaks. The tick marks on the x-axis correspond to theoretical 19-year peaks. In particular, notice how the 20th Century warming trend began at the trough of the 172-year cycle in 1920. And notice how the trend accelerated upward once the

The 57-Year EUWS Cycle

57-year trough passed during the 1970s. Finally, notice how global temperatures abruptly reversed during January 2007 – the same year and month that the 57-year and 172-year EUWS cycles peaked. This temperature reversal also coincided with the eruption of the sub-prime financial crisis – down to the month!

This chapter reveals how the 57-year EUWS cycle influences commodity prices, stock prices, economic depressions, war activity, and global climate. These amazing correlations continue as the EUWS cycles divide even further.

The Unified Cycle Theory

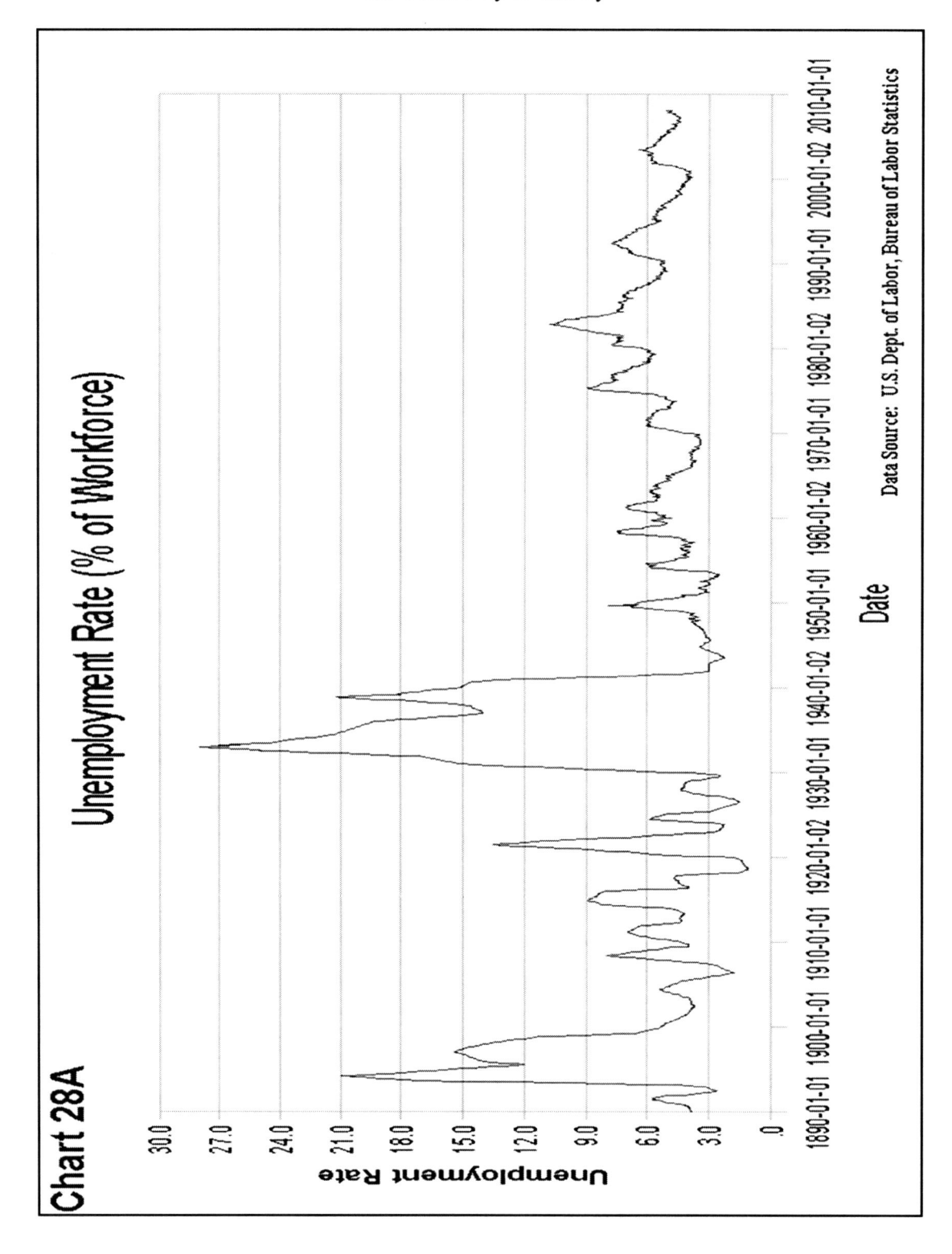

Chart 28A Unemployment Rate (% of Workforce)

Data Source: U.S. Dept. of Labor, Bureau of Labor Statistics

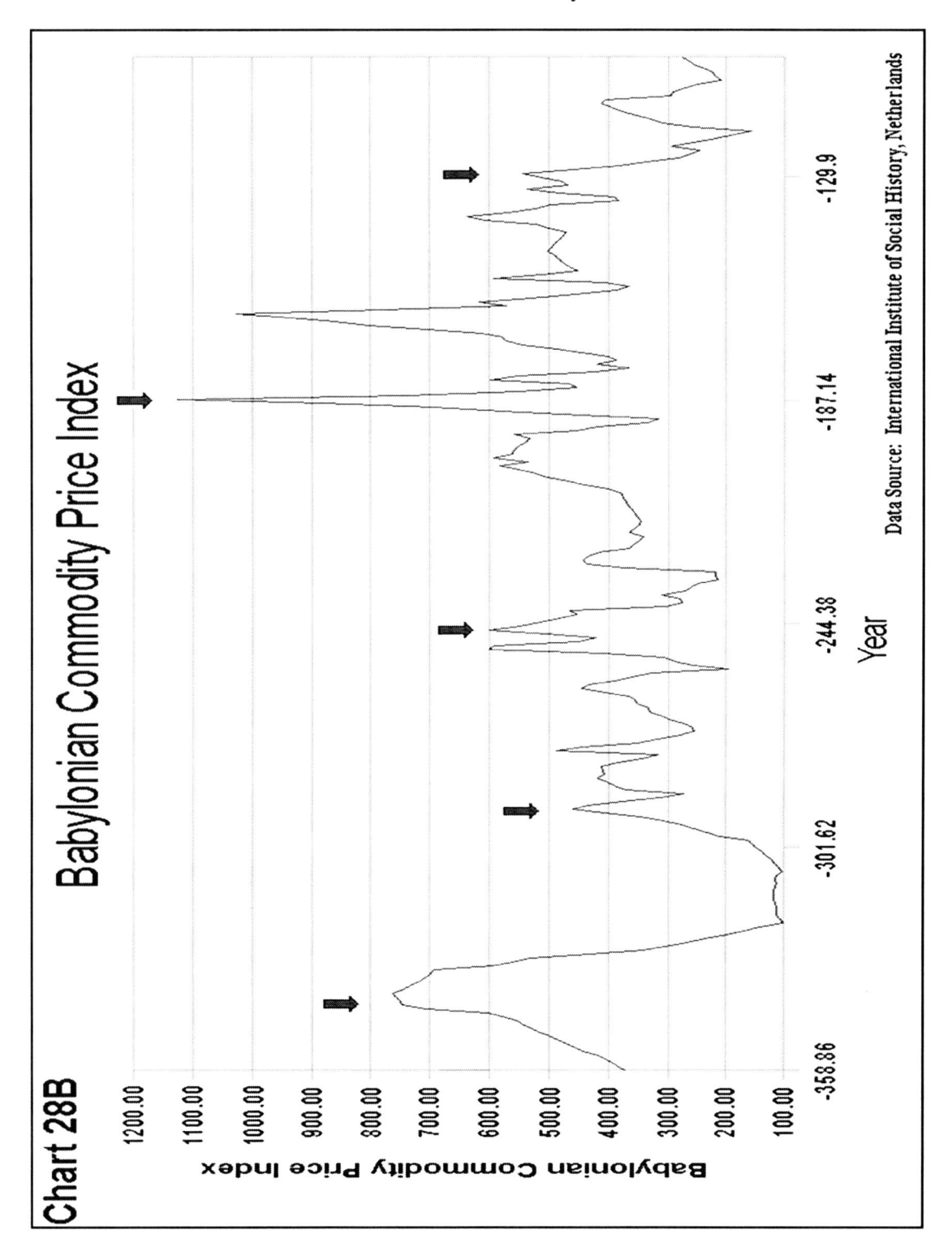

The 57-Year EUWS Cycle

Chart 28B — Babylonian Commodity Price Index

Data Source: International Institute of Social History, Netherlands

The Unified Cycle Theory

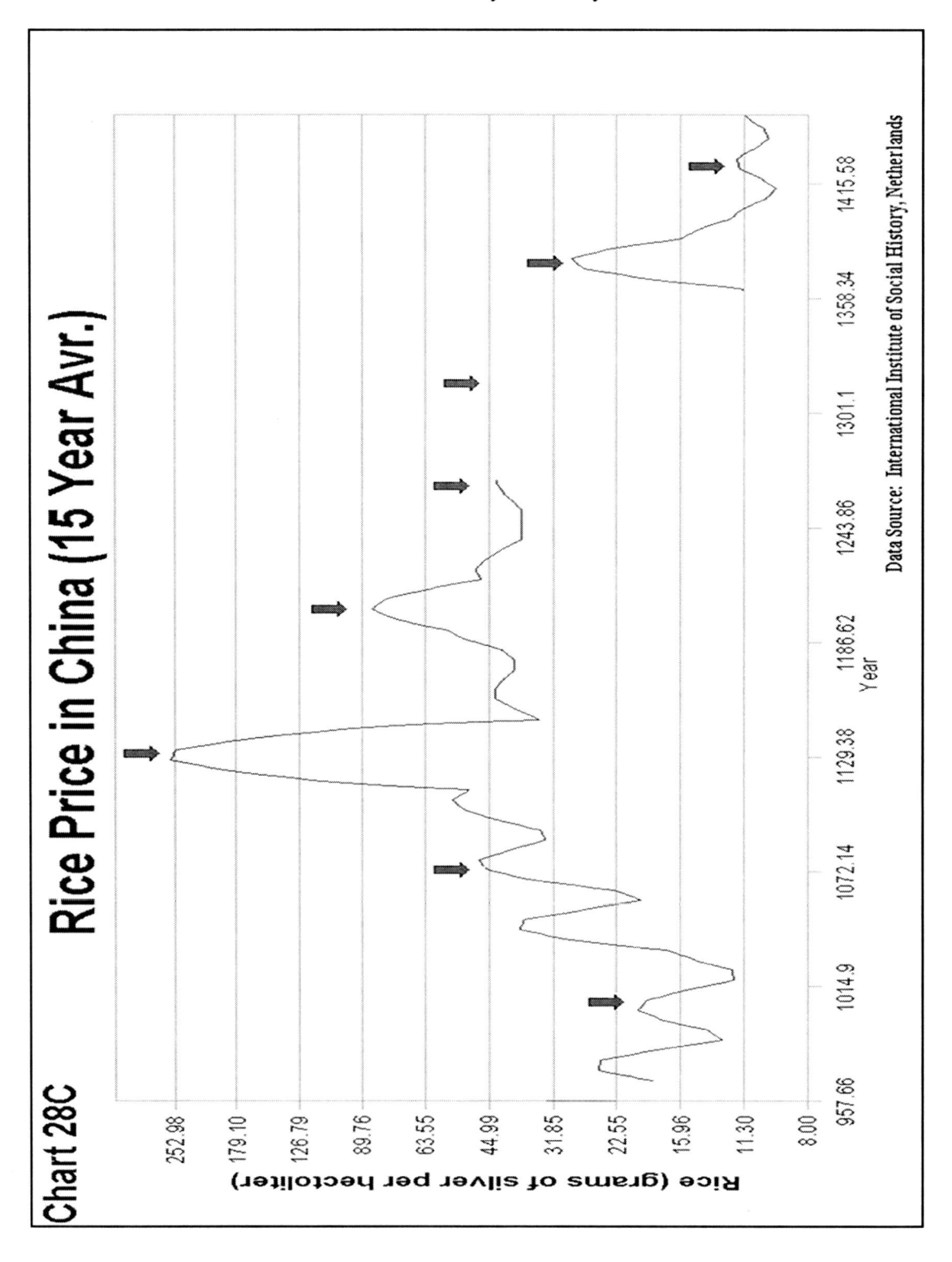

Chart 28C

Rice Price in China (15 Year Avr.)

Data Source: International Institute of Social History, Netherlands

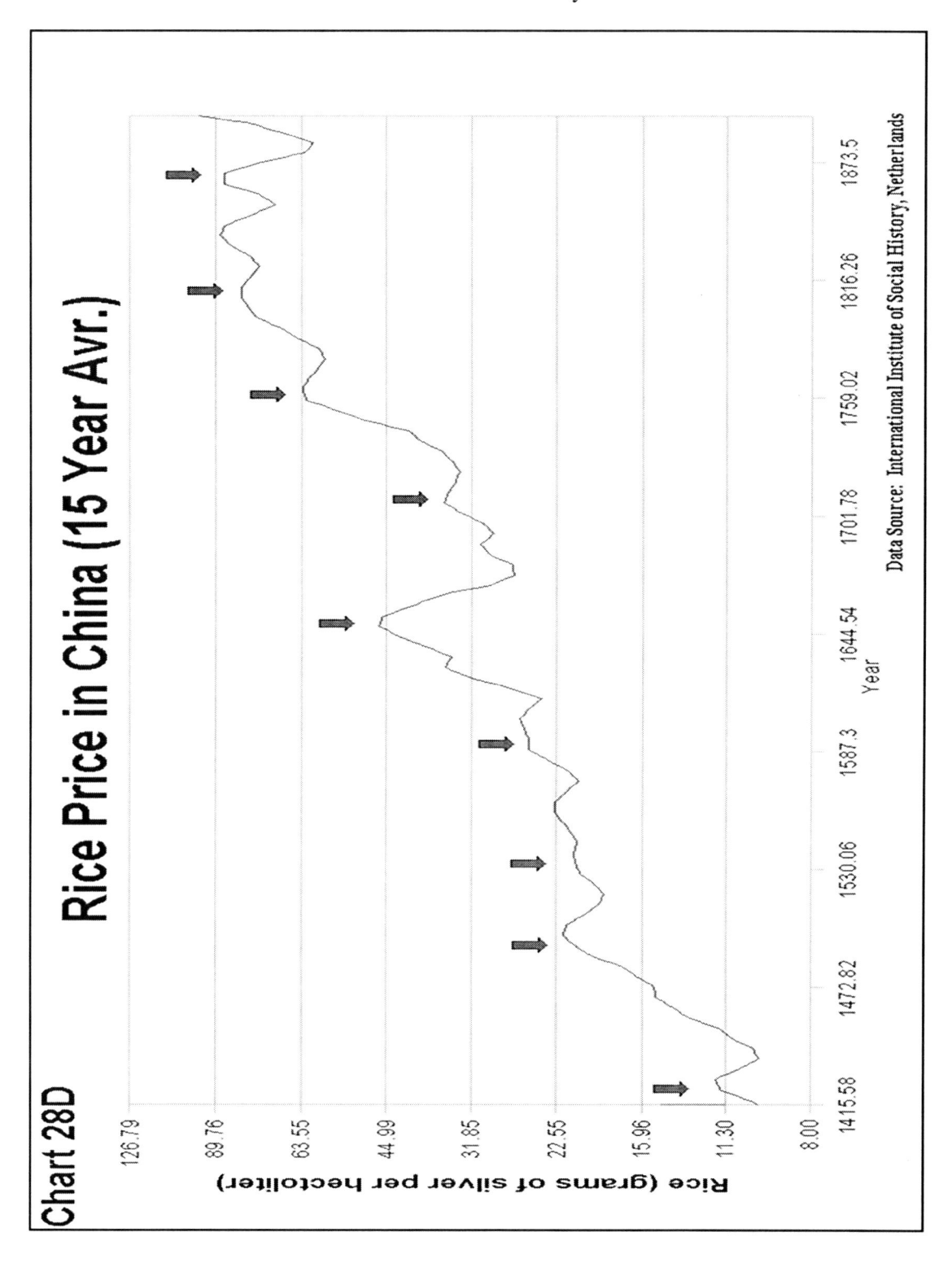

The 57-Year EUWS Cycle

Chart 28D: Rice Price in China (15 Year Avr.)

Y-axis: Rice (grams of silver per hectoliter)
X-axis: Year

Data Source: International Institute of Social History, Netherlands

The Unified Cycle Theory

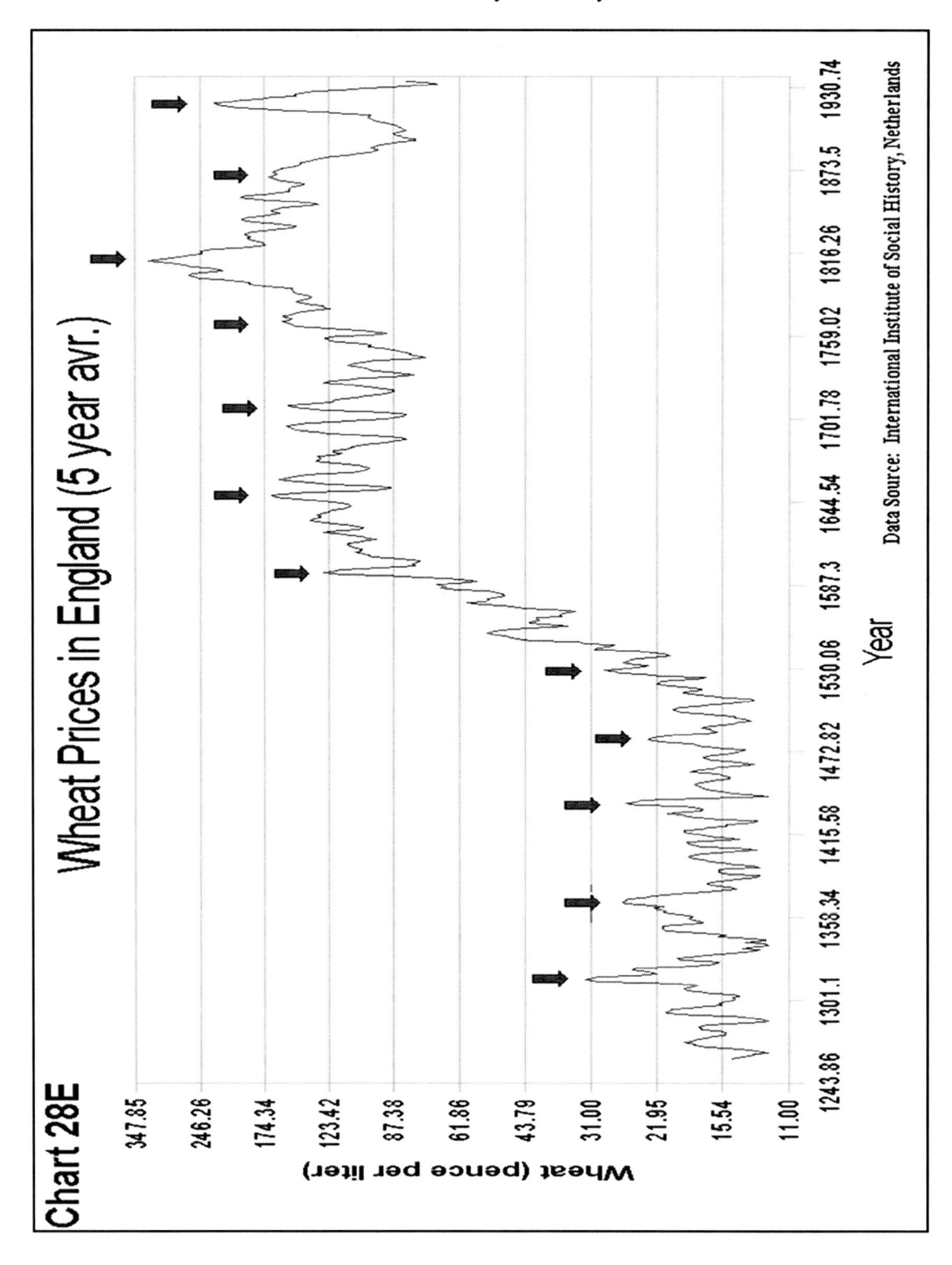

Chart 28E Wheat Prices in England (5 year avr.)

Data Source: International Institute of Social History, Netherlands

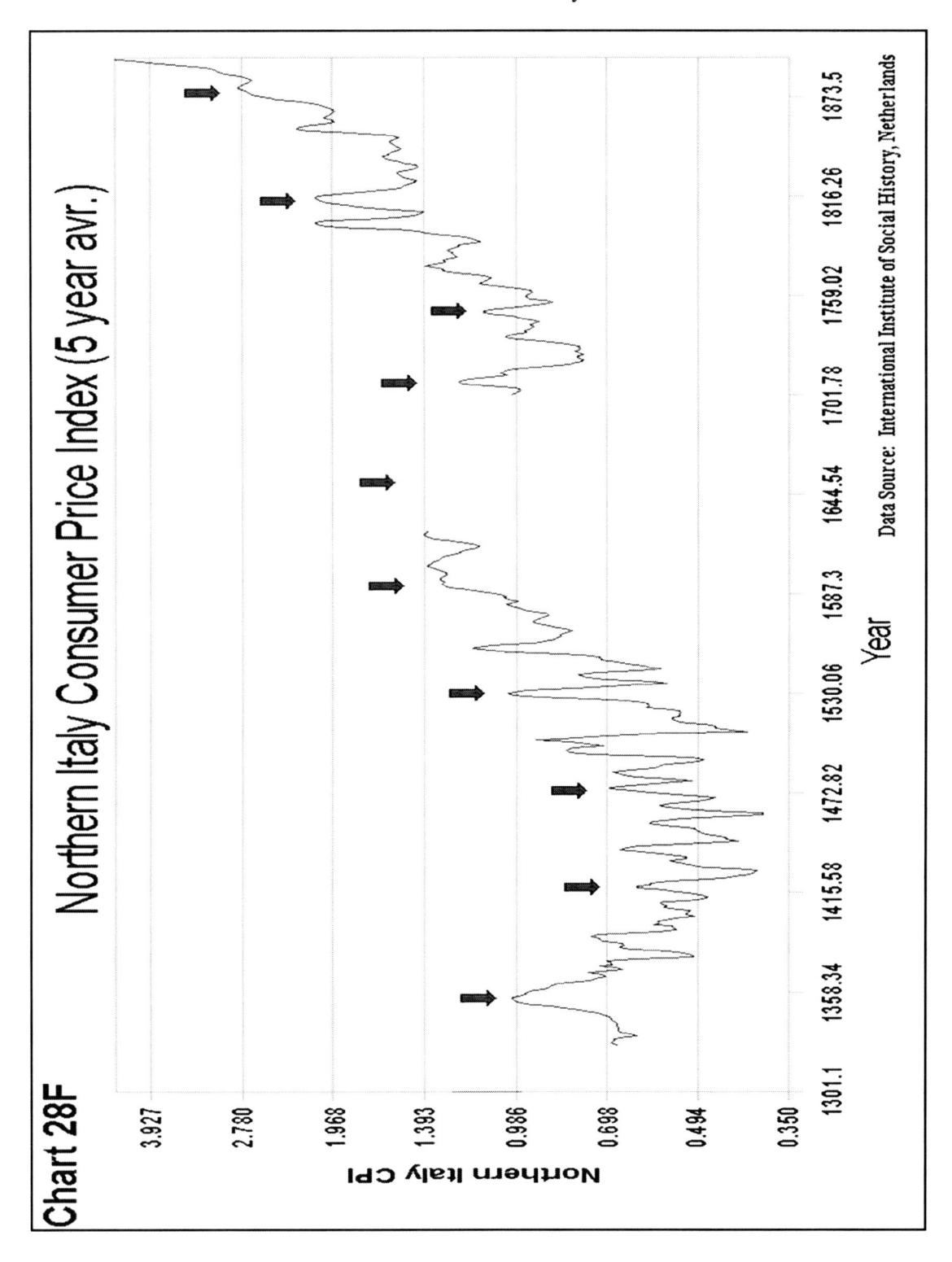

The 57-Year EUWS Cycle

Chart 28F — Northern Italy Consumer Price Index (5 year avr.)

Data Source: International Institute of Social History, Netherlands

The Unified Cycle Theory

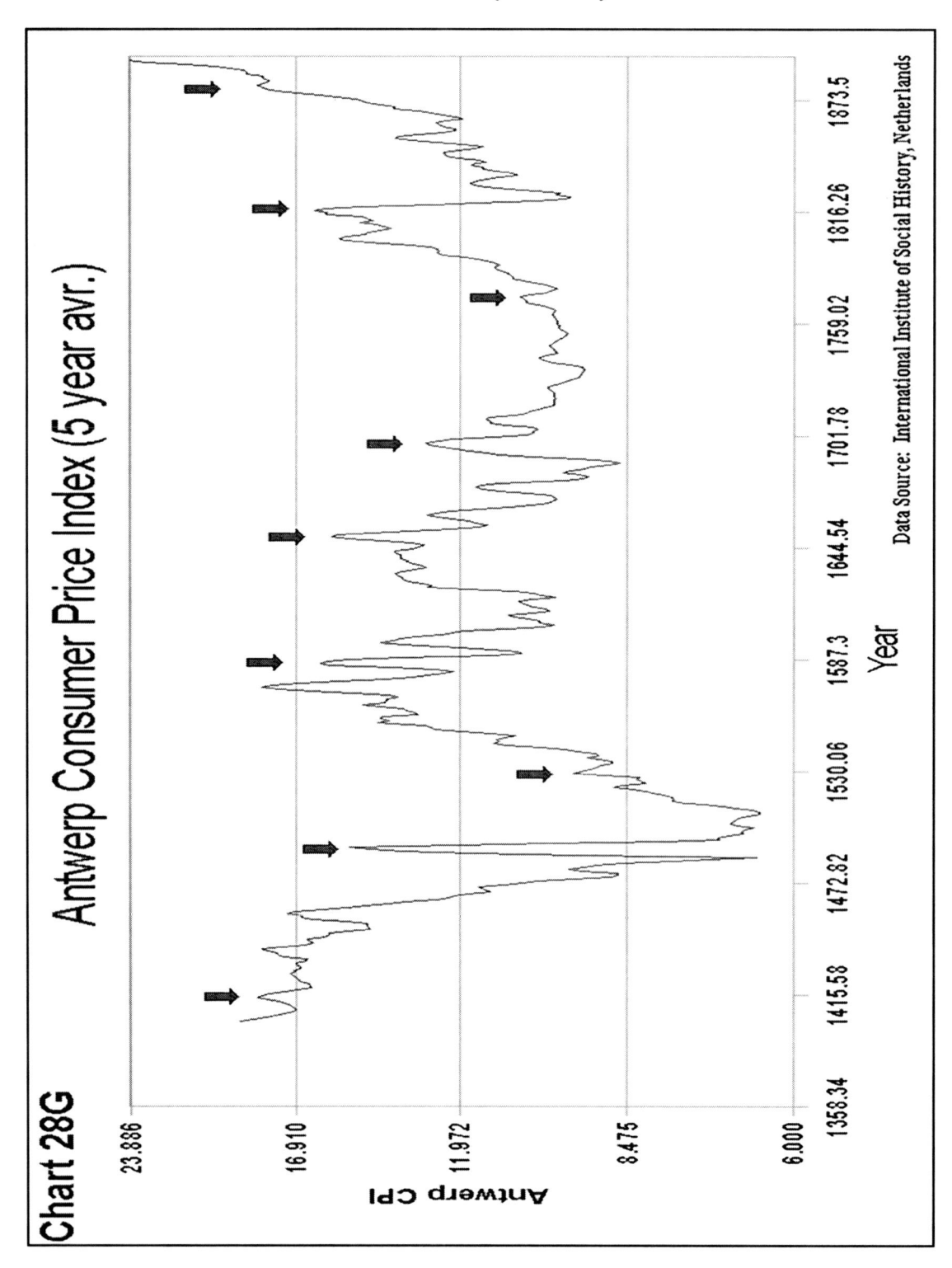

Chart 28G Antwerp Consumer Price Index (5 year avr.)

Data Source: International Institute of Social History, Netherlands

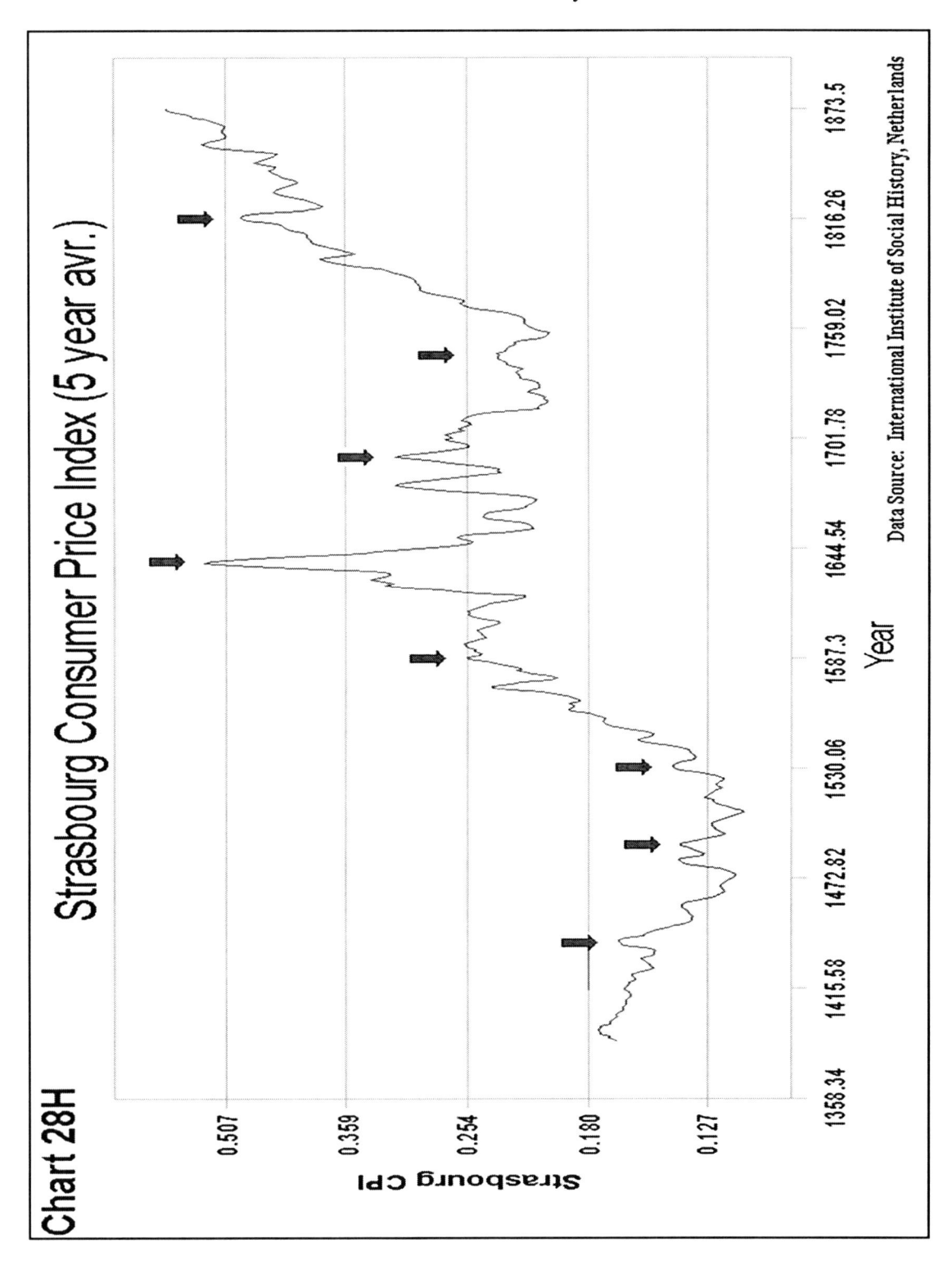

The 57-Year EUWS Cycle

Chart 28H — Strasbourg Consumer Price Index (5 year avr.)

Data Source: International Institute of Social History, Netherlands

The Unified Cycle Theory

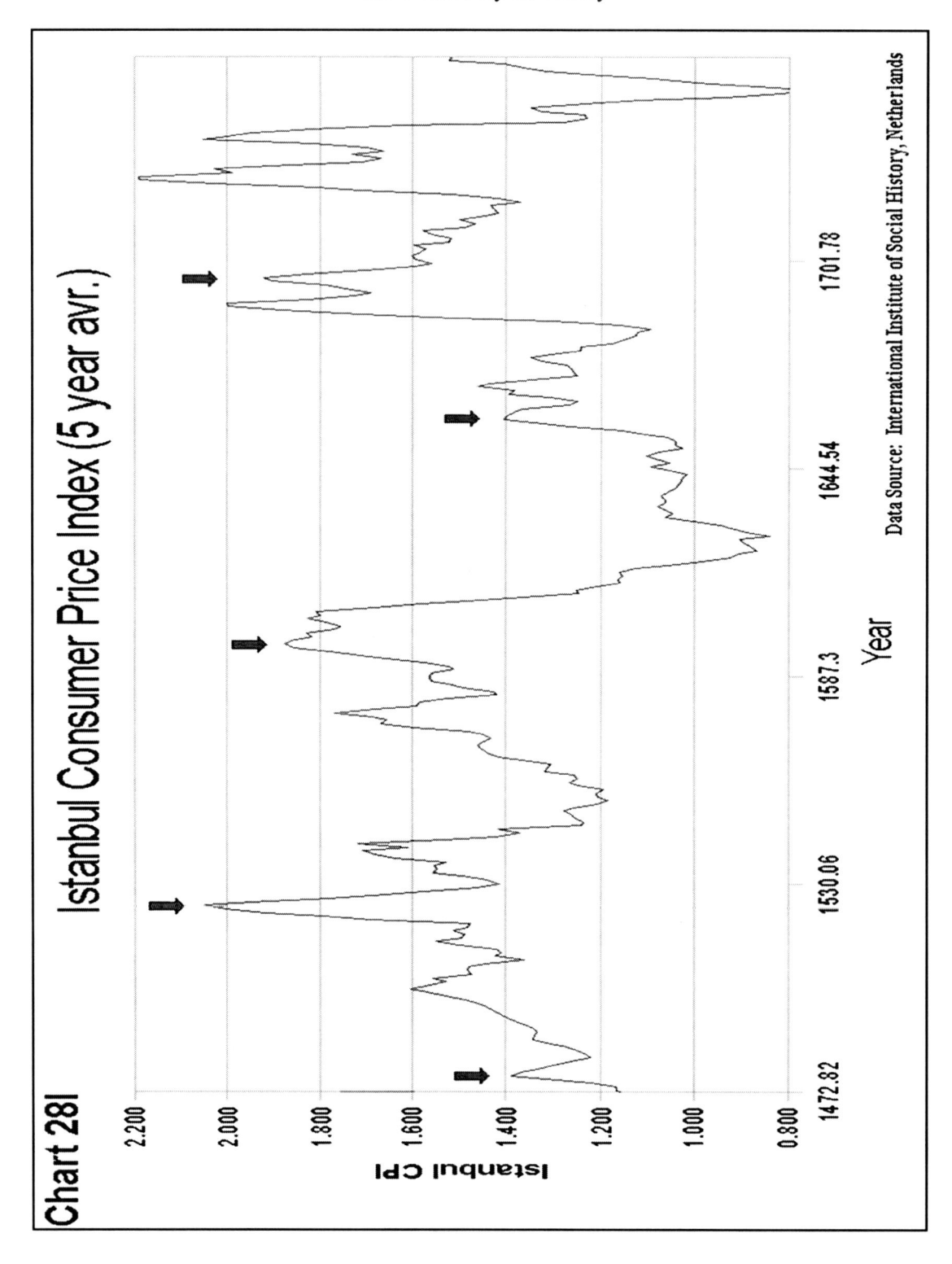

Chart 28I — Istanbul Consumer Price Index (5 year avr.)

Data Source: International Institute of Social History, Netherlands

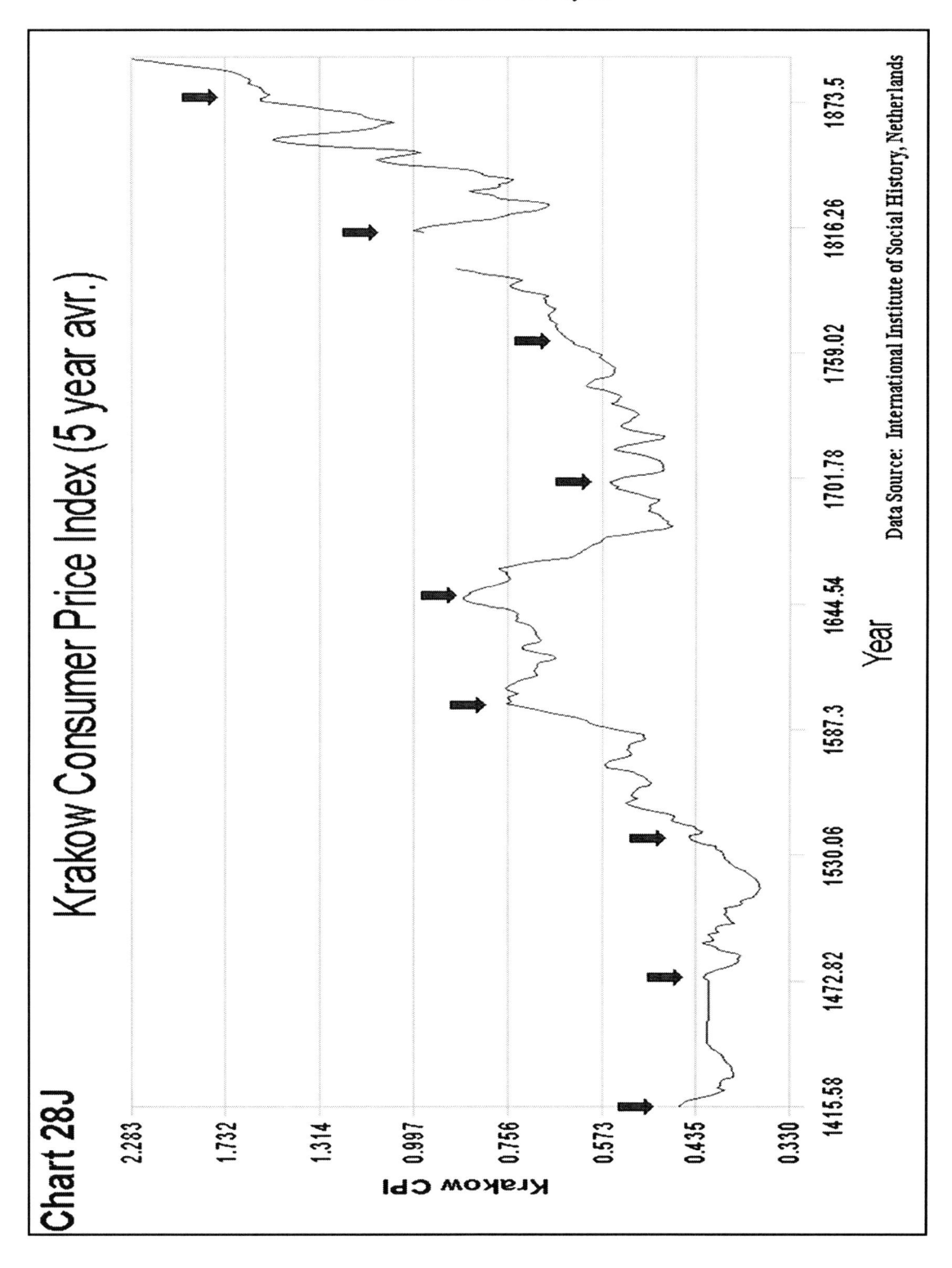

The 57-Year EUWS Cycle

Chart 28J — Krakow Consumer Price Index (5 year avr.)

Data Source: International Institute of Social History, Netherlands

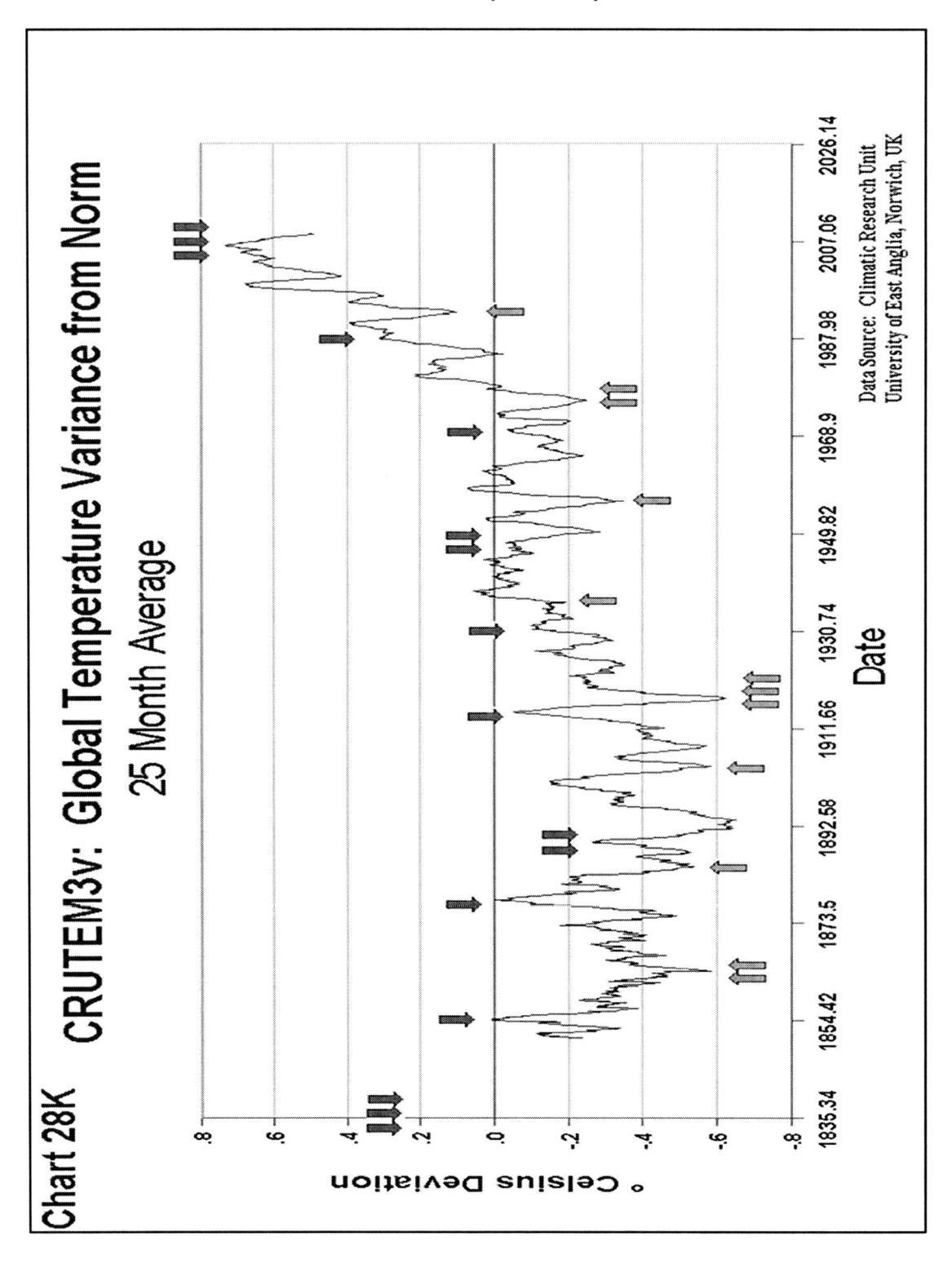

Chart 28K CRUTEM3v: Global Temperature Variance from Norm
25 Month Average

Data Source: Climatic Research Unit
University of East Anglia, Norwich, UK

Chapter 29
The 19-Year EUWS Cycle

After dividing the 57.24-year cycle by three, it yields the next frequency – the 19.08-year EUWS cycle. Financial panics stand out as the distinguishing feature of the 19-year cycle. A financial panic usually involves some type of deflation. Declining asset values prompt cautious individuals to withdrawal deposits from institutions perceived to be overextended. In pre-FDIC times, financial runs typically involved only commercial banks. In more recent times, financial runs involve hedge funds, mutual funds, investment banks, as well as commercial banks. In addition, declining asset prices immediately translate into depreciated corporate book values. Hence, stock prices often crash during a panic. At the end of the daisy-chain, financial turmoil typically results in reduced business spending, which translates into economic recession. **Table 29.1** contains theoretical peaks for all 19-year cycles during the last 300 years and their associated panics.

As they relate to the business community, the major difference between the 57-year cycle and the 19-year cycle amounts to the degree of impact. Downturns produced by the 57-year cycle trigger economic depressions. However, declines associated with the 19-year cycle simply trigger brief bear markets resulting in recessions that fail to morph into anything larger.

The panic cycle is one of the most identifiable of the EUWS cycles. Simon Kuznets discovered this cycle in the 1930s, originally reporting its period as a range between 15 and 20 years. Within two decades, Edward Dewey refined the cycle's length to 18.33 years, calling it a real estate cycle.

Table 29.1 – Recent Peaks of the 19-Year Financial Panic Cycle.

Theoretical Peaks: Decimal Date	Theoretical Peaks: Calendar Date	Date of Actual Top	Deviation from Theoretical	Events Associated with Panic Cycle
1720.86	Nov 1720	Jan 1720 June 1720	Direct hit Direct hit	Crash of the Mississippi Bubble, France Crash of the South Sea Bubble, England
1739.94	Dec 1739	1738 1741	1 Year 2 Years	Top in South Carolina wholesale prices Top in Philadelphia wholesale prices
1759.02	Jan 1759	1759 1763	Direct hit 4 Years	Top in South Carolina wholesale prices Top in Philadelphia wholesale prices
1778.10	Feb 1778	1782	4 Years	Top in South Carolina wholesale prices The Continental Currency collapses
1797.18	Mar 1797	1796	1 Year	Wholesale prices top in S.Carolina and Philadelphia; 4 year recession begins
1816.26	Apr 1816	1817	1 Year	Top in US stocks and commodities
1835.34	May 1835	May 1835	Direct hit	1835-1842 bear market in US stocks
1854.42	June 1854	1852	2 Years	Five year bear market begins
1873.50	July 1873	Aug 1873	Direct hit	Stock market crash, Panic of 1873
1892.58	Aug 1892	June 1892	Direct hit	Bear market, high unemployment
1911.66	Aug 1911	Dec 1909	2 Years	Eleven year bear market begins
1930.74	Sept 1930	Sept 1929	1 Year	Market crash, economic depression
1949.82	Oct 1949	--	(Skipped)	No panic in the United States
1968.90	Nov 1968	Dec 1968	Direct hit	Top of 1960s Go-Go bull market
1987.98	Dec 1987	Aug 1987	Direct hit	October 1987 stock market crash
2007.06	Jan 2007	Jan 2007 Oct 2007	Direct hit Direct hit	Subprime crisis erupts All major global stock markets peak

The 19-year panic cycle also appeared in commodity prices from ancient Babylonia. Unfortunately, occasional gaps in the data make this set of data less than perfect. Hence, it needs to be assigned a smaller weighting than other pieces of evidence. Data gaps were especially numerous from 360 BC to 300 BC. Nonetheless, Babylonian commodity prices displayed a series of tops regularly spaced about 19 years apart. See **Chart 29A**.

In spite of the limitations associated with data gaps, Babylonian commodity prices represent the

The 19-Year EUWS Cycle

only truly ancient commodity price series in existence. It contains enough accuracy to give a general picture of how ancient commodity prices fluctuated. And it produced the same 19-year beats that modern markets do today!

While ancient Babylonian data sits in the interesting category, extensive histories since medieval times solidify support for the 19-year panic cycle. Specifically, London consumer prices between the years 1260 and 1880 offer a 620-year continuous history, with accuracy within one year. (The Chinese rice price history could not be used for this analysis because it consisted of decade averages.)

Charts 29B, 29C, and 29D show the 19-year cycle's influence on consumer prices in London. These charts display a variety of EUWS features – already discussed extensively in previous chapters. London's commodity history simply reinforces all of these traits. To highlight these features, the charts display these features:

- Marking 57-year tops with double-arrows (the parent cycle of the 19-year frequency)
- Marking 19-year tops with single-arrows
- Marking 6.36-year tops with a dot (the child cycle of the 19-year frequency)

This chapter focuses on the 19-year cycle. However, every EUWS cycle faces interference from its parent cycle, its child cycle, and other non-EUWS cycles. In addition to interference, shifts appear quite frequently. But these shifts tend to occur in sub-cycle increments. As 57-year peaks come and go, prices change in a stair-step manner corresponding to 19-year and 6.36-year sub-cycles.

In spite of shifts, interference, and deviations, a fairly cohesive pattern emerges. The EUWS cycles tend to work together in harmony quite well. As a final note: In some cases, only one 6.36-year cycle appears in between 19-year peaks. This doesn't necessarily imply a skipped beat. Brief spikes, marking 6.36-year highs, could be masked by the one year average employed by this data series.

The charts also reveal the tendency of commodity markets to top anywhere from 6 to 27 years ahead of theoretical peaks of the 57-year cycle.

Next in the analysis comes a set of charts from Continental Europe. These charts cover consumer prices between the years 1663 and 1880. Except for Antwerp, they only contain single arrows, depicting actual tops of the 19-year panic cycle. Mostly, these charts show the same things as the London datasets. In particular, consumer prices tended to top 19 years ahead of theoretical peaks in the depression cycle – but the tops ranged anywhere from 6 to 27 years prior to the theoretical peaks.

Chart 29E shows the 19-year cycle in Antwerp prices. During the first portion of the chart, market tops repeat approximately midway through each cycle. Then, starting in 1797, they shift forward toward theoretical peaks. The chart also shows how the sunspot cycle interfered with visibility of the 19-year cycle. In **Chart 29E**, every significant spike upward corresponded to either a sunspot maximum or a 19-year theoretical peak. The 19-year EUWS cycle frequently wages a tug-of-war against the 11-year Schwabe sunspot cycle. To demonstrate the power of the sunspot cycle, review its correlation with various Antwerp CPI tops in **Table 29.2**.

Table 29.2 – Peaks in the Sunspot Cycle versus Antwerp CPI Tops.

Sunspot Peak	Antwerp CPI Peak	Deviation in Years
1615	1616	1
1626	1626	0
1639	1641	2
1649	1651	2
1660	1661	1
1675	1675	0
1685	1684	1
1693	1693	0
1705	1709	4
1718	1720	2
1727	1725	2
1738	1741	3
1750	1750	0
1761	1762	1
1770	1772	2
1778	1780	2
1788	1788	0
1805	1804	1
1816	1817	1
1830	1832	2
1837	1839	2
1848	1847	1
1860	1861	1
1870	1871	1

On average during this period, the sunspot cycle repeated once every eleven years. Assuming random behavior, the odds of any intermediate CPI top falling into a 1-year slot corresponding to an 11-year sunspot maximum equals 9%. Giving each peak a 2½-year leeway, 22 out of 24 sunspot cycles (92%) came within 2½ years of consumer price tops. The laws of probability show this should only happen randomly 45% of the time.

Using the probability mass function, here's the calculation for determining the probability of getting 22 successes in 24 trials, with the probability of a success equaling 45%....

$$\text{Probability} = (24!/22!(24-22)!) * (.45^{22} * .55^{2})$$

The 19-Year EUWS Cycle

Performing the math, that equates to 1 chance in 5 million that 22 CPI tops would randomly fall within 2½ years of a sunspot cycle peak. Obviously, the sunspot cycle exerted great influence on Antwerp CPI prices during this period. Always remember that geomagnetic cycles continually interfere with the visibility of short-term EUWS oscillations.

Next, **Chart 29F** shows Polish consumer prices between 1663 and 1880. Conflicts with Russia, Prussia, Spain, and France resulted in loss of data during the early part of the 19th Century. For this particular series, the 19-year cycle shifted forward for the initial four cycles and the 1835 cycle. The tops of all remaining cycles closely matched their 19-year theoretical peaks.

Each of these charts reflects different conditions in different counties. Each country controls its own military and monetary system. And the various regions get hit with different weather patterns. These outside factors cause the amplitudes of the oscillations to vary from country to country. Yet, in spite of these regional variations, actual CPI peaks and troughs remain fairly consistent, regardless of the country or the decade. This regularity indicates that the force behind these cycles distributes its power globally.

Between the years 1663 and 1880, Strasbourg consumer prices reflect this consistency. Similar to the other cities, the Strasbourg CPI tended to top ahead of the theoretical peaks during the first half of the chart, and then consumer prices shifted by matching theoretical peaks during the 1800s. See **Chart 29G**.

The consumer price analysis concludes with the Italian series shown in **Chart 29H**. The Northern Italy CPI chart depicts much of the same – price tops occurred ahead of 19-year cycle theoretical peaks until 1759. Starting in 1778, price tops closely coincided with theoretical peaks.

Now, attention moves to security prices. While commodity prices sometimes climax as much as ten years ahead of theoretical peaks of the 19-year cycle, equity market tops consistently match theory. The year 1720 provides a perfect example. Shortly after mankind invented equity shares, financial bubbles inflated in Holland, England, and France from 1717 to 1720. The most famous of these were the Mississippi Bubble in France and the South Sea Bubble in England. **Chart 29I** shows the South Sea inflation/deflation.

Several data gaps appear in this series. Nonetheless, **Chart 29I** closely approximates the path taken as the South Sea Bubble inflated until July 1720, and then crashed during the July-September quarter. This crash contained most of the elements of other great crashes. The sunspot cycle peaked two years earlier, the eclipse cycle came into play, and the bulk of the crash occurred during the seasonally weak month of September. However, perhaps most important of all, the theoretical peak of the 19-year panic cycle arrived on November 10, 1720. The top of the South Sea Bubble preceded that critical date by only four months!

Originally published in 1841, Charles Mackay's classic book *Extraordinary Popular Delusions and the Madness of Crowds* outlines many of the important elements of financial bubbles. These include the widespread use of leverage, great emotional involvement, and tremendous monetary stimulus from government. Mackay relayed this account as the South Sea Bubble burst:

> **"It would be needless and uninteresting to detail the various arts employed by the directors to keep up the price of stock. It will be sufficient to state that it finally rose to one thousand percent.... The bubble was then full-blown, and began to quiver and shake preparatory to its bursting.... During the whole month of August the stock fell, and on the 2nd of September it was quoted at seven hundred only. The state of things now became alarming. To prevent, if possible, the utter extinction of public confidence in their proceedings, the directors summoned a general court of the whole corporation, to meet in Merchant Tailors' Hall on the 8th of September. By nine o'clock in the morning, the room was filled to suffocation.... Several resolutions were passed at this meeting, but they had no effect upon the public. Upon the very same evening the stock fell to six hundred and forty, and on the morrow to five hundred and forty. Day after day, it continued to fall.... [By September 20] ... men were running to and fro in alarm and terror, their imaginations filled with great calamity, the form and dimensions of which nobody knew: 'Black it stood as night – Fierce as ten furies – terrible as hell.'"** [Mackay, 1980]

With credit-based monetary systems firmly established by the 19th Century, the regularity of bubbles strengthened. In the USA, a continuous series of panics (evenly spaced approximately 19 years apart) occurred from 1796 until 1929. A stock market proxy constructed from several sources clearly shows these evenly spaced panics.[Warren & Pearson, 1933], [Smith & Cole, 1935], [Macaulay, 1938], [Cowles, 1939], [S&P, 1986] See **Chart 29J**.

This index closely tracks a continuous stock market series developed by University of Rochester professor G. William Schwert.[Schwert, 1991] However, Schwert's index started with the year 1802.

All of the long-term bear markets in USA history began within two years of theoretical peaks of ten of the last eleven panic cycles. These theoretical peaks occurred in 1816, 1835, 1854, 1873, 1892, 1911, 1930, 1949, 1968, 1987, and 2007. The only miss came in 1949, which amounted to a skipped beat. Before further examining these financial panics, another probability session estimates the odds of this happening randomly.

For a 19-year cycle, the likelihood of a market top falling into any of its 19 annual slots equals 5.24%. Hence, the likelihood of the top coming within two years of a 19-year theoretical peak equals 21%. Using the probability mass function, the odds of getting 10 successes in 11 trials, with the probability of a success at 21%, equates to the following....

$$\text{Probability} = (11!/10!(11-10)!) * (.21^{10} * .79)$$

The 19-Year EUWS Cycle

Performing the math, that computes to about 1 chance in 7 million that these tops formed randomly! Since security markets initially developed 300 years ago, reoccurring financial panics, at 19-year intervals, emerge as another set of evidence strongly supportive of a EUWS component. A credit-based economy working in combination with the 19-year panic cycle creates a lethal duo for equity investors.

This analysis continues by examining each of these financial panics in more detail. Each of the following panics appears as a down-arrow in **Chart 29J**.

Cycle 1, March 1797 Theoretical Peak – The previous cycle ended with Shays' Rebellion in 1786-1787. Daniel Shays and his cohorts protested excessive land taxes and economic depression following the American Revolution; while many other protests took place during this period.[Encarta: Shays, 2008] The economy then recovered until 1796 – one year prior to a theoretical 19-year peak. According to the National Bureau of Economic Research, the United States endured a lengthy recession from 1796-1799.[Davis, 2005] Stocks failed to recover from that recession until about 1814.

Cycle 2, April 1816 Theoretical Peak – Even after hitting bottom in 1814, equities only managed to nudge higher until 1817. The 1817 top came one year after a theoretical peak. Two years after the 1817 top, the Panic of 1819 erupted. That signaled the bottom of the cycle. Stocks then moved sideways for another decade.

Cycle 3, May 1835 Theoretical Peak – Another up-phase began in 1828, and it lasted until the stock market reached its euphoric climax in May 1835. The May 1835 top coincided perfectly with a 19-year theoretical peak. Two years later, the Panic of 1837 erupted. The panic coincided with a brief bottom, only lasting a few months. Within a year, stocks resumed their decline, finally hitting bottom in 1842. According to the NBER, two economic contractions covered most of this period. The first one spanned the years 1836-1838. The second downturn lasted from 1839 to 1843.

Cycle 4, June 1854 Theoretical Peak – A stock market recovery began about one year ahead of the economic nadir in 1843. During the subsequent bull market, most of the advance occurred before 1845. The stock market then declined modestly until 1848, before reaching its next major top in 1852. That came two years ahead of the 1854 theoretical peak. Following the 1852 top, another set of panics followed. The Panic of 1854 came first. It only affected New York City banks – developing after financial difficulties hit the Knickerbocker Bank. Three years later, the troubles intensified with the arrival of the Panic of 1857. The 1857 crisis started with the failure of Ohio Life and Trust Company on August 24. By September 17, the fear intensified after New York banks lost $2 million of uninsured gold bullion after an ocean vessel sank, resulting in a stock market crash. On October 12, stocks finally hit bottom. By then, leading railroad stocks had lost almost 50% from their apexes three months earlier. The 1852-1857 bear market also resulted in two economic recessions. The first from 1853-1855, and the second lasted from 1856-1858.

Cycle 5, July 1873 Theoretical Peak – Once again, the stock market recovered ahead of the economy. A strong 16-year bull market followed the 1857 panic lows. Even the Civil War didn't deter the bull market. But in 1873, the good times came to an end. In that year, an important stock market top materialized one month after a theoretical 19-year peak. On cue, the bull market topped in early August 1873, and then crashed the following month. See **Chart 29K**.

By the close of trading on September 20, the panic intensified to the point where NYSE officials suspended trading for ten days. The History Channel Online gives this account:

> **"The Panic of 1873, which was triggered by over speculation, continued to wreak havoc on the nation's economy. Just two days after Jay Cooke & Co., one of the nation's most reputable brokerage firms, declared bankruptcy, the New York Stock Exchange decided to close down for ten days to wait out the worst of the crisis. The Secretary of the Treasury responded to these events by pumping $26 million of new currency into the economy, swelling the amount of paper money in circulation to $382 million."**[History.com: Panic, 2008]

After reopening for trading on September 30, 1873, the Treasury's money-pumping helped for a while. However, the bear market soon resumed. And the stock market continued lower until finally hitting bottom in 1877.

Cycle 6, August 1892 Theoretical Peak – Very similar to the previous cycle, from the 1877 low, most of the subsequent gains occurred during the initial four years of the ensuing bull market. After 1881, the stock market stagnated for eleven years. When the theoretical peak of the panic cycle arrived in August 1892, the stock market had already reached its summit five months earlier. After making its top, stocks drifted lower until the August 1 theoretical peak passed. Then the bottom fell out of the market. See **Chart 29L**. During the next year, the stock market lost nearly half of its value, ending with the Panic of 1893. After July 1893, a partial recovery took hold, but the market gave back all of those gains by 1896. Record high unemployment emerged as the distinguishing feature of the mid-1890s depression. After 1896, the stock market initiated another recovery.

Cycle 7, August 1911 Theoretical Peak – Each and every economic depression gives birth to the next cycle. And it so happened in 1896. As measured by the Dow Jones Industrial Average, a ten year bull market climaxed in January 1906. See **Chart 29M**. However, a divergence occurred. After the Panic of 1907, the more broadly based Cowles Commission Index recovered from all of the 1907 losses, and later hit its bull market zenith during November-December 1909. That top came within two years of the theoretical peak of the 1911 panic cycle. After that, a shallow (but lengthy) eleven year bear market followed. See **Chart 29J** again for the broad-market trend.

The lengthy bear market of the 1910s ended with intense deflation in 1920-21. And the deflation of the early 1920s paved the way for the great bull market of the Roaring '20s. Two other important events occurred around then. First, in December 1913 Congress authorized creation of the Federal Reserve banking system. The Fed became fully operational in 1914. Second, the 172-year EUWS cycle passed its theoretical low point during 1921.

Cycle 8, September 1930 Theoretical Peak – With these dual forces acting as propellants, one of the greatest bull markets in history began. See **Chart 29N**. While EUWS cycles acted as agents for building confidence and risk-taking, easy monetary conditions provided the credit-fuel for the speculation that stretched far beyond reasonable levels. From this standpoint, historians generally overlook the role played by the newly created Federal Reserve in promoting the speculative bubble of the '20s. From the 6.0% discount rate set on January 24, 1920, the Fed lowered rates four different times during the 1920s. By August 4, 1927, the Fed had pushed the discount rate down to an overly simulative 3.5%. By then the damage had been done. The mania had already captivated too many speculators. By 1927, with indirect support from the Federal Reserve, the banking system pushed broker and security loans in large numbers. In turn, easy security lending allowed an ever increasing number of speculators into the stock market.

Excessive credit creation always played a critical role in fueling history's great bubbles. Episodes of misallocated credit happened as far back as the Mississippi Bubble and the South Sea Bubble in 1720, it happened again in 1929. (And it continued happening throughout the last 50 years.) By 1928, the Federal Reserve governors finally became concerned about the speculation. However, their actions came too late to prevent the looming depression. Belatedly, on February 21, 1928 the Fed tightened credit conditions, albeit, very modestly by raising the discount rate to 4.0%. The Fed raised the discount rate two more times during 1928, and then stopped.

In spite of the Fed's efforts, the bull market continued. Broker-loans and credit continued to expand. Releasing the credit-genie from its bottle was easy. But putting the genie back in proved to be a more difficult task. The bull market finally crested on the first trading day of September 1929. At that time, the unbridled optimism of the 1920s shifted toward more realistic caution.

In spite of the carnage that took place during September, early October brought bargain-hunters into the market. Stock prices partially recovered. Then panic erupted. During the last half of October 1929, the stock market crashed in perhaps the most documented panic in recorded history. The crash financially devastated the nation. Soon afterwards, the USA sank into its deepest economic depression.

After a short-term rebound during the early months of 1930, the stock market turned lower again. It continued its downward path until 1932. At the low, equities had lost more than 90% of their August 1929 levels. After the 1932 bottom, the stock market recovery picked up momentum after the theoretical low point of the 6.36-year cycle in late 1933. And the recovery continued until the next important stock market top arrived on March 10, 1937. Not by accident, that top came about one month from the February 8, 1937 theoretical peak of the 6.36-year cycle. Then phase two of the depression began. Unemployment remained high until 1939. And the stock market finally reached its valley in 1942.

From its start in 1920, to its end in 1942, the entire cycle surrounding the bull market of the 1920s and its subsequent crash serves as a near perfect example of the inter-related workings of the 19-year panic cycle, the 6.36-year recession cycle, the 172-year cycle, and changing credit market conditions.

Cycle 9, October 1949 Theoretical Peak – In a rare occurrence for the EUWS cycles, the panic cycle skipped a beat in 1949. However, it could be argued that stocks topped three years ahead of their 1949 theoretical peak. Following World War II, the Dow Jones Industrial Average topped at 212.50 on May 29, 1946. After the war, most Americans expected a post-war depression. Using the War of 1812, the Civil War, and World War I as guidelines, it's understandable how this expectation arose.

Nonetheless, for an economic depression to develop, unstable sectors of the economy must be forced into debt liquidation. Even though World War II saddled the federal government with large debts, state governments, corporations, and households all enjoyed their most pristine balance sheets of the century during the late-1940s. When the federal government made no attempt to pay down its debt after World War II, the USA avoided the only means of triggering a depression.

Without substantial leveraged positions, forced liquidations and financial panic cannot develop. And that, more than anything else, distinguished the late 1940s from all earlier and all subsequent theoretical peaks of the 19-year panic cycle. The fact that the USA avoided panic during this cycle indicates that mankind has the ability to overcome the effects of this cycle. However, to accomplish this on a permanent basis, nations must essentially eliminate debt from their monetary systems, and convert to pay-as-you-go monetary systems.

Cycle 10, November 1968 Theoretical Peak – After the devastating financial effects of the Great Depression and the psychological scars associated with World War II, confidence slowly returned to the general population. And as confidence returned, consumers gradually took on debt again. As always, the willingness to take on debt initially created an economic boom. And the expanding economy then led the way to the Go-Go stock market of the 1960s. See **Chart 29O**.

Actually, this Dow Industrial chart fails to give proper justice to the stock market of 1967-68. It shows a triple top in 1966, 1968, and 1973. But every other stock index for this period shows the late-1968 top reached levels far above those registered in either 1966 or 1973. (Refer to **Chart 9E** to see the Value Line Index during the 1960s and 1970s.) Without a doubt, 1968 marked the top of the bull market for this cycle. And when the market climaxed on December 3, 1968, it only missed the panic cycle's theoretical peak by six calendar days (and three trading days)!

After the late-1968 top, except for the Dow Jones Industrial Average, all major market averages entered a six year bear market. The only major interruption to this lengthy bear market came with the counter trend rally from late-1970 to mid-1971. By December 1974, the bear market finally reached its end.

Cycle 11, December 1987 Theoretical Peak – The next bubble in the stock market formed slowly from 1975 to 1982. The severe 1981-82 recession brought an end to the high interest rate environment endured during the early 1980s. With interest rates finally declining, the tremendous bull market of the 1980s took off. See **Chart 29P**.

Like other multi-year advances, increased leverage became an integral part of the bull market of the mid-1980s. (Refer to **Chart 9B** to see how margin debt expanded during this period.) The EUWS cycles stamped their mark at every significant turning point within this panic cycle. The 1981-82 bear market started with a 6.36-year peak. The August 1982 low coincided with a theoretical 2.12-year low. The late-1983 high closely matched a 2.12-year peak. The 1984 low came slightly ahead of a theoretical low of the 6.36-year cycle. That's when the frothy portion of the bull market began. During the next down phase of the 2.12-year cycle, a brief pause interrupted the speculative fever. Then the blow-off phase carried the Dow Jones Industrial Average to its pinnacle on August 25, 1987. The top came four months ahead of the 19-year theoretical peak – which occurred on December 27. Two months after the top, on October 19, the stock market crashed with the largest one-day loss in the history of the New York Stock Exchange. After the crash, stocks drifted sideways for nearly a year before rising to form another top by early 1990. A significant downturn developed later that year. Economic recession followed in 1991. As the recession deepened, the stock market hit bottom. The bottom for the cycle arrived in early 1991 – close to a 6.36-year theoretical low.

Cycle 12, January 2007 Theoretical Peak – Early 1991 coincided with the start on the next great bull market. The bull market of the 1990s centered around financial and technology shares. Interruptions appeared quite frequently. The nation's largest bank, Citicorp, nearly failed after the 1990-91 recession. During 1994-95, the Orange County bankruptcy and the Latin American debt crisis triggered substantial fear within the banking community. During October 1997, the Southeast Asian currency crisis coincided with a large drop in the stock market. And in September 1998, the blowup of Long Term Capital Management hedge fund stirred fears of a systemic failure.

Yet, when in need of assurances, the Federal Reserve always took quick action aimed at calming the markets. These quick responses became known as the Greenspan-Put – implying that the stock market was always safe from serious downside risk because of the perceived benevolence of the Fed Chairman. But the Greenspan-Put created an unintended moral hazard from its frequent use. It reassured investors that the Fed would always be there when needed. This prompted institutional managers and market speculators to begin taking increasingly outrageous risks. The Internet Bubble of 1998-2000 provides one example. The housing bubble of 2002-2006 provides another.

The Internet Bubble climaxed during March 2000, and it never recovered. However, for other sectors of the stock market, an even greater bubble inflated after a brief bear market ended in 2002. See **Chart 29Q**.

After the Internet Bubble burst, economic recession followed. By late 2002 and early 2003, fear reached the point where systemic failure became a concern again. On cue, the Federal Reserve responded to the market's cry for help. This time, the Fed pushed short-term interest rates down to 50-year lows. It dropped its key fed-funds rate all the way down to 1% from late-2003 to mid-2004. Moral hazard intensified. Risk taking turned insane. The stock market resumed its ascent, and the great housing bubble of 2002-2006 inflated out of control.

But conditions soon changed. On January 24, 2007, another theoretical peak arrived for the 19-year panic cycle. And that's precisely the month when the sub-prime crisis struck and when most financial stocks topped. During January 2007, Markit's asset-backed security index for home equity loans began plunging in value. Scores of small and medium-sized mortgage companies became insolvent, declared bankruptcy, or merged with larger rivals. By June 2007, two large hedge funds operated by Bear Stearns & Company became insolvent. As the crisis intensified, Bear Stearns halted redemptions in both hedge funds. That decision unleashed a series of negative reactions, including panic from the executives of other brokerage firms who had loaned money to the failing hedge funds.

As the sub-prime crisis took hold in early 2007, Federal Reserve officials assured investors that the crisis was contained. Yet, by mid-August 2007, the panic spread to other sectors – including hedge funds, structured investment vehicles, other asset backed markets, and junk bonds. As it had done under Greenspan, the Federal Reserve Board, under its new chairman Ben Bernanke, began lowering interest rates aggressively at the first hint of trouble.

The stock market initially responded positively to the rate cut – as it had always done after other Fed rate cuts. It resumed its upward trend. However, by October 2007, investors had second thoughts. The sub-prime crisis continued to intensify, and the contagion spread to virtually every sector of the credit markets. Starting on October 9, 2007, equity markets around the global declined in concert with markets in the USA. During March 2008, equities continued sliding when the investment bank Bear Stearns became insolvent. On March 16, 2008, in order to keep the brokerage firm from failing, the Federal Reserve brokered its sale to JP Morgan Chase. Once again, the stock market responded positively to the Fed's intervention. But this time, a rather pathetic rally ensued.

The story doesn't end there. However, it does take us to the present. While the future remains unknown, a number of indicators show that the Fed has lost control of monetary conditions in the USA. Too much debt exists to easily resolve the current crisis. The Federal Reserve now suffers from impotence resulting from the consequences of its prior over-issuance of credit.

The January 24, 2007 theoretical peak represented not only a 19-year high, but a 172-year high. More than likely, the negative impact from the downturn of this 172-year cycle will result in the worst depression in the history of the USA. In spite of occasional counter-trend rallies, fear in the credit markets continues to spread.

As we approach yearend-2008, the Federal Reserve and our federal government face a dilemma. Do they fight to save the financial sector of the economy? Or do they fight to gain control over our collapsing currency? It's a tough choice. However, it's one that managers of all credit-based monetary systems must eventually face.

Monetary Legislation and the 19-Year Panic Cycle.

The 19-year EUWS cycle also influences the President and Congress. Following the peak of every panic cycle since 1816, Congress enacted major monetary legislation. As financial panics unfolded, the House and Senate passed new laws in an attempt to prevent future panics. Obviously, these legislative attempts failed miserably at halting panics. A permanent solution involves elimination of central banking and fractional reserve lending, and a return to lending based on free-market interest rates. For example, brokered loans from existing savings would result in non-inflationary lending. However, even a brokered loan system wouldn't completely eliminate the effects of the 19-year cycle, but it would definitely reduce its impact by constraining lending. As the following section shows, except for 1836 and 1873, Congress generally opted for monetary solutions that encouraged debt growth.

1816 Theoretical Peak - In 1816, Congress created the Second Bank of the United States. This institution served as a central bank. In 1819, after calling in notes that helped create the Panic of 1819, the Second Bank became unpopular among USA citizens. Some states tried to tax its branches to drive it out of their territories.

1835 Theoretical Peak – In 1836, Andrew Jackson blocked renewal of the unpopular Second Bank of the United States. This first experiment with central banking came to an end one year after a 172-year EUWS peak. That started the free banking era.

1854 Theoretical Peak – Near the theoretical bottom of this panic cycle, Congress passed the National Bank Act of 1863. It established national charters for banks, thus ending the free banking era.

1873 Theoretical Peak – The Coinage Act of 1873 demonetized silver. US mints ceased coining silver dollars.

1892 Theoretical Peak – In 1893, Congress repealed the Sherman Silver Purchase Act. Proponents hoped this law would boost the money supply, thus stimulating business activity. In 1900, the Gold Standard Act passed. As a result, bimetallism came to an end.

1911 Theoretical Peak – In 1913, the Federal Reserve Act passed. Congress reestablished central banking in the United States near the theoretical low of the 172-year cycle. By creating the Federal Reserve banking system, Congress hoped this lender-of-last-resort would prevent banking panics.

1930 Theoretical Peak – In 1933, Congress enacted the Glass-Steagall Act to separate commercial banking activities from investment banking. It also establishes the FDIC as a temporary federal agency. Also in 1933, Franklin Roosevelt ended the ability of citizens to exchange currency for gold. The end of convertibility allowed the United States Treasury to hold onto its dwindling gold reserves.

1949 Theoretical Peak – In 1945, immediately after World War II, the Bretton Woods Agreement established a new gold standard among central banks. As a result, central bank gold exchanges resumed. (This is the only major monetary legislation to be enacted prior to a panic cycle peak.) Next, Congress passed the Federal Deposit Insurance Act of 1950. In doing so, Congress established the FDIC as a permanent agency.

1968 Theoretical Peak – During 1971, as foreign central banks increasingly demanded gold for their swelling supply of dollars, Richard Nixon closed the gold window. As a consequence, the Bretton Woods Agreement, signed by world leaders during the previous cycle, ceased to function. Elimination of the Bretton Woods gold standard initiated a huge inflationary boom by allowing the Federal Reserve to create unlimited amounts of dollars.

1987 Theoretical Peak – After the savings and loan crisis of the 1980s, and the near-collapse of several large banks in 1990-1991, Congress passed the Federal Deposit Insurance Corporation Improvement Act of 1991. In doing so, Congress gave the FDIC more power to deal with the large number of failed banks at the time.

2007 Theoretical Peak – With the passage of another panic cycle peak, a new financial crisis has erupted. However, this time the peak also accompanies a 172-year maximum. Similar to the last 172-year peak, some type of central banking shakeup seems imminent. Could the Federal Reserve System come to an end? Very possibly, it could. Already, the Federal Reserve has come under attack for letting numerous bubbles inflate out of control, for its lack of oversight with banks, and for its liberal use of taxpayer's money to bail out banks. At a minimum, legislation modifying its mode of operation seems likely. After major peaks, such as a 172-year cycle, popular opinion normally turns strongly against institutions failing to deliver on their mission. Because of its recent failures, the Federal Reserve could become a victim of expanding negative sentiment.

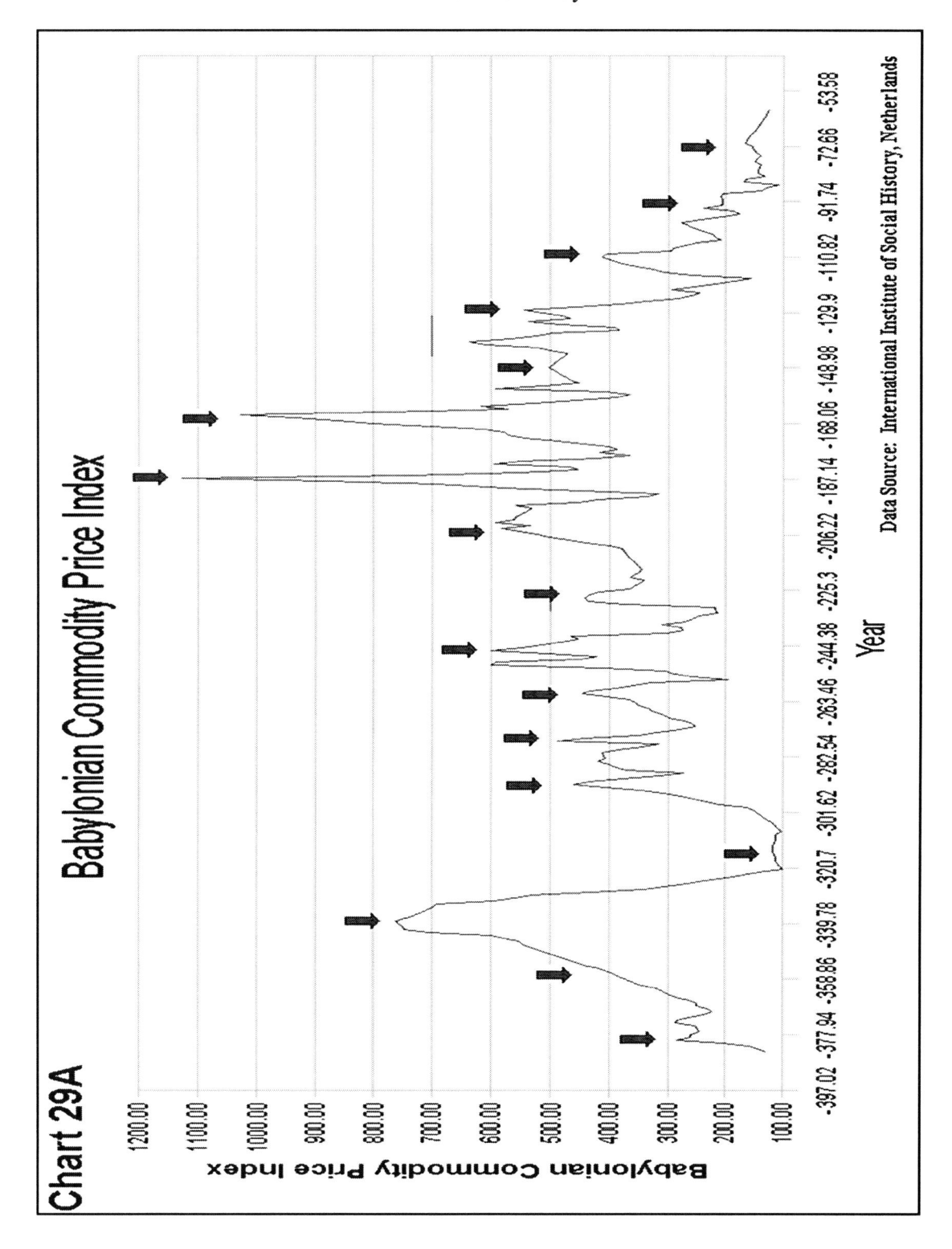

The 19-Year EUWS Cycle

Chart 29A
Babylonian Commodity Price Index

Data Source: International Institute of Social History, Netherlands

The Unified Cycle Theory

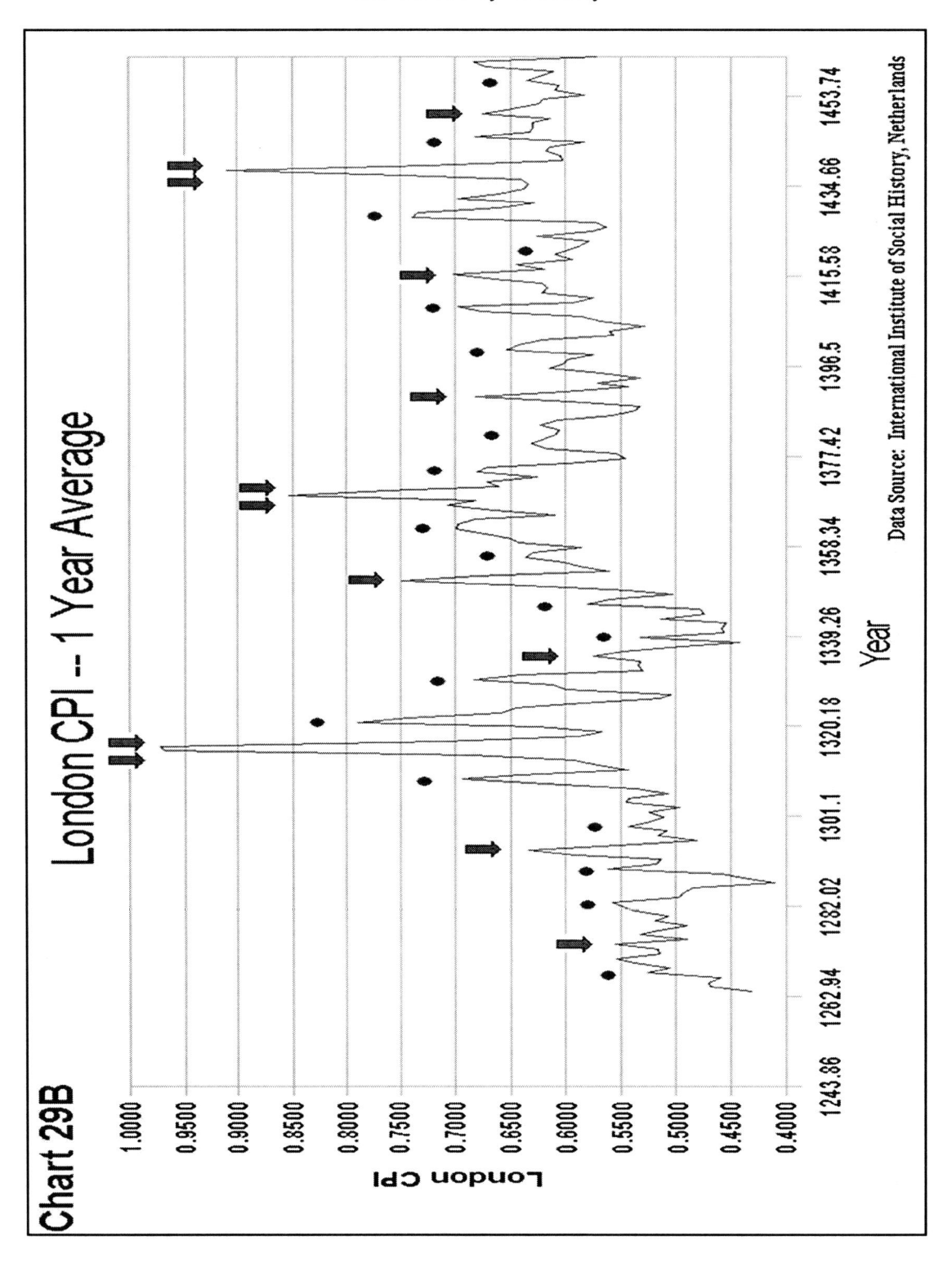

Chart 29B
London CPI -- 1 Year Average

Data Source: International Institute of Social History, Netherlands

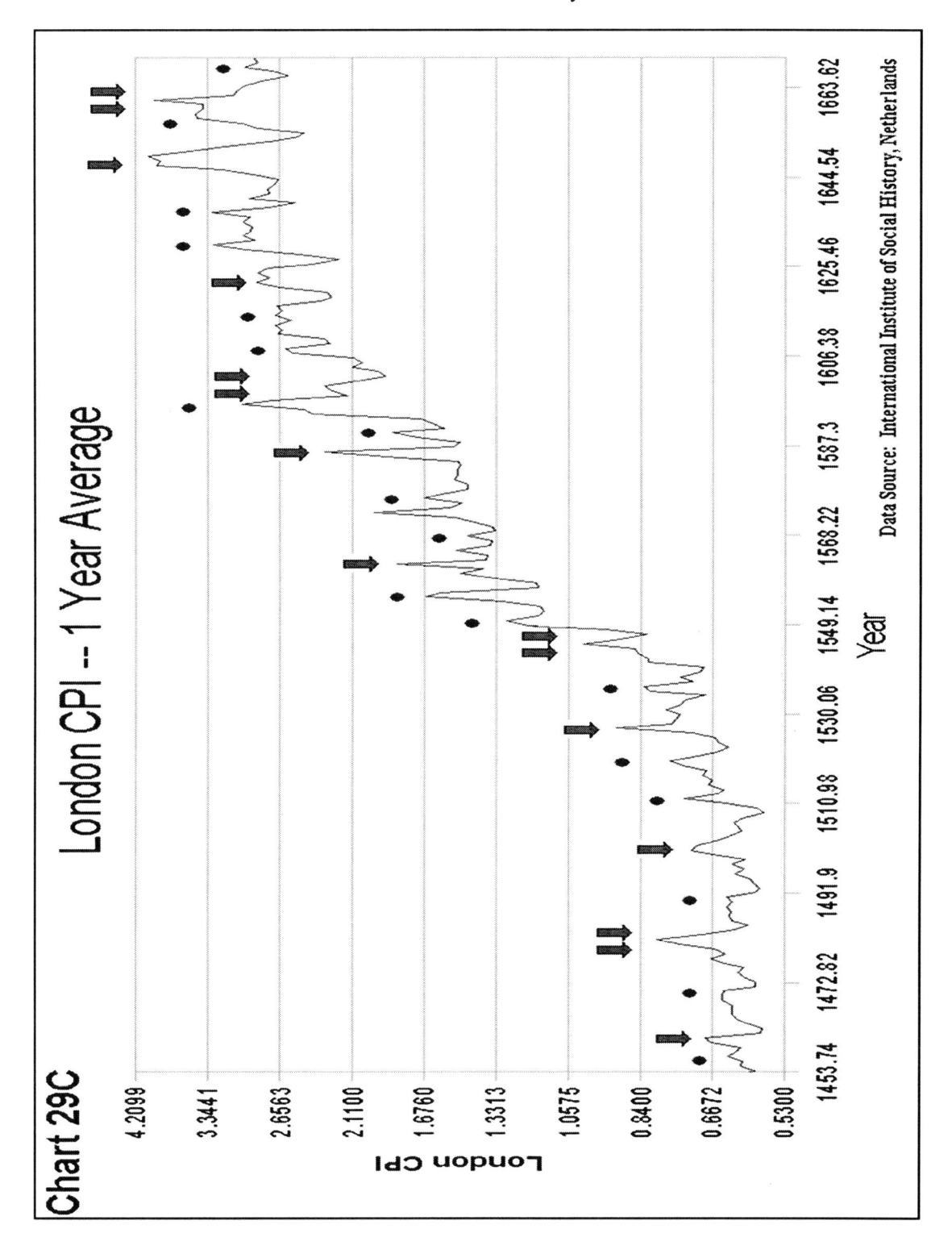

The 19-Year EUWS Cycle

The Unified Cycle Theory

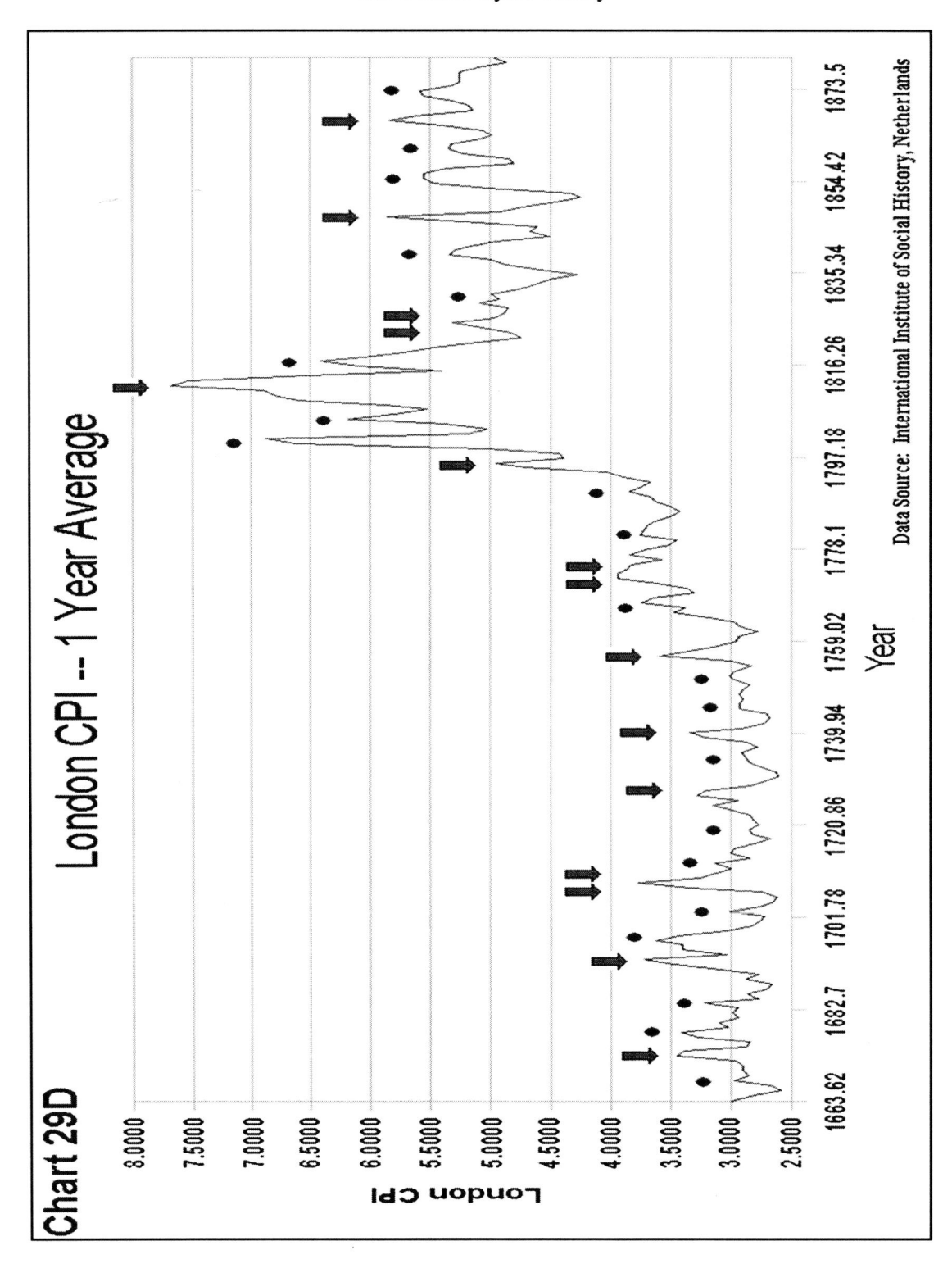

Chart 29D — London CPI -- 1 Year Average

Data Source: International Institute of Social History, Netherlands

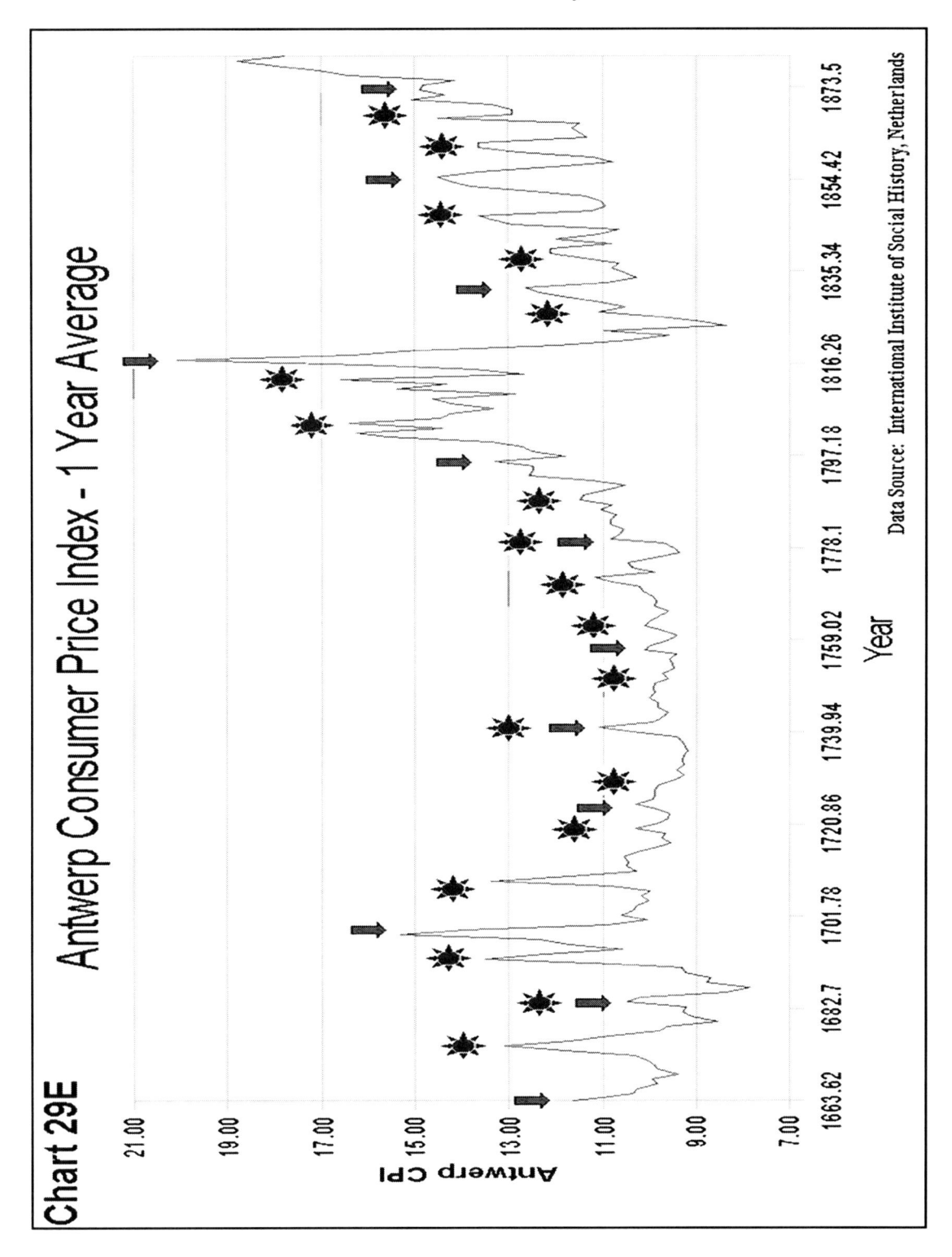

Chart 29E
Antwerp Consumer Price Index - 1 Year Average

The 19-Year EUWS Cycle

Data Source: International Institute of Social History, Netherlands

The Unified Cycle Theory

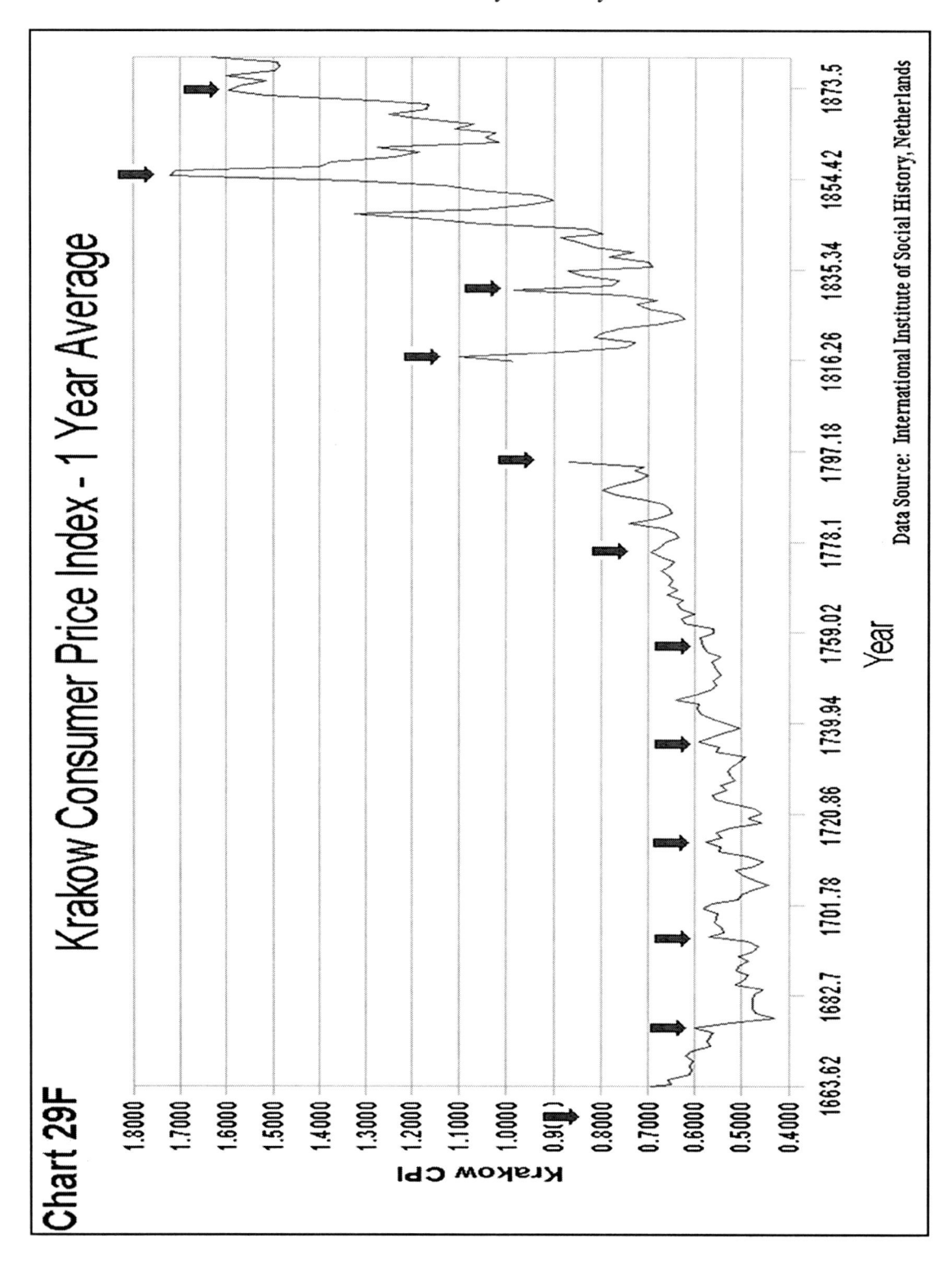

Chart 29F — Krakow Consumer Price Index - 1 Year Average

Data Source: International Institute of Social History, Netherlands

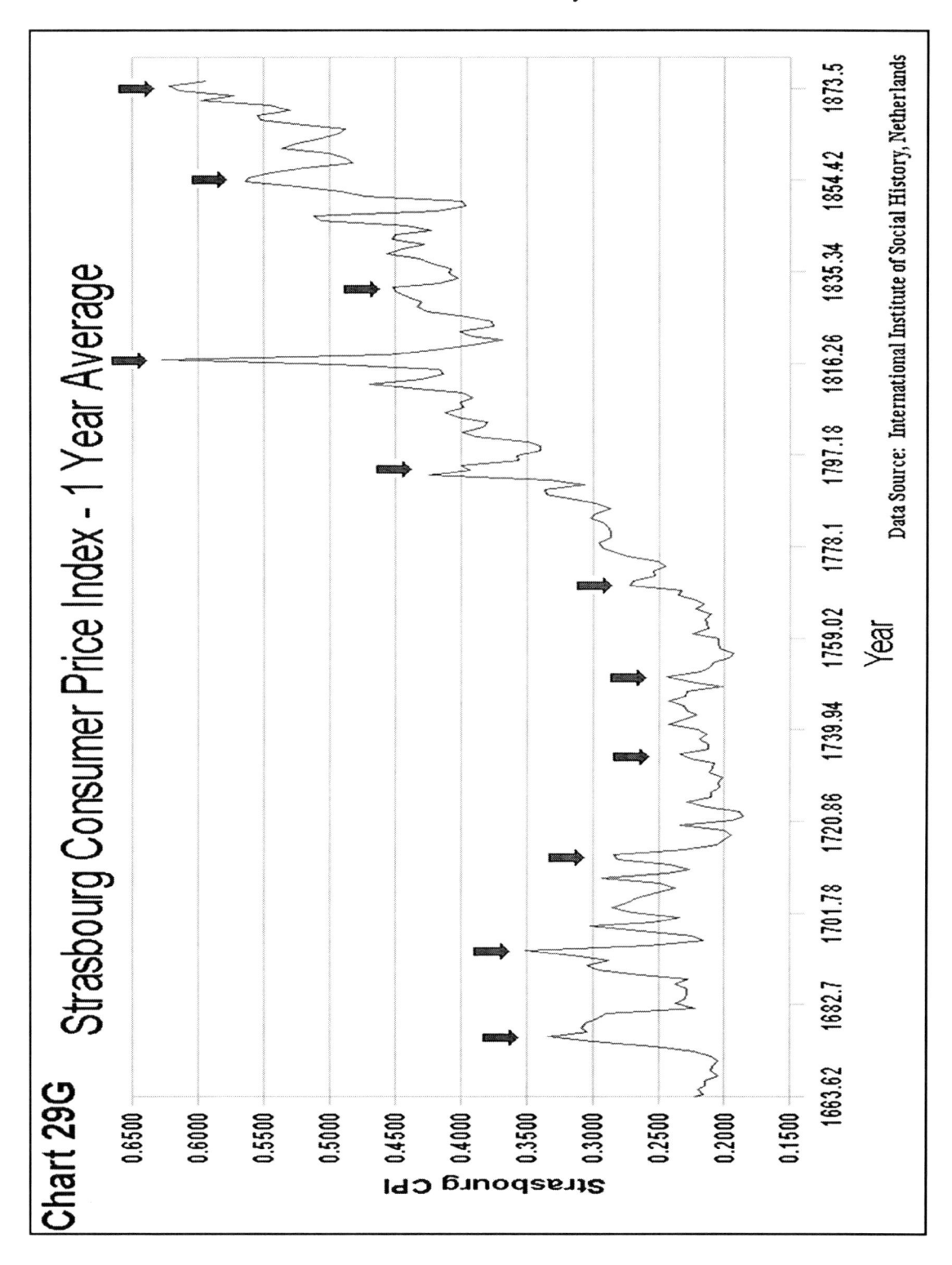

The 19-Year EUWS Cycle

Chart 29G: Strasbourg Consumer Price Index - 1 Year Average

Data Source: International Institute of Social History, Netherlands

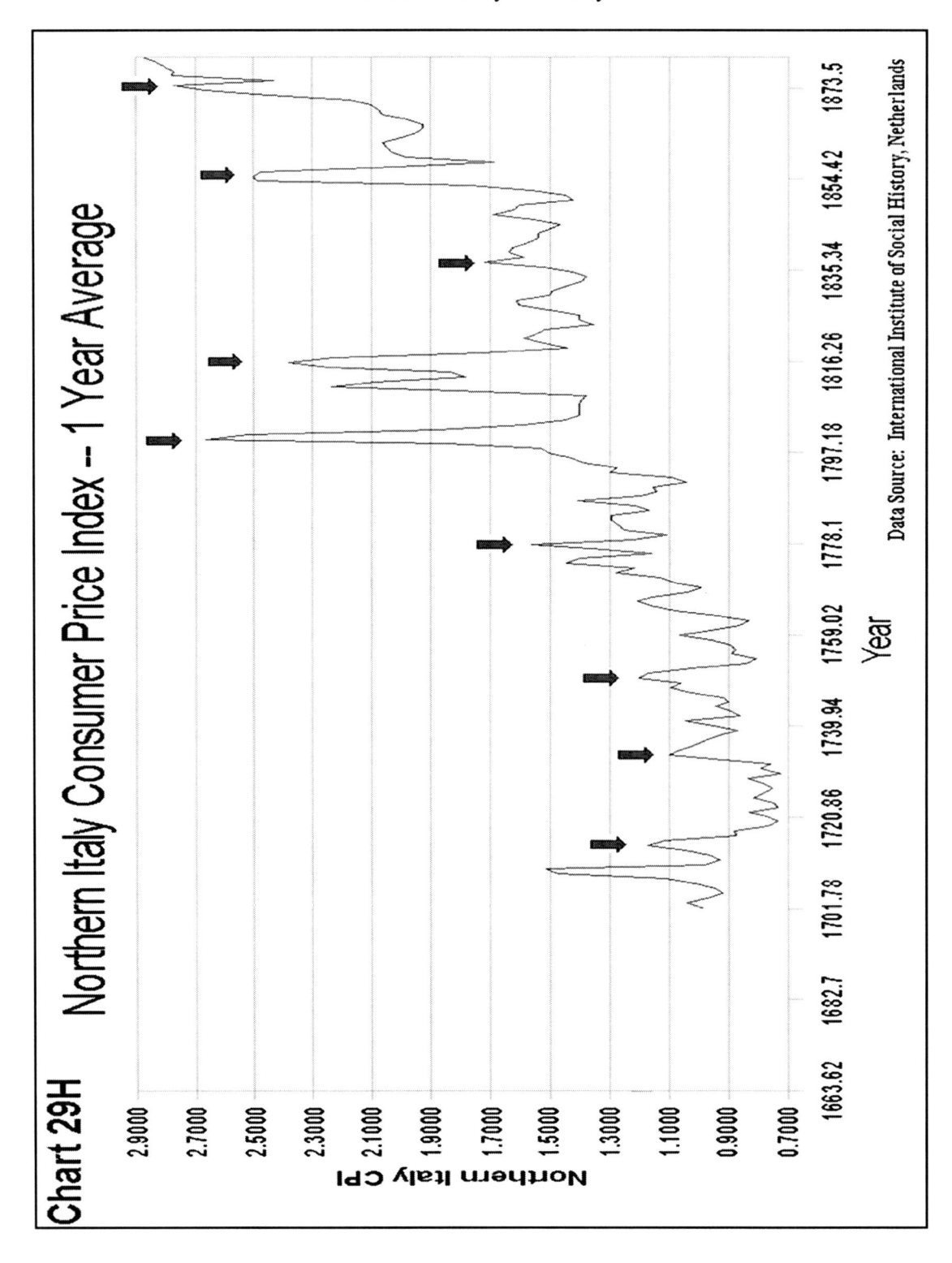

Chart 29H   Northern Italy Consumer Price Index – 1 Year Average

Data Source: International Institute of Social History, Netherlands

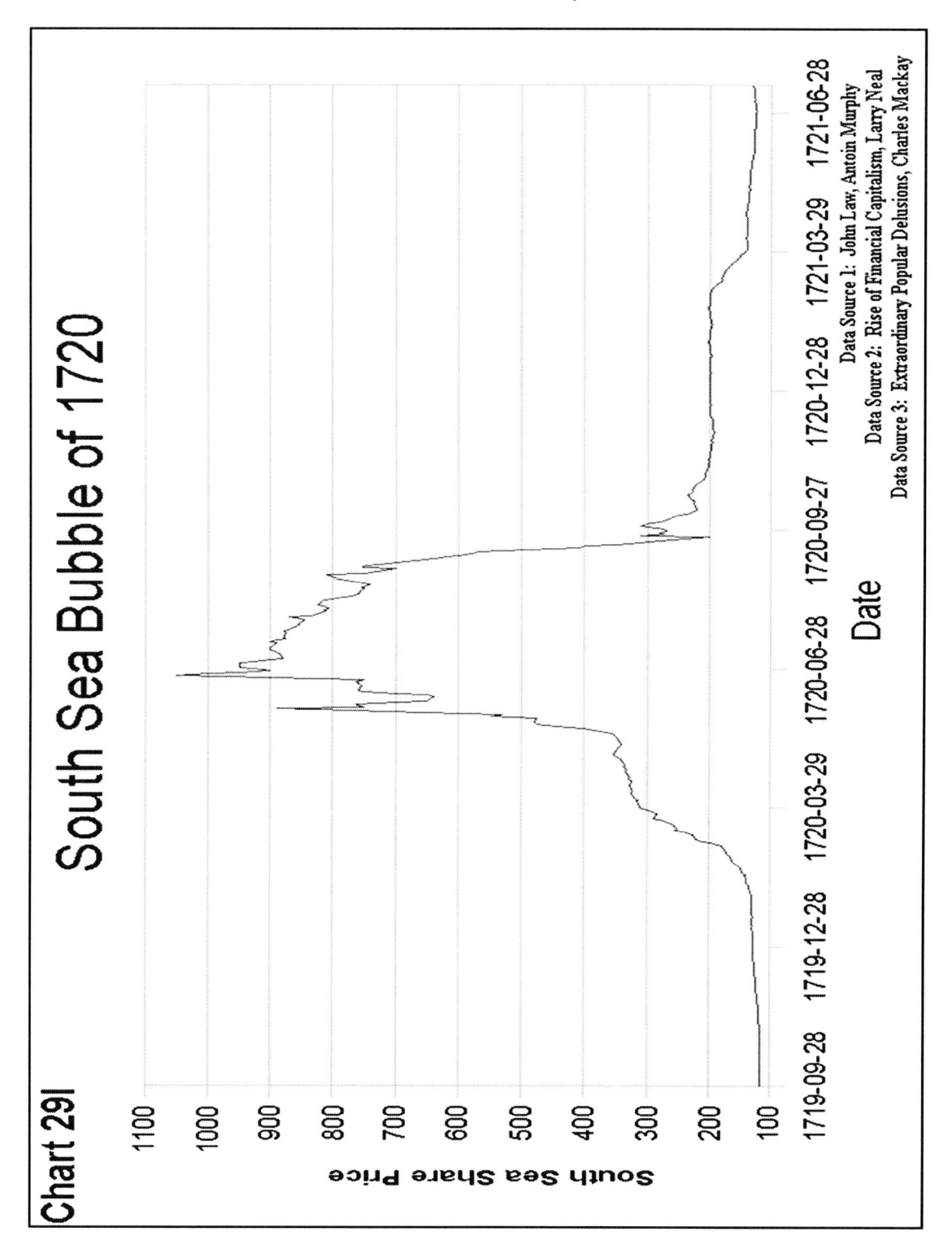

The 19-Year EUWS Cycle

Chart 29I

South Sea Bubble of 1720

South Sea Share Price vs **Date**

Data Source 1: John Law, Antoin Murphy
Data Source 2: Rise of Financial Capitalism, Larry Neal
Data Source 3: Extraordinary Popular Delusions, Charles Mackay

Chart 29J

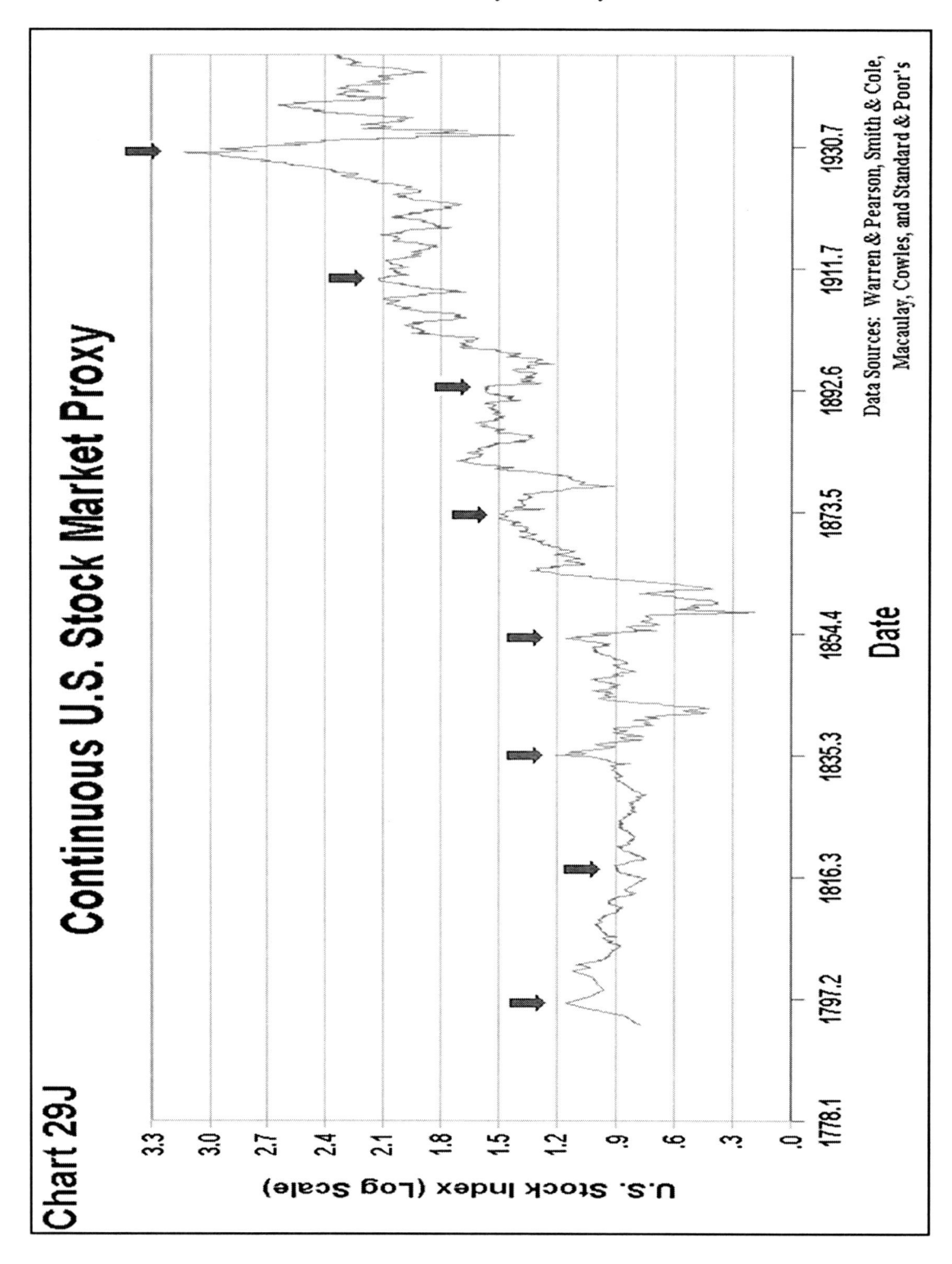

Continuous U.S. Stock Market Proxy

Data Sources: Warren & Pearson, Smith & Cole, Macaulay, Cowles, and Standard & Poor's

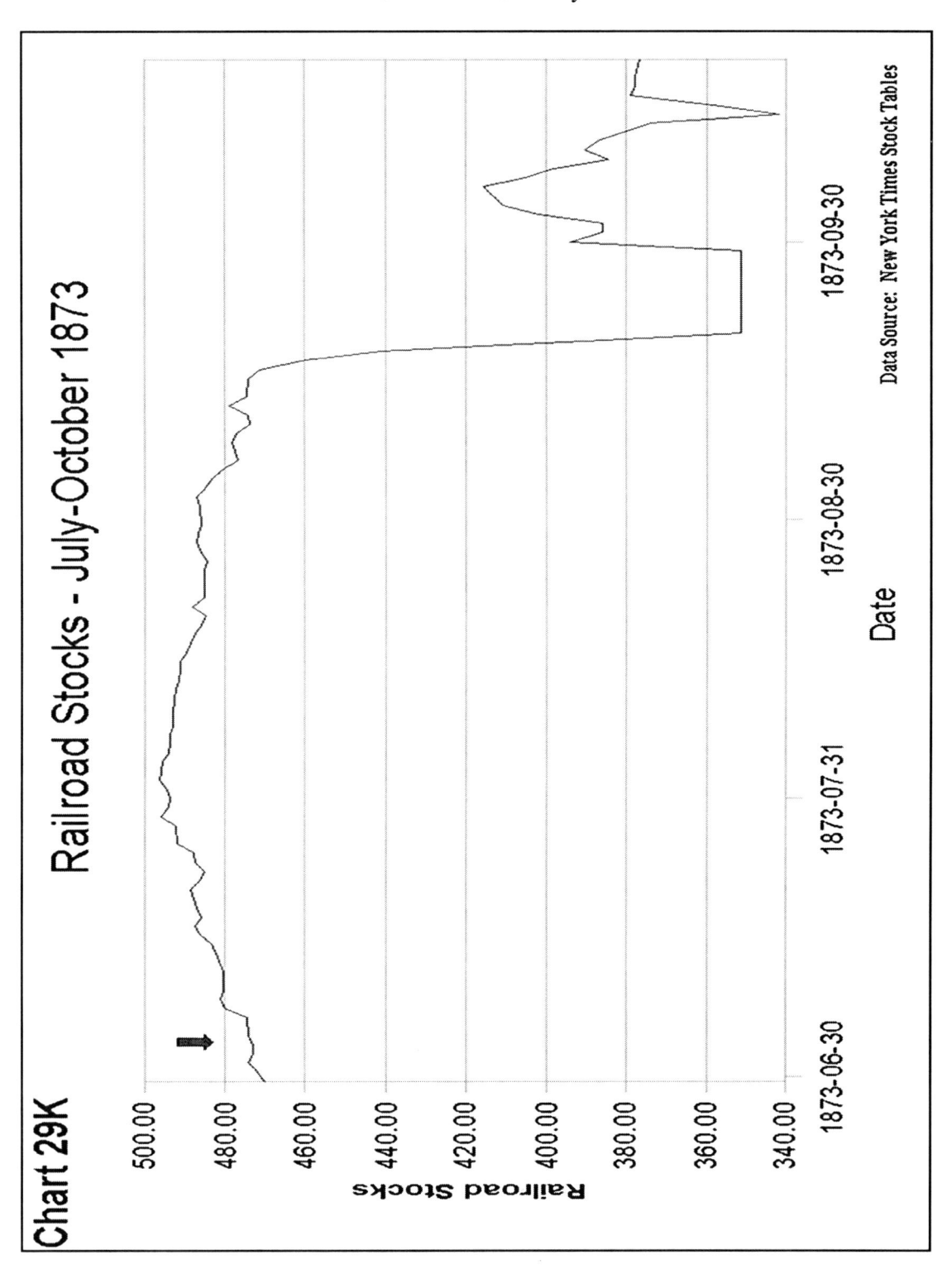

The 19-Year EUWS Cycle

Chart 29K

Railroad Stocks - July-October 1873

Data Source: New York Times Stock Tables

The Unified Cycle Theory

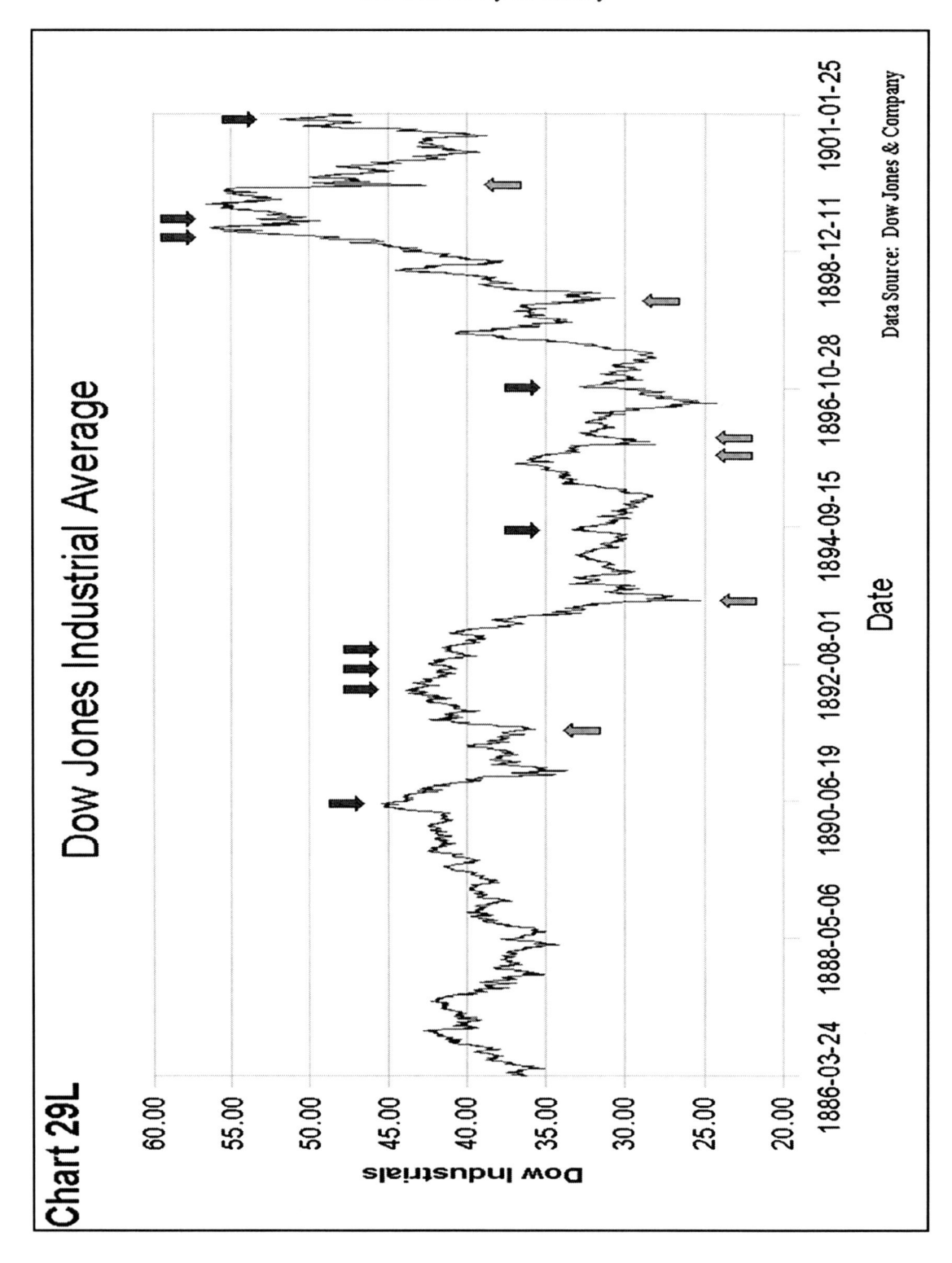

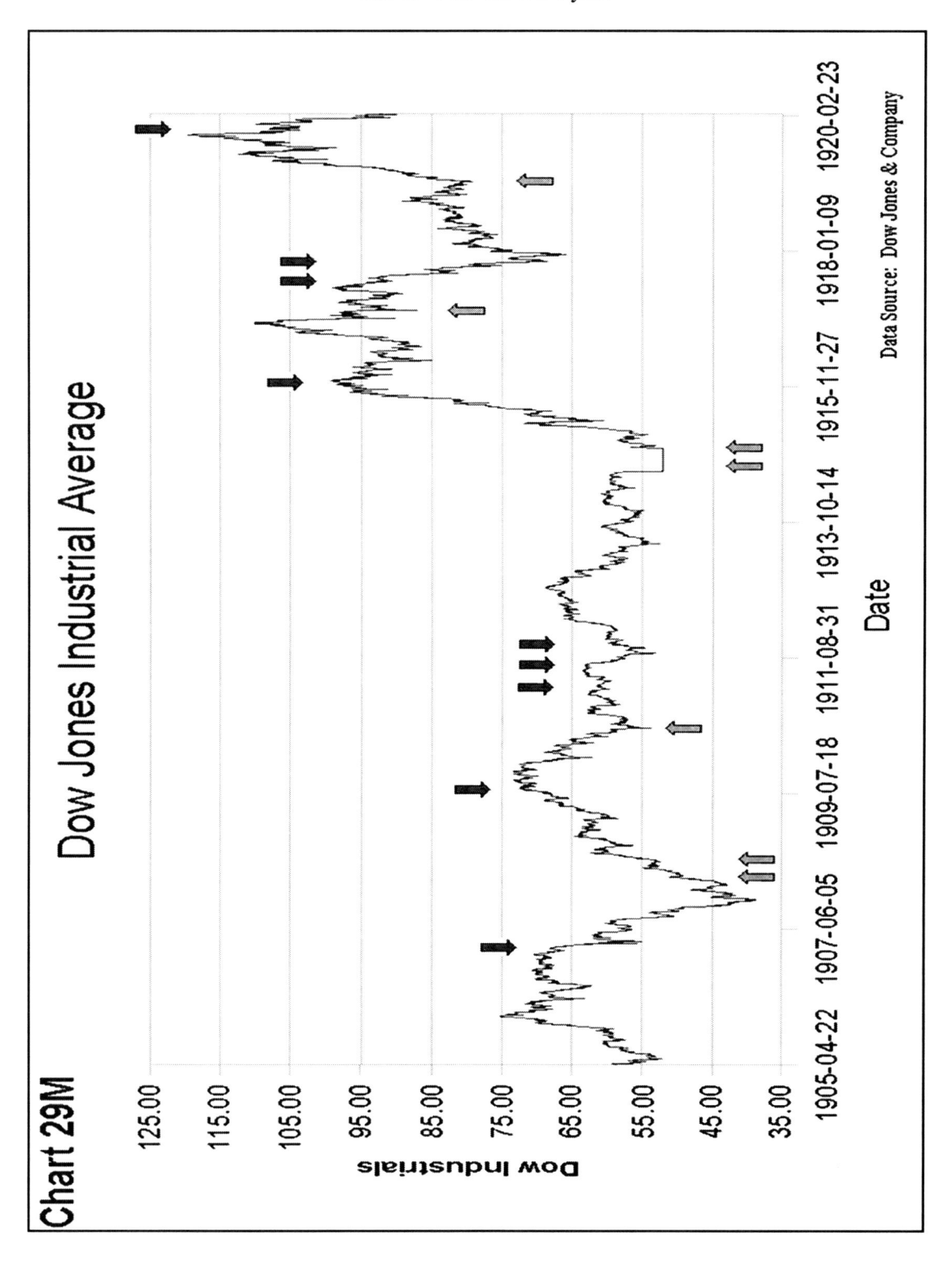

The 19-Year EUWS Cycle

Chart 29M

Dow Jones Industrial Average

Data Source: Dow Jones & Company

Chart 29N
Dow Jones Industrial Average

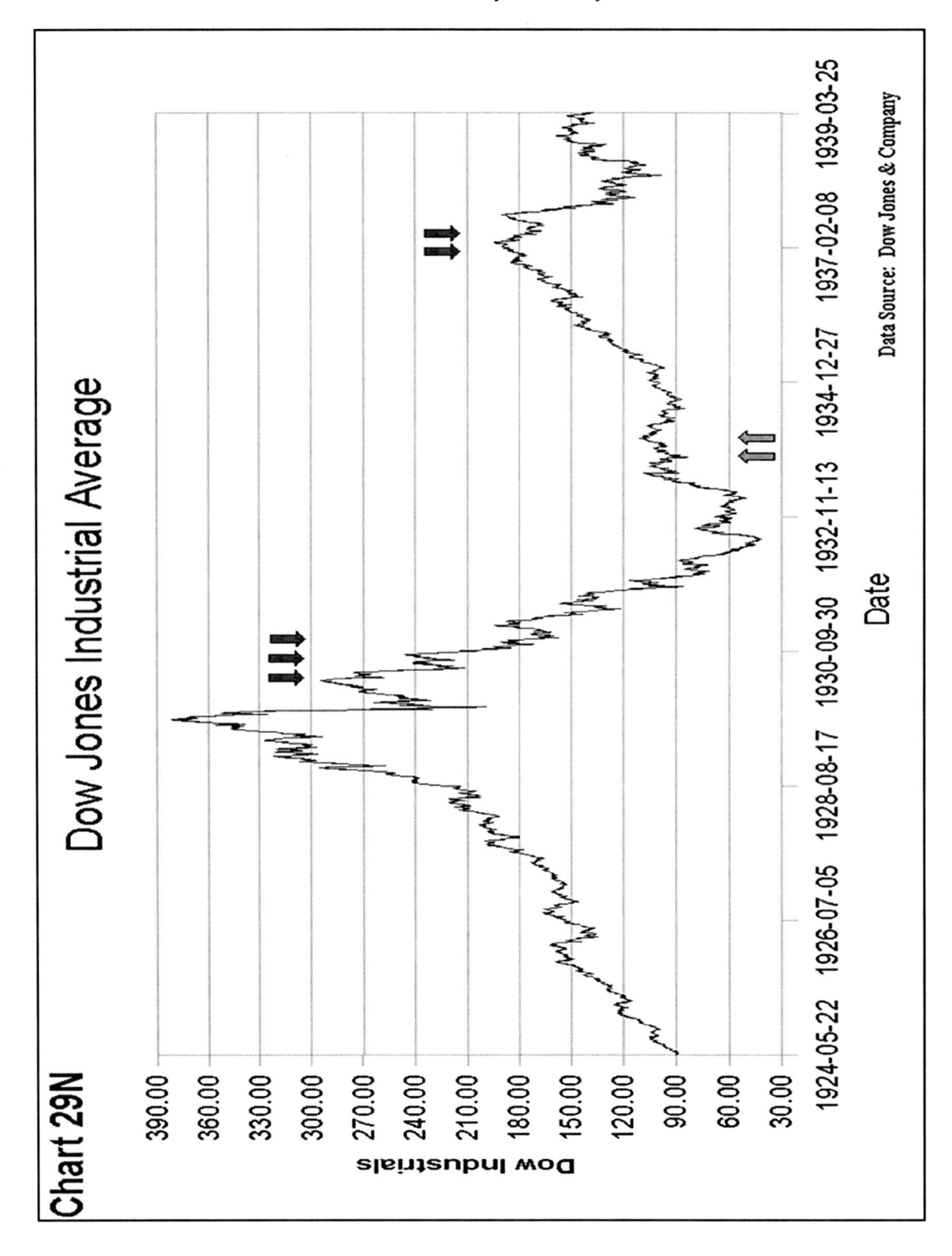

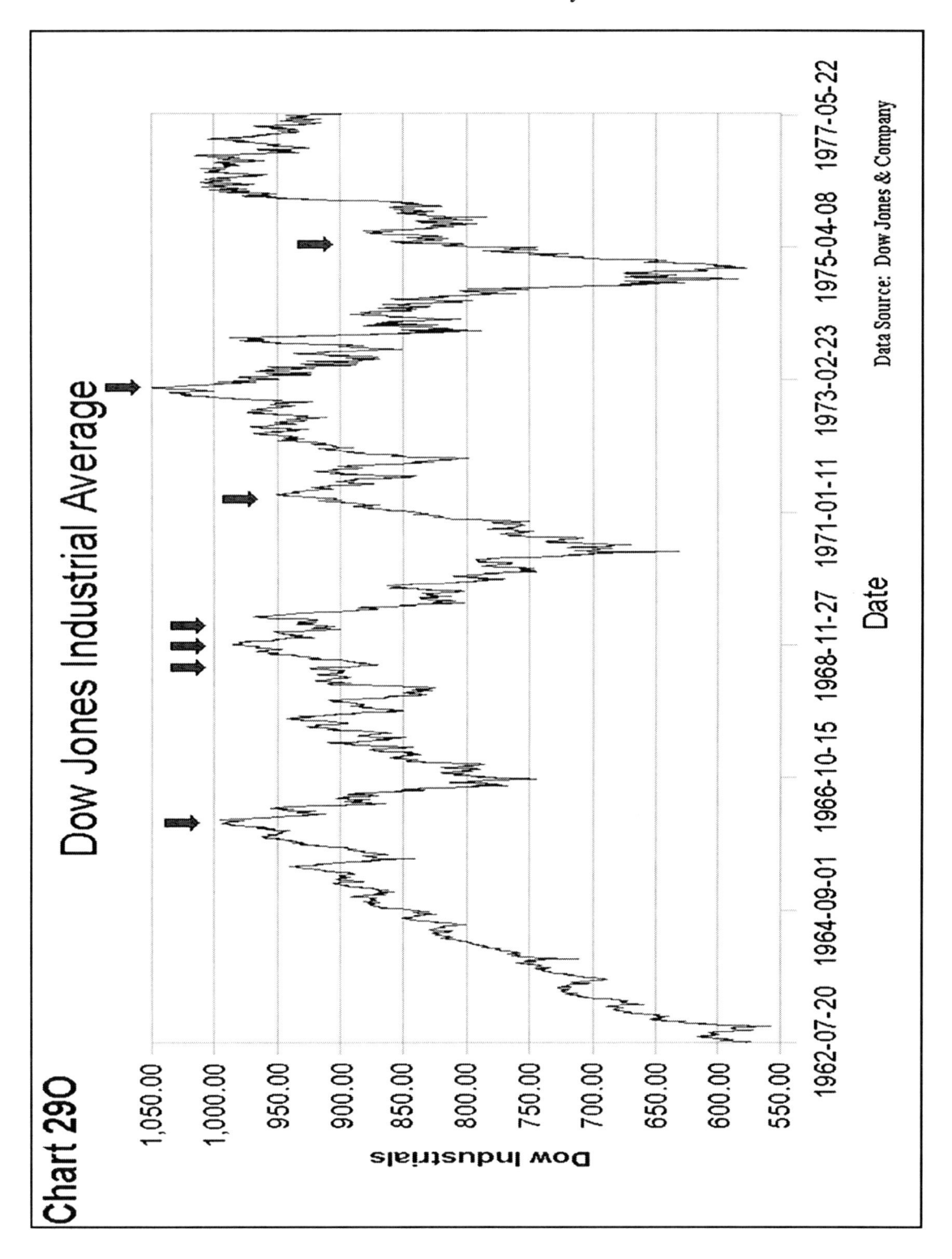

The 19-Year EUWS Cycle

Chart 290

Dow Jones Industrial Average

Data Source: Dow Jones & Company

The Unified Cycle Theory

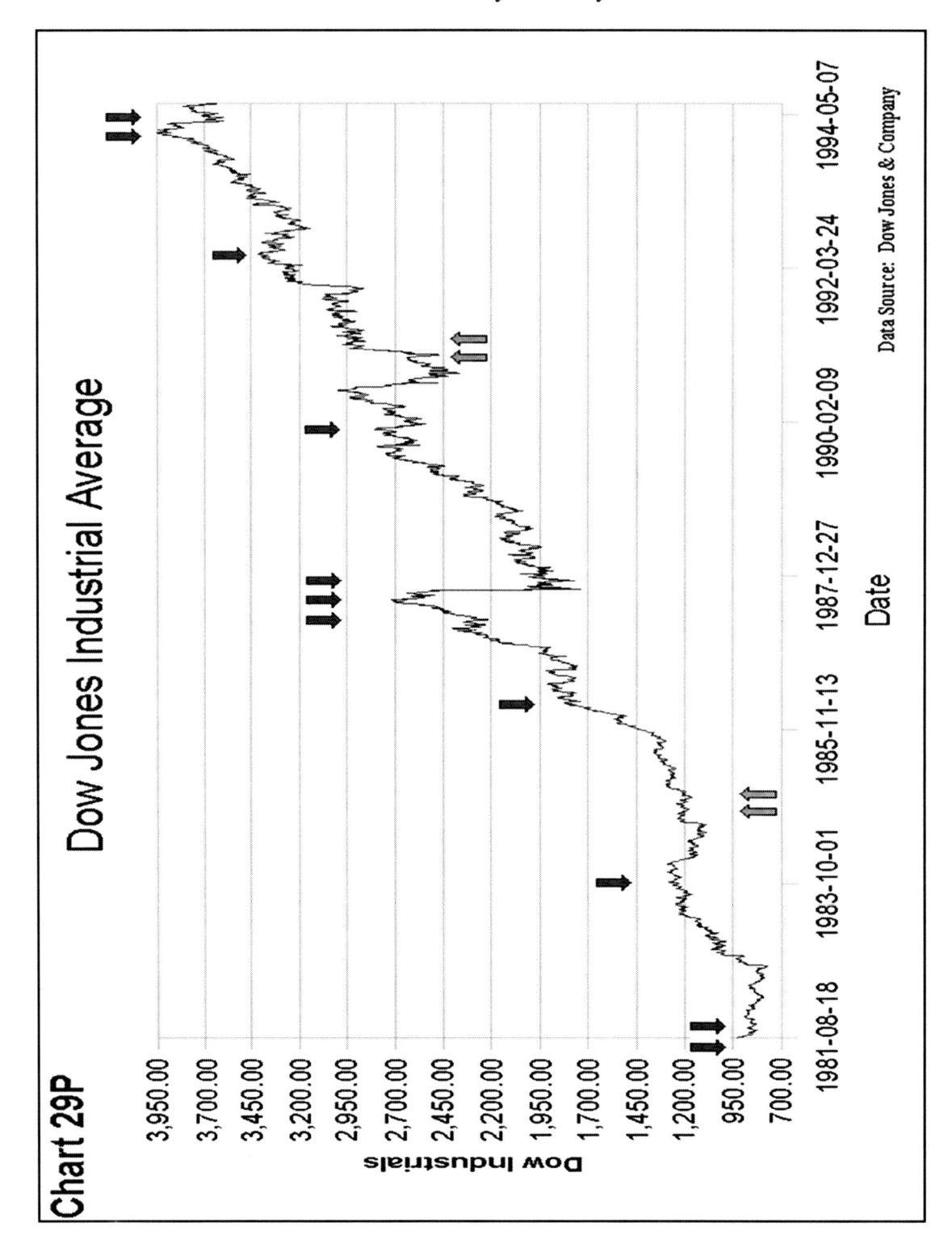

Chart 29P — Dow Jones Industrial Average

Data Source: Dow Jones & Company

388

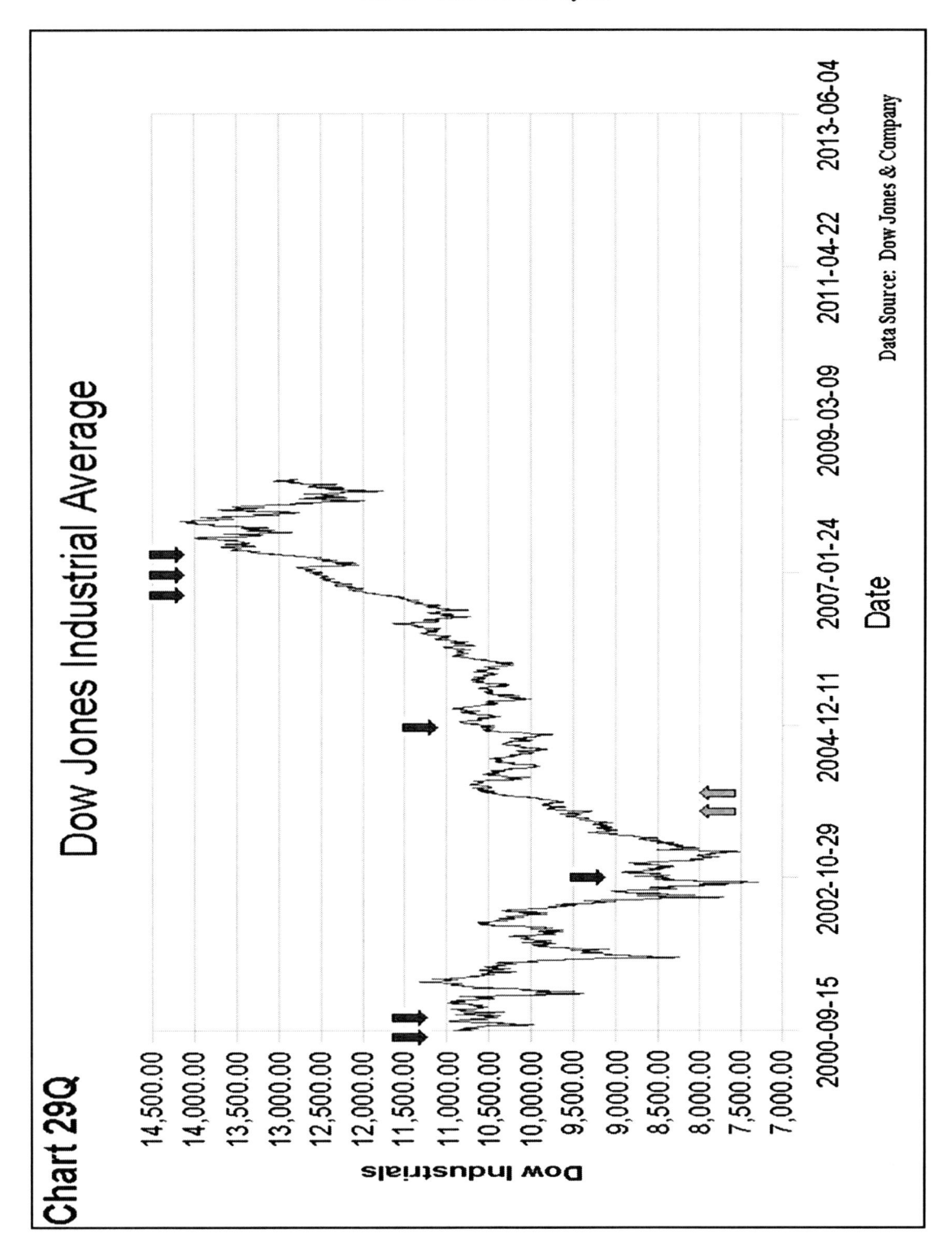

The 19-Year EUWS Cycle

Chapter 30
The 6.36-Year EUWS Cycle

The next cycle in the sequence comes from dividing the 19.08-year EUWS frequency by three – thus producing the 6.36-year cycle. This cycle correlates modestly with economic fluctuations. As previous chapters already showed, in general, commodity price movements have reflected economic conditions for centuries. In particular, review London CPI charts 15B, 15C, and 15D. In the United States, major commodity price movements since 1968 correlated with 6.36-year EUWS fluctuations fairly well.

Commodity price declines tend to act as coincident indicators of economic downturns. However, during strong inflationary periods, commodities can actually lag business conditions a few months. During the last seven cycles, commodity price tops in the USA came roughly at the same time as either the start of a recession or the start of a slowdown in economic growth. These commodity tops came approximately 6.36 years apart. The latest commodity top arrived during July 2008, during the same quarter that GDP started to contract.

From the perspective of the 57-year cycle, commodities normally climax ahead of theoretical peaks. However, for the 6.36-year business cycle the relationship reverses – equities turn first, while commodities lag. Because of its leading indicator status for the 6.36-year cycle, stock market activity becomes the most critical business barometer at this level. Attention now turns in that direction. A review follows for all 6.36-year equity market cycles since 1910.

The Unified Cycle Theory

Chart 30A employs a log scale for the Dow Jones Industrial Average from 1911 to 1962. Keep in mind, at this frequency the 11-year sunspot cycle substantially interferes with EUWS visibility. For that reason, the 6.36-year cycles fails to exhibit the same strength as the other EUWS cycles.

September 1, 1911 Theoretical Peak – For this cycle, the stock market topped in December 1909, 21 months ahead of its theoretical peak. This top also coincided with a 19-year EUWS peak. From the perspective of the more broadly based S&P Index, the stock market began an eleven year bear market at this time. The first part of the bear market ended with heavy selling on July 30, 1914 – associated with the start of World War I. To calm the markets, the NYSE closed the exchange for more than four months. This period of closure precisely coincided with the theoretical low point of the 6.36-year cycle. When the NYSE reopened in December 1914, the market began a two year counter-trend bull move.

January 10, 1918 Theoretical Peak – The war-rally ended during November 1916 – 14 months ahead of the next theoretical peak. However, the advancing portion of the sunspot cycle briefly interrupted the subsequent decline. As a result, the stock market formed a double-top. The first top came in November 1916 – associated with a peak of the 6.36-year cycle. The second top came in November 1919 – associated with a sunspot cycle maximum. After November 1919, with both cycles in sync to the downside, stocks moved lower. The combined negative force carried the long-term bear market to its ultimate low in August 1921. After World War I, strong deflation triggered a moderate depression. Share prices plummeted. All of this happened a few months after the theoretical low point for both the 6.36-year cycle and the 19-year cycle. But the 1921 deflation gave birth to a new bull market. In **Chart 30A**, see the two up-arrows above 1921 – signifying the trough of the 19-year cycle.

May 21, 1924 Theoretical Peak – Similar to the previous cycle, the actual top arrived ahead of the theoretical peak by 14 months. After March 1923, a period of consolidation lasted for more than a year. By June 1924, the uptrend resumed. Three factors contributed to the ensuing bull market. (1) The sunspot cycle turned upward. (2) The 19-year EUWS cycle advanced toward its peak. (3) Credit market stimulus from the newly created Federal Reserve provided speculators with ample funds for buying stocks.

September 30, 1930 Theoretical Peak – The bull market of the 1920s provided the nation with its first glimpse of what happens when a central bank gets credit-happy while responding to a crisis. In the process of fighting the deflation of 1921, the Federal Reserve helped inflate the Roaring '20s stock market bubble. The top came on the first trading day of September – midway between peaks in the sunspot cycle (in 1928) and the 6.36-year cycle (in 1930). In **Chart 30A**, see the two down-arrows above 1930 – signifying the top of the 19-year cycle (and the 6.36-year cycle). During the ensuing bear market, stocks lost 90% of their value. The crash bottomed on July 8, 1932. It arrived more than one year ahead of the theoretical low of the 6.36-year cycle. Then, another bull market began. However, it formed slowly. As the 1933 theoretical low approached, the choppy, sideway action stopped. And a new bull market gained momentum.

The 6.36-Year EUWS Cycle

February 8, 1937 Theoretical Peak – Three factors ignited the bull market of the mid-1930s. First, the 6.36-year cycle turned higher in 1933. Second, the sunspot cycle moved toward its maximum. Third, measured by book values and dividend yields, stocks offered their best bargains of the century. Once momentum turned, the market staged a powerful multi-year advance. However, on March 10, 1937 the Dow Industrials topped at 194.40 – occurring near the time of a sunspot maximum. Additionally, a 6.36-year theoretical peak arrived one month earlier. With both of these factors turning negative in early-1937, another nasty bear market followed. The subsequent decline wasn't as steep as the 1929-32 crash. However, it lasted longer. The bottom finally arrived five years later, on April 28, 1942. In **Chart 30A**, see the two up-arrows by 1940 signifying the bottom of the 19-year and 6.36-year cycles.

June 20, 1943 Theoretical Peak – After the 1942 low, stocks rocketed sharply higher. The stock market went straight up for almost one year. The first significant pullback came on July 14, 1943. Like the 1937 cycle, the actual top came one month from a theoretical peak of the 6.36-year cycle. However, the decline only lasted five months. At that point, after almost 30 years of near-perfect correlation with the 6.36-year cycle, the stock market deviated from expected patterns. After the brief decline, the bull market resumed. And it didn't stop until May 1946. As the theoretical low point of the 6.36-year cycle arrived, the stock market did succumb to a steep decline that lasted six months – until October 1946. But after the sell-off, instead of heading higher, stocks moved sideways for three years. Then, as the next theoretical peak approached, stocks began moving higher again.

October 29, 1949 Theoretical Peak – Similar to the 1943 cycle, market activity differed substantially from EUWS predictions. October 1949 coincided with a panic cycle peak. However, instead of panic, calm prevailed. Perhaps the best description amounts to the cycle simply skipping a beat. As already explained in previous chapters, the lack of cyclic visibility during this period coincided with an absence of financial leverage. And debt acts as a prerequisite for development of a financial panic. Go back and review **Chart 3B**. It shows that in 1949 mortgage debt equaled about 8% of the value of all residential homes – one of the lowest levels of the past century. Also by 1949, corporate debt and financial-sector debts sank toward lows for the century. Without debt to trigger forced selling, the stock market moved higher. Then, by September 1951, a top finally arrived, albeit two years late. However, instead of a sell-off, stocks moved sideways for two years – actually overshooting the theoretical low point by a year. But then, after two cycles of poor correlation, the 6.36-year cycle jumped back on track.

March 9, 1956 Theoretical Peak – Similar to the 1937 and 1943 cycles, the high arrived on April 6, 1956 – only one month after its theoretical peak. A significant bear market followed. But it only lasted one-and-a-half years. During 1958 and 1959, stocks again moved higher. Then another bear market set in. The bottom came about a year after the theoretical low for both the panic cycle and the 6.36-year cycle. As this cyclic bottom passed (which also marked a 19-year low), the market laid the foundation for yet another bubble to inflate in the never-ending series of financial speculations.

July 19, 1962 Theoretical Peak – Before the Go-Go market of the 1960s took hold, one final selloff developed. The stock market topped during November 1961, several months ahead of its theoretical peak. The decline started slowly, but then one month ahead of the theoretical peak, stocks sold off sharply under near-panic conditions during June 1962. Notice the final portion of **Chart 30A**. After building a base near its low, stocks began soaring again by late October 1962. The Go-Go market had begun. See the early portion of **Chart 30B**.

However, **Chart 30B** inadequately reflects the magnitude of the froth that transpired during the 1960s. By late 1968, the Dow Jones Industrial Average had regained everything lost during the 1966 bear market. In the process, the Dow formed a double-top. But **Chart 30C**, depicting the Value Line Index, paints a completely different picture. During the 1960s, secondary issues and low-priced stocks soared much more than the Dow. By late-1968, the Value Line Index rocketed 35% above its 1966 highs. From this perspective, the extensive bull market of the 1960s clearly peaked in 1968 – not 1966, as some argue. When the bull exhausted itself in late November 1968, the apex came within one week of its 6.36-year theoretical peak. In **Chart 30C**, notice the two down-arrows above 1968.

November 27, 1968 Theoretical Peak – The 1968 top also coincided with a 19-year panic cycle peak. As measured by the broad market averages, the bear market that followed was the largest in the post-WWII era. That's true from the standpoint of both time and magnitude. The ensuing bear market lasted almost the entire length of the 6.36-year period. After the market touched bottom in late-1974, it quickly shot higher – never looking back in the process.

April 8, 1975 Theoretical Peak – By mid 1975, the Dow Industrials had already recovered a good portion of the bear market losses. And the Value Line Index started its long road to recovery. However, the recovery slowed after early 1976, and the markets stalled until mid-1978. That marked the theoretical low point for three important EUWS cycles – those of 6.36 years, 19 years, and 57 years. In **Chart 30C**, notice the triple up-arrows at 1978. At that point, the greatest bull market in the history of the USA began. Except for brief selloffs that usually lasted less than a year, the stock market went virtually straight up for more than thirty years!

August 17, 1981 Theoretical Peak – The next interruption to the bull came during April 1981. That happened four months from its 6.36-year theoretical peak in August 1981. A mild bear market followed, ending during August 1982. The bull market reached new highs throughout 1983, followed by another mild downturn in 1984. The 1984 low point came close to the theoretical low of the 6.36-year cycle. In **Chart 30C**, see the up-arrow at 1984. After the 1984 low, stocks soared uninterrupted for three years.

December 27, 1987 Theoretical Peak – The next interruption also came four months ahead of a 6.36-year theoretical peak. After climaxing during August 1987, by late-October 1987 all major stock market averages had crashed by nearly 50%. Fears quickly spread of another economic depression engulfing the country. Kondratieff followers warned that the 58-year interval since the last great crash signaled the start of another depression. However, this time, neither the market nor the economy followed the path taken six decades earlier. In 1929, the economy deviated by sinking into depression one 19-year cycle ahead of schedule. In 1987, both the market and the economy quickly recovered from the crash. By doing so, the economic depression cycle reverted back to its normal theoretical peak – projected for January 2007. In the process, the economic recovery totally confounded Kondratieff followers. For the investing public, the majority believed that government and the Federal Reserve had finally learned how to control the business cycle. The relatively quick recovery from the 1987 crash emboldened investors with confidence that the Federal Reserve would always be there when needed. At that point, the Fed's interventions unintentionally gave birth to the moral hazard of intensified speculation and risk-taking. As the down-phase of the 6.36-year cycle ended, the stock market started to boom higher again in January 1991. In **Chart 30C**, see the up-arrow at 1991.

May 7, 1994 Theoretical Peak – From the early-1991 valley, the bull market resumed its uptrend until January 31, 1994. That happened about three months ahead of the next 6.36-year theoretical peak. From there, the market stalled for a year before zooming higher for two more years. The Southeast Asian currency crisis brought global stocks down sharply in 1997 – at the time of a theoretical low of the 6.36-year cycle. After the LTCM panic in 1998, stock market speculation became heavily concentrated in technology shares. The Internet Bubble inflated massively during 1999 and early 2000.

September 15, 2000 Theoretical Peak – As credit became increasingly easier during the '70s, the '80s, and then the '90s, a string of ever larger bubbles formed. First, the commodity market bubble of the 1970s burst just ahead of the 6.36-year peak in 1981. Then, the stock market bubble of the mid-1980s burst a few months before the 6.36-year peak in 1987. Next, leveraged trades in foreign currencies came crashing down after the 6.36-year peak in 1994. This started with debt crises throughout Latin America in 1994-95, and then spread to Southeast Asia in 1997, and finally ended with the Russian currency and bond market crash in 1998. During September 1998, the collapse of Long Term Capital Management served as the culminating event in this series of revolving currency crises. Even with credit conditions already easy, the Federal Reserve's policy became easier still. As one bubble deflated, two more sprang up – thus snaring in an ever increasing number of speculators. Technology stocks became the next Fed-inspired bubble. By March 2000, technology shares became so over-valued that the NASDAQ Composite Index lost 75% of its value over the next three years. However, even after the initial April 2000 selloff, technology stocks partially recovered, and the broader market fully recovered until September 6, 2000. That final top during September 2000 arrived nine days ahead of a 6.36-year EUWS theoretical peak. After the theoretical peak passed, all sectors of the market cratered, and the bear market of 2000-2002 ensued. As always, another bubble waited in the wings. This time, the bubble centered on structured finance and the housing market.

January 24, 2007 Theoretical Peak – In addition to structured finance and housing, the 2003-2007 credit expansion carried a wide range of bubbles in its swath. Bubbles spread to equity markets in developing countries, commodity markets, junk bonds, secondary stocks, Treasury bonds, and derivatives. Anywhere and everywhere, bubbles inflated. By the start of 2007, the granddaddy of all bubbles had formed. Enticed by super easy credit from the Federal Reserve and European central banks, and enhanced by Wall Street structured finance, the international flow of dollars extended to every nation around the globe. The financial party lasted until a critical turning point arrived on January 24, 2007. Not only did that date coincide with a 6.36-year theoretical peak, it marked the turning point for the 19-year panic cycle, the 57-year economic depression cycle, and the 172-year minor civilization cycle. On cue, the subprime crisis erupted and a downturn in housing prices began immediately. Between August 2007 and October 2008, repeated credit market lock-ups required a series of unprecedented central bank innervations, as well as budget breaking federal bailouts.

Immediately after the January 2007 downturn began, the Federal Reserve lost control of its great three-decade credit expansion. By making credit excessively easy during this portion of the 172-year cycle's up-phase, every sector of the USA economy became heavily indebted during the good times. Unfortunately, with the hard times just beginning, liquidity sits at historically low levels – just the opposite of what good planning would normally dictate.

With the start of a 172-year depressionary down-phase, a great unwinding must take place. Pushing more loans onto overly indebted consumers and businesses simply won't work. In this environment, the 6.36-year cycle will still make its presence felt. However, down-phases will be devastating, while 6.36-year up-cycles will be weak. In other words, the stock market should now follow the mirror image of what happened during the up-phase over the past thirty years.

The *Unified Cycle Theory* postulates that various physical cycles affect the way humans feel. As such, markets, cultures, and civilizations fluctuate to reflect these changing attitudes. Yet, markets and civilizations represent indirect measurements of human attitudes. They serve as good indicators because of their longevity. However, markets and civilization oscillations act as proxies for human sentiment trends – not true, direct measures.

Within the past fifty years, better, more direct measures have become available. Direct measurements include stock market surveys from Investors Intelligence, Market Vane, and Consensus. And polls from the University of Michigan, the Conference Board, and ABC News provide direct measurements of consumer attitudes.

Unfortunately, because these direct measures only came into existence within the last fifty years, they fail to provide the longevity needed to study how EUWS cycles affect human behavior. Nonetheless, as shown in **Chart 30D**, fluctuations in the 200 Week Average from *Investors' Intelligence* match predicted EUWS cycles even better than stock prices.

Compare **Chart 30D** with **Charts 30B and 30C**. In particular, notice how major bullish peaks (1968, 1987, and 2007) and major bearish troughs (1975 and 1996) occurred around the time of theoretical turning points of the 19-year EUWS cycle. Smaller swings in bullishness occurred in

The 6.36-Year EUWS Cycle

between, corresponding to 6.36-year turning points. Also notice how bullish opinions reached record levels during the weeks surrounding the 172-year theoretical peak on January 24, 2007.

Many ways exist for measuring human feelings, emotions, and sentiments. As lengthy histories build for these more direct means of measurement, analyzing EUWS cycles will become easier.

Chart 30A — Dow Jones Industrial Average

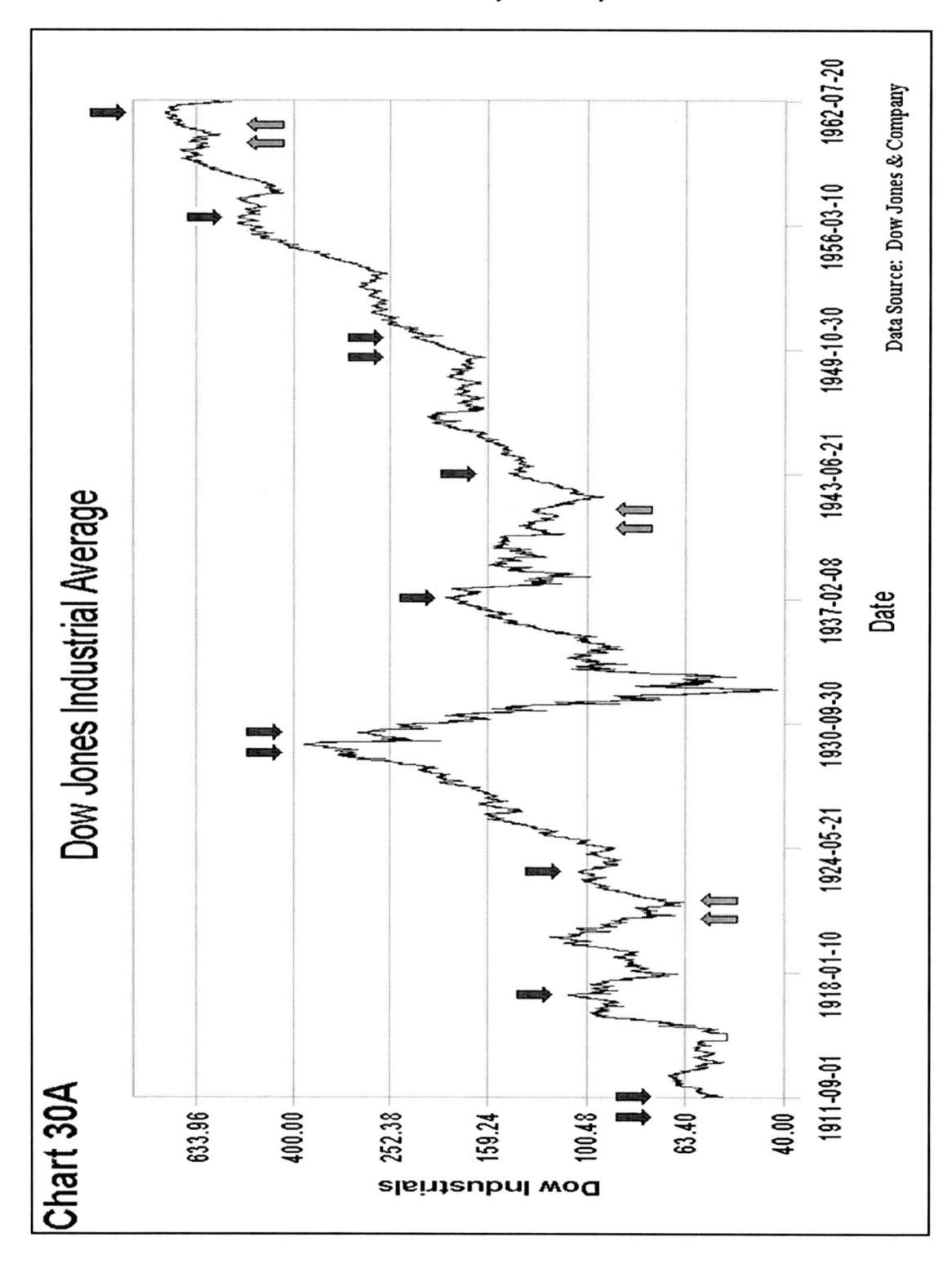

Data Source: Dow Jones & Company

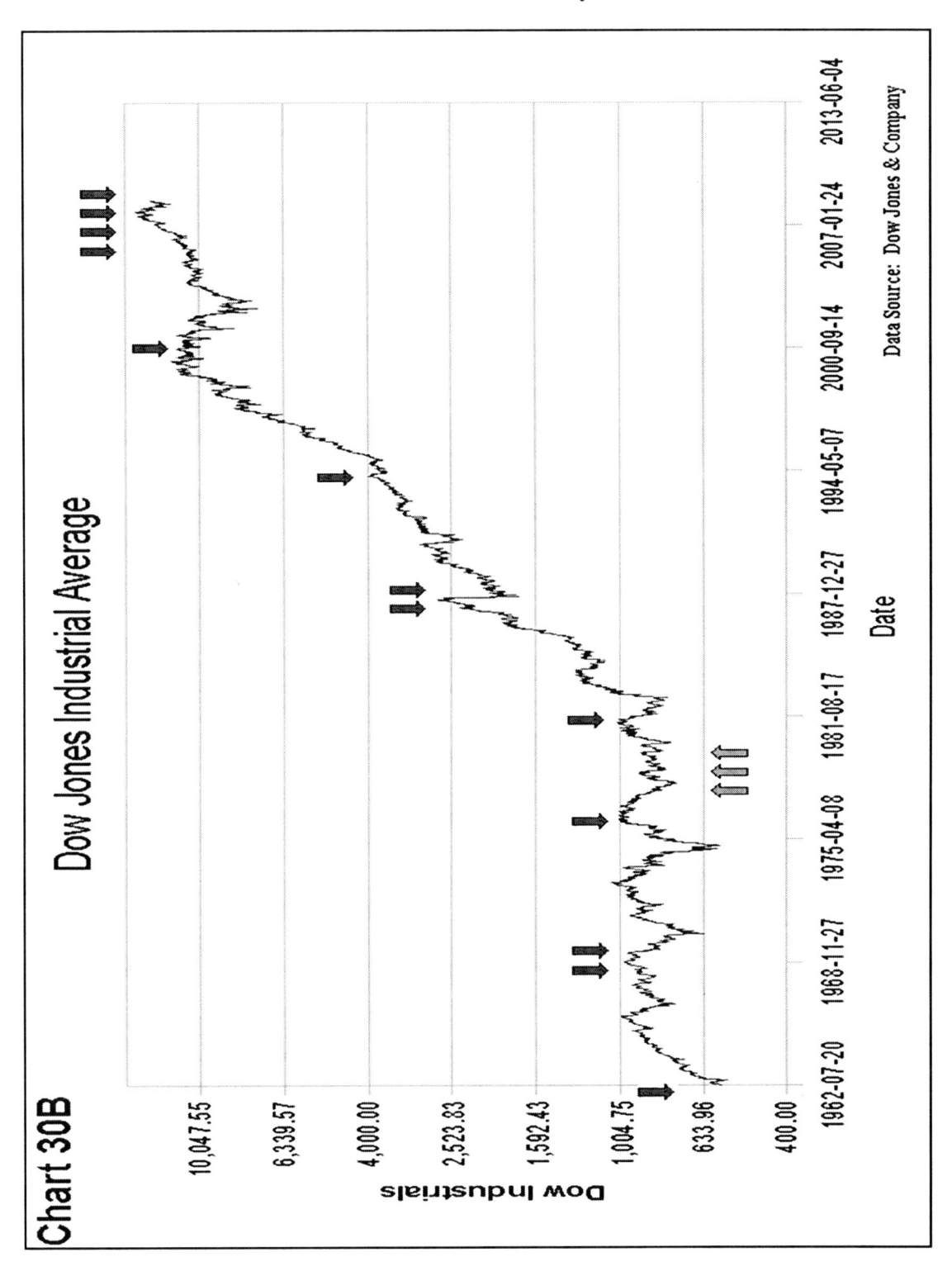

The 6.36-Year EUWS Cycle

The Unified Cycle Theory

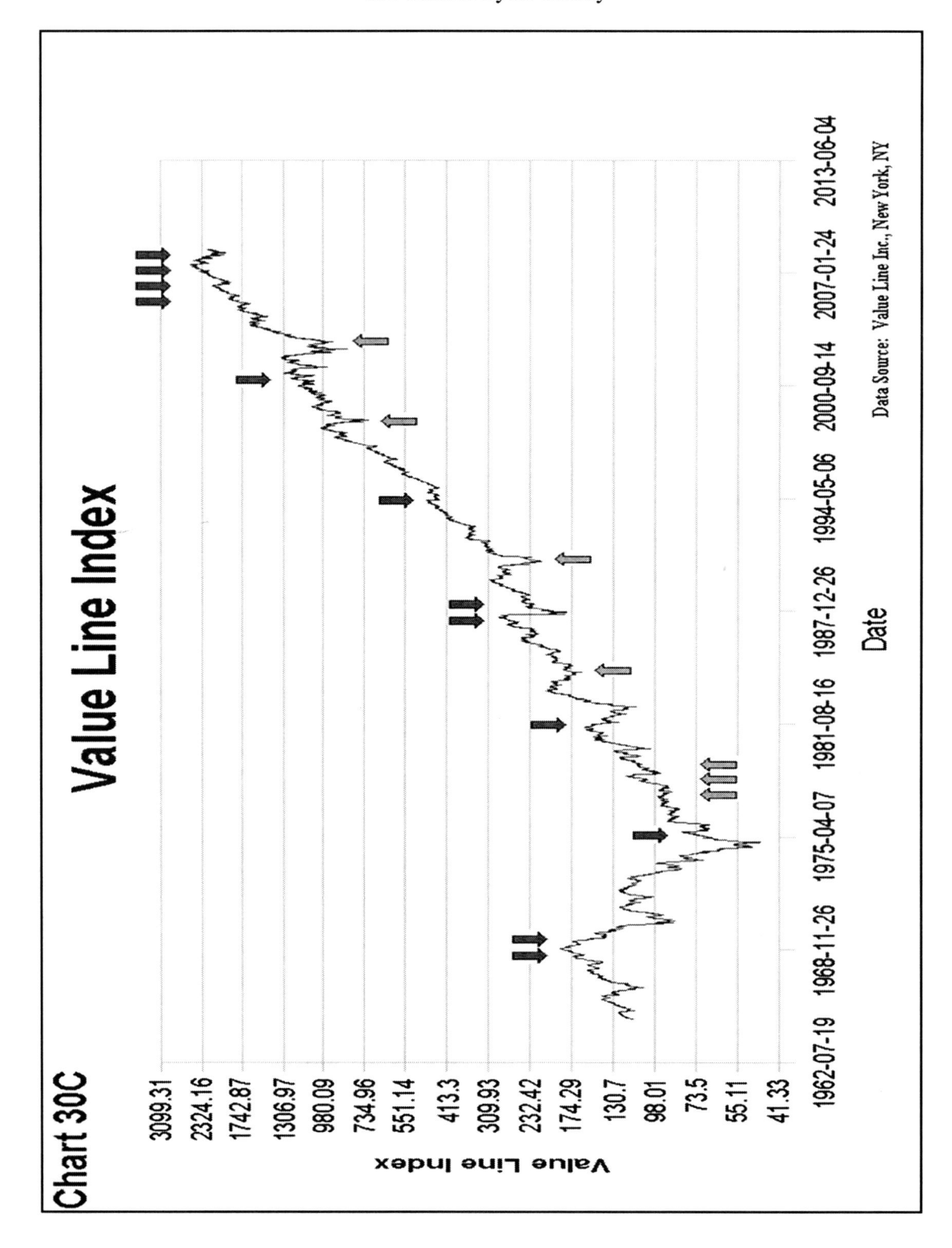

Chart 30C
Value Line Index

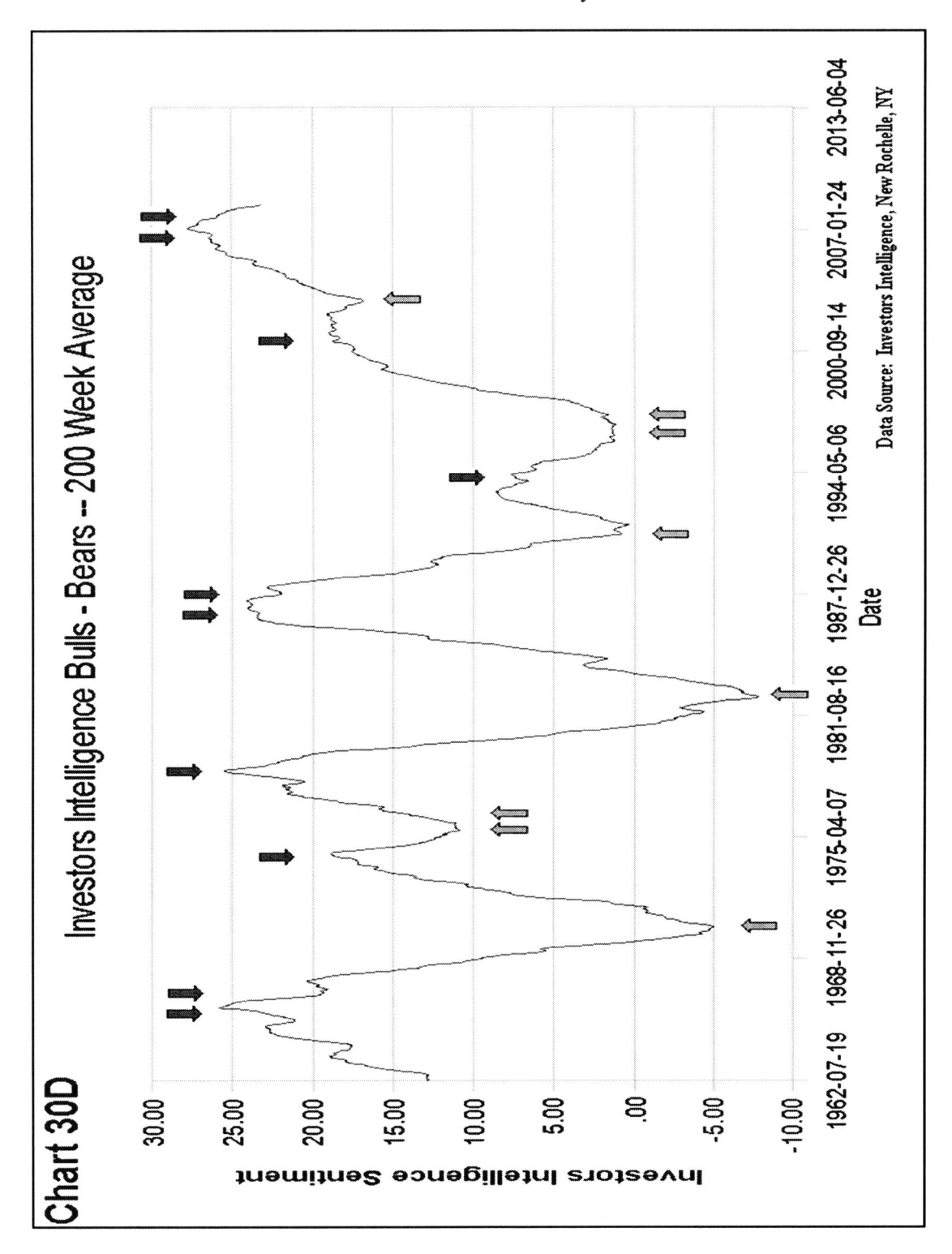

The 6.36-Year EUWS Cycle

401

Chapter 31
The 2.12-Year EUWS Cycle

Dividing by three, the 6.36-year cycle produces the 2.12-year EUWS cycle. This cycle reveals itself through intermediate-term stock market oscillations. Similar to other EUWS cycles under 19 years, the 2.12-year cycle behaves perfectly sometimes and erratically on other occasions. Nonetheless, whenever deviations occur, the fluctuations eventually return to theoretical turning points with good precision.

The list below provides dates for recent theoretical peaks of the 2.12-year cycle.

Feb 8, 1990	June 19, 1996	Oct 29, 2002
Mar 23, 1992	Aug 2, 1998	Dec 11, 2004
May 7, 1994	Sept 15, 2000	Jan 24, 2007

Chart 31A shows activity during a choppy period for stocks. As a general rule, choppy markets enhance the visibility of EUWS cycles. During the 1934-1949 timeframe, that certainly held true for the 2.12-year cycle.

During this fifteen year period, market tops and bottoms spaced themselves fairly evenly – about 2.12 years apart. Five of the six market lows came close to their theoretical low points. And three of the six tops matched their theoretical highs. For all three deviations, double tops formed.

The Unified Cycle Theory

In **Chart 31A**, notice the double-tops surrounding the theoretical peaks during March 1939, May 1941, and September 1947. This type of pattern, with two high points surrounding a mild trough at the theoretical peak, appears on occasion for all of the EUWS cycles. A moving average could be employed to remove these mild valleys, thus embellishing the cycles' aesthetics. With a moving average, actual tops and theoretical tops would have matched six for six.

However, in some cases, aesthetics can obscure knowledge. An important part of the *Unified Cycle Theory* relates to explaining all EUWS deviations. And moving averages hinder that effort.

A closer examination of the tops during March 1939, May 1941, and September 1947 reveals two 258-day sub-cycle highs surrounding each side of the 2.12-year theoretical peaks. Except for their different frequencies, this formation exactly matched what happened with the 57-year cycle in 1949. Specifically, two important 19-year sub-cycle tops (1930 and 1968) sandwiched a valley at the 1949 theoretical peak of the 57-year cycle.

Why cyclic tops occasionally split in this manner remains a mystery. While the TPD Principle accounts for these sub-cycle deviations, the precise reason for splits must be deferred to future research.

Chart 31B shows another interesting deviation. The period from 1949 to 1964 spanned the first phase of the great post-WWII bull market. Put bluntly, none of the EUWS cycles, including the 2.12-year cycle, performed well during this period.

During lengthy bull markets, EUWS cycles respond differently than during choppy markets. For the period between 1949 and 1964, the bull market deviated in the following ways.

First, all 2.12-year down-phases consisted of very brief declines. Declines only lasted a few weeks to a few months – instead of the expected span of slightly more than one year.

Second, instead of occurring midway through the cycle, the market established lows close to its theoretical peaks, or shortly thereafter. In essence, the highs and lows occurred at about the same time – near theoretical peaks.

Third, the cycles compressed together in bunches around 1952, 1956, and 1960. Previous chapters already discussed explanations – lack of leverage reduced cyclic visibility and interference from sunspot cycles.

Connecting 20th Century sunspot data with estimates from NOAA's Paleoclimatology Program reveals that, during the 1950s, sunspot activity reached its highest level in over 10,000 years! In recent times, the only previous sunspot maximum that approached this intensity occurred 171 years earlier, in 1789.

The 2.12-Year EUWS Cycle

Very possibly, record-breaking sunspot activity during the 1950s greatly disrupted operation of all shorter-term EUWS cycles. Regardless of the reasons, by the late 1960s, the bull market ended. And another lengthy period of choppy markets began. Along with it, normalcy returned to the 2.12-year oscillations. See **Chart 31C**.

Throughout this period, only four significant deviations occurred. Those deviations came when (1) the 1966 top arrived one sub-cycle early, (2) the 1974 low came one sub-cycle late, (3) the 1975 low advanced one sub-cycle early, and (4) the 1977 top arrived one sub-cycle early. The other nine turning points matched theoretical projections almost perfectly. After the well-behaved 1964-1979 period, another lengthy bull market began. And similar to 1949-1964, the bull move hampered EUWS visibility once again. Review **Chart 31D**.

Even with a strong bull market, the actual tops developed quite regularly – consistently hitting theoretical 2.12-year peaks very closely. The deviations resulted from shortened downturns. In fact, after the 1987 peak, the ensuing crash only covered two months. Immediately after the devastating panic, the bull market slowly regained its footing and marched higher for another 19 years.

Chart 31E shows the same bull market characteristic – brief downturns developing shortly after theoretical peaks of the 2.12-year cycle. That pattern continued until September 2000, when the technology crash briefly interrupted the long-term uptrend. As the '90s came to an end, the bull market reversed in three phases. Phase 1 came when the Dow Jones Industrial Average topped in January 2000 – one 258-day sub-cycle ahead of the theoretical peak of the 6.36-year cycle. Phase 2 came when NASDAQ and the Internet stocks climaxed in March 2000 – two 86-day sub-cycles ahead of the theoretical peak of the 6.36-year cycle. Phase 3 came when the rest of the markets topped on September 6, 2000, one week ahead of the theoretical peak of the 6.36-year cycle.

After the September 2000 top, normal behavior returned to the 2.12-year cycle. The latest top formed during October 2007, one 258-day cycle after the theoretical peaks of the 172-year cycle in January 2007! If the market continues to follow EUWS patterns, the developing bear market will turn into something unprecedented in the relatively brief 300-year history of financial markets.

Equity markets had never reached such extreme levels of leverage and valuation. For example, dividend yields on the Dow Jones Industrials stayed below 3% for the entire decade prior to the 2007 top. In contrast, throughout much of the 20th Century, equities typically yielded anywhere from 4% to 8%. Coming at the time of a theoretical peak of the 172-year cycle, this mixture of leverage and over-valuation will likely prove lethal. In this deleveraging environment, the longer-term EUWS cycles may obliterate visibility of 2.12-year up-phases. Expect downturns to be steep and lengthy, with upturns appearing weak and brief.

As mentioned in the last chapter, stock market cycles represent an indirect measurement of

EUWS oscillations. To view a more direct gauge, **Charts 31F and 31G** show a 20 week average of market sentiment from *Investors' Intelligence*. These charts show the 2.12-year EUWS cycle slightly clearer than major stock market indices.

Up to this point, all EUWS cycles involved multi-year frequencies. Now, the analysis moves down to EUWS cycles in the sub-annual category.

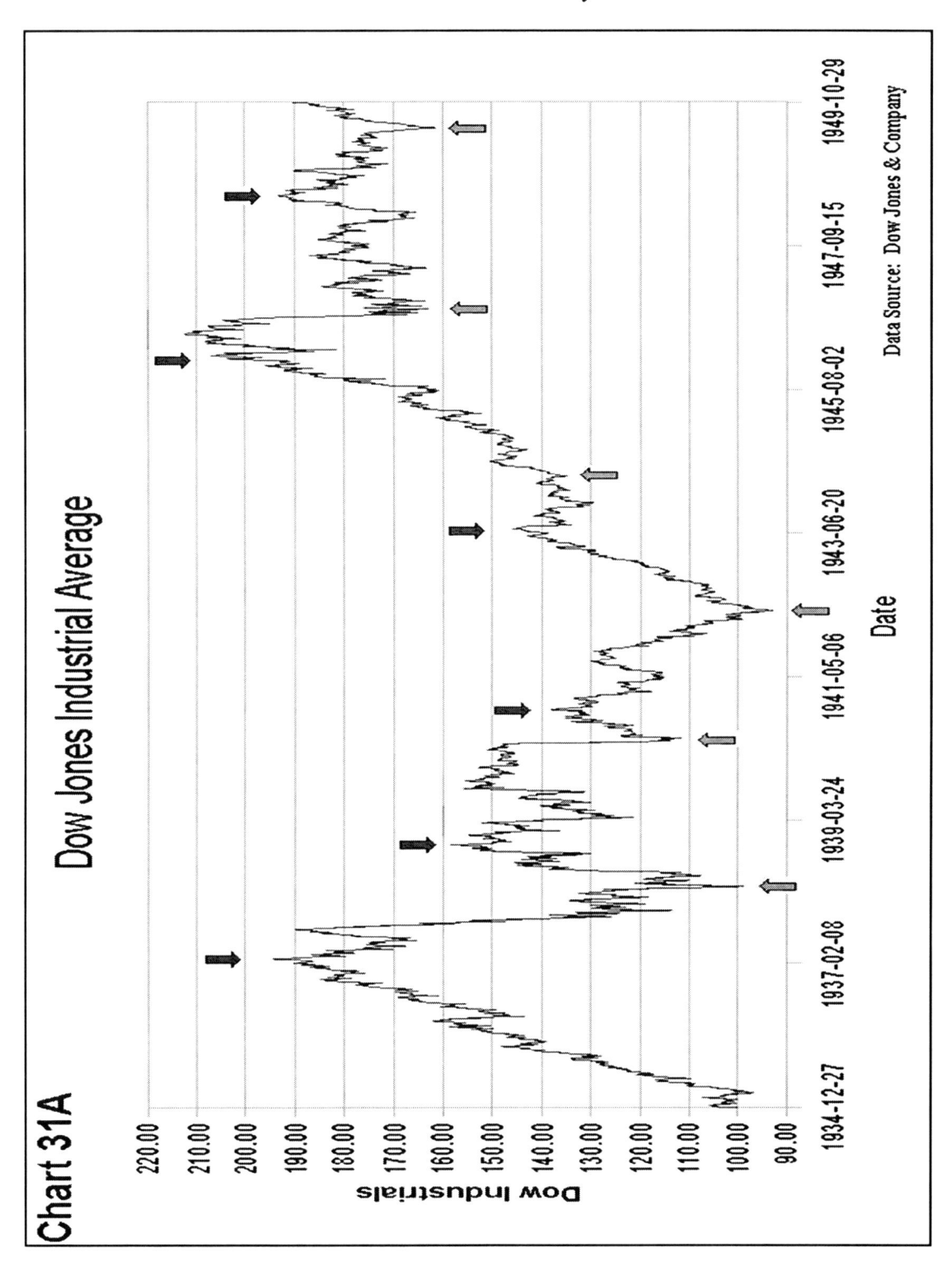

The 2.12-Year EUWS Cycle

Chart 31B

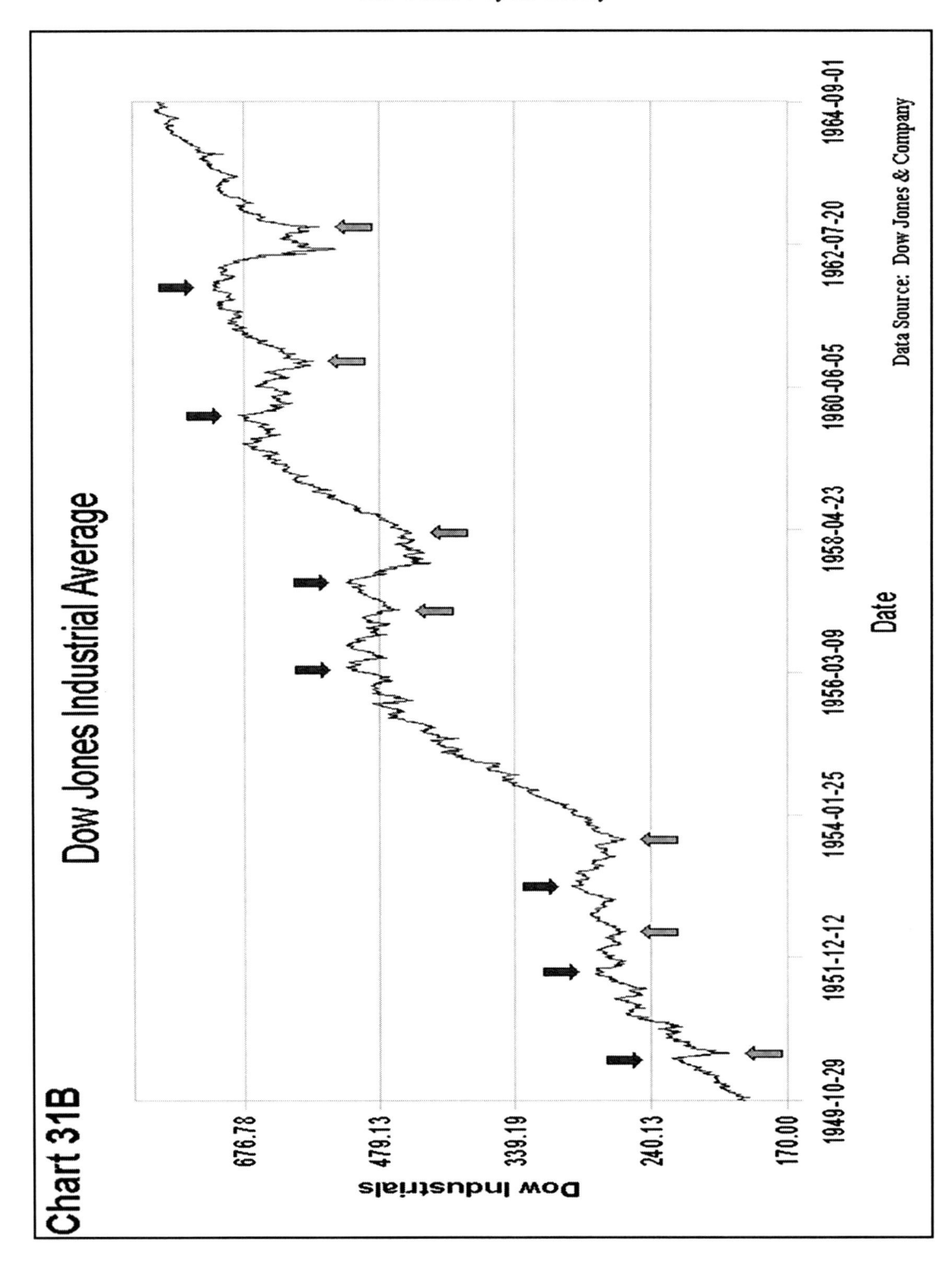

Dow Jones Industrial Average

Data Source: Dow Jones & Company

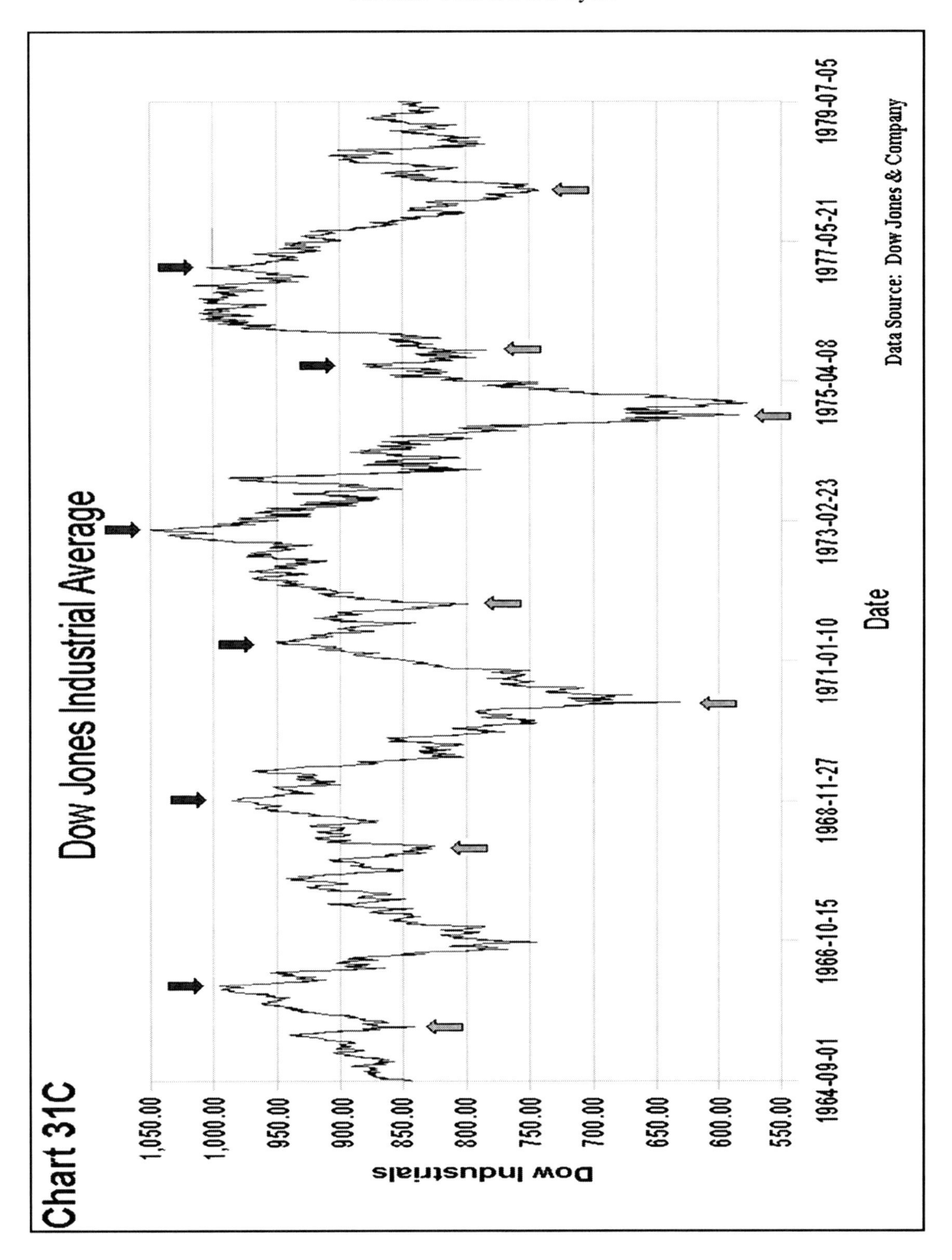

The 2.12-Year EUWS Cycle

The Unified Cycle Theory

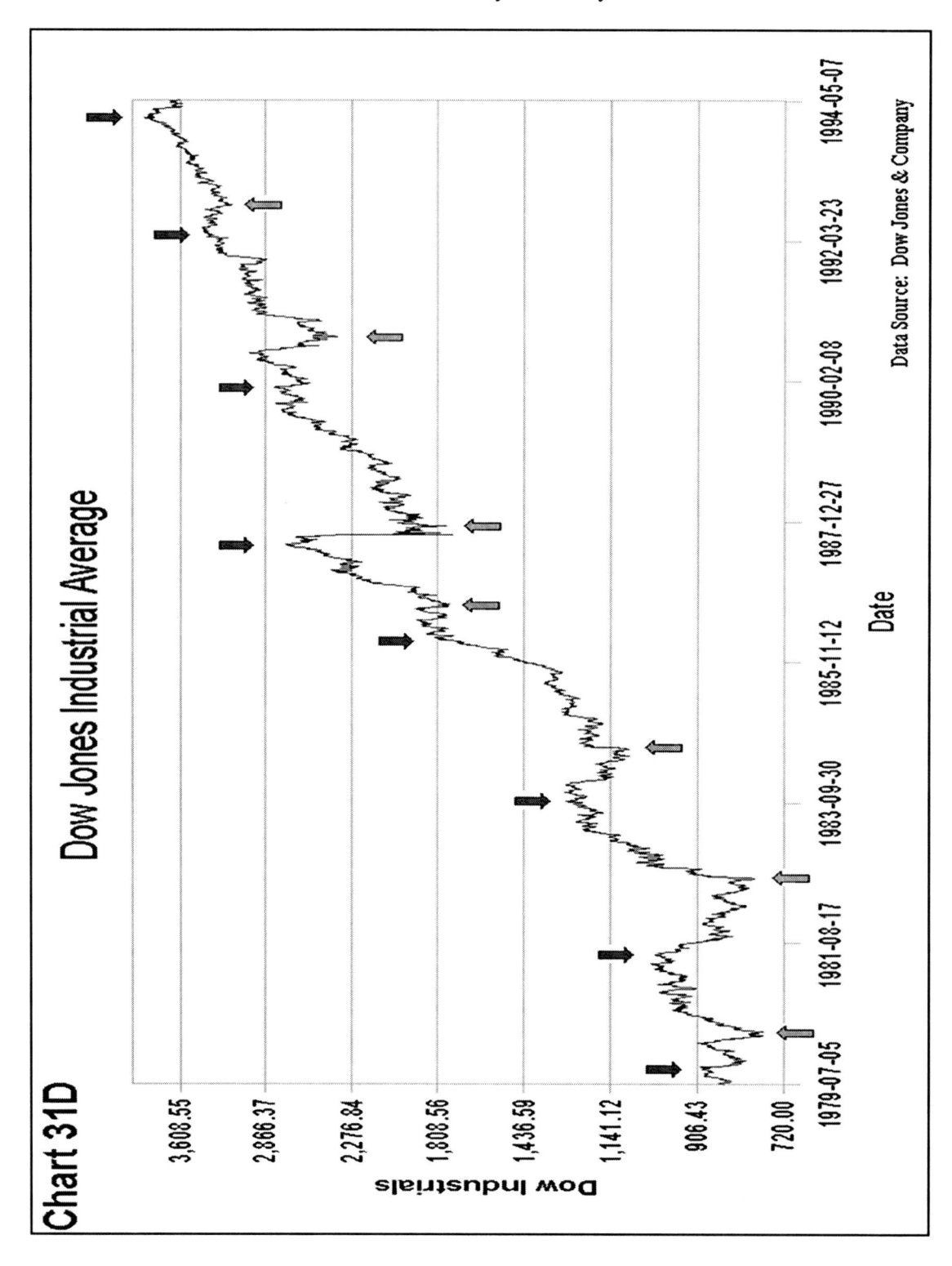

Chart 31D — Dow Jones Industrial Average

Data Source: Dow Jones & Company

410

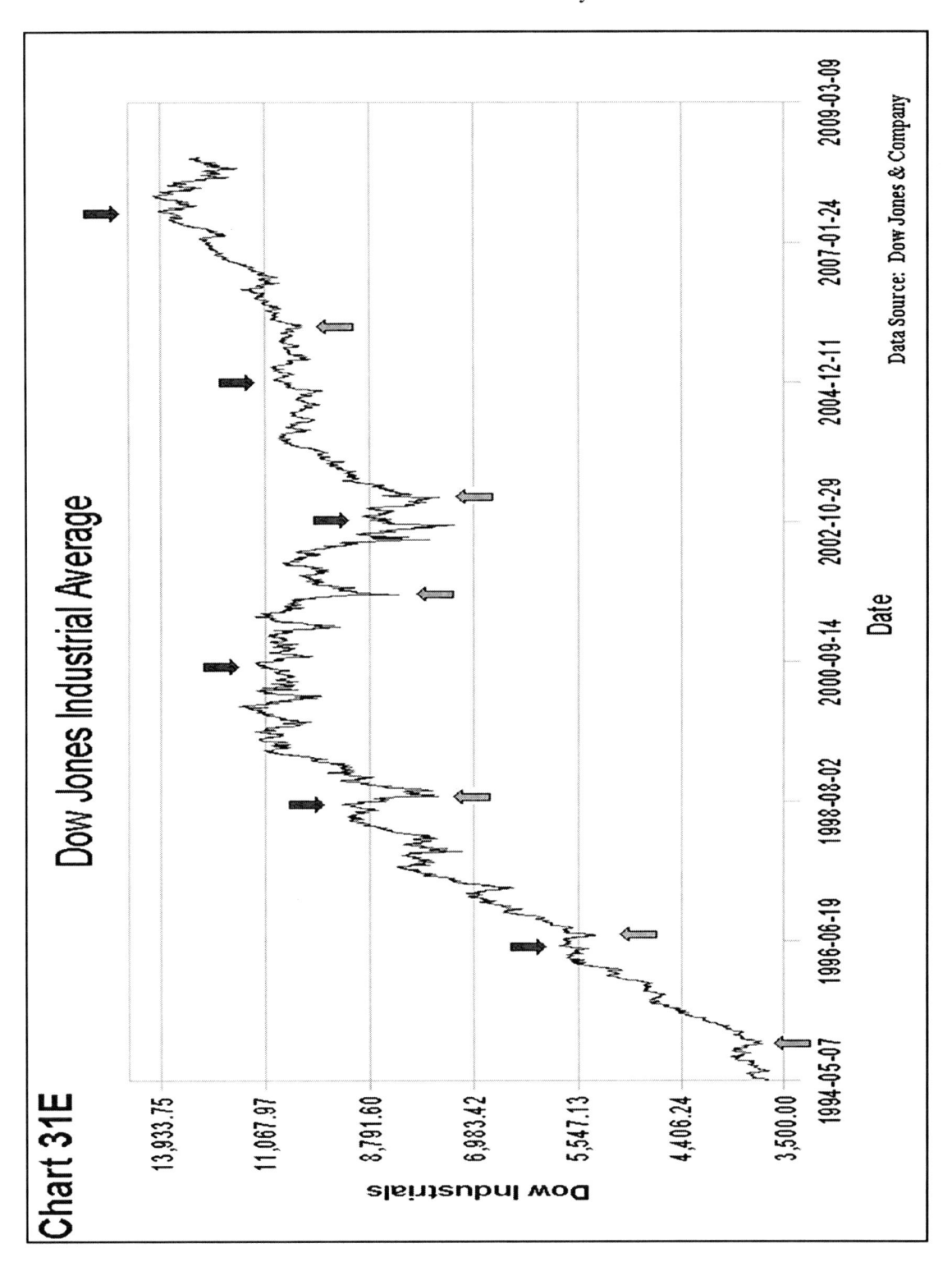

The 2.12-Year EUWS Cycle

Chart 31E

Dow Jones Industrial Average

Data Source: Dow Jones & Company

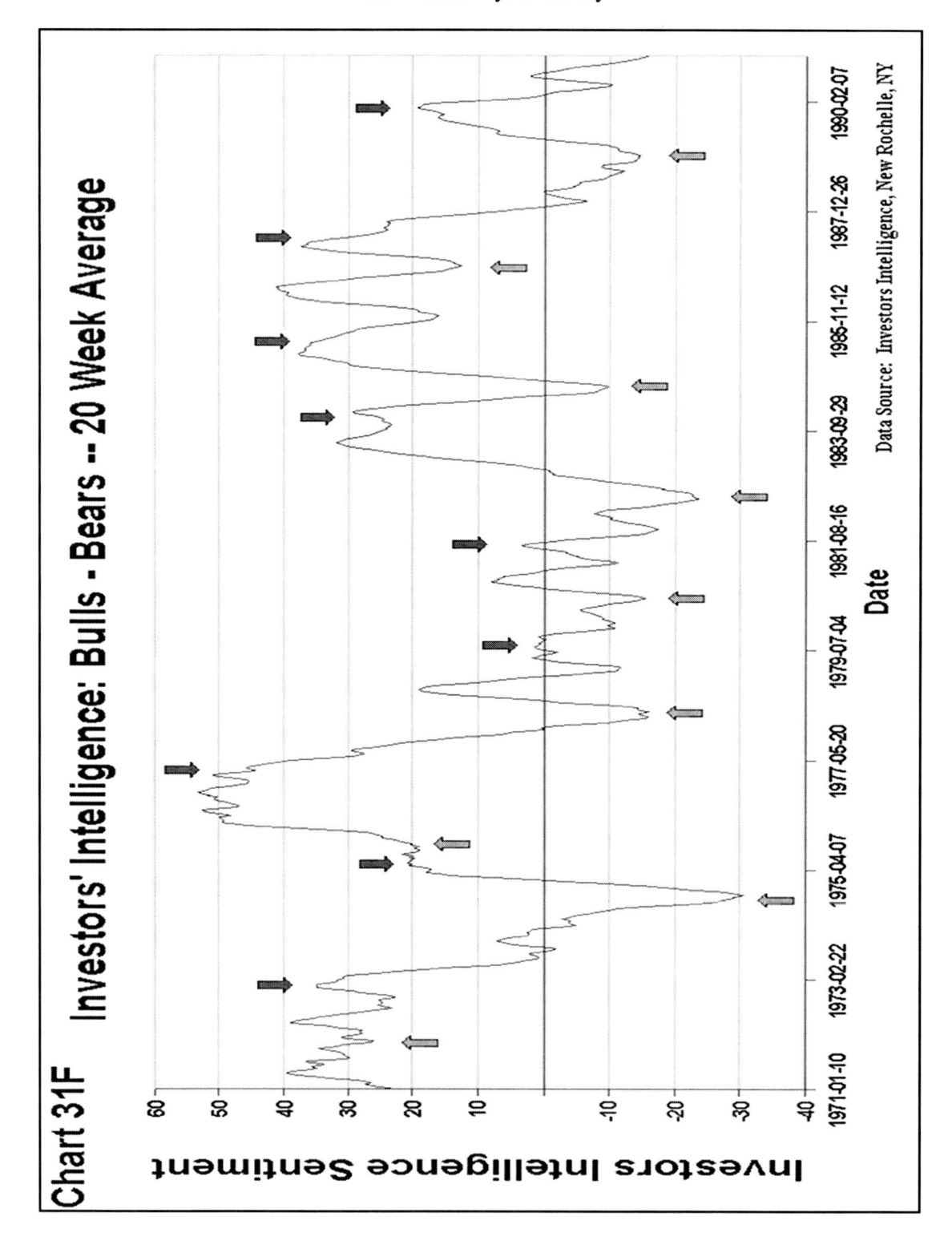

Chart 31F Investors' Intelligence: Bulls - Bears -- 20 Week Average

Data Source: Investors Intelligence, New Rochelle, NY

The 2.12-Year EUWS Cycle

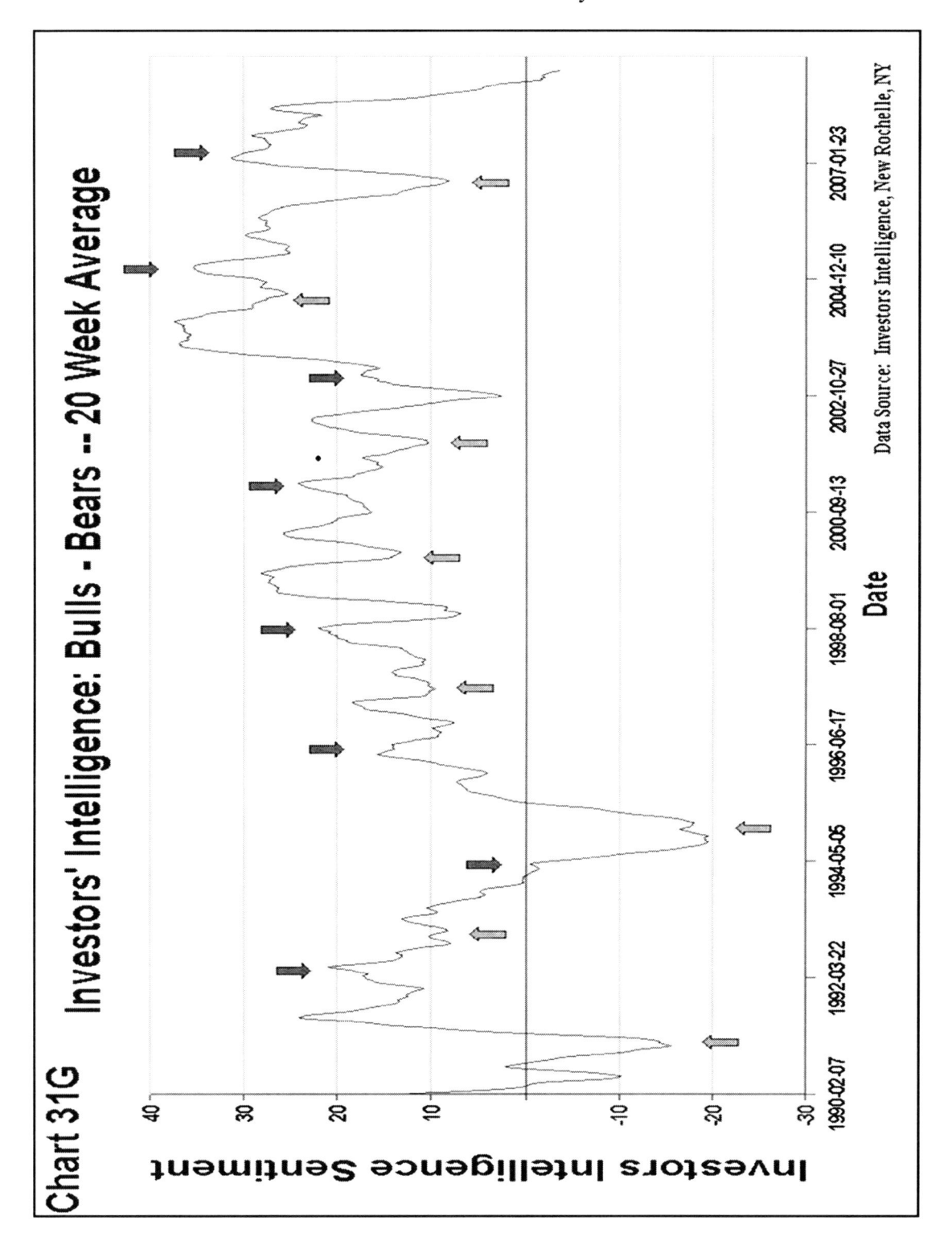

Chart 31G — Investors' Intelligence: Bulls - Bears -- 20 Week Average

Data Source: Investors Intelligence, New Rochelle, NY

Chapter 32
The 258.11-Day EUWS Cycle

Dividing 2.12-years by three produces the 258.11-day EUWS cycle. Compared to the longer-term EUWS cycles, this particular wave exhibits diminished influence. A couple of reasons account for this. First, because of its short duration, it produces small amplitudes. Second, the geomagnetic cycles tend to dominate at this level, thus masking its appearance. However, for the past ten years, the cycle exhibited substantial power. Review **Charts 32A and 32B**, showing all 258-day cycles between 1998 and 2008.

The 258-day cycle made especially noteworthy appearances at the 6.36-year tops in the years 2000 and 2007.

In 2000, the top formed in three stages. First, the Dow Jones Industrial Average topped on January 14 – fourteen days after a theoretical peak of the 258-day cycle. Second, the NASDAQ Index (and the Internet Bubble) began to implode around the time of the 86-day theoretical peak on March 26. Third, the S&P 500 Index completed a double-top on September 1 – closely coinciding with a theoretical peak of the 258-day cycle. For the next 10 oscillations, the 258-day cycle came close to direct hits.

Finally, as the long-term bull market reached its summit in 2007, the 258-day cycle modulated the three phases of the ensuing crash. A Wall Street adage states that credit market changes foreshadow stock market trends by six to nine months, and stock market swings precede economic oscillations by six to nine months. The 2007 bubble imploded in accordance with this adage, separated precisely by 258-day intervals.

First, the subprime crisis began within days of the January 24, 2007 theoretical peak of the 172-year cycle (which also coincided with a 258-day peak). Second, global equity markets topped on October 9, 2007, one day before the theoretical peak of the next 258-day cycle. Third, the global commodity bubble burst on July 3, 2008, one week after the succeeding 258-day cyclic peak. The bursting of the commodity bubble was significant because declining commodity prices typically indicate either the start of recession or the beginning of slow-growth.

Although it probably produces other effects, at this time, the 258-day EUWS cycle can only be associated with fluctuations in stocks and commodities.

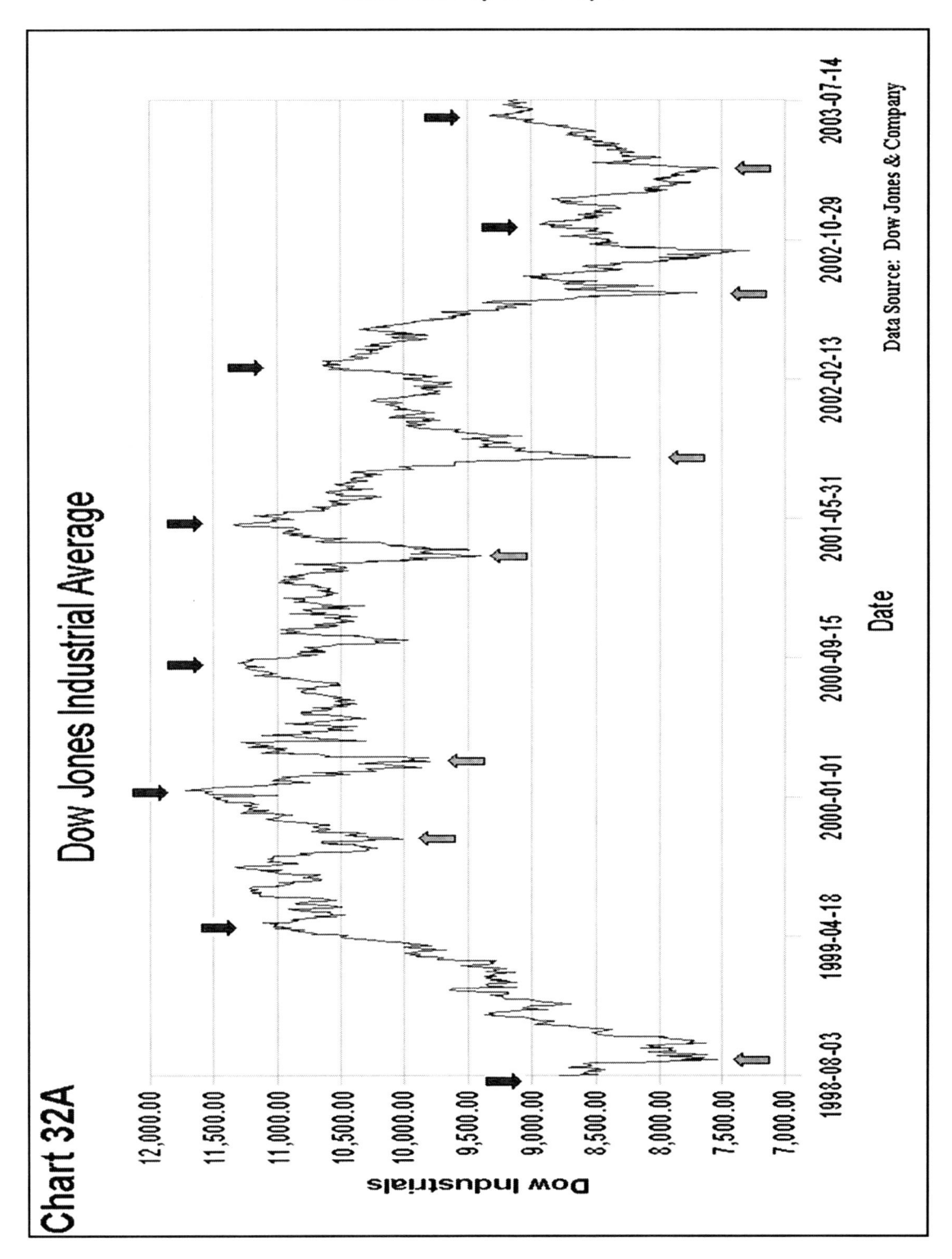

The 258.11-Day EUWS Cycle

Chart 32B

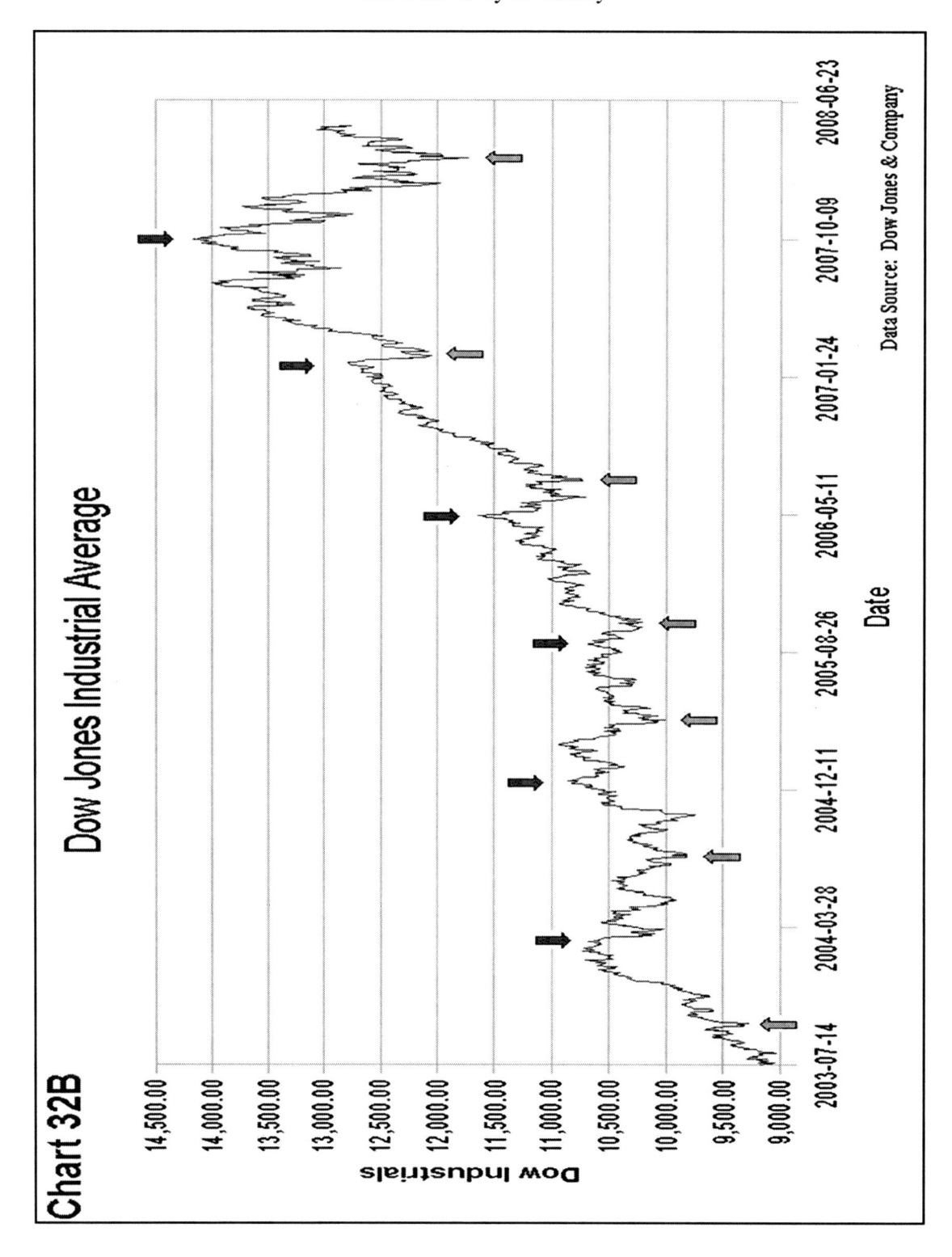

Dow Jones Industrial Average

Data Source: Dow Jones & Company

Chapter 33
The 86.04-Day EUWS Cycle

Dividing by three, the 258.11-day cycle produces the next frequency – the 86.04-day EUWS cycle. Hence, this cycle covers a period slightly less than three months. Identical to its parent cycle, the 86-day cycle exhibited limited visibility prior to 2000, and then suddenly emerged as a dominant force. See **Charts 33A through 33E**.

Since 2000, the 86-day cycle repeated 36 times with near perfection. Reduced sunspot activity during this period, working in combination with highly leveraged markets, probably caused the sudden increase in its visibility.

That concludes this chapter! Previous chapters already presented all of the important EUWS characteristics. Little remains to be told. Only one more chapter remains. However, it contains an important piece of evidence not found in any of the other EUWS cycles.

The Unified Cycle Theory

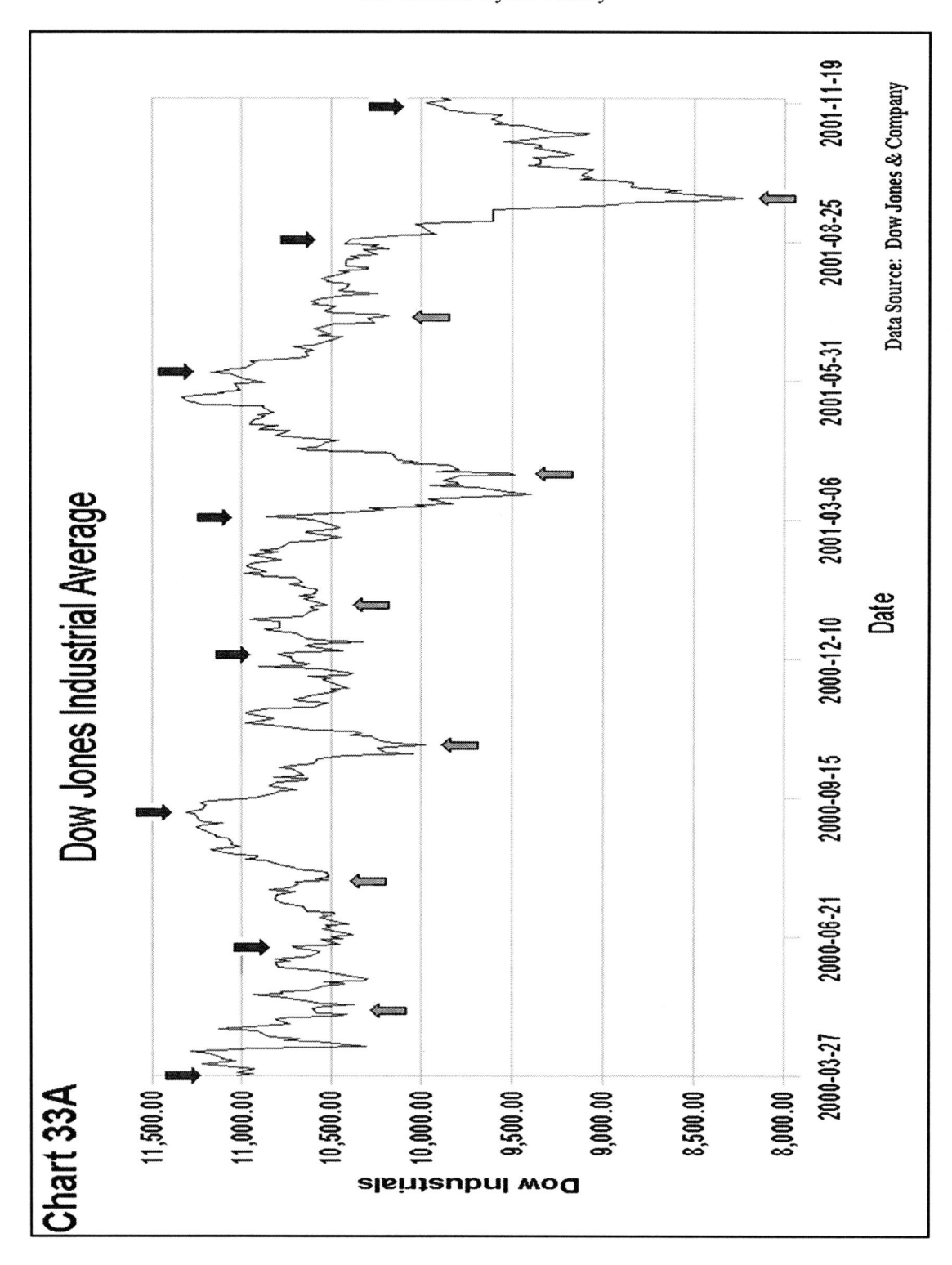

Chart 33A

Dow Jones Industrial Average

Data Source: Dow Jones & Company

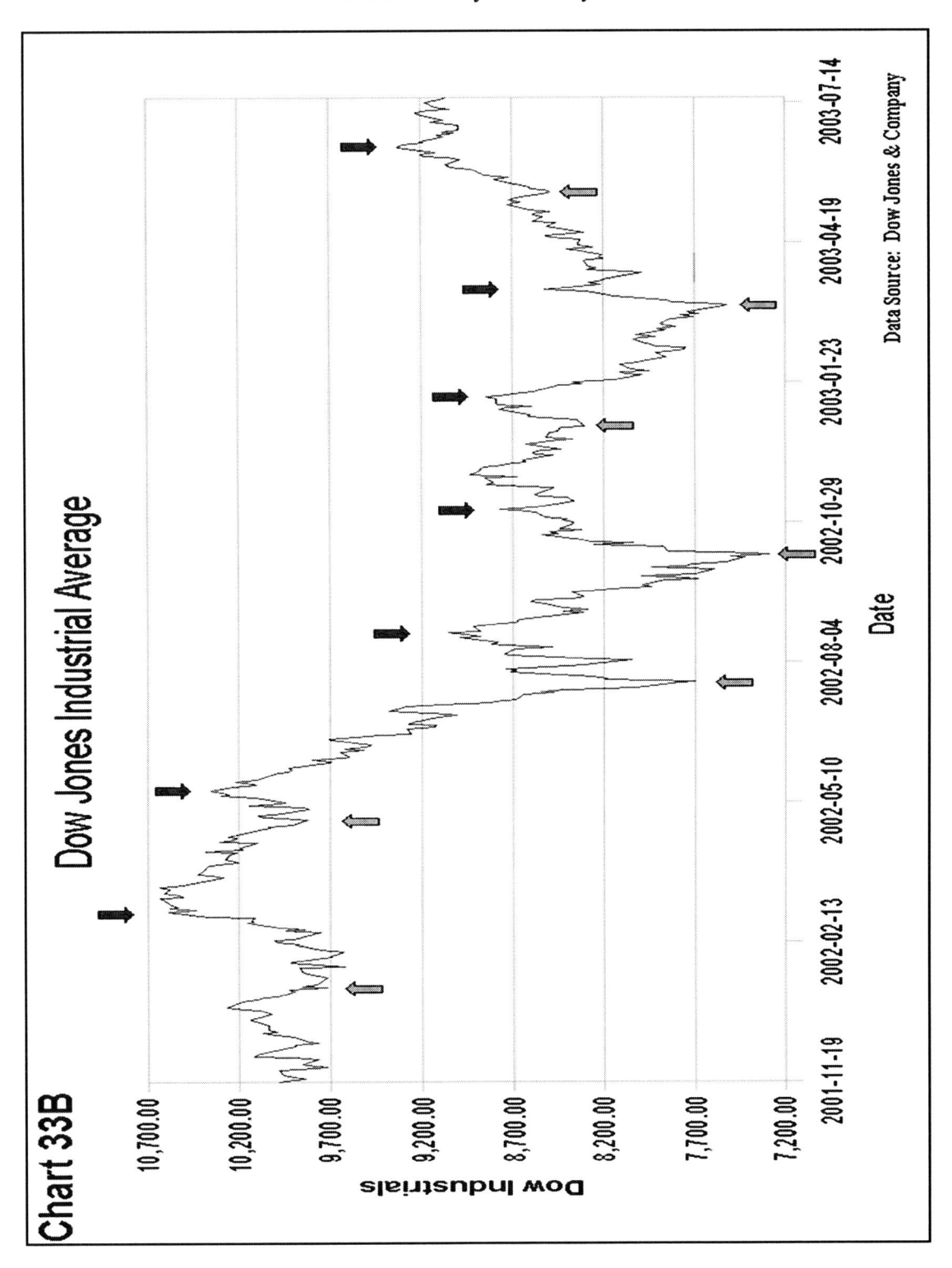

The 86.04-Day EUWS Cycle

The Unified Cycle Theory

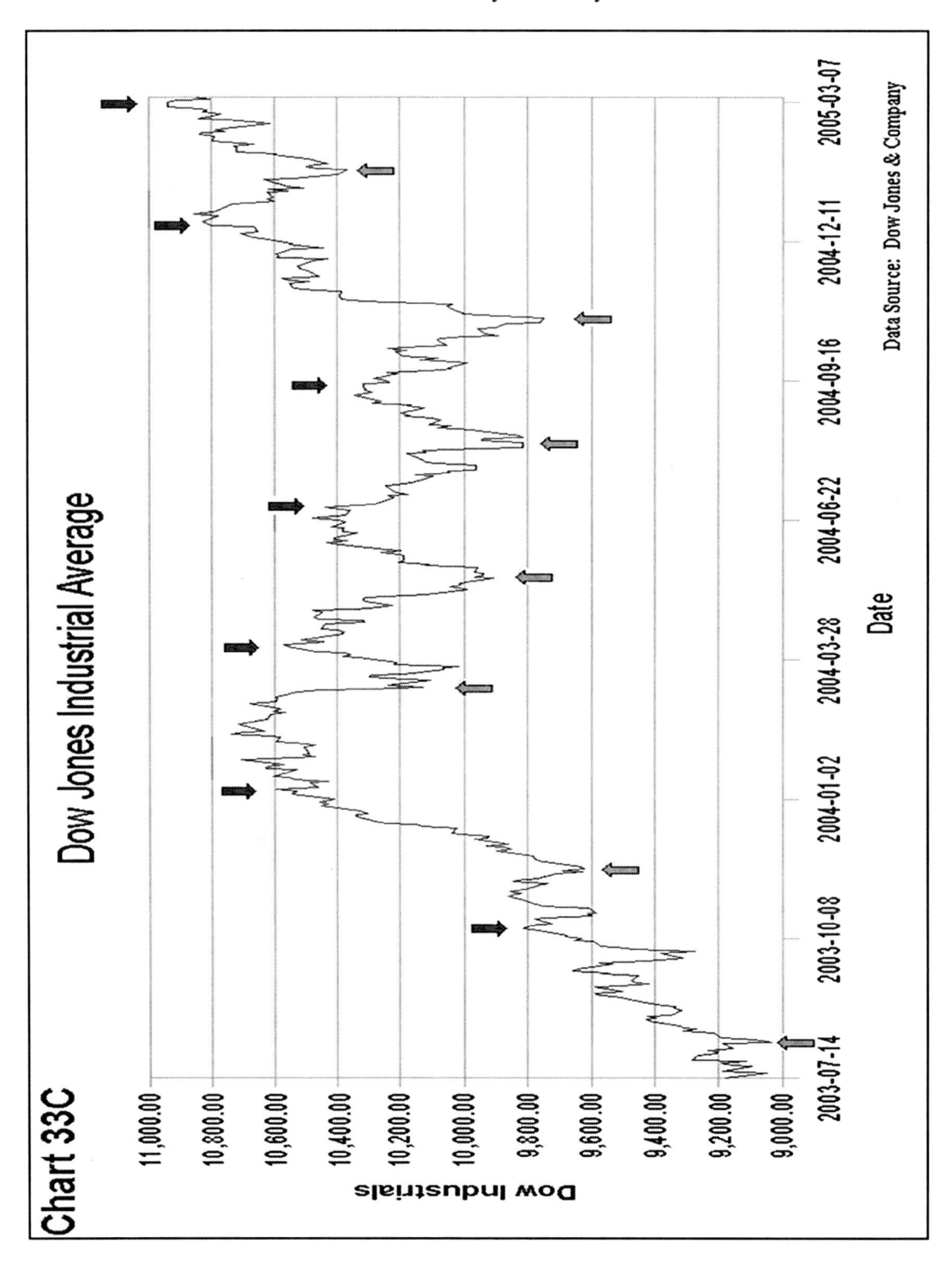

Chart 33C Dow Jones Industrial Average

Data Source: Dow Jones & Company

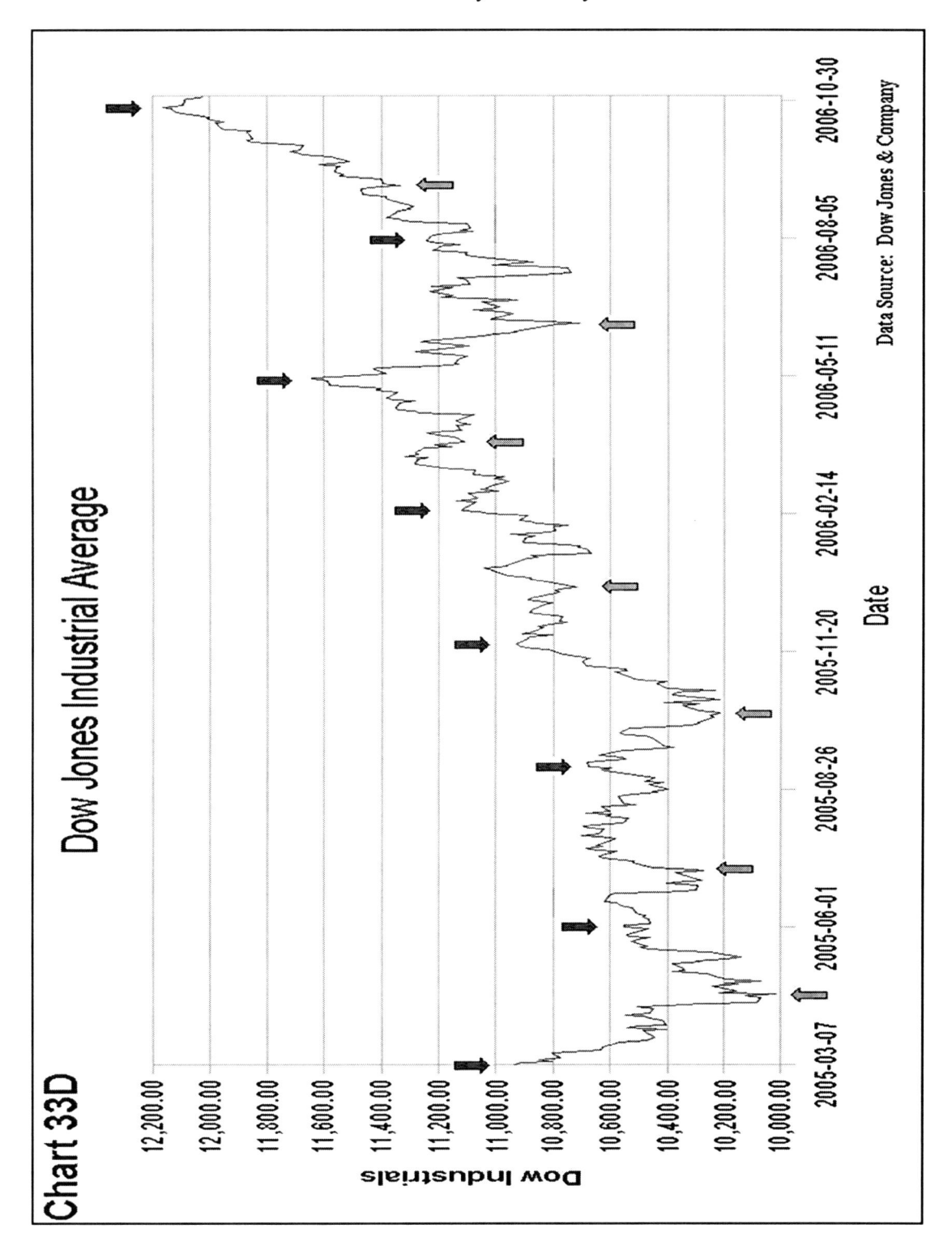

The 86.04-Day EUWS Cycle

Chart 33D — Dow Jones Industrial Average

Data Source: Dow Jones & Company

Chart 33E

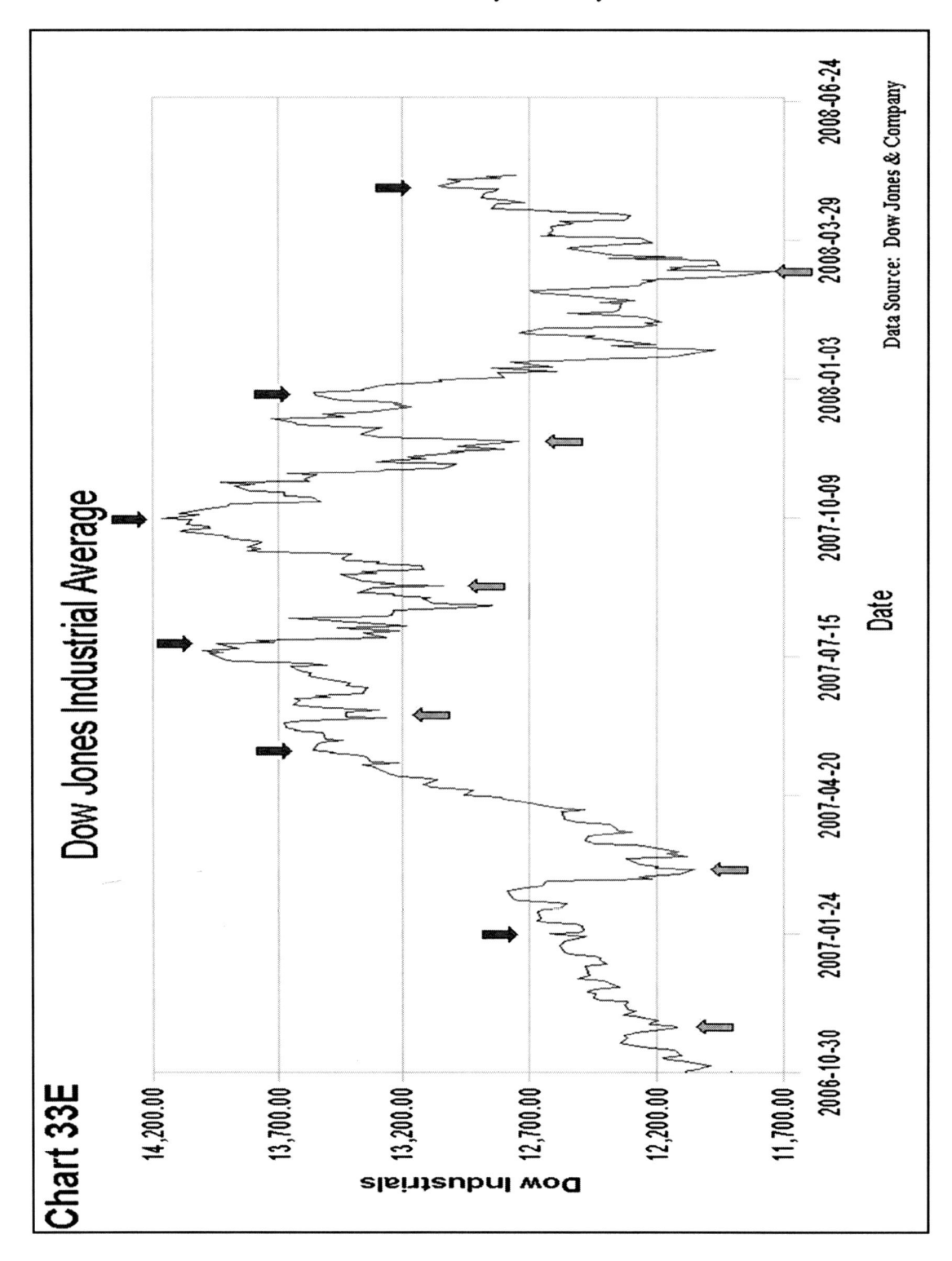

Dow Jones Industrial Average

Chapter 34
The 28.68-Day EUWS Cycle

By the time the EUWS cycles shrink to 28.68 days, their magnitude and significance almost become completely irrelevant. In addition, two geomagnetic-related cycles closely rival its frequency. The 29.53-day lunar cycle stands slightly above it, while the 27-day solar cycle resides slightly below it. Their comparable periods add to the complexity of determining which of these three cycles causes the roughly 29-day cycle in human behavior.

This particular situation resembles the comparable frequencies exhibited between Milankovitch cycles and EUWS cycles – in the range of 13-kyr to 1-myr. And that leads to the possibility that, somehow, the 28.68-day EUWS cycle modulates the others. At a minimum, this possibility deserves some consideration.

To start with, the widely reported 27-day solar cycle actually consists of different periods, depending on the area of the Sun being observed. The Stanford University Solar Center describes the Sun's differential rotation:

> **"Low latitudes rotate at a faster angular rate (approx. 14 degrees per day) than do high latitudes (approx. 12 degrees per day). For example, the equatorial rotation period is 27.7 days compared to 28.6 days at latitude 40 degrees."**[Stanford Solar, 2008]

With a high-latitude solar rotation of 30 days, a 40 degree latitude rotation of 28.6 days, and the equatorial rotation of 27.7 days, the Sun's average rotational speed (combining all latitudes) equals about 28.8 days. That comes exceedingly close to the 28.68-day EUWS frequency.

Another compelling discovery enhances the possibility of a link between EUWS cycles and solar activity – cycles in solar neutrinos. Neutrinos are one of the basic particles in our universe. They originate from nuclear reactions (similar to those that take place at the Sun's core) and from cosmic ray collisions with atoms. Additionally, they travel close to the speed of light, and about one million billion neutrinos pass through a human body every second! Dave Casper, a physicist at the University of California, Irvine, describes these tiny particles:

> **"Neutrinos are similar to the more familiar electron, with one crucial difference: neutrinos do not carry electric charge. Because neutrinos are electrically neutral, they are not affected by the electromagnetic forces which act on electrons. Neutrinos are affected only by a weak sub-atomic force of much shorter range than electromagnetism, and are therefore able to pass through great distances in matter without being affected by it."**[Casper, 2008]

Up to now, this book only discussed cycles related to gravity and the electromagnetic force. However, a team of physicists recently discovered that solar neutrinos bombard Earth in a frequency closely approximating the 28.68-day EUWS cycle. In an article entitled *Twenty-Eight Day Cycle Found in Solar Neutrinos*, Science Daily reviews the discoveries of a research team led by Stanford physicist Peter Sturrock:

> **"Their analysis is based on data collected at the Homestake neutrino detector in South Dakota over a 24-year period. Using advanced statistical procedures, they found clear evidence for a 28.4-day cycle. 'We estimate the probability that the cycle is due to chance to be about three parts in a hundred,' Walther said.... At the same time, the new analysis did not find any evidence for neutrino variations that correspond to the 11-year solar cycle and only weak evidence for two other proposed cycles: a 157-day periodicity that Eric Rieger of the Max Planck Institute in Germany found in the intensity of solar flares, and a 780-day 'quasi-biennial' periodicity that Kunitomo Sakurai from Kanagawa University reported finding in the Homestake data."**[Science Daily: Neutrinos, 1997]

The Science Daily article reveals two interesting details. First, the 28.4-day neutrino cycle falls within 1% of the 28.68-day EUWS cycle. Second, even though supporting evidence appeared weak, a possible 780-day neutrino cycle identified by Sakurai comes within 1% of the 2.12-year (774.33-day) EUWS cycle.

Because these neutrino cycles come tantalizing close to two EUWS cycles, it opens the possibility that a link exists. If so, then it becomes virtually certain that the weak nuclear force is involved somehow. That's because neutrinos primarily react via the weak nuclear force. Taken in sum, through all the frequencies reviewed in this book, various EUWS cycles closely correlate to cycles involving three of the four basic physical forces – electromagnetic, gravitational, and the weak nuclear force.

That completes the review of the first set of cycles in the 29-day range. The second cycle involves the 29.53-day lunar cycle. For the most part, scientists discount ideas that human behavior cycles related to the Moon originate from electromagnetism. That's because the Moon lacks a magnetic core. Nonetheless, in an article entitled *The Moon and the Magnetotail*, Tony Phillips of NASA's Goddard Space Flight Center describes an electromagnetic interaction between the Moon and the Earth:

> "**Behold the full moon. Ancient craters and frozen lava seas lie motionless under an airless sky of profound quiet. It's a serene, slow-motion world where even a human footprint may last millions of years. Nothing ever seems to happen there, right? Wrong. NASA-supported scientists have realized that something happens every month when the moon gets a lashing from Earth's magnetic tail.... 'Earth's magneto-tail extends well beyond the orbit of the moon and, once a month, the moon orbits through it,' says Tim Stubbs, a University of Maryland scientist working at the Goddard Space Flight Center. 'This can have consequences ranging from lunar dust storms to electrostatic discharges.' ... Our entire planet is enveloped in a bubble of magnetism, which springs from a molten dynamo in Earth's core. Out in space, the solar wind presses against this bubble and stretches it, creating a long magneto-tail in the downwind direction.... Anyone can tell when the moon is inside the magneto-tail. Just look: 'If the moon is full, it is inside the magneto-tail,' says Stubbs. 'The moon enters the magneto-tail three days before it is full and takes about six days to cross and exit on the other side.' It is during those six days that strange things can happen.... Fine particles of dust on the moon's surface can actually float off the ground when they become charged by electrons in Earth's magneto-tail. During the crossing, the moon comes in contact with a gigantic plasma sheet of hot charged particles trapped in the tail. The lightest and most mobile of these particles, electrons, pepper the moon's surface and give the moon a negative charge.... On the moon's day side this effect is counteracted to a degree by sunlight: UV photons knock electrons back off the surface, keeping the build-up of charge at relatively low levels. But on the night side, in the cold lunar dark, electrons accumulate and surface voltages can climb to hundreds or thousands of volts.**"[Phillips, 2008]

The Earth's magneto-tail may explain why lunar cycles in markets fade in and out. The magneto-tail becomes more heavily ionized during periods of high sunspot activity. For example, this happened early in the decades of 1980, 1990, and 2000.

The magneto-tail may explain why the gold market had such a strong lunar relationship during the early 1980s. And that may further explain why lunar cycle oscillations prevailed during other famous market crashes. Almost all great market crashes occurred during sunspot maxima.

The 1987 crash was an exception.

Charts 34A, 34B, and 34C span the time between 2002 and 2003. That coincided with a time of declining, yet relatively heavy sunspot activity. Hence, an enhanced magneto-tail existed at that time. In the charts, arrows mark theoretical peaks of the 28.68-day EUWS cycle, while circles signify full moons. Notice how stock market tops tended to coincide with both cycles. When their peaks were out of phase, a tug-of-war developed between the two. And when their peaks arrived in sync, the cycles tended to amplify stock market movements.

In summary, the 28.68-day EUWS frequency closely correlates with cycles in the Sun's rotation as well as cycles in solar neutrinos. These correlations open the possibility of a link between the EUWS cycles and the weak nuclear force.

Additionally, because the 28.68-day EUWS cycle closely approximates the frequency of the lunar cycle, interference between the two complicates studying human activity cycles in this range. More than likely, the lunar cycle in human activity results from interactions related to the Earth's magneto-tail.

That completes review of the EUWS cycles. Next, the spotlight shifts to the important question: What causes these well synchronized oscillations?

Chart 34A

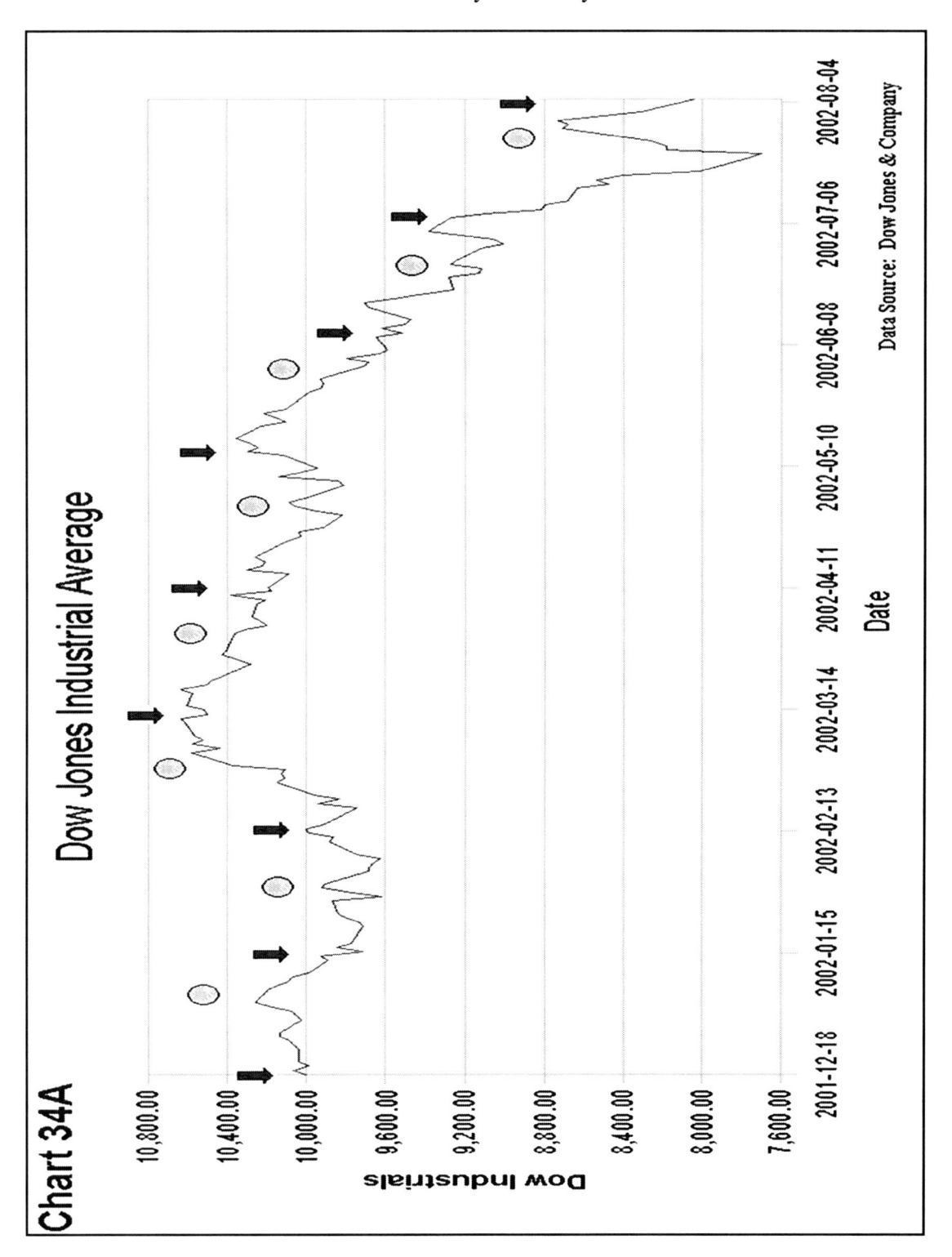

The 28.68-Day EUWS Cycle

Dow Jones Industrial Average

Data Source: Dow Jones & Company

The Unified Cycle Theory

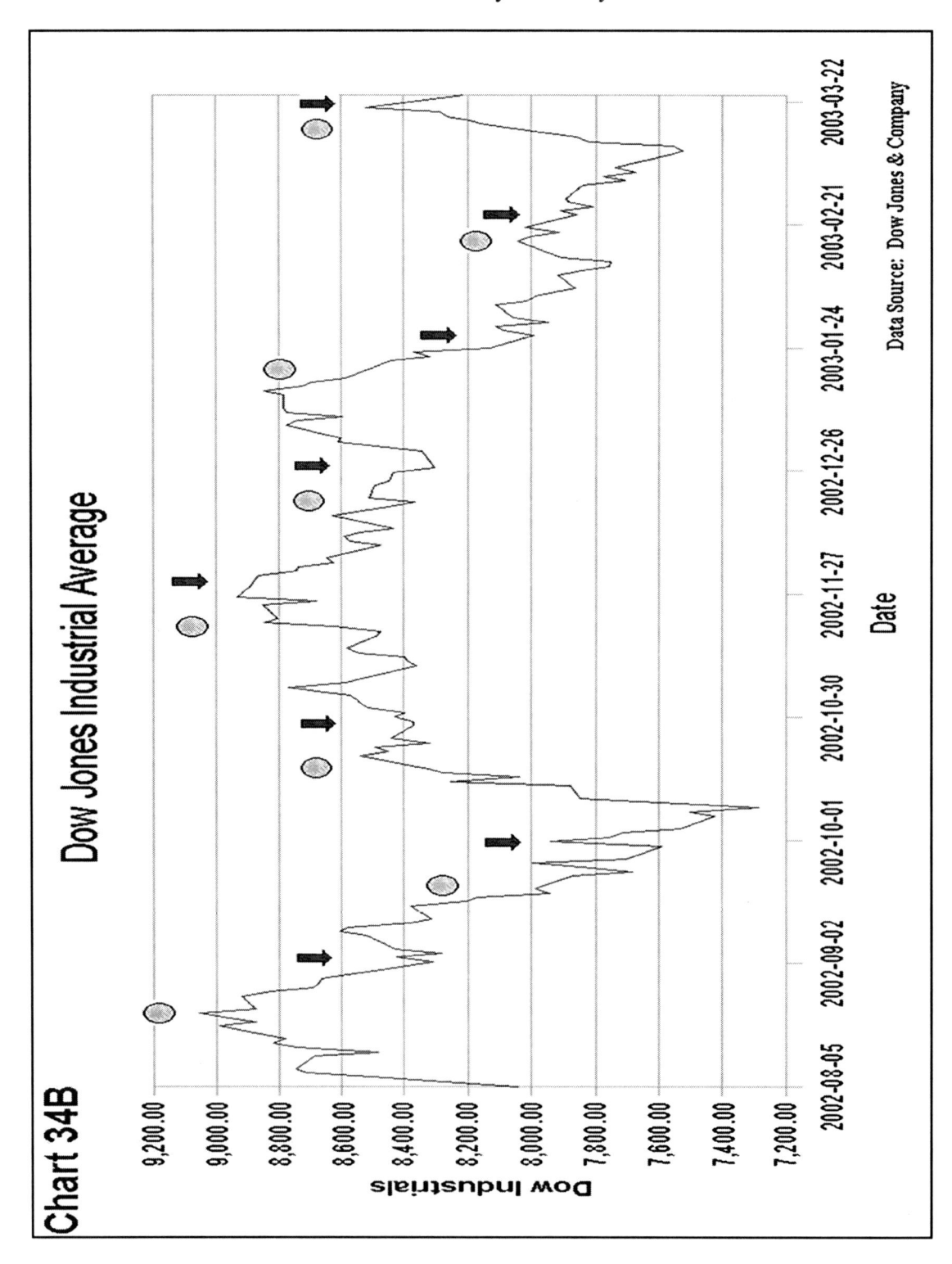

Chart 34B — Dow Jones Industrial Average

Data Source: Dow Jones & Company

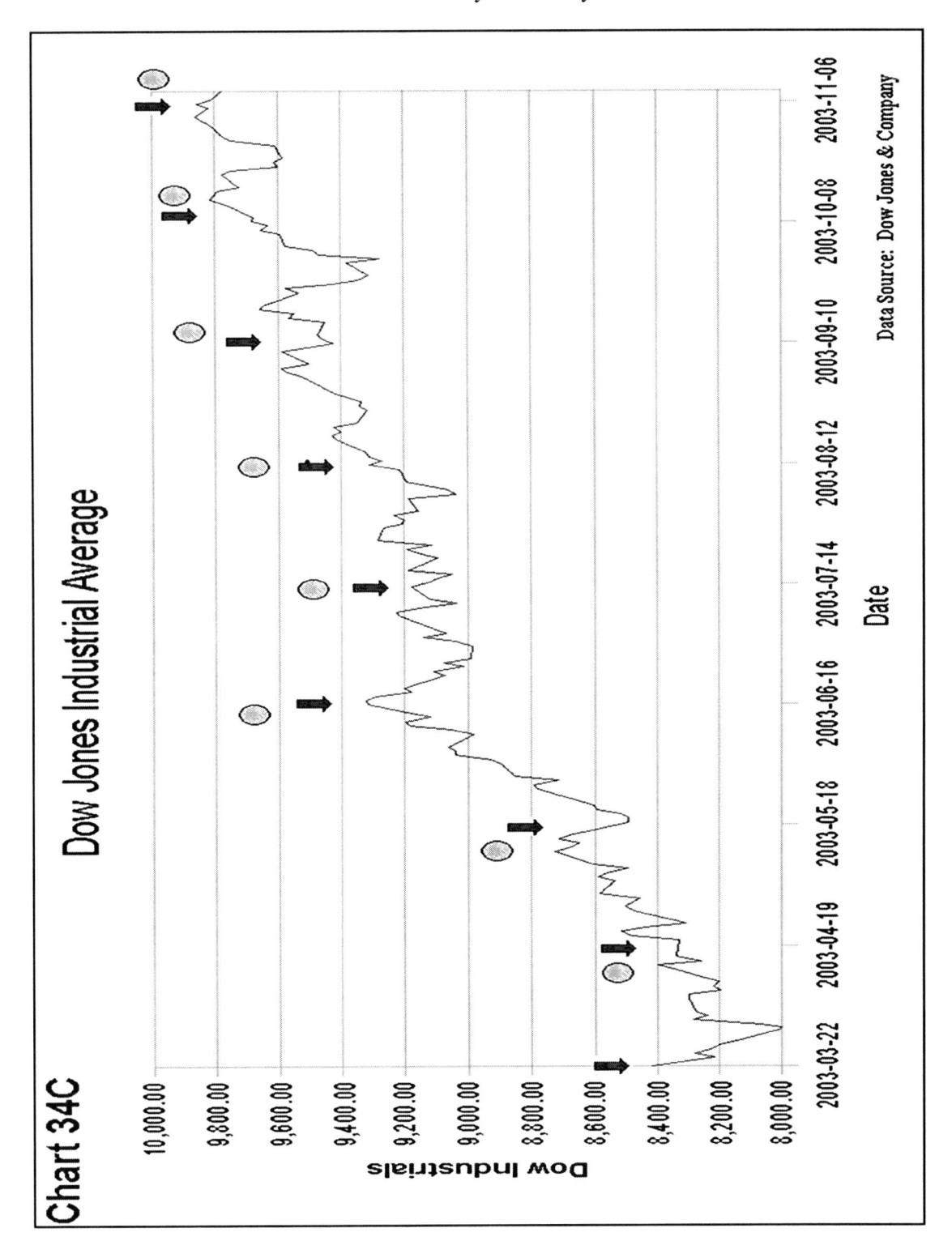

The 28.68-Day EUWS Cycle

Chapter 35
Summary, Analysis, and Conclusions

This chapter reviews pieces of evidence listed earlier in the book. An analysis follows, ending with a conclusion. In some cases, insufficient evidence disallows conclusions with a high degree of certainty. In all cases, confidence level estimates are assigned to each conclusion. In some cases, the review lists multiple conclusions when more than one solution appears possible.

Chapter 4, Item 1: A series of 25 consecutive EUWS cycles exists. Starting with the smallest, the frequency of each parent cycle equals the child's frequency multiplied by three. All EUWS cycles within the range of 28.68 days to 22.2-gyr show their presence in nature through a wide variety of ways – including impacts on geology, biology, and climate. In addition, each child cycle derives its theoretical turning points from its parent cycle.

Chapter 4, Item 1, Conclusion: Both the harmonic relationship of their frequencies and the consistency of their turning points indicate the EUWS cycles all originate from the same physical source. It's virtually inconceivable that such a close-knit group of cycles could originate from completely different sources. (99% confidence)

Chapter 6, Item 1: Sunspot activity affects the solar wind, which results in fluctuations in the Earth's magnetic field.

Chapter 6, Item 1, Conclusion: The solar cycle influences geomagnetic cycles. (99% confidence)

Chapter 6, Item 2: R.W. Kay reports that geomagnetic storms affect the behavior of mental patients. This happens with a delay of two weeks and in a seasonal pattern.

Chapter 6, Item 2, Conclusion: The delays and influences reported by R.W. Kay match the effects that geomagnetic cycles produced in the Dow Jones Industrial Average between 1885 and 2007. Geomagnetic cycles affect human behavior with a delay. (99% confidence)

Chapter 6, Item 3: The World Health Organization explained how magnetic fields affect the navigational abilities of a variety of organisms, and the health of humans. However, it does not provide evidence of mental health or emotional effects from magnetism.

Chapter 6, Item 3, Conclusion: This information does not add to the *Unified Cycle Theory*. However, it does show that electromagnetic fields affect human health. This opens the possibility that magnetic fields create a physical-mental link in humans. (25% confidence)

Chapter 6, Item 4: A research team led by Samuel McClure of Princeton found two areas of the brain that often operate in conflict with each other. The limbic system controls the emotional brain, while the pre-frontal cortex and posterior parietal cortex control the logical brain. When people succumb to instant gratification, the emotional brain over-rules the logical brain.

Chapter 6, Item 4, Conclusion: The behaviors described by the McClure team closely correspond to investor actions. Most people allow their emotional brains to prevail when investing in markets. Bull markets coincide with periods of confidence eventually turning into over-confidence. Near the end of bull markets, the majority move together in a herd as their emotional brains lure them into the illusion of easy money and quick profits, thus pushing prices to exorbitant levels. During bear markets, fear replaces confidence – especially for leveraged speculators. (95% confidence)

Chapter 7, Item 1: According to the Milankovitch Theory, gravity produces cycles in eccentricity, obliquity, and precession. In turn, these cycles combine to produce a 100,000 year cycle in ice-ages.

Chapter 7, Item 1, Conclusion: Eccentricity, obliquity, and precession probably influence global climate; however, they impact ice-age cycles far less than the Milankovitch Theory suggests. Instead, EUWS cycles work with Milankovitch cycles to produce climate oscillations. (99% confidence)

Chapter 7, Item 2: As reflected by the daily and seasonal cycles, gravity and motion cycles oscillate in a manner that replicates a sine wave.

Summary, Analysis, and Conclusions

Chapter 7, Item 2, Conclusion: Gravity and motion cycles reveal their identity in climate cycles with a sine wave fingerprint. (99% confidence)

Chapter 7, Item 3: Geomagnetic cycles develop from three sources – the Sun, the Earth's annual rotation, and the Moon. The sun produces a 27-day cycle, the 11-year Schwabe cycle, the 87.8-year Gleissberg cycle, and the 208-year Suess cycle. The Earth's rotation produces a semi-annual geomagnetic cycle. And the Moon produces a 173-day eclipse cycle (as its plane with the Earth and Sun oscillates), and a 29.53-day lunar cycle (from full moon to new moon and back to a full moon). Each of these geomagnetic cycles functions as leading indicators for the stock market.

Chapter 7, Item 3, Conclusion: Fluctuations in geomagnetism affect human behavior via our electrochemical nervous systems and emotional brains. (99% confidence)

Chapter 7, Item 4: A study by P.T. Nastos *et al.* showed that symptoms of panic and depression appeared in a seasonal pattern, with a maximum during September and October. This pattern closely matches the seasonal pattern of markets.

Chapter 7, Item 4, Conclusion: The study by the Nastos team provides additional evidence that fluctuations in geomagnetism affect human behavior. (99% confidence)

Chapter 8, Item 1: When markets fail to reverse direction at theoretical turning points of EUWS cycles, the deviations tend to occur at theoretical turning points of competing cycles and/or child cycles. In this way, the various cycles reach a compromise – allowing a best fit that partially satisfies all of the cycles. Market crashes in 1929 (stocks), 1980 (silver), 1987 (stocks), and 1990 (Japanese stocks) reveal the conflicts among cycles, as well as the compromises made to achieve a best fit for all cycles involved.

Chapter 8, Item 1, Conclusion: The TPD Principle describes how cycles deviate from their expected turning points, in a non-random manner. By deviating in this manner, cycles achieve "best fit" harmony with their competitors. (95% confidence)

Chapter 10, Item 1: Geomagnetic fluctuations related to the 11-year cycle, the seasonal cycle, the eclipse cycle, and the lunar cycle were compared to stock market trends between the years 1885 and 2007. For all four cycles, the stock market correlated inversely to the geomagnetic cycles, with the geomagnetic cycles leading changes in the stock market.

Chapter 10, Item 1, Conclusion: If the stock market oscillated independently from geomagnetism, then it would be highly unlikely for each pair of cycles to correlate inversely. All 4 cycles showed the same negative correlation, with geomagnetism leading changes in stocks. This research conclusively links geomagnetic cycles to human behavior cycles. (99% confidence)

Chapter 11, Item 1: The Milankovitch Theory fails to account for at least eleven different climate issues.

Chapter 11, Item 1, Conclusion: Climatologists over-rate how Milankovitch cycles impact global climate. The EUWS cycles fill in the void left by cycles in eccentricity, obliquity, and precession. (99% confidence)

Chapter 11, Item 2: The Greenland ice-core record shows that temperature changes either coincide with or precede changes in carbon-dioxide. Sometimes, temperature changes occur one thousand years ahead of corresponding CO_2 moves.

Chapter 11, Item 2, Conclusion: This shows that atmospheric carbon-dioxide exerts a limited impact, if any, on global temperatures. As part of a classic cause-and-effect error related to correlations, the scientific community has grossly over-rated the impact that atmospheric CO_2 exerts on global climate. (99% confidence)

Chapter 12, Item 1: The trough of the 22.2-gyr EUWS cycle closely coincides with the Big Bang, while its peak coincides with the 2.7-gyr Event on Earth.

Chapter 12, Item 1, Conclusion: Because the EUWS cycles all share the same source, and because of the sheer size of the 22.2-gyr cycle, these cycles cannot possibly originate from galactic cosmic rays from inside the Milky Way Galaxy. The Big Bang could have released these waves in some manner. (10% confidence) However, the ability of an explosion, such as the Big Bang, to modulate itself seems unlikely. If the Big Bang failed to create the EUWS cycles, then their sheer size implies they must originate from outside our known universe. (90% confidence)

Chapter 12, Item 2: Assuming the EUWS cycles originate from outside our universe, the existence of other universes becomes likely. If they operate as open systems, then mergers and exchanges would create mismatches in matter and energy inside each individual universe.

Chapter 12, Item 2, Conclusion: With universes operating as open systems, prior to a big crunch, matter does not necessarily equal anti-matter. As such, a subsequent big bang would produce a universe with unequal portions of matter and anti-matter. Hence, a multiverse consisting of open universes solves the problem of matter not equaling anti-matter inside our own observable universe. (75% confidence)

Chapter 13, Item 1: The Great Unconformity at Frenchman Mountain in Nevada created a 1.15-gyr gap in the geological record from 0.55 Ga to 1.70 Ga. This coincided almost perfectly with a theoretical low point in a model primarily controlled by the 22.2-gyr, 7.39-gyr, and 2.46-gyr cycles.

Chapter 13, Item 1, Conclusion: A geological unconformity represents the exact opposite of a

Summary, Analysis, and Conclusions

tectonic upheaval. Unconformities occur at major EUWS troughs along with extremely cold climates, leaving limited geological trails. During these EUWS troughs, reduced tectonic activity provides less sediment to produce good, permanent geological records. For the limited sediment produced during these cold periods, ice sheets quickly erode all remaining geological evidence. (99% confidence)

Chapter 14, Item 1: Paleorecords show continual oscillations in Oxygen-18 levels closely corresponding to the EUWS cycles.

Chapter 14, Item 1, Conclusion: EUWS cycles cause fluctuations in global temperature. (99% confidence)

Chapter 14, Item 2: A wide variety of geological data shows precise matches with theoretical turning points of the EUWS cycles. The geological indicators include the age of juvenile continental crust, deformed granitoid zircon, orogenic gold deposits, delta Carbon-13 levels, global mantle plume events, and volcanic flood basalts.

Chapter 14, Item 2, Conclusion: Near major theoretical peaks, the EUWS cycles cause great tectonic upheavals, including substantially greater volcanic activity. During these major up-phases, Earth attempts to expand. (99% confidence)

Chapter 14, Item 3: Meteorites tend to strike Earth heavily twice during every 274-myr cycle. While not a direct match, the meteorite cycle acts as a secondary harmonic of the 274-myr cycle. Oscillations in universal clumpiness correspond to EUWS cycles in three ways. (1) in big-crunch big-bang cycles, (2) with two meteorite-strike cycles every 274-myr, and (3) through star formation rates coinciding with every other 2.46-gyr cycle.

Chapter 14, Item 3, Conclusion: Possible explanations for the association between oscillations in universal clumpiness of matter and EUWS cycle troughs include...... 1. A universal membrane exists as theorized by Steinhardt and Turok. As the membrane oscillates, it causes expansions and contractions matching EUWS cycles. (10% confidence) 2. The EUWS cycles cause fluctuations in gravitational forces. (15% confidence) 3. The EUWS cycles produce a sequence of magnetic waves that knock matter around. (25% confidence) 4. A combination of gravitational forces, electromagnetic forces, and another basic undiscovered force, causes the clumpiness resulting from the EUWS cycle. (50% confidence)

Chapter 14, Item 4: A close examination of the appearance of new gene families shows that evolution occurs in cycles corresponding to the EUWS frequencies of 274-myr, 821-myr, 2.46-gyr, 7.39-gyr, and 22.2-gyr.

Chapter 14, Item 4, Conclusion: The EUWS cycles regulate life on Earth. Even though massive environmental changes occur during up-phases of the EUWS cycles, all types of species adapt, survive, and multiply. During EUWS down-phases, organisms lose their ability to adapt and evolve. During down-phases, species must either survive dramatic environmental changes in their current form – or become extinct. This strange behavior indicates that EUWS cycles activate an evolutionary switch in all living organisms. (80% confidence)

Chapter 15, Item 1: From the geological timescale, geological periods closely correspond to theoretical peaks of the 30.4-myr and 91.3-myr cycles.

Chapter 15, Item 1, Conclusion: This adds to the evidence that EUWS cycles produce alterations in Earth's geology. (99% confidence)

Chapter 15, Item 2: Relatively reliable temperature proxies for the past 500 million years show five global temperature oscillations corresponding precisely to theoretical periods of the 91.3-myr EUWS cycle.

Chapter 15, Item 2, Conclusion: EUWS cycles act as the primary cause of fluctuations in global temperature. (99% confidence)

Chapter 15, Item 3: Deposits of buried organic carbon drop off sharply after theoretical peaks of the EUWS cycles.

Chapter 15, Item 3, Conclusion: Down-phases of the EUWS cycles contribute to the extinction of species. (99% confidence) As a corollary, EUWS cycles modulate the global carbon cycle. (99% confidence)

Chapter 15, Item 4: For the past 600 million years, the number of genera within phylum porifera (sponges) and phylum brachiopoda (lamp shells) oscillated with the 91.3-myr EUWS cycle. Going back 2 billion years, 20 cycles of significant generic change occur with a frequency of 91.3-myr, corresponding to the EUWS.

Chapter 15, Item 4, Conclusion: The EUWS cycles regulate life on Earth. (99% confidence)

Chapter 15, Item 5: Meteorites regularly strike Earth corresponding to the 91.3-myr EUWS cycle.

Chapter 15, Item 5, Conclusion: EUWS cycles cause matter to clump periodically. (99% confidence)

Chapter 15, Item 6: Fluctuations in zircon ages from Western Australia corresponding to the 91.3-myr EUWS cycle.

Chapter 15, Item 6, Conclusion: Near theoretical peaks, the EUWS cycles produces major alterations in Earth's geology. (99% confidence)

Summary, Analysis, and Conclusions

Chapter 16, Item 1: Global climate fluctuated in a 30.4-myr cycle for 17 cycles over the past 520 million years. Additionally, detailed records from Zachos *et al* show the Late Paleocene Thermal Maximum occurred at 54.95 Ma – almost exactly equaling the 54.959 Ma EUWS theoretical peak.

Chapter 16, Item 1, Conclusion: EUWS cycles act as the primary modulator of global temperature. (99% confidence)

Chapter 16, Item 2: For the past 600 million years, the number of genera – of cephalopods (mollusks) and of all genera surviving less than 45 million years – oscillated in cycles of 30.4-myr. Their turning points corresponded to the 30.4-myr EUWS cycle. And this reoccurred for 17 cycles.

Chapter 16, Item 2, Conclusion: The EUWS cycles modulate life on Earth. (99% confidence)

Chapter 17, Item 1: During the Phanerozoic, all but three borders for geological epochs closely correspond to theoretical peaks of the 10.1-myr cycle. 29 of the 32 breakpoints correlated with 10.1-myr EUWS theoretical peaks (with 11 doubling as 30.4-myr peaks). Only three epochs failed to match with one of the 10.1-myr maxima.

Chapter 17, Item 1, Conclusion: Near theoretical peaks, the EUWS cycles produces major alterations in Earth's geology. (99% confidence)

Chapter 17, Item 2: Over the last 200 million years, global temperatures fluctuated in 20 cycles of 10.1-myr. Furthermore, every third cycle exhibited greater amplitude, reflecting the presence of the 30.4-myr cycle.

Chapter 17, Item 2, Conclusion: EUWS cycles act as the primary modulator of global temperature. (99% confidence)

Chapter 17, Item 3: After 70 Ma, global temperature estimates achieved a high degree of accuracy. The TPD Principle didn't even come into play with the initial five theoretical peaks of the 10.1-myr cycle. From 65.10 Ma to 14.40 Ma, global temperatures hit maxima precisely at 10.1-myr peaks. And even the slight miss at 4.26 Ma, may have been a direct hit – depending on the data series used.

Chapter 17, Item 3, Conclusion: EUWS cycles act as the primary modulator of global temperature. (99% confidence)

Chapter 18, Item 1: Geological ages tend to be separated by 3.38-myr intervals. Of course, that corresponds to a EUWS frequency. Around major EUWS troughs; however, geological ages often skip a beat by producing a geological age equaling a multiple of the 3.38-myr cycle.

Chapter 18, Item 1, Conclusion: A skipped beat in the geological-age sequence equates to a minor unconformity. Unconformities occur at major EUWS troughs along with colder than normal climates, leaving limited geological trails. During these EUWS troughs, reduced tectonic activity provides less sediment to produce good geological records. For the limited sediment produced during these periods, ice sheets quickly erode all remaining geological evidence. By producing limited sediment and then erasing what remains, nature prevents geologists from properly identify every geological age. In this way, unconformities produce skipped beats in the geological timescale. (99% confidence)

Chapter 18, Item 2: Over the last 38 million years, global temperatures fluctuated in 11 cycles of 3.38-myr.

Chapter 18, Item 2, Conclusion: EUWS cycles act as the primary modulator of global temperature. (99% confidence)

Chapter 19, Item 1: Over the last 18 million years, global temperatures fluctuated in 16 cycles of 1.13-myr.

Chapter 19, Item 1, Conclusion: EUWS cycles act as the primary modulator of global temperature. (99% confidence)

Chapter 19, Item 2: Over the past 5 million years, equatorial temperatures in the Eastern Pacific Ocean coincided perfectly with 1.13-myr theoretical peaks at 4.260, 3.133, and 2.006 Ma. For the 880 Ka cycle, the TPD Principle came into play. For that cycle, the sea temperature peaked one 376-kyr cycle early, at 1.255 Ma.

Chapter 19, Item 2, Conclusion: EUWS cycles act as the primary modulator of global temperature. (99% confidence)

Chapter 20, Item 1: Over the last 8 million years, global temperatures fluctuated in 21 cycles of 376-kyr. Over the last 5 million years, Eastern Pacific Ocean temperatures oscillated in 14 cycles of 376-kyr.

Chapter 20, Item 1, Conclusion: EUWS cycles act as the primary modulator of global temperature. (99% confidence)

Chapter 21, Item 1: Over the last 2.9 million years, the 125-kyr EUWS frequency repeated 23 times in global temperature proxies. Over the last 5 million years, the 125-kyr EUWS frequency repeated 42 times in Eastern Pacific Ocean temperatures.

Chapter 21, Item 1, Conclusion: EUWS cycles act as the primary modulator of global temperature. (99% confidence)

Summary, Analysis, and Conclusions

Chapter 21, Item 2: During the down-phase of the last 125-kyr cycle (between 128.6 Ka and 66 Ka), a number of species of large animals became extinct. Furthermore, Homo-Sapiens almost became extinct near the trough of this 125-kyr cycle. However, after the theoretical trough passed, sometime around 66 Ka, human populations multiplied throughout the world. Dramatic environmental changes only affect the survival of species when climate changes from warm-to-cold. Dramatic climate changes from cold-to-warm enhance evolution.

Chapter 21, Item 2, Conclusion: The EUWS cycles modulate life on Earth. (99% confidence) Within individual organisms, the EUWS cycles accomplish this by turning evolutionary switches off-and-on that affect the ability of organisms to mutate, adapt, and survive. (80% confidence)

Chapter 22, Item 1: Oscillations of the 41.0-kyr obliquity cycle and the 41.7-kyr EUWS cycle combine to produce alternating periods of dominance for the 41-kyr frequency.

Chapter 22, Item 1, Conclusion: The *Unified Cycle Theory* covers a variety of physical cycles. The EUWS cycles and the Milankovitch cycles represent two sets of cycles under that umbrella. Interference between the 41.0-kyr obliquity cycle and the 41.7-kyr EUWS cycle periodically reduces their visibility. (90% confidence)

Chapter 22, Item 2: Milankovitch cycles tend to separate themselves with frequencies roughly equal to EUWS cycles. Two of these Milankovitch frequencies equal secondary harmonics of EUWS cycles – the 80-to-85-kyr year beat of the 100-kyr cycle and the 21-kyr precession cycle. This close link between Milankovitch cycles and EUWS cycles opens the possibility that EUWS cycles modulate gravitational cycles. In addition, the Earth tends to contract and expand in phase with EUWS cycles – suggesting cycles of compression and expansion related to fluctuating gravity.

Chapter 22, Item 2, Conclusion: If this proves to be true, then EUWS cycles may consist of a basis physical force that potentially unifies the 4 known physical forces. (10% confidence) The EUWS cycles cause fluctuations in the gravitational force. (15% confidence)

Chapter 22, Item 3: The EUWS cycles exhibit many electromagnetic characteristics. These characteristics include spiked reversals (rather than sine wave reversals), widely variable periods, an association with galactic cosmic ray cycles, and a correlation with evolution (electromagnetism affects all organisms).

Chapter 22, Item 3, Conclusion: The EUWS cycles produce a sequence of magnetic waves that knock matter around. (25% confidence)

Chapter 22, Item 4: Over the past 2.1 million years, Eastern Pacific Ocean temperatures fluctuated in phase with the 41.7-kyr EUWS frequency for 50 cycles.

Chapter 22, Item 4, Conclusion: The EUWS cycles act as the primary modulator of global temperature. (99% confidence)

Chapter 22, Item 5: Over the last 800 thousand years, the amplitude of Earth's paleointensity oscillated for 20 cycles in phase with the 41.7-kyr EUWS frequency.

Chapter 22, Item 5, Conclusion: The EUWS cycles act as the primary modulators for long-term cycles in geomagnetism. (99% confidence) Note: The Gleissberg and Schwabe solar cycles act as the primary short-term regulators of geomagnetism. (99% confidence)

Chapter 23, Item 1: Over the last 615 thousand years, the 13.9-kyr cycle repeated 44 times in Eastern Pacific Ocean temperatures.

Chapter 23, Item 1, Conclusion: The EUWS cycles act as the primary modulator of global temperature. (99% confidence)

Chapter 24, Item 1: For 1.5 cycles, the 4636-year EUWS cycle correlates with reconstructed sunspot activity.

Chapter 24, Item 1, Conclusion: For periods in excess of 300 years, the EUWS cycles act as the primary modulator of sunspot activity. (95% confidence)

Chapter 24, Item 2: Over the last 50 thousand years, the 4636-year EUWS frequency appeared for 10 cycles in the Greenland ice core.

Chapter 24, Item 2, Conclusion: The EUWS cycles act as the primary modulator of global temperature. (99% confidence)

Chapter 25, Item 1: Over the last 10 thousand years, the 1545-year EUWS cycle correlated with reconstructed sunspot activity for 6 cycles.

Chapter 25, Item 1, Conclusion: For periods in excess of 300 years, the EUWS cycles act as the primary modulator of sunspot activity. (95% confidence)

Chapter 25, Item 2: Over the last 50 thousand years, the 1545-year EUWS frequency appeared for 32 cycles in the Greenland ice core.

Chapter 25, Item 2, Conclusion: The EUWS cycles act as the primary modulator of global temperature. (99% confidence)

Chapter 25, Item 3: Over the last 6500 years, the 1545-year EUWS cycle coincided with 4 dark-ages in human civilization.

Summary, Analysis, and Conclusions

Chapter 25, Item 3, Conclusion: The EUWS cycles regulate human emotions. Collective emotions alternate between periods of cooperation, trust, and confidence and periods of individualism, distrust, and fear. (99% confidence)

Chapter 26, Item 1: For nearly 6000 years, civilizations in China, Meso-America, Egypt, Greece, Rome, and Europe have come and gone in cycles of 515 years.

Chapter 26, Item 1, Conclusion: The EUWS cycles regulate human emotions. Collective emotions alternate between periods of cooperation, trust, and confidence and periods of individualism, distrust, and fear. (99% confidence)

Chapter 26, Item 2: Over the last 9,500 years, the 515-year EUWS cycle correlates with reconstructed sunspot activity for 22 cycles.

Chapter 26, Item 2, Conclusion: For periods in excess of 300 years, the EUWS cycles act as the primary modulator of sunspot activity. (95% confidence)

Chapter 26, Item 3: Over the last 9,500 years, the 515-year EUWS cycle correlates with the climate proxy from the Greenland ice cores.

Chapter 26, Item 3, Conclusion: The EUWS cycles act as the primary modulator of global temperature. (99% confidence)

Chapter 27, Item 1: Peaks of the 172-year cycle coincide with the bursting of massive financial bubbles. The last two peaks of the 172-year cycle came in 1835 and 2007. For the United States, these two years coincided with the start of two of the greatest financial panics in its history. The only other panic to rival these two came with the great crash from 1929 to 1932.

Chapter 27, Item 1, Conclusion: The EUWS cycles regulate human emotions. Collective emotions alternate between periods of cooperation, trust, and confidence and periods of individualism, distrust, and fear. (99% confidence)

Chapter 27, Item 2: Ancient records of commodity prices from Babylonia, China, and England show evidence of a 172-year cycle corresponding to the EUWS frequency.

Chapter 27, Item 2, Conclusion: The EUWS cycles regulate human emotions. Collective emotions alternate between periods of cooperation, trust, and confidence and periods of individualism, distrust, and fear. (99% confidence)

Chapter 27, Item 3: Over the past 11,000 years, sunspot activity has fluctuated in cycles of 172-years that become somewhat masked by the 87.8-year Gleissberg solar cycle.

Chapter 27, Item 3, Conclusion: The 172-year EUWS cycles shares control over sunspot activity with the Gleissberg cycle. For frequencies under 100 years, internal solar cycles dominate sunspot activity, with minor influence from EUWS cycles. (95% confidence)

Chapter 27, Item 4: Over the last 4,000 years, the 172-year EUWS cycle correlates with the climate proxy from the Greenland ice cores for 23 cycles.

Chapter 27, Item 4, Conclusion: The EUWS cycles act as the primary modulator of global temperature. (99% confidence)

Chapter 28, Item 1: On a global scale since 1663, economic depressions started very close to theoretical peaks of the 57-year cycle for 7 consecutive cycles. The only significant deviation came in 1929, when a global depression started one 19-year sub-cycle early.

Chapter 28, Item 1, Conclusion: The EUWS cycles regulate human emotions. Collective emotions alternate between periods of cooperation, trust, and confidence and periods of individualism, distrust, and fear. (99% confidence)

Chapter 28, Item 2: Ancient records of commodity prices from Babylonia, China, England, Italy, Belgium, France, Germany, Turkey, and Poland show evidence of a 57-year cycle corresponding to the EUWS frequency.

Chapter 28, Item 2, Conclusion: The EUWS cycles regulate human emotions. Collective emotions alternate between periods of cooperation, trust, and confidence and periods of individualism, distrust, and fear. (99% confidence)

Chapter 28, Item 3: Measured on a monthly basis, global temperatures have fluctuated in 57-year cycles since 1850. Furthermore, global temperatures turned significantly lower during January 2007, which corresponded to a 172-year EUWS peak.

Chapter 28, Item 3, Conclusion: The EUWS cycles act as the primary modulator of global temperature. (99% confidence)

Chapter 29, Item 1: In the USA and England, financial panics – corresponding to the 19-year EUWS cycle – repeated for 15 out of 16 cycles since 1720. (The 1949 cycle skipped a beat.)

Chapter 29, Item 1, Conclusion: The EUWS cycles regulate human emotions. Collective emotions alternate between periods of cooperation, trust, and confidence and periods of individualism, distrust, and fear. (99% confidence)

Chapter 29, Item 2: Ancient records of commodity prices from Babylonia, England, Italy, Belgium, France, Germany, Turkey, and Poland show evidence of a 19-year cycle corresponding to the EUWS frequency.

Summary, Analysis, and Conclusions

Chapter 29, Item 2, Conclusion: The EUWS cycles regulate human emotions. Collective emotions alternate between periods of cooperation, trust, and confidence and periods of individualism, distrust, and fear. (99% confidence)

Chapter 29, Item 3: In the USA, major monetary legislation reoccurs in a 19-year cycle, slightly lagging theoretical peaks of the 19-year EUWS cycles. As each new panic erupted, Congress tried to "fix" the monetary system to prevent future panics. Obviously, every one of these Congressional fixes failed.

Chapter 29, Item 3, Conclusion: Government laws cannot control emotional cycles. At best, a properly designed monetary system can limit damage during a panic cycle. In the end, the EUWS cycles regulate human emotions. Collective emotions alternate between periods of cooperation, trust, and confidence and periods of individualism, distrust, and fear. (99% confidence)

Chapter 30, Item 1: Since 1911, the Dow Jones Industrial Average fluctuated in unison with the 6.36-year EUWS frequency for 15 cycles.

Chapter 30, Item 1, Conclusion: The EUWS cycles regulate human emotions. Collective emotions alternate between periods of cooperation, trust, and confidence and periods of individualism, distrust, and fear. (99% confidence)

Chapter 30, Item 2: For the past 45 years, investor sentiment toward the stock market oscillated in unison with the 6.36-year EUWS cycle. Larger sentiment shifts come at 19-year EUWS peaks and troughs.

Chapter 30, Item 2, Conclusion: The EUWS cycles regulate human emotions. Collective emotions alternate between periods of cooperation, trust, and confidence and periods of individualism, distrust, and fear. (99% confidence)

Chapter 31, Item 1: Since 1934, the Dow Jones Industrial Average fluctuated in unison with the 2.12-year EUWS frequency for 34 cycles. The correlation mostly disappeared during the late-1940s and the 1950s, but reappeared after 1960.

Chapter 31, Item 1, Conclusion: The EUWS cycles regulate human emotions. Collective emotions alternate between periods of cooperation, trust, and confidence and periods of individualism, distrust, and fear. (99% confidence)

Chapter 32, Item 2: Since 1998, the Dow Jones Industrial Average fluctuated in unison with the 258-day EUWS frequency for 14 cycles. The correlation was virtually perfect.

Chapter 32, Item 2, Conclusion: The EUWS cycles regulate human emotions. Collective emotions alternate between periods of cooperation, trust, and confidence and periods of individualism, distrust, and fear. (99% confidence)

Chapter 33, Item 1: Since 2000, the Dow Jones Industrial Average fluctuated in unison with the 86-day EUWS frequency for 33 cycles. The correlation was virtually perfect.

Chapter 33, Item 1, Conclusion: The EUWS cycles regulate human emotions. Collective emotions alternate between periods of cooperation, trust, and confidence and periods of individualism, distrust, and fear. (99% confidence)

Chapter 34, Item 1: With periods of roughly one month, the 28.68-day EUWS cycle competes with the 27-day solar cycle and the 29.53-year lunar cycle. All three cycles influence the stock market. Their interaction creates interference – making visibility of the 28.68-day cycle somewhat obscure.

Chapter 34, Item 1, Conclusion: The 28.68-day EUWS cycle, the 27-day solar cycle, and the 29.53-day lunar cycle share control over market activity in the monthly range. (99% confidence)

Chapter 34, Item 2: The 28.68-day EUWS frequency may regulate cycles in solar neutrinos. Weak evidence of a neutrino–EUWS relationship also appears for the 2.12-year frequency.

Chapter 34, Item 2, Conclusion: The EUWS cycles modulate formation of neutrinos in the Sun's core. (50% confidence)

Type of Force that regulates the EUWS cycles: Even though the EUWS cycles exhibit characteristics linking them to magnetism, gravity, and the weak nuclear force, more than likely, the source modulator of the EUWS cycles comes from another force. The variable periods and the spiked tops eliminate gravity. And for cycles less than fifty years, inconsistencies in expected Geomagnetic AA oscillations eliminate an electromagnetic source. The short-term EUWS cycles impact human behavior more extensively than geomagnetism. Yet, the short-term EUWS cycles show no visible influence on short-term Geomagnetic AA oscillations. If the EUWS cycles were truly magnetic, they should alter geomagnetism – similar to the way the Schwabe solar cycle leaves its imprint on Geomagnetic AA. But they don't. Therefore, instead of being electromagnetic waves, the EUWS cycles seem to behave as modulators of electromagnetic waves.

The source of the EUWS cycles remains puzzling. The EUWS cycles show their presence by their recurring impact on our universe, Earth's geology, global climate, and life. Yet, they leave limited evidence revealing their nature. Perhaps, their mystery indicates a link to the other ambiguous components of our universe – dark matter and dark energy. If the EUWS cycles behave as propagators of gravity, then they could easily create the gravitational illusions we perceive as dark matter and dark energy. Ultimately, scientists will discover the source. The *Unified Cycle Theory* provides the first major step in that direction.

Summary, Analysis, and Conclusions

In spite of the mystery surrounding their origin, **the existence of the Extra-Universal Wave Series cycles seems certain beyond reasonable doubt**. Pure and simple, these cycles left their trail in all of the major sciences, throughout all timescales, and in a consistent and harmonic manner that reduces the chances of these events being coincidence to virtually none. Furthermore, an origin from outside our observable universe seems highly likely. Yet, the precise manner in which the EUWS cycles give life and death to universes, galaxies, stars, planets, and species remains a secret of nature – at least, for now.

Chapter 36
Summary of the Unified Cycle Theory

1) **In our universe, two types of cycles dominate the movement of matter and energy:**

   - Gravitational cycles.
   - The Extra Universal Wave Series (EUWS) cycles.

2) **The force that modulates the EUWS cycles remains unknown. However, the EUWS cycles display characteristics that validate their existence and distinguish them from other classes of cycles:**

   - The frequency of each child-cycle exactly equals its parent-cycle's frequency divided by 3.
   - The magnitudes of EUWS cycles rise in proportion to their frequencies.
   - To varying degrees, 25 EUWS cycles have been identified with frequencies between 28.68 days and 22.2 billion years.
   - The turning points for each child cycle derive from the parent's theoretical turning points.
   - The EUWS cycles impact the formation of universes, stars, and planets.
   - The EUWS cycles affect human emotions.
   - The 22.2-gyr EUWS cycle is too large to originate within our universe. Hence, its origin lies outside our universe. This implies the existence of a multiverse.

3) **In addition to gravitational and EUWS cycles, on Earth, geomagnetic cycles also exert substantial influence. Important geomagnetic cycles include:**

   - An 87.8-year cycle related to the Gleissberg solar cycle.
   - An 11-year cycle related to the Schwabe solar cycle.
   - A 27-day cycle related to the 27-day solar cycle.
   - A semi-annual cycle related to Earth's position relative to the Sun and MW galactic center.
   - A 173.31-day cycle caused by the eclipse cycle.
   - A 28.68-day cycle associated with the Moon's passage through Earth's magneto-tail.

4) **As observed in markets, the Turning Point Distribution (TPD) Principle describes how cycles reverse:**

   - All market movements, involving all time periods, reside on the list of possible cyclic turning points.
   - The theoretical turning points of the geomagnetic cycles and the EUWS cycles hold the highest probabilities of being achieved – with a small margin of error.
   - In most instances, the various geomagnetic and EUWS cycles work in conflict. Some point higher, while others point lower.
   - The longer-term cycles determine the major trends.
   - Around the time of major inflection points, the major cycles yield their dominance to the short-term cycles. Short-term cycles play the major role in timing reversals.
   - Market tops usually deviate from their norms in a way that fits the collective pattern of all the cycles – both long-term and short-term.
   - Deviations aren't random. Instead, they center on <u>second most likely scenarios</u>, which equate to theoretical turning points of competing cycles.

5) **Emotions drive markets. Geomagnetic and EUWS cycles modulate human emotions:**

   - Geomagnetic and EUWS cycles affect the human nervous system, which causes the "emotional brain" to produce oscillations of confidence and fear – in phase with the physical cycles.
   - At tops, people first become fearful, and then stocks fall in response. At bottoms, people first become confident, and then stocks rise. In short, collective emotional swings act as the stock market's engine.
   - Financial manias coincide with periods when emotions run higher than normal, when the majority moves together in a herd, and when prices rise to exorbitant levels.
   - The use of leverage intensifies emotional involvement in markets because speculators risk losing everything when they purchase on margin. Because markets oscillate from cycles of confidence and fear, increased leverage results in increased emotions. Leisurely accumulation during a bull market, versus emotional, forced

Summary of the Unified Cycle Theory

selling during a bear market accounts for the difference needed to affect cycle visibility.

6) **On Earth, ice-age cycles result from:**

   - Gravitational cycles in eccentricity, obliquity, and precession play a minor role.
   - EUWS cycles in the range of 1.13-myr to 41.7-kyr act as primary modulators.
   - About every 1 to 2 million years, ice-ages alternate between 100-kyr and 41-kyr periods. This happens because of the enhancement/canceling effect produced by combining the amplitudes of the 41.0-kyr obliquity cycle and the 41.7-kyr EUWS cycle.

7) **The EUWS cycles left their imprints throughout the histories of our universe and Earth:**

   - The trough of the 22.2-gyr cycle coincided with the Big Bang.
   - The peak of the 22.2-gyr cycle coincided with the 2.7 Ga Event on Earth.
   - Cycles in Oxygen-18 correspond to all EUWS frequencies, indicating that EUWS cycles control global temperatures. Global temperature fluctuations are visible for all EUWS frequencies between 57.24-years and 2.46-gyr.
   - The carbon cycle correlates with all EUWS frequencies.
   - Geological timescales moderately correlate with EUWS cycles in the range of 3.38-myr to 821-myr.
   - Mantle plume events correlate with EUWS peaks in the range of 821-myr to 7.4-gyr.
   - Geological unconformities correlate with EUWS troughs in the range of 91.3-myr to 7.4-gyr.
   - The EUWS cycles regulate cycles in evolution and extinctions for frequencies between 30.4-myr and 7.4-gyr. During up-phases, organisms easily adapt to all environmental changes, even abrupt ones. During down-phases, organisms lose their ability to adapt and evolve. During down-phases, species must either survive dramatic environmental changes in their current form – or become extinct. The EUWS cycles accomplish this by triggering "evolutionary switches" in all organisms.
   - The human population shrank to near-extinction at the trough of the last 125-kyr cycle.

8) **The EUWS cycles modulate the formation of universes as follows:**

   - A multiverse exists. Each constituent universe within this structure operates as an open system – occasionally exchanging matter and energy with other universes in the system. In some cases, individual universes merge to form larger universes.
   - Because they operate as open systems, <u>matter never equals anti-matter</u> inside any universe.

- EUWS contractions in the trillion year range cause universes to implode. These contractions produce big crunches. Because individual universes do not contain offsetting elements in the correct proportions to totally annihilate each other, big crunches may approach singularity, but they never fully achieve it.
- Even though big crunches fail to achieve singularity, their super-condensed black holes eventually explode into big bangs. Big bangs develop once the appropriate trillion-year range EUWS cycle reaches a trough, and then begins to expand.

9) **The EUWS cycles exhibit a few characteristics linking them to gravity:**

- Milankovitch cycles tend to separate themselves with frequencies roughly equal to EUWS cycles. Two Milankovitch frequencies equal secondary harmonics of EUWS cycles – the 80-to-85-kyr year beat of the 100-kyr cycle and the 21-kyr precession cycle.
- Earth tends to contract and expand in phase with EUWS cycles – suggesting alternating gravitational cycles cause these periods of compression and expansion.
- The Big Bang occurred slightly after a 22.2-gyr trough, during the initial expansionary phase of the 22.2-gyr cycle. Associated with this initial expansionary phase, a relaxation of gravitational forces may have allowed a compressed universal black-hole to explode into the Big Bang.

10) **The EUWS cycles exhibit many characteristics linking them to electromagnetism:**

- Reversals usually appear as spikes, rather than smooth sine waves associated with gravity.
- On occasion, EUWS cycles deviate – displaying widely variable periods. These deviations closely resemble the variability of magnetic solar cycles.
- Some of the large EUWS cycles show a weak link to galactic cosmic rays.
- The EUWS cycles regulate evolution. This implies an electromagnetic component to these cycles because electromagnetism affects all organisms.

11) **Short-term EUWS cycles exhibit a mild link to the weak nuclear force:**

- The 28.68-day EUWS frequency may regulate cycles in solar neutrinos. Weak evidence of a neutrino–EUWS relationship also emerges for the 2.12-year frequency.
- Neutrinos primarily react via the weak nuclear force.

12) **EUWS cycles affect mankind in many ways:**

- The human population shrank to near-extinction at the trough of the last 125-kyr cycle.
- The 1545-year EUWS cycle correlates with dark-ages in history.
- The 515-year EUWS cycle controlled the rise and fall of numerous civilizations

over the past 6000 years.
- The last two 172-year EUWS theoretical peaks coincided with two of the three greatest financial speculations in USA history. Ancient records of commodity prices from Babylonia, China, and England show evidence of a 172-year cycle corresponding to the EUWS frequency.
- On a global basis, the last seven 57-year theoretical peaks (with one cycle deviating by a 19-year sub-cycle) coincided with periods of economic depression. Ancient records of commodity prices from Babylonia, China, England, Italy, Belgium, France, Germany, Turkey, and Poland show evidence of a 57-year cycle corresponding to the EUWS frequency.
- In the USA and England, 15 of the last 16 peaks of the 19-year EUWS cycle corresponded to financial panics. Ancient records of commodity prices from Babylonia, England, Italy, Belgium, France, Germany, Turkey, and Poland show evidence of a 19-year cycle corresponding to the EUWS frequency.
- In the USA, major monetary legislation reoccurs every 19 years, slightly lagging theoretical peaks of the 19-year EUWS cycle. After panic erupts, Congress tries to "fix" the monetary system to prevent future panics.
- Since 1911, the Dow Jones Industrial Average fluctuated in unison with the 6.36-year EUWS frequency for 15 cycles. For the past 45 years, investor sentiment toward the stock market oscillated in unison with the 6.36-year EUWS cycle.
- Since 1934, the Dow Jones Industrial Average correlated with the 2.12-year EUWS frequency for 34 cycles. The correlation became obscure during the late-1940s and the 1950s, but reappeared after 1960.
- Since 1998, the Dow Jones Industrial Average fluctuated in parallel with the 258-day EUWS frequency for 14 cycles. The correlation has been virtually perfect.
- Since 2000, the Dow Jones Industrial Average fluctuated in unison with the 86-day EUWS frequency for 33 cycles. The correlation has been virtually perfect.

13) The *Unified Cycle Theory* consolidates all major cycles into classes by their originating forces. In addition, the EUWS cycles unify many of the major sciences by modulating fluctuations for the following ... the multiverse, universes, Earth's geology, global temperatures, evolution, economies, markets, and society. From this perspective, the *Unified Cycle Theory* unifies both cycles and sciences.

Chapter 37
Predictions Based on the EUWS Cycles

1) Coinciding with the 172-year EUWS peak on January 24, 2007, a major global economic depression began. It started with the sub-prime meltdown in the USA during January 2007. It became global when virtually every major stock market reached all-time highs during October 2007. It entered its deflationary phase when global commodity markets crashed starting in July 2008. And it reached the crisis stage when numerous international banks began failing during September-November 2008. The developing depression could last a decade, or longer. As a prerequisite to hitting bottom, housing prices must stabilize and financial systems must be cleansed of trillions of dollars of leverage.

2) Coinciding with the 172-year EUWS peak on January 24, 2007, global temperatures should cool during most of the 21st Century. In fact, global temperatures began cooling moderately during 2007. This cooling should continue until the initial down-phase ends in 2035. See **Chart 37A**. This prediction will serve as a crucial test of the Unified *Cycle Theory*. Most current climate models project rapidly increasing global temperatures based on atmospheric emissions of industrial CO_2. Global climate conditions during the next 30 years should allow scientists to calculate the amount of temperature variation attributable to CO_2 emissions and the amount attributable to EUWS cycles.

3) Starting 268 million years ago, a new global cooling phase began. The cooling will reach an extreme with a pair of "Snowball Earth" episodes. **Chart 37B** puts this

prediction into perspective – showing all major events on Earth since our universe first formed.

Up arrow 1 – The Big Bang erupted at 13.73 Ga, near the 22.2-gyr cycle's theoretical low.

Up arrow 2 – The Sun and Earth formed around 4.54 Ga, coinciding with an 821-myr theoretical low.

Down arrow 3 – Global warming occurred at 2.7 Ga. At that time, Earth achieved record-breaking expansion as seen in the massive volcanic flood basalts created then. The 2.7 Ga Event coincided with theoretical peaks of the 22.2-gyr and 7.39-gyr EUWS cycles.

Down arrow 4 – This coincided with a global climate maximum at 268 Ma. At that time, the 2.46-gyr EUWS cycle peaked.

Vertical line between arrows 4 and 5 – The vertical line represents the present. The current plateau period will end within 6 million years.

Up arrow 5 – The next "Snowball Earth" epoch will hit bottom in 140-myr AP.

Up arrow 6 – A "Super Snowball Earth" epoch will end at 960-myr AP. At that time, the 7.39-gyr cycle will hit its theoretical low.

That concludes the explanation of the *Unified Cycle Theory*. At a minimum, I hope this book sparks intensified investigation into how the EUWS cycles affect mankind and our universe. The existence of the EUWS cycles means that many existing theories, in many different sciences, must be modified to account for them. The existence of these cycles means that we must alter of views of the universe and the way human populations act in response to them. Because of their existence, cycles certainly deserve more attention than they have received in the past.

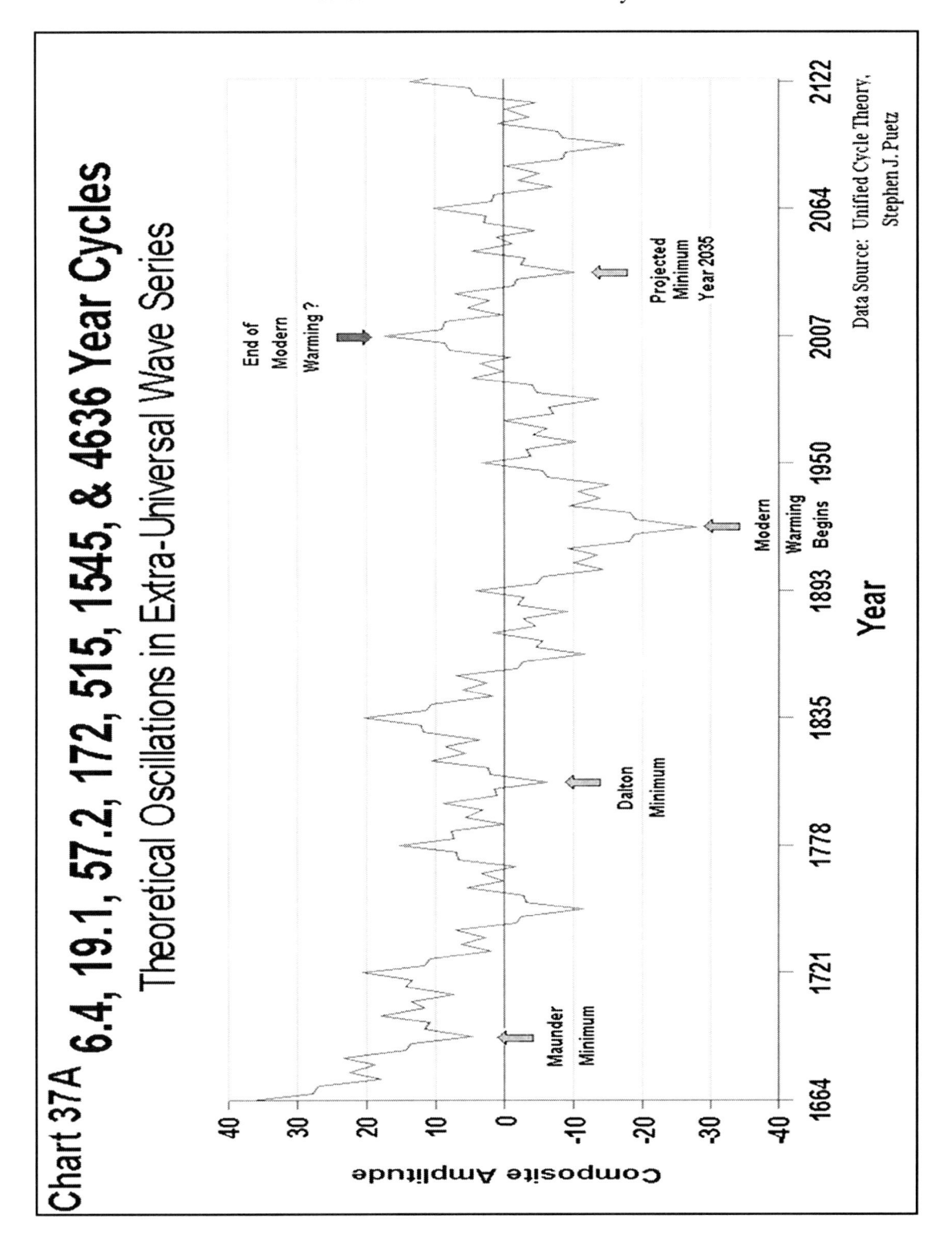

Predictions Based on the EUWS Cycles

Chart 37A **6.4, 19.1, 57.2, 172, 515, 1545, & 4636 Year Cycles**
Theoretical Oscillations in Extra-Universal Wave Series

Data Source: Unified Cycle Theory, Stephen J. Puetz

457

Chart 37B   .030, .091, .274, .821, 2.46, 7.39, & 22.2 Gyr Cycles

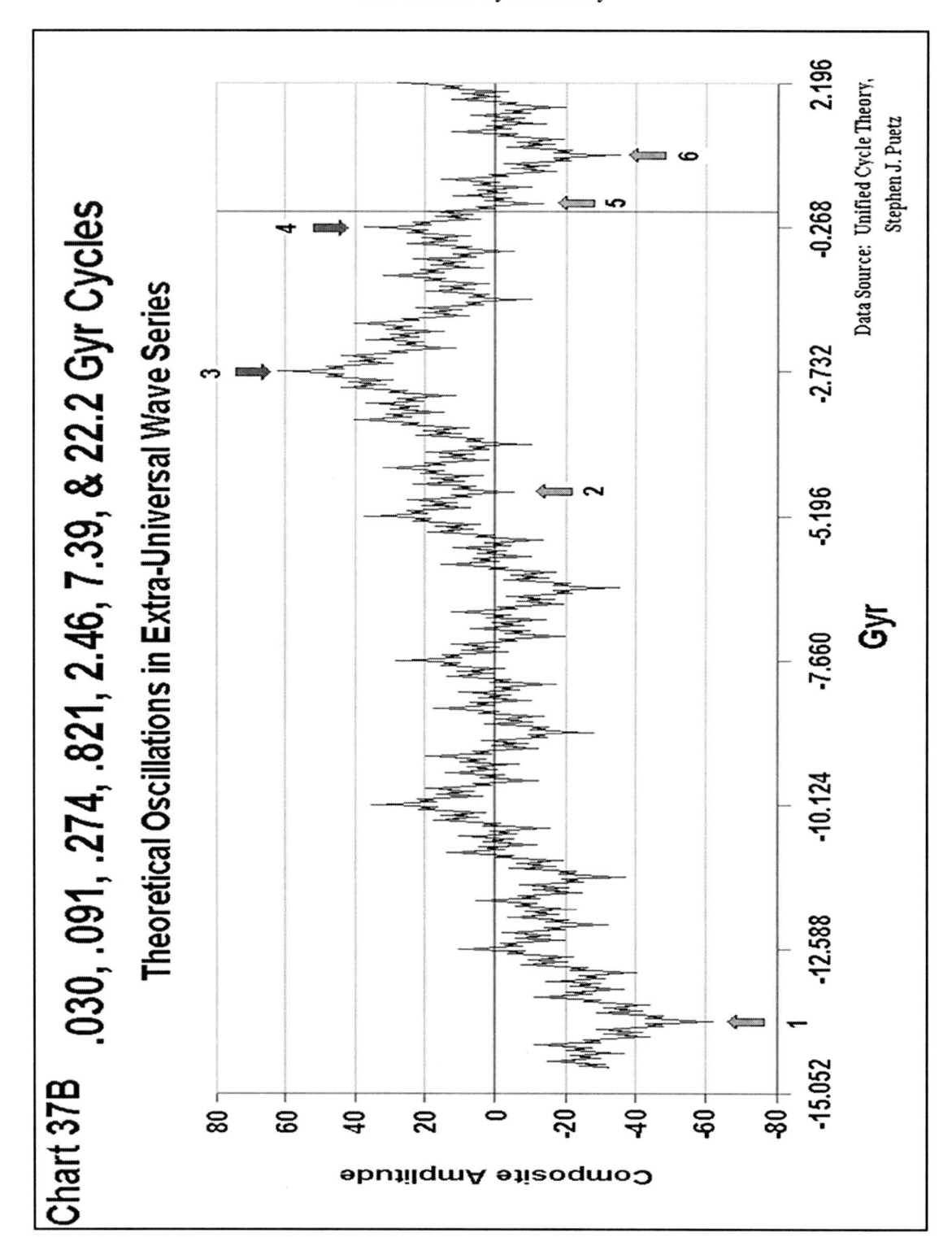

Theoretical Oscillations in Extra-Universal Wave Series

Data Source: Unified Cycle Theory, Stephen J. Puetz

Appendix – EUWS Cycle Peaks &Troughs

For the tables without troughs listed, the theoretical low resides midway between the peaks. Tables A1 through A25 contain theoretical turning points for all Extra-Universal Wave Series cycles between 28-7 days and 22.2 billion years.

Table A1 – 22.176-Gyr EUWS Cycle. Theoretical Peaks and Troughs.

Peak	Trough	Peak	Trough	Peak	Trough
91.436 Ga	80.348 Ga	24.908 Ga	13.820 Ga	41.620 GAP	52.708 GAP
69.260 Ga	58.172 Ga	2.732 Ga	8.356 GAP	63.796 GAP	74.884 GAP
47.084 Ga	35.996 Ga	19.444 GAP	30.532 GAP	85.972 GAP	97.060 GAP

Table A2 – 7.392-Gyr EUWS Cycle. Theoretical Peaks and Troughs.

Peak	Trough	Peak	Trough	Peak	Trough
47.084 Ga	43.388 Ga	24.908 Ga	21.212 Ga	2.732 Ga	0.964 Gvr AP
39.692 Ga	35.996 Ga	17.516 Ga	13.820 Ga	4.660 Gvr AP	8.356 Gvr AP
32.300 Ga	28.604 Ga	10.124 Ga	6.428 Ga	12.052 Gvr AP	15.748 Gvr AP

Table A3 – 2.464-Gyr EUWS Cycle. Theoretical Peaks and Troughs.

Peak	Trough	Peak	Trough	Peak	Trough
15.052 Ga	13.820 Ga	7.660 Ga	6.428 Ga	0.268 Ga	0.964 Gvr AP
12.588 Ga	11.356 Ga	5.196 Ga	3.964 Ga	2.196 Gvr AP	3.428 Gvr AP
10.124 Ga	8.892 Ga	2.732 Ga	1.500 Ga	4.660 Gvr AP	5.892 Gvr AP

Table A4 – 821.331-Myr EUWS Cycle. Theoretical Peaks and Troughs.

Peak	Trough	Peak	Trough	Peak	Trough
5.196 Ga	4.785 Ga	2.732 Ga	2.321 Ga	0.268 Ga	0.143 Gvr AP
4.375 Ga	3.964 Ga	1.911 Ga	1.500 Ga	0.553 Gvr AP	0.964 Gvr AP
3.553 Ga	3.143 Ga	1.089 Ga	0.679 Ga	1.375 Gvr AP	1.785 Gvr AP

Table A5 – 273.777-Myr EUWS Cycle. Theoretical Peaks and Troughs.

Peak	Trough	Peak	Trough	Peak	Trough
5.196 Ga	5.059 Ga	2.458 Ga	2.321 Ga	0.280 Gvr AP	0.417 Gvr AP
4.922 Ga	4.785 Ga	2.184 Ga	2.047 Ga	0.553 Gvr AP	0.690 Gvr AP
4.648 Ga	4.511 Ga	1.911 Ga	1.774 Ga	0.827 Gvr AP	0.964 Gvr AP
4.375 Ga	4.238 Ga	1.637 Ga	1.500 Ga	1.101 Gvr AP	1.238 Gvr AP
4.101 Ga	3.964 Ga	1.363 Ga	1.226 Ga	1.375 Gvr AP	1.512 Gvr AP
3.827 Ga	3.690 Ga	1.089 Ga	0.952 Ga	1.649 Gvr AP	1.785 Gvr AP
3.553 Ga	3.416 Ga	0.815 Ga	0.679 Ga	1.922 Gvr AP	2.059 Gvr AP
3.279 Ga	3.143 Ga	0.542 Ga	0.405 Ga	2.196 Gvr AP	2.333 Gvr AP
3.006 Ga	2.869 Ga	0.268 Ga	0.131 Ga	2.470 Gvr AP	2.607 Gvr AP
2.732 Ga	2.595 Ga	.006 GAP	0.143 GAP	2.744 Gvr AP	2.881 Gvr AP

Appendix – EUWS Cycle Peaks &Troughs

Table A6 – 91.259-Myr EUWS Cycle. Theoretical Peaks and Troughs.

Peak	Trough	Peak	Trough	Peak	Trough
3.827 Ga	3.781 Ga	2.549 Ga	2.504 Ga	1.272 Ga	1.226 Ga
3.736 Ga	3.690 Ga	2.458 Ga	2.412 Ga	1.180 Ga	1.135 Ga
3.644 Ga	3.599 Ga	2.367 Ga	2.321 Ga	1.089 Ga	1.044 Ga
3.553 Ga	3.508 Ga	2.276 Ga	2.230 Ga	0.998 Ga	0.952 Ga
3.462 Ga	3.416 Ga	2.184 Ga	2.139 Ga	0.907 Ga	0.861 Ga
3.371 Ga	3.325 Ga	2.093 Ga	2.047 Ga	0.815 Ga	0.770 Ga
3.279 Ga	3.234 Ga	2.002 Ga	1.956 Ga	0.724 Ga	0.679 Ga
3.188 Ga	3.143 Ga	1.911 Ga	1.865 Ga	0.633 Ga	0.587 Ga
3.097 Ga	3.051 Ga	1.819 Ga	1.774 Ga	0.542 Ga	0.496 Ga
3.006 Ga	2.960 Ga	1.728 Ga	1.682 Ga	0.450 Ga	0.405 Ga
2.914 Ga	2.869 Ga	1.637 Ga	1.591 Ga	0.359 Ga	0.314 Ga
2.823 Ga	2.778 Ga	1.546 Ga	1.500 Ga	0.268 Ga	0.222 Ga
2.732 Ga	2.686 Ga	1.454 Ga	1.409 Ga	0.177 Ga	0.131 Ga
2.641 Ga	2.595 Ga	1.363 Ga	1.317 Ga	0.085 Ga	0.040 Ga

Table A7 – 30.420-Myr EUWS Cycle. Theoretical Peaks and Troughs.

Peak	Trough	Peak	Trough	Peak	Trough
2002 Ma	1987 Ma	1333 Ma	1317 Ma	663 Ma	648 Ma
1971 Ma	1956 Ma	1302 Ma	1287 Ma	633 Ma	618 Ma
1941 Ma	1926 Ma	1272 Ma	1257 Ma	603 Ma	587 Ma
1911 Ma	1895 Ma	1241 Ma	1226 Ma	572 Ma	557 Ma
1880 Ma	1865 Ma	1211 Ma	1196 Ma	542 Ma	526 Ma
1850 Ma	1835 Ma	1180 Ma	1165 Ma	511 Ma	496 Ma
1819 Ma	1804 Ma	1150 Ma	1135 Ma	481 Ma	466 Ma
1789 Ma	1774 Ma	1120 Ma	1104 Ma	450 Ma	435 Ma
1758 Ma	1743 Ma	1089 Ma	1074 Ma	420 Ma	405 Ma
1728 Ma	1713 Ma	1059 Ma	1044 Ma	390 Ma	374 Ma
1698 Ma	1682 Ma	1028 Ma	1013 Ma	359 Ma	344 Ma
1667 Ma	1652 Ma	998 Ma	983 Ma	329 Ma	314 Ma
1637 Ma	1622 Ma	968 Ma	952 Ma	298 Ma	283 Ma
1606 Ma	1591 Ma	937 Ma	922 Ma	268 Ma	253 Ma
1576 Ma	1561 Ma	907 Ma	892 Ma	237 Ma	222 Ma
1546 Ma	1530 Ma	876 Ma	861 Ma	207 Ma	192 Ma
1515 Ma	1500 Ma	846 Ma	831 Ma	177 Ma	161 Ma
1485 Ma	1469 Ma	815 Ma	800 Ma	146 Ma	131 Ma
1454 Ma	1439 Ma	785 Ma	770 Ma	116 Ma	101 Ma
1424 Ma	1409 Ma	755 Ma	739 Ma	85 Ma	70 Ma
1393 Ma	1378 Ma	724 Ma	709 Ma	55 Ma	40 Ma
1363 Ma	1348 Ma	694 Ma	679 Ma	25 Ma	9 Ma

Appendix – EUWS Cycle Peaks &Troughs

Table A8 – 10.140-Myr EUWS Cycle. Theoretical Peaks and Troughs.

Peak	Trough	Peak	Trough	Peak	Trough
561.95 Ma	556.88 Ma	369.30 Ma	364.23 Ma	176.64 Ma	171.57 Ma
551.81 Ma	546.74 Ma	359.16 Ma	354.09 Ma	166.50 Ma	161.43 Ma
541.67 Ma	536.60 Ma	349.02 Ma	343.95 Ma	156.36 Ma	151.29 Ma
531.53 Ma	526.46 Ma	338.88 Ma	333.81 Ma	146.22 Ma	141.15 Ma
521.39 Ma	516.32 Ma	328.74 Ma	323.67 Ma	136.08 Ma	131.01 Ma
511.25 Ma	506.18 Ma	318.60 Ma	313.53 Ma	125.94 Ma	120.87 Ma
501.11 Ma	496.04 Ma	308.46 Ma	303.39 Ma	115.80 Ma	110.73 Ma
490.97 Ma	485.90 Ma	298.32 Ma	293.25 Ma	105.66 Ma	100.59 Ma
480.83 Ma	475.76 Ma	288.18 Ma	283.11 Ma	95.52 Ma	90.45 Ma
470.69 Ma	465.62 Ma	278.04 Ma	272.97 Ma	85.38 Ma	80.31 Ma
460.55 Ma	455.48 Ma	267.90 Ma	262.83 Ma	75.24 Ma	70.17 Ma
450.41 Ma	445.35 Ma	257.76 Ma	252.69 Ma	65.10 Ma	60.03 Ma
440.27 Ma	435.21 Ma	247.62 Ma	242.55 Ma	54.96 Ma	49.89 Ma
430.14 Ma	425.07 Ma	237.48 Ma	232.41 Ma	44.82 Ma	39.75 Ma
420.00 Ma	414.93 Ma	227.34 Ma	222.27 Ma	34.68 Ma	29.61 Ma
409.86 Ma	404.79 Ma	217.20 Ma	212.13 Ma	24.54 Ma	19.47 Ma
399.72 Ma	394.65 Ma	207.06 Ma	201.99 Ma	14.40 Ma	9.33 Ma
389.58 Ma	384.51 Ma	196.92 Ma	191.85 Ma	4.26 Ma	0.81 Myr AP
379.44 Ma	374.37 Ma	186.78 Ma	181.71 Ma	5.88 Myr AP	10.95 Myr AP

Table A9 – 3.380-Myr EUWS Cycle. Theoretical Peaks and Troughs.

Peak	Trough	Peak	Trough	Peak	Trough
163.118 Ma	161.428 Ma	105.66 Ma	103.97 Ma	48.199 Ma	46.509 Ma
159.738 Ma	158.048 Ma	102.28 Ma	100.59 Ma	44.819 Ma	43.129 Ma
156.358 Ma	154.668 Ma	98.899 Ma	97.209 Ma	41.439 Ma	39.749 Ma
152.978 Ma	151.288 Ma	95.519 Ma	93.829 Ma	38.059 Ma	36.369 Ma
149.598 Ma	147.908 Ma	92.139 Ma	90.449 Ma	34.679 Ma	32.989 Ma
146.218 Ma	144.528 Ma	88.759 Ma	87.069 Ma	31.299 Ma	29.609 Ma
142.838 Ma	141.148 Ma	85.379 Ma	83.689 Ma	27.919 Ma	26.229 Ma
139.458 Ma	137.768 Ma	81.999 Ma	80.309 Ma	24.539 Ma	22.849 Ma
136.078 Ma	134.388 Ma	78.619 Ma	76.929 Ma	21.160 Ma	19.470 Ma
132.698 Ma	131.008 Ma	75.239 Ma	73.549 Ma	17.780 Ma	16.090 Ma
129.318 Ma	127.628 Ma	71.859 Ma	70.169 Ma	14.400 Ma	12.710 Ma
125.938 Ma	124.248 Ma	68.479 Ma	66.789 Ma	11.020 Ma	9.330 Ma
122.558 Ma	120.869 Ma	65.099 Ma	63.409 Ma	7.640 Ma	5.950 Ma
119.179 Ma	117.489 Ma	61.719 Ma	60.029 Ma	4.260 Ma	2.570 Ma
115.799 Ma	114.109 Ma	58.339 Ma	56.649 Ma	0.880 Ma	0.810 Mvr AP
112.419 Ma	110.729 Ma	54.959 Ma	53.269 Ma	2.500 Mvr AP	4.190 Mvr AP
109.039 Ma	107.349 Ma	51.579 Ma	49.889 Ma	5.880 Mvr AP	7.570 Mvr AP

Table A10 – 1.127-Myr EUWS Cycle. Theoretical Peaks and Troughs.

Peak	Trough	Peak	Trough	Peak	Trough
34.679 Ma	34.116 Ma	22.286 Ma	21.723 Ma	9.893 Ma	9.330 Ma
33.553 Ma	32.989 Ma	21.160 Ma	20.596 Ma	8.766 Ma	8.203 Ma
32.426 Ma	31.863 Ma	20.033 Ma	19.470 Ma	7.640 Ma	7.076 Ma
31.299 Ma	30.736 Ma	18.906 Ma	18.343 Ma	6.513 Ma	5.950 Ma
30.173 Ma	29.609 Ma	17.780 Ma	17.216 Ma	5.386 Ma	4.823 Ma
29.046 Ma	28.483 Ma	16.653 Ma	16.090 Ma	4.260 Ma	3.696 Ma
27.919 Ma	27.356 Ma	15.526 Ma	14.963 Ma	3.133 Ma	2.570 Ma
26.793 Ma	26.229 Ma	14.400 Ma	13.836 Ma	2.006 Ma	1.443 Ma
25.666 Ma	25.103 Ma	13.273 Ma	12.710 Ma	0.880 Ma	0.316 Ma
24.539 Ma	23.976 Ma	12.146 Ma	11.583 Ma	0.247 Mvr AP	0.810 Mvr AP
23.413 Ma	22.849 Ma	11.020 Ma	10.456 Ma	1.374 Mvr AP	1.937 Mvr AP

Appendix – EUWS Cycle Peaks & Troughs

Table A11 – 375,552-Year EUWS Cycle. Theoretical Peaks and Troughs.

Peak	Trough	Peak	Trough	Peak	Trough
11.020 Ma	10.832 Ma	6.889 Ma	6.701 Ma	2.757 Ma	2.570 Ma
10.644 Ma	10.456 Ma	6.513 Ma	6.325 Ma	2.382 Ma	2.194 Ma
10.269 Ma	10.081 Ma	6.137 Ma	5.950 Ma	2.006 Ma	1.819 Ma
9.893 Ma	9.705 Ma	5.762 Ma	5.574 Ma	1.630 Ma	1.443 Ma
9.517 Ma	9.330 Ma	5.386 Ma	5.199 Ma	1.255 Ma	1.067 Ma
9.142 Ma	8.954 Ma	5.011 Ma	4.823 Ma	0.880 Ma	0.692 Ma
8.766 Ma	8.579 Ma	4.635 Ma	4.447 Ma	0.504 Ma	0.316 Ma
8.391 Ma	8.203 Ma	4.260 Ma	4.072 Ma	0.129 Ma	0.059 Mvr AP
8.015 Ma	7.827 Ma	3.884 Ma	3.696 Ma	0.247 Mvr AP	0.435 Mvr AP
7.640 Ma	7.452 Ma	3.509 Ma	3.321 Ma	0.622 Mvr AP	0.810 Mvr AP
7.264 Ma	7.076 Ma	3.133 Ma	2.945 Ma	0.998 Mvr AP	1.186 Mvr AP

Table A12 – 125,184-Year EUWS Cycle. Theoretical Peaks and Troughs.

Peak	Trough	Peak	Trough	Peak	Trough
4009.3 Ka	3946.7 Ka	2507.1 Ka	2444.5 Ka	1004.9 Ka	942.3 Ka
3884.1 Ka	3821.5 Ka	2381.9 Ka	2319.3 Ka	879.7 Ka	817.1 Ka
3758.9 Ka	3696.4 Ka	2256.7 Ka	2194.1 Ka	754.5 Ka	691.9 Ka
3633.8 Ka	3571.2 Ka	2131.6 Ka	2069.0 Ka	629.3 Ka	566.8 Ka
3508.6 Ka	3446.0 Ka	2006.4 Ka	1943.8 Ka	504.2 Ka	441.6 Ka
3383.4 Ka	3320.8 Ka	1881.2 Ka	1818.6 Ka	379.0 Ka	316.4 Ka
3258.2 Ka	3195.6 Ka	1756.0 Ka	1693.4 Ka	253.8 Ka	191.2 Ka
3133.0 Ka	3070.4 Ka	1630.8 Ka	1568.2 Ka	128.6 Ka	66.0 Ka
3007.8 Ka	2945.2 Ka	1505.6 Ka	1443.0 Ka	3.4 Ka	59.2 Kvr AP
2882.7 Ka	2820.1 Ka	1380.4 Ka	1317.9 Ka	121.8 Kvr AP	184.3 Kvr AP
2757.5 Ka	2694.9 Ka	1255.3 Ka	1192.7 Ka	246.9 Kvr AP	309.5 Kvr AP
2632.3 Ka	2569.7 Ka	1130.1 Ka	1067.5 Ka	372.1 Kvr AP	434.7 Kvr AP

Table A13 – 41,728-Year EUWS Cycle. Theoretical Peaks and Troughs.

Peak	Trough	Peak	Trough	Peak	Trough
2215.0 Ka	2194.1 Ka	1380.5 Ka	1359.6 Ka	545.9 Ka	525.0 Ka
2173.3 Ka	2152.4 Ka	1338.7 Ka	1317.9 Ka	504.2 Ka	483.3 Ka
2131.6 Ka	2110.7 Ka	1297.0 Ka	1276.1 Ka	462.4 Ka	441.6 Ka
2089.8 Ka	2069.0 Ka	1255.3 Ka	1234.4 Ka	420.7 Ka	399.8 Ka
2048.1 Ka	2027.2 Ka	1213.5 Ka	1192.7 Ka	379.0 Ka	358.1 Ka
2006.4 Ka	1985.5 Ka	1171.8 Ka	1150.9 Ka	337.3 Ka	316.4 Ka
1964.6 Ka	1943.8 Ka	1130.1 Ka	1109.2 Ka	295.5 Ka	274.7 Ka
1922.9 Ka	1902.1 Ka	1088.4 Ka	1067.5 Ka	253.8 Ka	232.9 Ka
1881.2 Ka	1860.3 Ka	1046.6 Ka	1025.8 Ka	212.1 Ka	191.2 Ka
1839.5 Ka	1818.6 Ka	1004.9 Ka	984.0 Ka	170.3 Ka	149.5 Ka
1797.7 Ka	1776.9 Ka	963.2 Ka	942.3 Ka	128.6 Ka	107.7 Ka
1756.0 Ka	1735.1 Ka	921.4 Ka	900.6 Ka	86.9 Ka	66.0 Ka
1714.3 Ka	1693.4 Ka	879.7 Ka	858.9 Ka	45.2 Ka	24.3 Ka
1672.5 Ka	1651.7 Ka	838.0 Ka	817.1 Ka	3.4 Ka	17.4 Kvr AP
1630.8 Ka	1609.9 Ka	796.3 Ka	775.4 Ka	38.3 Kvr AP	59.2 Kvr AP
1589.1 Ka	1568.2 Ka	754.5 Ka	733.7 Ka	80.0 Kvr AP	100.9 Kvr AP
1547.4 Ka	1526.5 Ka	712.8 Ka	691.9 Ka	121.8 Kvr AP	142.6 Kvr AP
1505.6 Ka	1484.8 Ka	671.1 Ka	650.2 Ka	163.5 Kvr AP	184.3 Kvr AP
1463.9 Ka	1443.0 Ka	629.3 Ka	608.5 Ka	205.2 Kvr AP	226.1 Kvr AP
1422.2 Ka	1401.3 Ka	587.6 Ka	566.8 Ka	246.9 Kvr AP	267.8 Kvr AP

Appendix – EUWS Cycle Peaks &Troughs

Table A14 – 13,909-Year EUWS Cycle. Theoretical Peaks and Troughs.

Peak	Trough	Peak	Trough	Peak	Trough
531,982 BP	525,027 BP	323,342 BP	316,387 BP	114,702 BP	107,747 BP
518,072 BP	511,118 BP	309,432 BP	302,478 BP	100,793 BP	93,838 BP
504,163 BP	497,208 BP	295,523 BP	288,568 BP	86,883 BP	79,929 BP
490,254 BP	483,299 BP	281,614 BP	274,659 BP	72,974 BP	66,019 BP
476,344 BP	469,390 BP	267,704 BP	260,750 BP	59,065 BP	52,110 BP
462,435 BP	455,480 BP	253,795 BP	246,840 BP	45,155 BP	38,201 BP
448,526 BP	441,571 BP	239,886 BP	232,931 BP	31,246 BP	24,291 BP
434,616 BP	427,662 BP	225,976 BP	219,022 BP	17,337 BP	10,382 BP
420,707 BP	413,752 BP	212,067 BP	205,112 BP	3,427 BP	3,537 AP
406,798 BP	399,843 BP	198,158 BP	191,203 BP	10,482 AP	17,437 AP
392,888 BP	385,934 BP	184,249 BP	177,294 BP	24,391 AP	31,346 AP
378,979 BP	372,024 BP	170,339 BP	163,385 BP	38,301 AP	45,255 AP
365,070 BP	358,115 BP	156,430 BP	149,475 BP	52,210 AP	59,165 AP
351,160 BP	344,206 BP	142,521 BP	135,566 BP	66,119 AP	73,074 AP
337,251 BP	330,296 BP	128,611 BP	121,656 BP	80,029 AP	86,983 AP

Table A15 – 4636.44-Year EUWS Cycle. Theoretical Peaks and Troughs.

Peak	Trough	Peak	Trough	Peak	Trough
77,610 BP	75,292 BP	45,155 BP	42,837 BP	12,700 BP	10,382 BP
72,974 BP	70,656 BP	40,519 BP	38,201 BP	8,064 BP	5,746 BP
68,338 BP	66,019 BP	35,882 BP	33,564 BP	3,427 BP	1,109 BP
63,701 BP	61,383 BP	31,246 BP	28,928 BP	1,209 AP	3,527 AP
59,065 BP	56,746 BP	26,610 BP	24,291 BP	5,846 AP	8,164 AP
54,428 BP	52,110 BP	21,973 BP	19,655 BP	10,482 AP	12,800 AP
49,792 BP	47,474 BP	17,337 BP	15,018 BP	15,118 AP	17,437 AP

Table A16 – 1545.48-Year EUWS Cycle. Theoretical Peaks and Troughs.

Peak	Trough	Peak	Trough	Peak	Trough
26,610 BP	25,837 BP	15,791 BP	15,018 BP	4,973 BP	4,200 BP
25,064 BP	24,291 BP	14,246 BP	13,473 BP	3,427 BP	2,655 BP
23,519 BP	22,746 BP	12,700 BP	11,927 BP	1,882 BP	1,109 BP
21,973 BP	21,200 BP	11,155 BP	10,382 BP	336 BP	436 AP
20,428 BP	19,655 BP	9,609 BP	8,837 BP	1,209 AP	1,982 AP
18,882 BP	18,109 BP	8,064 BP	7,291 BP	2,755 AP	3,527 AP
17,337 BP	16,564 BP	6,518 BP	5,746 BP	4,300 AP	5,073 AP

Table A17 – 515.16-Year EUWS Cycle. Theoretical Peaks and Troughs.

Peak	Trough	Peak	Trough	Peak	Trough
6578 BC	6321 BC	3487 BC	3230 BC	397 BC	139 BC
6063 BC	5806 BC	2972 BC	2715 BC	118	375
5548 BC	5291 BC	2457 BC	2200 BC	633	890
5033 BC	4775 BC	1942 BC	1684 BC	1148	1406
4518 BC	4260 BC	1427 BC	1169 BC	1663	1921
4003 BC	3745 BC	912 BC	654 BC	2178	2436

Table A18 – 171.72-Year EUWS Cycle. Theoretical Peaks and Troughs.

Peak	Trough	Peak	Trough	Peak	Trough
3316 BC	3230 BC	1427 BC	1341 BC	461	547
3144 BC	3058 BC	1255 BC	1169 BC	633	719
2972 BC	2886 BC	1083 BC	998 BC	805	890
2801 BC	2715 BC	912 BC	826 BC	976	1062
2629 BC	2543 BC	740 BC	654 BC	1148	1234
2457 BC	2371 BC	568 BC	482 BC	1320	1406
2285 BC	2200 BC	397 BC	311 BC	1491	1577
2114 BC	2028 BC	225 BC	139 BC	1663	1749
1942 BC	1856 BC	53 BC	32	1835	1921
1770 BC	1684 BC	118	204	2007	2092
1599 BC	1513 BC	289	375	2178	2264

Appendix – EUWS Cycle Peaks &Troughs

Table A19 – 57.24-Year EUWS Cycle. Theoretical Peaks from 2801 BC to 2293.

2801 BC	1942 BC	1083 BC	225 BC	633	1491
2743 BC	1885 BC	1026 BC	168 BC	690	1549
2686 BC	1828 BC	969 BC	110 BC	747	1606
2629 BC	1770 BC	912 BC	53 BC	805	1663
2572 BC	1713 BC	854 BC	3	862	1720
2514 BC	1656 BC	797 BC	60	919	1778
2457 BC	1599 BC	740 BC	118	976	1835
2400 BC	1541 BC	683 BC	175	1033	1892
2343 BC	1484 BC	625 BC	232	1091	1949
2286 BC	1427 BC	568 BC	289	1148	2007
2228 BC	1370 BC	511 BC	347	1205	2064
2171 BC	1312 BC	454 BC	404	1262	2121
2114 BC	1255 BC	397 BC	461	1320	2178
2056 BC	1198 BC	339 BC	518	1377	2236
1999 BC	1141 BC	282 BC	576	1434	2293

Table A20 – 19.08-Year EUWS Cycle. Theoretical Peaks from 187 BC to 2083.

Nov 10, 187 BC	June 18, 194	Jan 23, 576	Aug 30, 957	Apr 8, 1339	Nov 10, 1720
Dec 9, 168 BC	July 17, 213	Feb, 22, 595	Sept 28, 976	May 5, 1358	Dec 9, 1739
Jan 7, 148 BC	Aug 14, 232	Mar 22, 614	Oct 27, 995	June 2, 1377	Jan 7, 1759
Feb 5, 129 BC	Sept 11, 251	Apr 22, 633	Nov 25, 1014	July 1, 1396	Feb 5, 1778
Mar 9, 110 BC	Oct 11, 270	May 19, 652	Dec 25, 1033	July 30, 1415	Mar 9, 1797
Apr 8, 91 BC	Nov 10, 289	June 18, 671	Jan 23, 1053	Aug 30, 1434	Apr 8, 1816
May 5, 72 BC	Dec 9, 308	July 17, 690	Feb 22, 1072	Sept 28, 1453	May 5, 1835
June 2, 53 BC	Jan 7, 328	Aug 14, 709	Mar 22, 1091	Oct 27, 1472	June 4, 1854
July 1, 34 BC	Feb 5, 347	Sept 11, 728	Apr 22, 1110	Nov 25, 1491	July 3, 1873
July 30, 15 BC	Mar 9, 366	Oct 11, 747	May 19, 1129	Dec 25, 1510	Aug 1, 1892
Aug 30, 3	Apr 8, 385	Nov 10, 766	June 18, 1148	Jan 23, 1530	Sept 1, 1911
Sept 28, 22	May 5, 404	Dec 9, 785	July 17, 1167	Feb 22, 1549	Sept 30, 1930
Oct 27, 41	June 2, 423	Jan 7, 805	Aug 14, 1186	Mar 22, 1568	Oct 29, 1949
Nov 25, 60	July 1, 442	Feb 5, 824	Sept 11, 1205	Apr 22, 1587	Nov 27, 1968
Dec 25, 79	July 30, 461	Mar 9, 843	Oct 11, 1224	May 19, 1606	Dec 27, 1987
Jan 23, 99	Aug 30, 480	Apr 8, 862	Nov 10, 1243	June 18, 1625	Jan 24, 2007
Feb 22, 118	Sept 28, 499	May 5, 881	Dec 9, 1262	July 17, 1644	Feb 23, 2026
Mar 22, 137	Oct 27, 518	June 2, 900	Jan 7, 1282	Aug 14, 1663	Mar 24, 2045
Apr 22, 156	Nov 25, 537	July 1, 919	Feb 5, 1301	Sept 11, 1682	Apr 22, 2064
May 19, 175	Dec 25, 556	July 30, 938	Mar 9, 1320	Oct 11, 1701	May 21, 2083

Appendix – EUWS Cycle Peaks &Troughs

Table A21 – 6.36-Year EUWS Cycle. Theoretical Peaks from 1580 to 2070.

Dec 11, 1580	Aug 14, 1663	Apr 20, 1746	Dec 27, 1828	Sept 1, 1911	May 7, 1994
Apr 22, 1587	Dec 25, 1669	Aug 30, 1752	May 5, 1835	Jan 10, 1918	Sep 15, 2000
Sept 1, 1593	May 3, 1676	Jan 7, 1759	Sept 13, 1841	May 21, 1924	Jan 24, 2007
Jan 7, 1600	Sept 11, 1682	May 19, 1765	Jan 23, 1848	Sept 30, 1930	June 4, 2013
May 19, 1606	Jan 22, 1689	Sept 28, 1771	June 4, 1854	Feb 8, 1937	Oct 14, 2019
Sept 28, 1612	June 2, 1695	Feb 5, 1778	Oct 13, 1860	June 20, 1943	Feb 23, 2026
Feb 5, 1619	Oct 11, 1701	June 18, 1784	Feb 22, 1867	Oct 29, 1949	July 4, 2032
June 18, 1625	Feb 20, 1708	Oct 27, 1790	July 3, 1873	Mar 9, 1956	Nov 12, 2038
Oct 27, 1631	July 1, 1714	Mar 9, 1797	Nov 12, 1879	July 19, 1962	Mar 24, 2045
Mar 9, 1638	Nov 10, 1720	July 17, 1803	Mar 23, 1886	Nov 27, 1968	Aug 2, 2051
July 17, 1644	Mar 21, 1727	Nov 25, 1809	Aug 1, 1892	April 8, 1975	Dec 12, 2057
Nov 25, 1650	July 30, 1733	Apr 8, 1816	Dec 11, 1898	Aug 17, 1981	Apr 22, 2064
Apr 6, 1657	Dec 9, 1739	Aug 17, 1822	Apr 22, 1905	Dec 27, 1987	Aug 31, 2070

Table A22 – 2.12-Year EUWS Cycle. Theoretical Peaks from 1881 to 2019.

Dec 25, 1881	Apr 22, 1905	Aug 16, 1928	Dec 12, 1951	Apr 8, 1975	Aug 2, 1998
Feb 7, 1884	June 5, 1907	Sept 30, 1930	Jan 24, 1954	May 21, 1977	Sep 15, 2000
Mar 23, 1886	July 18, 1909	Nov 12, 1932	Mar 9, 1956	July 4, 1979	Oct 29, 2002
May 5, 1888	Sept 1, 1911	Dec 26, 1934	Apr 22, 1958	Aug 17, 1981	Dec 11, 2004
Jun 18, 1890	Oct 14, 1913	Feb 8, 1937	June 4, 1960	Sep 30, 1983	Jan 24, 2007
Aug 1, 1892	Nov 27, 1915	Mar 24, 1939	July 19, 1962	Nov 12, 1985	Mar 9, 2009
Sept 14, 1894	Jan 10, 1918	May 6, 1941	Aug 31, 1964	Dec 27, 1987	Apr 22, 2011
Oct 27, 1896	Feb 23, 1920	June 20, 1943	Oct 14, 1966	Feb. 8, 1990	June 4, 2013
Dec 11, 1898	Apr 7, 1922	Aug 2, 1945	Nov 27, 1968	Mar. 23, 1992	July 19, 2015
Jan 24, 1901	May 21, 1924	Sept 15, 1947	Jan 10, 1971	May 7, 1994	Aug 31, 2017
Mar 9, 1903	July 4, 1926	Oct 29, 1949	Feb 22, 1973	June 19, 1996	Oct 14, 2019

Table A23 – 258.11-Day EUWS Cycle. Theoretical Peaks from 1964 to 2019.

Aug 31, 1964	Nov 7, 1973	Jan 15, 1983	Mar 23, 1992	May 31, 2001	Aug 7, 2010
May 16, 1965	July 23, 1974	Sept 30, 1983	Dec 6, 1992	Feb 13, 2002	Apr 22, 2011
Jan 29, 1966	Apr 8, 1975	June 14, 1984	Aug 21, 1993	Oct 29, 2002	Jan 5, 2012
Oct 14, 1966	Dec 22, 1975	Feb 27, 1985	May 7, 1994	July 14, 2003	Sept 19, 2012
June 29, 1967	Sept 5, 1976	Nov 12, 1985	Jan 20, 1995	Mar 28, 2004	June 4, 2013
Mar 13, 1968	May 21, 1977	July 28, 1986	Oct 5, 1995	Dec 11, 2004	Feb 18, 2014
Nov 27, 1968	Feb 3, 1978	Apr 12, 1987	June 19, 1996	Aug 26, 2005	Nov 3, 2014
Aug 12, 1969	Oct 19, 1978	Dec 27, 1987	Mar 4, 1997	May 11, 2006	July 19, 2015
Apr 27, 1970	July 4, 1979	Sept 10, 1988	Nov 17, 1997	Jan 24, 2007	Apr 2, 2016
Jan 10, 1971	Mar 18, 1980	May 26, 1989	Aug 2, 1998	Oct 10, 2007	Dec 16, 2016
Sept 25, 1971	Dec 1, 1980	Feb 8, 1990	Apr 17, 1999	June 24, 2008	Aug 31, 2017
June 9, 1972	Aug. 17, 1981	Oct 24, 1990	Dec 31, 1999	Mar 9, 2009	May 16, 2018
Feb 22, 1973	May 3, 1982	July 9, 1991	Sept 15, 2000	Nov 22, 2009	Jan 29, 2019

Table A24 – 86.04-Day EUWS Cycle. Theoretical Peaks from 1991 to 2015.

Oct 3, 1991	Oct 5, 1995	Oct 6, 1999	Oct 8, 2003	Oct 10, 2007	Oct 11, 2011
Dec 28, 1991	Dec 30, 1995	Dec 31, 1999	Jan 2, 2004	Jan 4, 2008	Jan 5, 2012
Mar 23, 1992	Mar 25, 1996	Mar 26, 2000	Mar 28, 2004	Mar 30, 2008	Mar 31, 2012
June 17, 1992	June 19, 1996	June 20, 2000	June 22, 2004	Jun 24, 2008	June 25, 2012
Sept 11, 1992	Sept 13, 1996	Sept 15, 2000	Sept 16, 2004	Sep 18, 2008	Sept 19, 2012
Dec 6, 1992	Dec 8, 1996	Dec 10, 2000	Dec 11, 2004	Dec 13, 2008	Dec 14, 2012
Mar 2, 1993	Mar 4, 1997	Mar 6, 2001	Mar 7, 2005	Mar 9, 2009	Mar 10, 2013
May 27, 1993	May 29, 1997	May 31, 2001	June 1, 2005	Jun 3, 2009	June 4, 2013
Aug 21, 1993	Aug 23, 1997	Aug 25, 2001	Aug 26, 2005	Aug 28, 2009	Aug 30, 2013
Nov 15, 1993	Nov 17, 1997	Nov 19, 2001	Nov 20, 2005	Nov 22, 2009	Nov 24, 2013
Feb 9, 1994	Feb 11, 1998	Feb 13, 2002	Feb 14, 2006	Feb 16, 2010	Feb 18, 2014
May 7, 1994	May 8, 1998	May 10, 2002	May 11, 2006	May 13, 2010	May 15, 2014
Aug 1, 1994	Aug 2, 1998	Aug 4, 2002	Aug 5, 2006	Aug 7, 2010	Aug 9, 2014
Oct 26, 1994	Oct 27, 1998	Oct 29, 2002	Oct 30, 2006	Nov 1, 2010	Nov 3, 2014
Jan 20, 1995	Jan 21, 1999	Jan 23, 2003	Jan 24, 2007	Jan 26, 2011	Jan 28, 2015
Apr 16, 1995	Apr 17, 1999	Apr 19, 2003	Apr 21, 2007	Apr 22, 2011	Apr 24, 2015
July 11, 1995	July 12, 1999	July 14, 2003	July 16, 2007	July 17, 2011	July 19, 2015

Appendix – EUWS Cycle Peaks &Troughs

Table A25 – 28.68-Day EUWS Cycle. Theoretical Peaks from 2004 to 2011.

July 21, 2004	Sept 24, 2005	Nov 28, 2006	Feb 1, 2008	Apr 7, 2009	June 11, 2010
Aug 18, 2004	Oct 23, 2005	Dec 27, 2006	Mar 1, 2008	May 5, 2009	July 9, 2010
Sept 16, 2004	Nov 20, 2005	Jan 24, 2007	Mar 30, 2008	June 3, 2009	Aug 7, 2010
Oct 15, 2004	Dec 19, 2005	Feb 22, 2007	Apr 27, 2008	July 2, 2009	Sept 5, 2010
Nov 12, 2004	Jan 17, 2006	Mar 23, 2007	May 26, 2008	July 30, 2009	Oct 3, 2010
Dec 11, 2004	Feb 14, 2006	Apr 21, 2007	June 24, 2008	Aug 28, 2009	Nov 1, 2010
Jan 9, 2005	Mar 15, 2006	May 19, 2007	July 22, 2008	Sep 26, 2009	Nov 30, 2010
Feb 7, 2005	Apr 13, 2006	June 17, 2007	Aug 20, 2008	Oct 24, 2009	Dec 28, 2010
Mar 7, 2005	May 11, 2006	July 16, 2007	Sept 18, 2008	Nov 22, 2009	Jan 26, 2011
Apr 5, 2005	June 9, 2006	Aug 13, 2007	Oct 16, 2008	Dec 21, 2009	Feb 24, 2011
May 4, 2005	July 8, 2006	Sept 11, 2007	Nov 14, 2008	Jan 18, 2010	Mar 24, 2011
June 1, 2005	Aug 5, 2006	Oct 10, 2007	Dec 13, 2008	Feb 16, 2010	Apr 22, 2011
June 30, 2005	Sept 3, 2006	Nov 7, 2007	Jan 10, 2009	Mar 17, 2010	May 21, 2011
July 29, 2005	Oct 2, 2006	Dec 6, 2007	Feb 8, 2009	Apr 14, 2010	June 19, 2011
Aug 26, 2005	Oct 30, 2006	Jan 4, 2008	Mar 9, 2009	May 13, 2010	July 17, 2011

References

Albrecht, A.; Magueijo, J. [1999]. *A Time Varying Speed of Light as a Solution to Cosmological Puzzles*. Physical Review D59 (1999) 043516.

All Empires: Teotihuacan [2008]. *Teotihuacan*. All Empires Online History Community. *Teotihuacan*. http://www.allempires.com/article/index.php?q=Meso-American_Civilizations

Alley, R.B. [2004], GISP2 Ice Core Temperature and Accumulation Data, IGBP PAGES/World Data Center for Paleoclimatology. *Data Contribution Series #2004-013*. NOAA/NCDC Paleoclimatology Program, Boulder CO, USA.

Alvarez, L.W.; Alvarez, F.; Michel, H.V. [1980]. *Alvarez* Science, v208, 1095-1108.

Ambou, M.G. [2008]. *Gravitational Explanation of the Glacial Periods and Calculation of the Precession of Mercury's Orbit*. http://www.ambou.net/mariano/glacial.html

Ambrose, S.H. [1998]. *Late Pleistocene Human Population Bottlenecks, Volcanic Winter, and Differentiation of Modern Humans*. Journal of Human Evolution 34 (6): 623–651. doi: 10.1006/jhev.1998.0219.

Archer, C., Vance, D. [2006]. *Coupled Fe and S Isotope Evidence for Archean Microbial Fe(III) and Sulfate Reduction*. Geology; March 2006; v. 34; no. 3; p. 153-156; DOI:

10.1130/G22067.1.

Asmerom, Y.; Polyak, V.; Burns, S.; Rasmussen, J. [2007]. *Solar forcing of Holocene Climate: New Insights from a Speleothem Record, Southwestern United States.* Geology,35, 1-4.

Barnola, J.M., Raynaud, D., Lorius, C., Barkov, N.I. [2003]. Laboratoire de Glaciologie et de Géophysique de l'Environnement, CNRS, BP96, 38402 Saint Martin d'Heres Cedex, France. *Historical Carbon Dioxide Record From the Vostok Ice Core.* http://cdiac.esd.ornl.gov/trends/co2/vostok.htm

Berger, A.; Loutre, M.F.; Melice, J.L. [2006]. *Equatorial Insolation: From Precession Harmonics to Eccentricity Frequencies.* Climate of the Past Discussions, 2, 131–136.

Bernanke, B.S. [2007]. *The Economic Outlook.* Board of Governors of the Federal Reserve System, Testimony before the Joint Economic Committee, U.S. Congress on March 28, 2007. http://www.federalreserve.gov/newsevents/testimony/Bernanke20070328a.htm

Bonanno, A.; Schlattl, H.; Patern, L. [2002]. *The Age of the Sun and the Relativistic Corrections in the EOS.* Astronomy and Astrophysics 390: 1115-1118, doi:10.1051/0004-6361:20020749.

Bond, G., Showers, W., Cheseby, M., Lotti, R., Almasi, P., deMenocal, P., Priore, P., Cullen, H., Hajdas, I., Bonani, G. [1997]. *A Pervasive Millennial-Scale Cycle in North Atlantic Holocene and Glacial Climates.* Science 278, 1257–1266.

Bond, G.; Kromer, B.; Beer, J.; Muscheler, R.; Evans, M.N.; Showers, W.; Hoffmann, S.; Lotti-Bond, R.; Hajdas, I.; Bonani, G. [December 2001]. *Persistent Solar Influence on North Atlantic Climate during the Holocene.* Science, Dec. 7, 2001, 294, 5549, Research Library Core, pg. 2130

Bradt, S. [2004]. *Brain Takes Itself Over -- Immediate vs. Delayed Gratification.* Harvard University Gazette, October 21, 2004.

Breiterman, C. [2004]. *Considering the Earth as an Open System.* Journal of Earth System Science Education, (Article), JESSE-03-400-05, 2004

Britannica: Aegean [2008]. *Aegean Civilizations.* Retrieved September 18, 2008 from Encyclopædia Britannica Online: http://www.britannica.com/EBchecked/topic/6965/Aegean-civilization

Britannica: Amratian [2008]. *Amratian Culture.* Retrieved September 15, 2008 from Encyclopedia Britannica Online: www.britannica.com/EBchecked/topic/21823/Amratian-culture

Britannica: Aztec [2008]. *Aztec.* Retrieved September 14, 2008, from Encyclopedia Britannica Online: www.britannica.com

Britannica: Byzantine [2008]. *Byzantine Empire*. Retrieved September 15, 2008 from Encyclopedia Britannica Online:
http://www.britannica.com/EBchecked/topic/87186/Byzantine-Empire

Britannica: Egypt [2008]. *Ancient Egypt*. Retrieved September 15, 2008 from Encyclopedia Britannica Online: www.britannica.com/EBchecked/topic/180468/ancient-Egypt

Britannica: Five & Ten [2008]. *Five Dynasties & Ten Kingdoms*. Retrieved September 15, 2008, from Encyclopedia Britannica Online:
www.britannica.com/EBchecked/topic/208994/Five-Dynasties

Britannica: Gerzean [2008]. *Gerzean culture*. Retrieved September 15, 2008 from Encyclopedia Britannica Online: www.britannica.com/EBchecked/topic/231949/Gerzean-culture

Britannica: Greek [2008]. *Ancient Greek Civilization*. Retrieved September 18, 2008 from Encyclopedia Britannica Online: http://www.britannica.com/EBchecked/topic/244231/ancient-Greece

Britannica: Han [2008]. *Han Dynasty*. Retrieved September 14, 2008, from Encyclopedia Britannica Online: www.britannica.com/EBchecked/topic/253872/Han-dynasty

Britannica: Maya [2008]. *Maya*. Retrieved September 14, 2008, from Encyclopedia Britannica Online: www.britannica.com/EBchecked/topic/370759/Maya

Britannica: Olmec [2008]. *Olmec*. Retrieved September 14, 2008, from Encyclopedia Britannica Online: www.britannica.com/EBchecked/topic/427846/Olmec

Britannica: Ottoman [2008]. *Ottoman Empire*. Retrieved September 16, 2008 from Encyclopedia Britannica Online:
http://www.britannica.com/EBchecked/topic/434996/Ottoman-Empire

Britannica: Qin [2008]. *Qin Dynasty*. Retrieved September 14, 2008, from Encyclopedia Britannica Online: www.britannica.com/EBchecked/topic/111681/Qin-dynasty

Britannica: Rome [2008]. *Ancient Rome*. Retrieved September 15, 2008 from Encyclopedia Britannica Online: http://www.britannica.com/EBchecked/topic/507905/ancient-Rome

Britannica: Spain [2008]. *Spain*. Retrieved September 21, 2008 from Encyclopedia Britannica Online: http://www.britannica.com/EBchecked/topic/ 557573/Spain

Britannica: Tang [2008]. *Tang Dynasty*. Retrieved September 15, 2008, from Encyclopedia Britannica Online: www.britannica.com/EBchecked/topic/582301/Tang-dynasty

Britannica: Teotihuacan [2008]. *Teotihuacan Civilization*. Retrieved September 14, 2008, from

Encyclopedia Britannica Online: www.britannica.com/EBchecked/topic/587669/Teotihuacan-civilization

Britannica: Zhou [2008]. *Zhou Dynasty.* Retrieved September 14, 2008, from Encyclopedia Britannica Online: www.britannica.com/EBchecked/topic/114678/Zhou-dynasty

Cal Davis Cosmology Group [2008]. *Einstein's Biggest Blunder.* Retrieved August 1, 2008 from Univ. of California Davis
http://www.physics.ucdavis.edu/Cosmology/albrecht/ein_transcript.htm

Casper, D. [2008]. *What's a Neutrino?* Retrieved June 17, 2008 from University of California, Irvine. http://www.ps.uci.edu/~superk/neutrino.html

Cavosie, A.J.; Wilde, A.; Liu, D.; Weiblen, P.W.; Valley, J.W. [2004]. *Internal Zoning and U–Th–Pb Chemistry of Jack Hills Detrital Zircons: A Mineral Record of Early Archean to Mesoproterozoic (4348–1576 Ma) Magmatism.* Precambrian Research 135 (2004) 251–279

Chew, S.C. [2006]. *The Recurring Dark Ages: Ecological Stress, Climate Changes, and System Transformation.* AltaMira Press. ISBN 13: 978-0759104525

China Guide [2008]. *History of China.* Retrieved September 8, 2008 from China Guide. http://www.travelchinaguide.com/intro/history/

Christl, M.; Mangini, A.; Holzkamper, S.; Spotl, C. [2003]. *Evidence for a Link Between the Flux of Galactic Cosmic Rays and Earth's Climate During the Past 200,000 Years.* Journal of Atmospheric and Solar-Terrestrial Physics 66 (2004) 313–322.

Coe, M.D. [2002]. *The Maya.* 6th edition, New York: Thames & Hudson, pp. 151-155. ISBN: 0-500-28066-5.

Cohn, M. [1989]. *Health and Rise of Civilization.* New Haven, Yale University Press.

Condie, K.C. [2003]. *Supercontinents and Superplume Events: Distinguishing Signals in the Geologic Record.* Elsevier, Physics of the Earth and Planetary Interiors, 146 (2004) 319–332

Condie, K.C. [1998]. *Episodic Continental Growth and Supercontinents: A Mantle Avalanche Connection?* Earth & Planetary Science Letters, Vol. 163, pp 97-108, November 1998.

Condie, K.C. [2003]. *What on Earth Happened 2.7 Billion Years Ago?* Geophysical Research Abstracts, Vol. 5, 01269, 2003

Condie, K.C. [2008]. *What on Earth Happened 2.7 Billion Years Ago?* Retrieved August 22,

2008 from Department of Earth and Environmental Science, New Mexico Tech, Socorro, New Mexico. http://www.ees.nmt.edu/Geol/classes/erth468/dl1.pdf

Cowles, A. and Associates [1939]. *Common Stock Indexes*. 2nd edition. Cowles Commission Monograph, Bloomington, Indiana. Principia Press.

CRU [2008]. *Climatic Research Unit, Data: Temperature*. University of East Anglia, Norwich, UK. http://www.cru.uea.ac.uk/cru/data/temperature/crutem3gl.txt

Cycles Research Institute [2008]. *Cycles*. Retrieved April 5, 2008 from Cycles Research Institute. http://www.cyclesresearchinstitute.org/cycles.html

Dalrymple, G.B. [1991]. *The Age of the Earth*. California: Stanford University Press. ISBN: 0-8047-1569-6.

Davis, J. [2005]. *An Improved Annual Chronology of U.S. Business Cycles since the 1790's*. The Vanguard Group & National Bureau of Economic Research. NBER Working Paper No.11157, issued in February 2005.

Del Peloso, E.F.; da Silva, L.; Porto de Mello, G.F.; Arany-Prado, L.I. [2005]. *The Age of the Galactic Thin Disk from Th/Eu Nucleocosmochronology: Extended Sample*. Proceedings of the International Astronomical Union (2005), 1: 485–486 Cambridge University Press.

Dewey, E.R. [1951]. *The 57-year Cycle in International Conflict*. Cycles, 2, 1, 4-6.

Dewey, E.R.; Dakin, E.F. [1947]. *Cycles: The Science of Prediction*. Henry Holt & Co., New York.

Ding, G., Kang, J., Liu, Q., Shi, T., Pei, G., Li, Y. [2006]. *Insights into the Coupling of Duplication Events and Macroevolution from an Age Profile of Animal Transmembrane Gene Families*. http://www.pubmedcentral.nih.gov/articlerender.fcgi?tool=pubmed&pubmedid=16895434

Dodson, A. [2008]. *The End of Civilization*. Retrieved September 15, 2008 from British Broadcasting Company. http://www.bbc.co.uk/history/ancient/egyptians/egypt_end_01.shtml

Eddy, John A. [2003]. *The Maunder Minimum*. Science, Vol. 192, pp 1189-1202.

Encarta: Archaic [2008]. *Ancient Greece: The Archaic Age (750-480 BC)*. Retrieved September 18, 2008 from Microsoft Encarta Online Encyclopedia. http://encarta.msn.com © 1997-2008 Microsoft Corporation.

Encarta: Byzantine [2008]. *Byzantine Empire*. Retrieved September 18, 2008 from Microsoft Encarta Online Encyclopedia. http://encarta.msn.com © 1997-2008 Microsoft Corporation.

Encarta: Classical [2008]. *Ancient Greece: The Classical Age (480-323 BC).* Retrieved September 18, 2008 from Microsoft Encarta Online Encyclopedia. http://encarta.msn.com © 1997-2008 Microsoft Corporation.

Encarta: Europe [2008]. *Europe: Arrival of Indo-Europeans.* Retrieved September 16, 2008 from Microsoft Encarta Online Encyclopedia. http://encarta.msn.com © 1997-2008 Microsoft Corporation.

Encarta: Germany [2008]. *Bismarck's Domestic Policies.* Retrieved September 24, 2008 from Microsoft Encarta Online Encyclopedia. http://encarta.msn.com © 1997-2008 Microsoft Corporation.

Encarta: Olmec [2008]. *Olmec.* Retrieved September 16, 2008, from Microsoft Encarta Online Encyclopedia. http://encarta.msn.com © 1997-2008 Microsoft Corporation.

Encarta: Ottoman [2008]. *Ottoman Empire.* Retrieved September 18, 2008 from Microsoft Encarta Online Encyclopedia. http://encarta.msn.com © 1997-2008 Microsoft Corporation.

Encarta: Rome [2008]. *Ancient Rome.* Retrieved September 18, 2008 from Microsoft Encarta Online Encyclopedia. http://encarta.msn.com © 1997-2008 Microsoft Corporation.

Encarta: Shays [2008]. *Shays' Rebellion.* Retrieved September 26, 2008 from Microsoft Encarta Online Encyclopedia. http://encarta.msn.com © 1997-2008 Microsoft Corporation.

Encarta: Spain [2008]. *Spain.* Retrieved September 20, 2008 from Microsoft Encarta Online Encyclopedia. http://encarta.msn.com © 1997-2008 Microsoft Corporation.

European Space Agency [2008]. *GAIA Project.* European Space Agency, 8-10 Rue Mario Nikis, Paris, France. GAIA Science Advisory Group.

Evansville [2008]. *Chinese Chronology.* Retrieved September 5, 2008 from the University of Evansville, Evansville, Indiana. http://eawc.evansville.edu/chronology/chpage.htm

Federal Reserve Board [2008]. *Flow of Funds Z.1.* http://www.federalreserve.gov/

Feynman, J. [1982]. *Geomagnetic and Solar Wind Cycles, 1900 –1975.* J. Geophys. Res., 87, 6153– 6162.

Fischer, A.G.; Arthur, M.A. [1977]. *Secular Variations in the Pelagic Realm.* The Society of Economic Paleontologists and Mineralogists, Special Publications no. 25, p. 19-50.

Foundation for the Study of Cycles [2008]. *Cycles Magazine.* 2929 Coors Blvd NW Suite 307; Albuquerque, NM 87120. http://www.foundationforthestudyofcycles.org

Frakes, L.A., Francis, J.E., Syktus, J.I. [1992]. *Climate Modes of the Phanerozoic*. Cambridge University Press. ISBN-13: 9780521021944; ISBN-10: 0521021944.

Georgia Perimeter College [2008]. *Structure of the Universe*. Retrieved July 8, 2008 from Georgia Perimeter College. http://facstaff.gpc.edu/~wlahaise/forces-c.pdf

Goldfarb, R.J.; Groves, D.I.; Gardoll, S. [2001]. *Orogenic Gold and Geologic Time: A Global Synthesis*. Ore Geologic Review 18, 1-75.

Gradstein, F.M.; Ogg, J.G.; Smith, A.G. [2005]. *A Geologic Time Scale 2004*. Cambridge University Press. ISBN: 0-521-78673-8.

Grimal, N. [1992]. *A History of Ancient Egypt*. Blackwell Publishers Ltd, Malden, MA. ISBN: 0-631-19396-0.

Gupta, A.K.; Das, M.; Anderson, D.M. [2005] *Solar Influence on the Indian Summer Monsoon during the Holocene*. Geophysical Research Letters, Vol. 32, L17703

Guyodo, Y.; Valet, J.P. [1999]. *Global Changes in Intensity of the Earth's Magnetic Field during the Past 800 Kyr*. Nature, 399, 249-252.

Hassan, F. [2008]. *The Fall of the Egyptian Old Kingdom*. Retrieved September 15, 2008 from British Broadcasting Company.
http://www.bbc.co.uk/history/ancient/egyptians/apocalypse_egypt_01.shtml

Hawking, S. [1988]. *A Brief History of Time*. Bantam Books, ISBN 13: 978-0553053401

Hawking, S. [1998]. *Physics Colloquiums – Inflation: An Open and Shut Case*. Retrieved July 12, 2008 from http://www.hawking.org.uk/text/physics/inflate.html

Helbling, T.; Terrones, M. [2003]. *Real and Financial Effects of Bursting Asset Price Bubbles*. International Monetary Fund,
http://www.imf.org/external/pubs/ft/weo/2003/01/pdf/chapter2.pdf

Hinshaw, G.; Weiland, J.L.; Hill, R.S.; Odegard, N.; Larson, D.; Bennett, C.L.; Dunkley, J.; Gold, B.; Greason, M.R.; Jarosik, N.; Komatsu, E.; Nolta, M.R.; Page, L.; Sperge, D.N.; Wollack, E.; Halpern, M.; Kogut, A.; Limon, M.; Meyer, S.S.; Tucker, G.S.; & Wright, E.L. [2008]. *Five-Year Wilkinson Microwave Anisotropy Probe (WMAP1) Observations: Data Processing, Sky Maps, & Basic Results*. Astrophysical Journal Supplement Series, arXiv:0803.0732.

Hirsch, J.A. [2008]. *Stock Trader's Almanac, 2008*. John Wiley & Sons, Inc., ISBN: 978-0-470-10985-4.

Historical Statistics of the United States [2005]. *National Wealth, by Type of Asset.* Cambridge University Press. doi:10.1017/ ISBN-9780511132971. Ce209-310

History.com [2008]. *Panic Closes NYSE.* Retrieved September 25, 2008 from The History Channel Online:
http://www.history.com/this-day-in-history.do?action=tdihArticleCategory&id=6108

Hooker, R. [2008]. *Bureaucrats and Barbarians, the Greek Dark Ages.* Retrieved July 30, 2008 from Washington State University.
http://www.wsu.edu:8080/~dee/MINOA/DARKAGES.HTM

Hunan China [2008]. *The Warring States Period of Chinese History.* Retrieved September 8, 2008 from Travel Hunan China. http://www.travelhunanchina.com/WarringStates.php

IISH: Prices and Wages [2008]. *List of Datafiles.* Retrieved March 10, 2008, from International Institute of Social History, Amsterdam, Netherlands. http://www.iisg.nl/hpw/data.php#world

Investors Intelligence [2008]. *The Advisory Sentiment Index.* Investors' Intelligence, 30 Church St., New Rochelle, NY 10802. http://www.investorsintelligence.com

Ipedia.net [2008]. *Information about Blytt Sernander, The Sequence.* Retrieved August 11, 2008 from Ipedia.net. http://www.ipedia.net/information/Blytt-Sernander

Jackson, M. [2003]. *Out, Damned Spot!* The IRM Quarterly, Spring 2003, Vol. 13, No.1

John Hopkins University [2008]. *Chronos – Cyclostratigraphy Online Database & Research Center.* Retrieved July 14, 2008, from John Hopkins University.
http://www.jhu.edu/~lhinnov1/chronoscyclostrat/aimandscope/millennialcycles.html

Karner, D.B.; Muller, R.A. [2000]. *A Causality Problem for Milankovitch.* Science, June 23, 2000: Vol. 288. no. 5474, pp. 2143 – 2144 DOI: 10.1126/science.288.5474.2143

Kasting, J.F.; Howard, M.T. [2006]. *Atmospheric Composition and Climate on the Early Earth.* Philosophical Transactions of the Royal Society B (2006) 361, 1733–1742, doi:10.1098/rstb.2006.1902

Katz, M.E.; Wright, J.D.; Miller, K.G.; Cramer, B.S.; Fennel, K.; and Falkowski, P.G. [2005]. *Biological Overprint of the Geological Carbon Cycle.* Marine Geology. 217:323-338.

Kay, R.W. [1994]. *Geomagnetic Storms: Association with Incidence of Depression as Measured by Hospital Admission.* The British Journal of Psychiatry 164: 403-409.

Keeley, L.H. [1996]. *War Before Civilization.* Oxford University Press, ISBN: 0-19-511912-6.

Keshavan, M.S.; Gangadhar, B.N.; Gautam, R.U.; Ajit, V.B.; Kapur, R.L., [1981]. *Convulsive Threshold in Humans and Rats and Magnetic Field Changes: Observations during Total Solar Eclipse*. Neuroscience Letter. 1981 March 10;22 (2):205-8 7231811.

Kondratieff, N. [1924]. *On the Notion of Economic Statics, Dynamics, and Fluctuations.*

Kopp, R.E.; Kirschvink, J.L.; Hilburn, I.A.; Nash, C.Z. [2005]. *The Paleoproterozoic Snowball Earth: A Climate Disaster Triggered by the Evolution of Oxygenic Photosynthesis*. Perspective, communicated by Paul F. Hoffman, Harvard University, Cambridge, MA, June 14, 2005 (received for review April 8, 2004)

Lannin, P.; Edmonds, S. [2006]. *U.S. Duo Win Physics Nobel for backing up Big Bang*. Reuters, October 3, 2006, Stockholm, Sweden.

Lawrence, K.T. [2006]. *Eastern Equatorial Pacific 5 Myr Alkenone SST and Paleoproductivity Reconstruction*. IGBP PAGES/World Data Center for Paleoclimatology. Data Contribution Series #2006-044. NOAA/NCDC Paleoclimatology Program, Boulder CO, USA.

Lear, C.H.; Rosenthal, Y.; Coxall, H.K.; Wilson, P.A. [2004]. *Late Eocene to Early Miocene Ice Sheet Dynamics and the Global Carbon Cycle*. Paleoceanography, Vol. 19, PA4015.

Lee, Y.K. [2002]. *Building the Chronology of Early Chinese History*. The Journal of Archaeology for Asia and the Pacific, Vol.41, 2002.

Macaulay, F.R. [1938]. *The Movements of Interest Rates, Bond Yields, and Stock Prices in the United States since 1856*. National Bureau of Economic Research, New York, NY.

Maca-Meyer, N.; Gonzalez, A.M.; Larruga, J.M.; Flores, C.; Cabrera, V.M. [2003]. *Major Genomic Mitochondrial Lineages Delineate Early Human Expansions*." BMC Genetics 2: 13. PMID 11553319.

Mackay, C. [Reprinted in 1980], Harmony Books. *Extraordinary Popular Delusions and the Madness of Crowds*. ISBN: 0-517-53919-5.

Marziali, C. [2008]. *Extinction by Asteroid a Rarity*. University of Southern California, Los Angeles, California. October 7, 2008 news release.

McClure, S.M.; Laibson, D.; Lowenstein, G.; Cohen, J. [2004]. *Separate Neural Systems Value Immediate and Delayed Monetary Rewards*. Science, 306.

McCulloch, M.T.; Bennett, V.C. [1994]. *Progressive Growth of the Earth's Continental Crust and Depleted Mantle – Geochemical Constraints*. Geochim. Cosmochim. Acta 58, 4717-4738.

Mercurio, E. [2001]. *The Effects of Galactic Cosmic Rays on Weather and Climate on Multiple*

Time Scales. Proceedings of the Seventeenth Annual Pacific Climate Workshop, March 2001.

Muller, R.A. [2008]. *Diversity Data Parsed in 171 Different Ways.* Department of Physics, University of California at Berkeley. http://muller.lbl.gov/papers/genera_book_7June08.pdf

Muller, R.A.; MacDonald, G.J. [2002]. *Origin of the 100 Kyr Glacial Cycle: Eccentricity or Orbital Inclination?* Department of Physics, University of California, Berkeley and San Diego

Murphy, A.E. [1997]. *John Law: Economic Theorist and Policy-Maker.* Oxford University Press, Oxford. ISBN-10: 0-198-28649-0.

Nador, A.; Lantos, M.; Toth-Makk, A.; Thamo-Bozso, E. [2002]. *Milankovitch-Scale Multi-Proxy Records from Fluvial Sediments of the last 2.6 Ma, Pannonian Basin, Hungary.* Quaternary Science Reviews.

NASA, Earth Observatory [2001]. *The Sun's Chilly Impact on Earth.* NASA News Archive, December 6, 2001.
http://earthobservatory.nasa.gov/Newsroom/NasaNews/2001/200112065794.html

NASA, Solar Physics [2008]. *The Solar Wind.* Retrieved June 1, 2008 from the Marshall Space Flight Center. http://solarscience.msfc.nasa.gov/SolarWind.shtml

Nastos, P.T.; Paliatsos, A.G.; Tritakis, V.P.; Bergiannaki. A. [2006] *Environmental Discomfort and Geomagnetic Field Influence on Psychological Mood in Athens*. Indoor and Built Environment 2006; 15; 365, DOI: 10.1177/1420326X06067372

National Association of Realtors [2008]. *Existing-Home Sales and Prices.*
http://www.realtor.org/research/research/ehsdata

National Science Foundation [2008]. *The Paleobiology Database.* Retrieved July 25, 2008 from the National Science Foundation.
http://flatpebble.nceas.ucsb.edu/cgi-bin/bridge.pl?action=startScale

National Snow and Ice Data Center, University of Colorado at Boulder & the World Data Center for Paleoclimatology. [2008]. *GISP2 Data.* NOAA/NGDC Paleoclimatology Program, Boulder CO, USA.

National Weather Service, Weather Forecast Office [2008]. *Historical Records and Past Events – Daily Records and Normals.* Sioux City, Iowa, Temperature. Retrieved August 29, 2008 from the Weather Forecast Office. http://www.crh.noaa.gov/fsd/climate/archive.php

Neal, L. [1990]. *The Rise of Financial Capitalism: International Capital Markets in the Age of Reason.* Cambridge University Press, New York. ISBN-10: 0-521-45738-6.

Newman, P.; NASA [2008]. *Imagine the Universe: Electromagnetic Spectrum.* Retrieved July 2, 2008 from the Goddard Space Flight Center.
http://imagine.gsfc.nasa.gov/docs/science/know_l1/emspectrum.html

NOAA, National Weather Service [2008]. *The Sun and Sunspots.* Retrieved June 5, 2008 from the Weather Forecast Office. http://www.crh.noaa.gov/fsd/astro/sunspots.php

NOAA, NGDC [2008]. *International Sunspot Numbers.* Retrieved February 3, 2008 from National Oceanic and Atmospheric Administration.
ftp://ftp.ngdc.noaa.gov/STP/SOLAR_DATA/SUNSPOT_NUMBERS

Pederson, J.L. [2000]. *Holocene Paleolakes of Lake Canyon, Colorado Plateau: Paleoclimate and Landscape Response from Sedimentology and Allostratigraphy.* GSA Bulletin; January 2000; v. 112; no. 1; p. 147–158.

Peristykh, A.N.; Damon, P.E. [2003]. *Persistence of the Gleissberg 88-Year Solar Cycle Over the Last 12,000 Years: Evidence from Cosmogenic Isotopes.* Journal of Geophysical Research, Vol. 108, No. A1, 1003.

Perry, C.A. [2007]. *Evidence for a Physical Linkage Between Galactic Cosmic Rays and Regional Climate Time Series.* Advances in Space Research, Vol. 40, doi:10.1016/j.asr2007.02.079, p.353-364.

Phillips, T. [2008]. *The Moon and the Magnetotail.* Retrieved August 20, 2008 from NASA, Goddard Space Flight Center.
http://www.nasa.gov/topics/moonmars/features/magnetotail_080416.html

Poon, L. [2008]. *History of China.* Retrieved September 7, 2008 from University of Maryland, College Park, Maryland. http://www-chaos.umd.edu/history/welcome.html

Poore, R.Z.; Dowsett, H.J.; Verardo, S.; Quinn, T.M. [2003]. *Millennial to Century-Scale Variability in Gulf of Mexico Holocene Climate Records.* Paleoceanography, Vol. 18, No. 2, 1048.

Prechter, R.R. Jr. [2003]. New Classics Library. *Pioneering Studies in Socionomics.* ISBN: 0-932750-56-7

Priem, H.N.A. [1997]. *CO_2 and Climate: A Geologist's View.* Dept. of Earth Sciences, Utrecht University, and Global Institute for the Study of Natural Resources, The Hague, The Netherlands.

Raup, D.M.; Sepkoski, J.J. Jr. [1984]. *Periodicity of Extinctions in the Geologic Past.* Proceedings of the National Academy of Sciences, USA, Vol. 81, pp. 801-805.

Raymo, M.E. [1997]. *The Timing of Major Climate Terminations.* Paleoceanography 12 (4), 577-585.

Redford, D.B. [1992]. *Egypt, Canaan, and Israel in Ancient Times.* Princeton University Press, 1992, p. 16. ISBN: 0-691-00086-7.

Rial, J.A.; Anaclerio, C.A. [2000]. *Understanding Nonlinear Responses of the Climate System to Orbital Forcing.* Quaternary Science Reviews 19 (2000) 1709-1722.

Rohde, R.A.; Muller, R.A. [2005]. *Cycles in Fossil Diversity.* Nature, 434: 208-210.

Sneppen, K.; Bak, P.; Flyvbjerg, H.; Jensen, M.H. [1995]. *Evolution as a Self-Organized Critical Phenomenon.* Proceedings of the National Academy of Sciences, USA, Vol. 92, pp. 5209-5213.

Rowland, S. [2008]. *Frenchman Mountain and the Great Unconformity.* Retrieved November 28, 2008, from Geosciences Department, University of Nevada Las Vegas. http://geoscience.unlv.edu/pub/rowland/Virtual/geology.html

Sarnthein, M.; van Kreveld, S.; Erlenkeuser, H.; Grootes, P. M.; Kucera, M.; Pflaumann, U.; Schulz, M. [2003]. *Centennial-to-Millennial-Scale Periodicities of Holocene Climate and Sediment Injections off the Western Barents Shelf, 75°N.* Boreas, Vol. 32, pp. 447–461. Oslo. ISSN 0300-9483.

Scarre, C. [1995]. *The Penguin Historical Atlas of Ancient Rome.* Penguin Books, London. ISBN: 0-14-051329-9.

Schwert, G.W. [1991]. *Indexes of United States Stock Prices from 1802 to 1987.* Journal of Business, 64 (July 1991) 442. Summarized in The C.F.A. Digest, 21 (Winter 1991) 3-5.

Science Daily: Neutrinos [1997]. *28-Day Cycle Found in Solar Neutrinos.* http://www.sciencedaily.com/releases/1997/12/971223070932.htm

Sharma, M. [2002]. *Variations in Solar Magnetic Activity During the Last 200,000 years: Is there a Sun-Climate Connection?* Earth and Planetary Science Letters, Vol. 199, Issues 3-4, June 10, 2002, pp 459-472.

Shaviv, N.J. [2002]. *Cosmic Rays Diffusion in the Dynamic Milky Way: Model, Measurement, and Terrestrial Effects.* Racah Institute of Physics, Hebrew University, Jerusalem, Israel.

Singer, S.F.; Avery, D.T. [2007]. *Unstoppable Global Warming: Every 1,500 Years.* Rowman & Littlefield Publishers. ISBN: 0-74-255117-2

Smith, W.B.; Cole, A.H. [1935]. *Fluctuations in American Business, 1790-1860.* Harvard University Press, Cambridge, Massachusetts.

Solanki, S.K. et al. [2005]. *11,000 Year Sunspot Number Reconstruction,* IGBP PAGES/World Data Center for Paleoclimatology. Data Contribution Series #2005-015. NOAA/NGDC Paleoclimatology Program, Boulder CO, USA.

Sonett, C.P.; Suess, H.E. [1984]. *Correlation of Bristlecone Pine Ring Widths with Atmospheric 14C Variations: A Climate-Sun Relation.* Nature. 307(5947): 141-143.

Spotts, P.N. [2002]. *The Big Bang (One More Time).* The Christian Science Monitor, May 9, 2002.

St. Andrews Press [2008]. *The End to a Mystery?* Press Office, University of St. Andrews, Scotland, UK

Stager, J.C.; Cumming, B.; Meeker, L.D. [1997]. *An 11,400-Year High-Resolution Diatom Record from Lake Victoria, East Africa.* Quaternary Research 47: 81-89.

Standard & Poor's Corp. [1986]. *Security Price Index Record, 1986.* Standard and Poor's Corp., New York, NY.

Stanford Solar [2008]. *Differential Rotation.* Retrieved August 15, 2008 from Stanford University, Stanford Solar Center. http://solar-center.stanford.edu/gloss.html

Stinson, G.S.; Dalcanton, J.J.; Quinn, T.; Kaufmann, T.; Wadsley, J. [2007]. *Breathing in Low Mass Galaxies: A Study of Episodic Star Formation.* Astronomy Department, University of Washington.

Stuiver, M.; Grootes, P.M.; Braziunas, T.F. [1995]. *The GISP2 18O Climate Record of the Past 16,500 Years and the Role of the Sun, Ocean, and Volcanoes.* Quaternary Research 44:341-354.

Stuiver, M.; Brauzanias, T.F. [1993]. *Modeling Atmospheric 14-C Influences and Radiocarbon Ages of Marine Samples Back to 10,000 BC.* Radiocarbon 35(1):231-249.

Sturrock, P.A.; Walther, G.; Wheatland, M.S. [1997]. *Search for Periodicities in the Homestake Solar Neutrino Data,* The Astrophysical Journal, 491: 409-413.

Toffler, A. [1980]. *The Third Wave.* Bantam Books. ISBN: 0-553-24698-4

Tuccille, J. [2004]. *Kingdom: The Story of the Hunt Family of Texas.* Beard Books, ISBN: 1587982269.

UCL: Egyptian Chronology [2008]. *Chronology*. Retrieved September 15, 2008 from University College London. http://www.digitalegypt.ucl.ac.uk/chronology/index.html

U.S. Dept. of Commerce [2008]. *Personal Income Series*. U.S. Department of Commerce: Bureau of Economic Analysis. http://research.stlouisfed.org/fred2/data/PINCOME.txt

U.S. Geological Survey [2000]. *The Sun and Climate*. U.S. Department of the Interior, USGS Fact Sheet FS-095-00, August 2000.

Usoskin, I.G.; Solanki, S.K.; Kovaltsov, G.A. [2007]. *Grand Minima and Maxima of Solar Activity: New Observational Constraints*. Astron.Astrophys. 471: 301–309, doi:10.1051/0004-6361:2007770.

Valley, J.W. [2005]. *A Cool Early Earth?* Scientific American, October 2005, pp 58-65.

Valley, J.W.; Lackey, J.S.; Cavosie, A.J.; Clechenko, C.C.; Spicuzza, M.J.; Basei, A.S.; Bindeman, I.N.; Ferreira, V.P.; Sial, A.N.; King, E.M.; Peck, W.H.; Sinha, A.K.; Wei, C.S. [2005]. *4.4 Billion Years of Crustal Maturation: Oxygen Isotope Ratios of Magmatic Zircon*. Contr. Mineral Petrol. DOI 10.1007/s00410-005-0025-8.

Van Flandern, T. [2008]. *The Speed of Gravity - What the Experiments Say*. Meta Research, University of Maryland Physics, Army Research Lab, 6327 Western Ave. NW, Washington, DC 20015-2456. Retrieved October 29, 2008. http://arts.sdhs.sandi.net/speed_of_gravity.htm

Varadi, F.; Runnegar, B.; Ghil, M. [2002]. *Successive Refinements in Long-Term Integrations of Planetary Orbits*. The Astrophysical Journal.

Veizer, J. [2008]. Department of Earth Sciences, University of Ottawa, Canada. *Isotope Data, Precambrian Database – Version 1.1(b)*. http://www.science.uottawa.ca/~veizer/isotope_data/

Veizer, J. [2008]. *Isotope Data, Phanerozoic Database – d18O-Update2004*. Department of Earth Sciences, University of Ottawa, Canada. http://www.science.uottawa.ca/~veizer/isotope_data/

Von Rad, U., et al. [1999]. *Multiple Monsoon-Controlled Breakdowns of Oxygen-Minimum Conditions During the Past 30,000 Years Documented in Laminated Sediments off Pakistan*. Palaeo., 152: 129-161.

Wade, B.S.; Palike, H. [2004]. *Oligocene Climate Dynamics*. Paleoceanography, Vol. 19, PA4019.

Warren, G.F.; Pearson, F.A. [1933]. *Prices*. John Wiley & Sons, New York, NY.

Wheelock, D.C. [2004]. *Monetary Policy and Asset Prices: A Look Back at Past U.S. Stock Market Booms*. Review -- Federal Reserve Bank of St. Louis,

http://www.allbusiness.com/north-america/united-states-missouri-metro-areas/1126202-1.html

Whitmire, D.P.; Matese, J.J. [1985]. *Periodic Comet Showers and Planet X*. Nature, 313: 36 – 38. DOI: 10.1038/313036a0.

Wikipedia: Kondratieff [2008]. *Kondratiev Wave*. Retrieved September 24, 2008 from Wikipedia. http://en.wikipedia.org/wiki/Kondratieff_wave

Wikipedia: Olmec [2008]. *Olmec*. Retrieved September 16, 2008, from Wikipedia. http://en.wikipedia.org/wiki/Olmec#cite_note-0

Wilkins, J. [2008]. *Companion to Energy: Milankovitch Cycles*. Retrieved August 18, 2008 from Ohio State University
http://www.physics.ohio-state.edu/~wilkins/energy/Companion/E16.7.pdf.xpdf

Williams, M.H. [2008]. *Depopulation of the City of Rome*. Retrieved July 29, 2008, from San Francisco State University, Department of History.
http://bss.sfsu.edu/mwilliams/hist329/lectures/hist329L9.pdf

World Health Organization [1987]. *Environmental Health Criteria for Magnetic Fields*. International Programme on Chemical Safety.
http://www.inchem.org/documents/ehc/ehc/ehc69.htm#SubSectionNumber:1.4.2

Zachos, J.; Pagani, M.; Sloan, L.; Thomas, E.; Billups, K. [2001]. *Trends, Rhythms, and Aberrations in Global Climate 65 Ma to Present*. DOI: 10.1126/science.1059412. Science 292, 686 (2001).

Zahorchak, M. [1983]. *Climate: The Key to Understanding Business Cycles by Raymond Wheeler*. Tide Press. ISBN 10: 0912931000 ISBN 13: 978-0912931005

Zhao, H.S.; Li, B. [2008]. *Dark Fluid: Towards a Unification of Empirical Theories of Galaxy Rotation, Inflation and Dark Energy*. Astrophysical Journal Letters, arXiv:0804.1588v1.

Zhivotovsky, L.A.; Rosenberg, N.A.; Feldman, M.W. [2003]. *Features of Evolution and Expansion of Modern Humans, Inferred from Genomewide Microsatellite Markers*. The American Journal of Human Genetics 72:1171–1186, 2003

Breinigsville, PA USA
22 September 2009
224385BV00003B/1/P